MEMS & Microsystems

Design and Manufacture

Tai-Ran Hsu
Microsystems Design and Packaging Laboratory
Department of Mechanical and Aerospace Engineering
San Jose State University

Tata McGraw-Hill Publishing Company Limited
NEW DELHI

McGraw-Hill Offices
New Delhi New York St Louis San Francisco Auckland Bogotá Caracas Kuala Lumpur Lisbon London Madrid Mexico City Milan Montreal San Juan Santiago Singapore Sydney Tokyo Toronto

Tata McGraw-Hill

MEMS & MICROSYSTEMS: DESIGN AND MANUFACTURE

Cover images:
Top picture: The miniaturized version of ENIAC-the world's first digital computer (Courtesy of the School of Engineering and Applied Science, University of Pennsylvania, Philadelphia, PA)
Bottom left picture: The "Micro car" (Courtesy of the Denso Corporation, Aichi, Japan)
Bottom center picture: A micro optical switch (Courtesy of Lucent Technologies, Murray Hill, NJ)
Bottom right picture: 8 & 12-inch silicon wafers (Courtesy of MEMC Electronic Materials Inc., St. Peters, Missouri)

Tata McGraw-Hill Edition 2002

Third reprint 2004
RLLYCRAIRALRD

Reprinted in India by arrangement with The McGraw-Hill Companies, Inc., New York

Sales territories: India, Nepal, Bangladesh, Sri Lanka and Bhutan

Library of Congress Cataloging-in-Publication Data

Hsu, Tai-Ran.
MEMS & microsystems: design and manufacture / Tai-Ran Hsu.—1st ed.
p. cm.
ISBN 0-07-239391-2
1. Microelectronics. 2. Microelectromechanical systems. 3. Microelectronic packaging.
I. Title.
TK7874.H794 2002
621.381—dc21

2001044737
CIP

ISBN 0-07-048709-X

Published by Tata McGraw-Hill Publishing Company Limited,
7-West Patel Nagar, New Delhi 110 00, and printed at
Krishna Offset, Delhi 110 032

The McGraw-Hill Companies

To:

My wife, Grace Su-Yong,
who has backed me up for decades
with infinite encouragement, support, and love.

My three loving children, Jean, May, and Leigh,
for their support and stimulating discussions on cyberspace
communication technology and molecular biology.

My students who motivated me to write this book.

老子天下篇：（公元前450年）

一尺之錘

日取其半

萬世不絕

An excerpt from a chapter of *"Under Heaven"* by Lao-Tze, 450 B.C.

"A foot-long stick,
cut in half every day.
The last cutting to eternity."

An ancient Chinese wisdom that inspired my belief in the fantasy land of the nanoworld.

CONTENTS

Chapter 5 **Thermofluid Engineering and Microsystem Design 163**

Chapter 6 **Scaling Laws in Miniaturization 215**

PREFACE

The technological advances of microsystem engineering in the past decade have been truly impressive in both pace of development and number of new applications. Microsystem engineering involves the design, manufacture, and packaging of microelectromechanical systems (MEMS) and peripherals. Applications of microsystems in the aerospace, automotive, biotechnology, consumer products, defense, environmental protection and safety, healthcare, pharmaceutical, and telecommunications industries prompted many experts to account for a staggering $82 billion in revenue for the microsystems and related products in the year 2000.

The strong demand for MEMS and microsystems by a rapidly growing market has generated strong interest, as well as a need for engineering educators to offer courses on this subject in their respective institutions. Likewise, many practicing engineers have shown similar interest and a desire to acquire the necessary knowledge and experience in the design and manufacture of microsystems. These individuals have faced great difficulty in realizing their goals and objectives because of a lack of comprehensive books that synergistically integrate the wide spectrum of disciplines in science and engineering that microsystem engineering draws on. A book that provides the reader with a broad perception of microsystem technology and the methodologies for the design, manufacture, and packaging of microsystem products is thus very much in demand. Such demand motivated me to develop this textbook.

The objective of this book is to provide upper division undergraduate and entry-level graduate students in mechanical, electrical, manufacturing, and related engineering disciplines with the necessary fundamental knowledge and experience in the design, manufacture, and packaging of microsystems. Emphasis has been placed on the application of students' knowledge and experience acquired in previous years to the design and manufacture of microsystems. The layout and the organization of the related topics, and the many design examples presented in this book, will usher practicing engineers too into the field of microsystems engineering.

The book is intended as a textbook for classes covered by one 15-week long semester at both the undergraduate and graduate levels. Students are expected to possess prerequisite knowledge in college mathematics, physics, and chemistry, as well as in engineering subjects such as fundamental material science, electronics, and machine design.

The book consists of eleven chapters:

Chapter 1 begins with an overview of microsystems and the evolution of microfabrication that led to the production of microsystems. It concludes with a preview of the current and potential markets for various types of microsystems.

Chapter 2 presents working principles for currently available microsensors, actuators and motors, valves, pumps, and fluidics used in microsystems.

Chapter 3 offers an overview of engineering science topics applicable to microsystems design and fabrication.

Chapter 4 includes engineering mechanics topics that are relevant to microsystem design and packaging. Topics in this chapter include mechanics of deformable solids and mechanical vibration theories. Also presented are the basic formulations of thermomechanics and fracture mechanics of interfaces of thin films that are common in microstructures. It concludes with an outline of the finite element method for stress analysis.

Chapter 5 deals with the application of thermofluid engineering principles in microsystems design. It begins with an overview of fluid mechanics theories and the capillary effect in microfluid flows. Also presented is the rarefaction effect of gases in submicrometer and nanoscales. Modified formulations for heat conduction in submicrometer scale systems are presented.

Chapter 6 relates to the scaling laws that are used extensively in the conceptual design of microdevices and systems. Students will learn both the positive and negative consequences of scaling down certain physical quantities that are pertinent to microsystems.

Chapter 7 deals with materials used for common microcomponents and devices. Active and passive substrates as well as packaging materials are presented in the chapter. Other materials—such as piezoresistives, piezoelectric, and polymer—for microsystems are described.

Chapter 8 presents an overview of microfabrication processes for micromanufacturing.

Chapter 9 covers the three common micromanufacturing techniques of bulk manufacturing, surface micromachining, and the LIGA process.

Chapters 10 and 11 offer essential elements involved in the design and packaging of microsystems. The use of CAD and the finite element method in these endeavors is presented. Selected case studies and examples in the design and packaging of micro pressure sensors and fluidics demonstrate the applications of these methodologies in product design and packaging.

To cover these diversified subjects in the textbook in a 15-week long semester with 3-hour-per-week lectures presents a major challenge to the instructor. For this reason, "Suggestions to Instructors," based on my personal experience in teaching this course to both levels of students, is offered in the subsequent section. Selections of subjects for teaching a MEMS or microsystem course in either a 15-week-long semester or a 10-week-long quarter are proposed in that section.

The task of preparing a book of this breadth cannot possibly be accomplished by a single person's effort. I enjoyed communicating with able professionals like Professor Ali Beskok on nano fluid dynamics and Dr. S. Krishnamoorthy on the design and modeling of capillary electrophoresis in microfluidics. It was also a great pleasure to have dedicated and capable students like Ta-jen Tai, Valerie Barker, Matthew Smith, Jeanette Wood, Jacob Griego, and Yen-chang Hu, who contributed by either filling in detailed computations or checking the numerical accuracy of a number of examples that I developed for the book. I also appreciate the help that I received from Mindy Kwan in digitizing many photographs and diagrams used in the book. The contributions of valuable photographs and drawings by various sectors of the MEMS industry in this country and abroad are appreciated as acknowledged in the book. Development of this book was a part of a project on the development of an undergraduate curriculum stem on mechatronics during 1995–1998, in which I played a principal role. Support of this project by the Division of Undergraduate Education of the National Science Foundation (NSF) is gratefully acknowledged.

I am grateful for indefatigable efforts of many reviewers, who went above and beyond the call of duty in making this work as comprehensive and lucid as possible.

Norman Tien, *Cornell Univ.*
Robert S. Keynton, *Univ. of Louisville*
Imin Kao, *State Univ. of New York at Stony Brook*
Dennis Polla, *Univ. of Minnesota*
Zeynep Celik-Butler, *Southern Methodist Univ.*
Ashok Srivastava, *Louisiana State Univ.*
Ryszard Pryputniewicz, *Worcester Polytechnic Inst.*
Michael Y. Wang, *Univ. of Maryland*
Michael Histand, *Colorado State Univ.*
Liwei Lin, *Univ. of California at Berkeley*
Masood Tabib-Azar, *Case Western Reserve Univ.*
George F. Watson, copy-editor

I am especially indebted to Cheah Choo Lek of Ngee Ann Polytechnic of Singapore, for his meticulous and thoughtful reading of the manuscript and many suggestions for improvement. Last, but not the least, I wish to acknowledge the excellent services provided by the editorial and production staff at McGraw-Hill. The support and able assistance of Jonathan Plant are greatly appreciated.

Tai-Ran Hsu,
San Jose, California

SUGGESTIONS TO INSTRUCTORS

This book is written as a textbook for upper division undergraduate and entry-level graduate students. It is also intended to be a reference book that will usher practicing engineers into the growing field of microsystems engineering. Its content is designed for 15-week-long semester classes at both undergraduate and graduate levels. With significant omission of materials, the book can also be used for classes covering a 10-week long quarter.

The textbook will be more effective for students in the classes with the following academic background and experience.

1. Undergraduate students in good upper division academic standing with sound knowledge of college mathematics, physics, and chemistry.
2. Students who have satisfied the prerequisite courses, or their equivalent, in fundamentals of electrical and electronic engineering, material sciences, engineering mechanics, and machine design.
3. Students with working experience in mathematics software such as MathCAD or MatLAB.

Teaching a course on MEMS with extreme diversity of subjects such as in this course in a 15-week-long semester or 10-week-long quarter with 3-hour-per-week lectures presents a major challenge to instructors. The problem may be further compounded by the likelihood of having students with diversified academic backgrounds and experiences in the same class. The following Schedules A and B offer suggested topics that the instructor may cover in either one semester or in one quarter.

Schedule A | For 15-week-long semesters

Week no.	Undergraduate classes	Graduate classes
1	Chap. 1	Chap. 1
2	Chap. 2	Chap. 2
3	Chap. 3	Secs. 3.5, 3.8, and 3.9. Assign the rest of the chapter as reading material. Secs. 4.1 and 4.2.
4	Secs. 4.1, 4.2, and 4.3	Secs. 4.3 and 4.4.
5	Secs. 4.4, 4.6, and 4.7	Remainder of Chap. 4 and Secs. 5.1 and 5.2
6	Secs. 5.1, 5.2, 5.4, and 5.6	Secs. 5.3 to 5.7
7	Secs. 5.8 and 5.9	Secs. 5.8 to 5.10
8	Chap. 6	Chap. 6 and Secs. 7.1 to 7.3
9	Secs. 7.1 to 7.5	Secs. 7.4 to 7.11
10	Secs. 7.6 to 7.11	Secs. 8.1 to 8.5
11	Secs. 8.1 to 8.5	Secs. 8.6 to 8.10
12	Secs. 8.6 to 8.10	Chap. 9, Secs. 10.1 and 10.2
13	Chap. 9	Secs. 10.3 to 10.6
14	Secs. 10.1 to 10.4, 10.6, 10.7, and 10.9	Secs. 10.7 to 10.9
15	Secs. 11.1 to 11.8, and 11.10	Chap. 11

Schedule B | For 10-week-long quarters

Week no.	Undergraduate classes	Graduate classes
1	Chap. 1 and 2 (reading and comprehension)	Chaps. 1 and 2 (reading and comprehension)
2	Secs. 3.5, 3.8, and 3.9	Chap. 3 (reading)
	Secs. 4.1, 4.2, 4.3.1, 4.3.2, 4.3.3, and 4.3.6	Chap. 4 (skip Secs. 4.3.1 and 4.3.2)
3	Secs. 4.4, 4.7, 5.4, and 5.6	Secs. 5.1, 5.4 to 5.10
4	Secs. 5.8, 5.9, and 5.10	Chap. 6
5	Secs. 7.1 to 7.9	Chap. 7
6	Secs. 7.10 and 7.11, and 8.1 to 8.3	Secs. 8.1 to 8.5
7	Secs. 8.4 to 8.10	Secs. 8.6 to 8.10, 9.1, and 9.2
8	Chap. 9	Secs. 9.3, 9.4, and 10.1 to 10.4
9	Secs. 10.1, 10.2, 10.4, 10.7, and 10.9	Secs. 10.5 to 10.9, and 11.1 to 11.3
10	Secs. 11.1 to 11.6	Secs. 11.4 to 11.8 and 11.10

Instructors, of course, may use their own discretion in selecting the topics other than those suggested in the above tables to suit their specific preferences and schedules.

The author would like to encourage the use of the blackboard for examples offered by the textbook, rather than have these examples swiftly disseminated to the students by view graphs or slides. Design projects on carefully chosen topics will be a valuable experience for students in learning topics such as specific microfabrication processes in depth. These projects may be assigned to groups of two or three students. Due credit, e.g., 20 percent of the total mark, needs be assigned to the projects. Extra time slots, or a "Project Day," may be reserved for the presentation of these projects for peer review among all students in the class. These presentations will benefit all students in the class as they learn from each other through presentations followed by active discussion and questioning. Instructors may incorporate ideas and results of these projects as additional materials for their lectures in future years.

ABOUT THE AUTHOR

Tai-Ran Hsu received his B.S. degree from the National Cheng–Kung University, Taiwan, China, an M.S. degree from the University of New Brunswick, Fredericton, N.B., Canada, and a Ph.D. degree from McGill University, Montreal, Canada. All his degrees were in mechanical engineering.

He is currently a Professor of Mechanical and Aerospace Engineering at San Jose State University (SJSU), San Jose, California. He joined SJSU as the Chair of the department in 1990—the position that he held until 1998. He served in a similar capacity at the University of Manitoba, Winnipeg, Canada before joining SJSU. Prior to his academic career, he worked extensively as a design engineer with heat exchangers, steam power plants, large steam turbines, and nuclear reactor fuel systems for major industries in both the United States and Canada. He has published four books on the finite element method and computer-aided design, and over 70 journal papers in thermomechanics with applications ranging from nuclear reactors to microelectronics packaging, His other research activities involved laser holographic interferometry for high precision metrology and computer-aided design. He was a pioneer in developing and installing an undergraduate mechatronics curriculum stem at the SJSU in 1997, which was the first such curriculum stem in this country. His current research activities include microsystems design and packaging as well as engineering education that involves teaching of microsystems and nano- technologies to both undergraduate and graduate students.

He is actively involved in national and international conferences and symposia on microelectronics packaging, mechatronics and MEMS education and microsystems technology with prominent professional societies. He has also served as Guest Editors for special sections in mechatronics and MEMS packaging in the IEEE Transactions of CPMT and Advanced Packaging.

Chapter 1

Overview of MEMS and Microsystems

CHAPTER OUTLINE

1.1 | MEMS AND MICROSYSTEMS

The term MEMS is an abbreviation of microelectromechanical system. A MEMS contains components of sizes in 1 micrometer (μm) to 1 millimeter (mm), (1 mm = 1000 μm). A MEMS is constructed to achieve a certain engineering function or functions by electromechanical or electrochemical means.

The core element in MEMS generally consists of two principal components: a sensing or actuating element and a signal transduction unit. Figure 1.1 illustrates the functional relationship between these two components in a microsensor.

Figure 1.1 I MEMS as a microsensor.

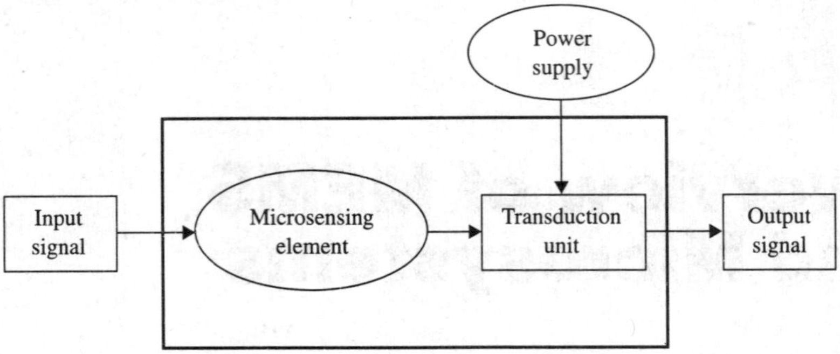

Micro sensors are built to sense the existence and the intensity of certain physical, chemical, or biological quantities, such as temperature, pressure, force, sound, light, nuclear radiation, magnetic flux, and chemical compositions. Microsensors have the advantages of being sensitive and accurate with minimal amount of required sample substance. They can also be mass produced in batches with large volumes.

There are many different types of microsensors developed for a variety of applications, and they are widely used in industry. Common sensors include biosensors, chemical sensors, optical sensors, and thermal and pressure sensors. Working principles of these sensors will be presented in Chapter 2. In a pressure sensor, an input signal such as pressure from a source is sensed by a microsensing element, which may include simply a silicon diaphragm only a few micrometers thick as illustrated in Figures 1.22c and 2.7. The deflection of the diaphragm induced by the applied pressure is converted into a change of electrical resistance by micropiezoresistors that are implanted in the diaphragm. The piezoresistors constitute a part of the transduction unit. The change of electrical resistance in the resistors induced by the change of the crystal structure geometry can be further converted into corresponding voltage changes by a micro Wheatstone bridge circuit also attached to the sensing element as another part of the transduction unit (see Fig. 2.8). The output signal of this type of microsensor is thus in the voltage change corresponding to the input pressure. Typical micro pressure sensors are shown as packaged products in Figure 1.2.

There are many other types of microsensors that are either available in the marketplace or being developed. They include chemical sensors for detecting chemicals or toxic gases such as CO, CO_2, NO, O_3, and NH_3, etc. either from exhaust from a combustion or a fabrication process, or from the environment for air quality control.

Biomedical sensors and biosensors will have significant share in the micro sensor market in the near future. Micro biomedical sensors are mainly used for diagnostic analyses. Because of their miniature size, these sensors typically require a minute amount of sample and can produce results significantly faster than the traditional

Figure 1.2 | Micro pressure sensor packages.

biomedical instruments. Moreover, these sensors can be produced in batches, resulting in low unit cost. Another cost saving is that most of these sensors are disposable; manual labor involving cleaning and proper treatment for reuse can be avoided. Biosensors are extensively used in analytical chemistry and biomedicine as well as genetic engineering.

Working principles for many of these microsensors will be presented in Section 2.2. Chapters 4 and 5 will present the design methodologies for these sensors.

Figure 1.3 illustrates the functional relationship between the actuating element and the transduction unit in a microactuator. The transduction unit converts the input power supply into the form such as voltage for a transducer, which functions as the actuating element.

Figure 1.3 | MEMS as a microactuator.

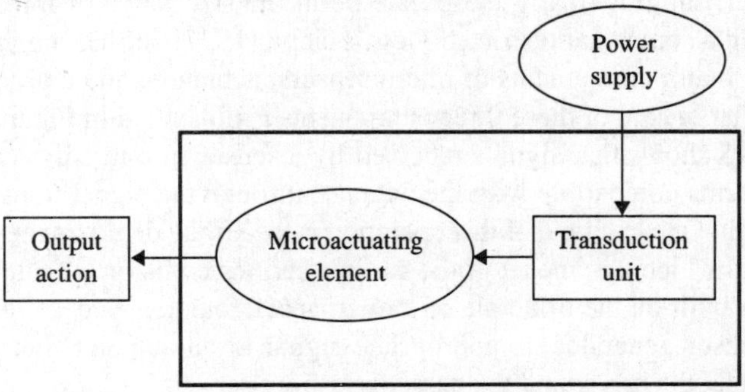

There are several ways by which microdevice components can be actuated, as will be described in Section 2.3. One popular actuation method involves electrostatic forces generated by charged parallel conducting plates, or electrodes separated by a dielectric material such as air. The arrangement is similar to that of capacitors. The application of input voltage to the plates (i.e., the electrodes in a capacitor) can result in electrostatic forces that prompt relative motion of these plates in normal direction of aligned plates or parallel movement for misaligned plates. These motions are set to accomplish the required actions. Electrostatic actuation is used in many microactuators. One such application is in a microgripper, as illustrated in Figure 1.4. Chapter 2 will include several other actuation methods and other types of devices.

Figure 1.4 | MEMS using electrostatic actuation.

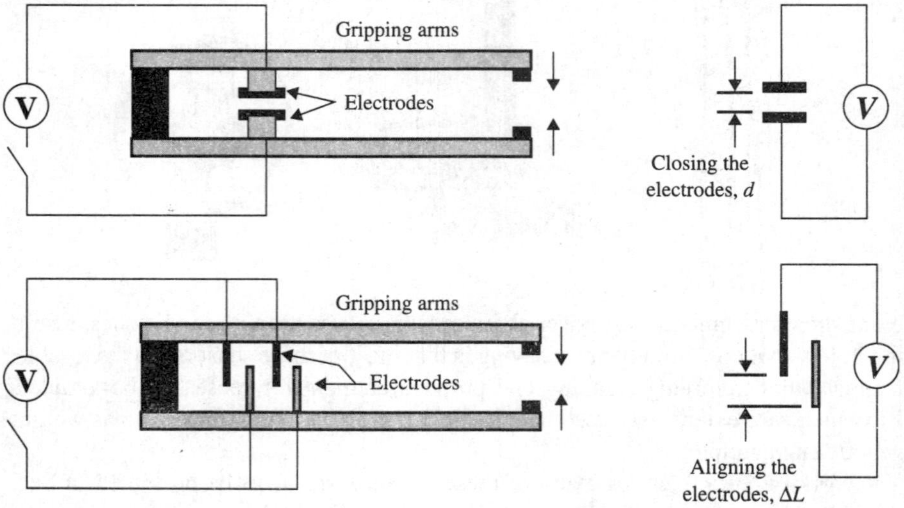

A *microsystem* is an engineering system that contains MEMS components that are designed to perform specific engineering functions. Despite the fact that many MEMS components can be produced in the size of micrometers, microsystems are typically in "mesoscales." (There is no clear definition of what constitutes the mesoscale. It implies a scale that is between micro and macro scales. Conventional wisdom, however, suggests that a mesoscale be in the size range of millimeters to a centimeter.) In a somewhat restricted view, Madou [1997] defines a microsystem to include three major components of micro sensors, actuators, and a processing unit. Functional relationship of these three components is illustrated in Figure 1.5.

Figure 1.5 shows that signals received by a sensor in a microsystem are converted into forms compatible with the actuator through the signal transduction and processing unit. One example of this operation is the airbag deployment system in an automobile, in which the impact of the car in a serious collision is "felt" by a micro inertia sensor built on the principle of a microaccelerometer, as described in Chapter 2. The sensor generates an appropriate signal to an actuator that deploys the airbag to protect the driver and the passengers from serious injuries.

Figure 1.6 shows a micro inertia sensor produced for this purpose. The sensor, which contains two microaccelerometers, is mounted to the chassis of the car. The accelerometer on the left measures the deceleration in the horizontal (x) direction, whereas the unit on the right measures the deceleration in the y direction. Both these accelerometers are mounted on the same integrated circuit (IC) chip with

Figure 1.5 | Components of a microsystem.

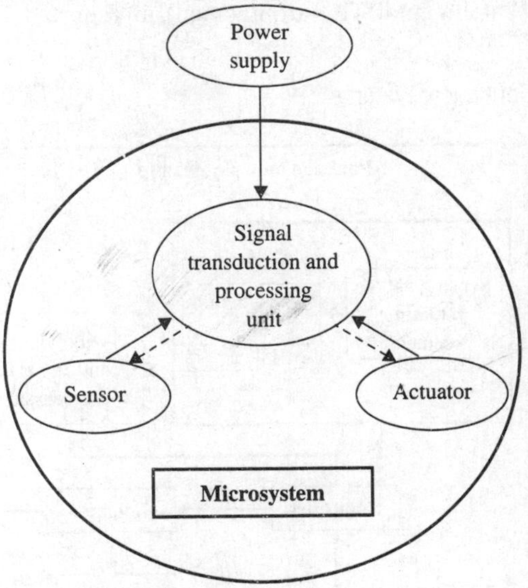

signal transduction and processing units. The entire chip has an approximate size of 3 × 2 mm, with the microaccelerometers taking about 10 percent of the overall chip area. The working principle of this type of microaccelerometer will be explained in Section 2.5.

Most microsystems are designed and constructed to perform single functions such as illustrated above. There is a clear trend in the industry to incorporate signal

Figure 1.6 | An Analog Devices ADXL276/ADXL 250 microaccelerometer

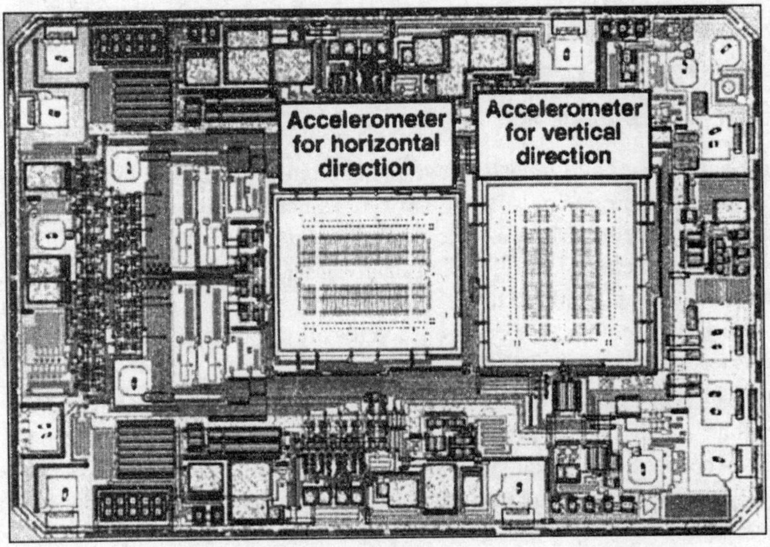

(Courtesy of Analog Devices, Norwood, Massachusetts.)

processing and closed-loop feedback control systems in a microsystem to make the integrated system "intelligent." The arrangement in Figure 1.7 illustrates such a possibility.

Figure 1.7 | Intelligent microsystems.

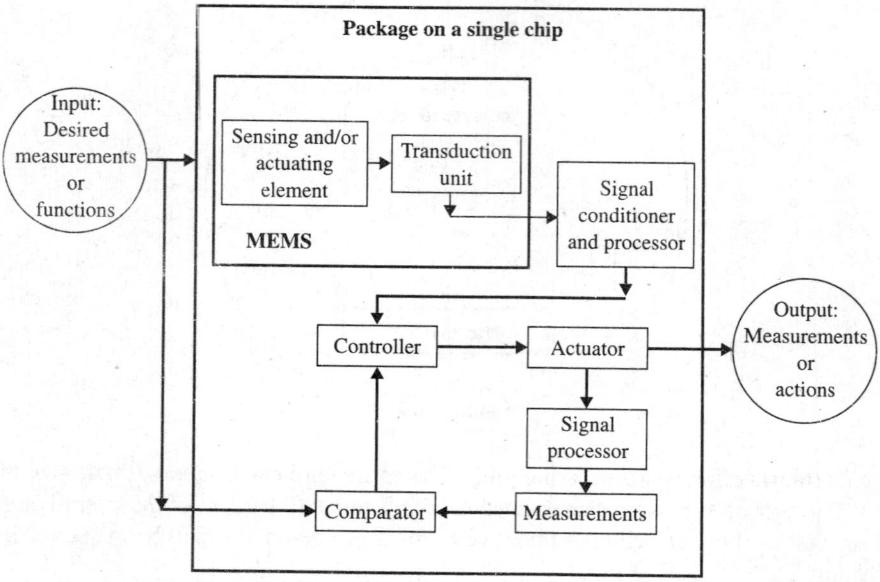

Many microsystems have been built on the "lab-on-a-chip" concept. The entire unit can be contained in a silicon chip of the size less than 0.5 × 0.5 mm. Figure 1.8 shows such an example for two different designs of microaccelerometers, or inertia sensors, used in air bag deployment systems of automobiles.

1.2 | TYPICAL MEMS AND MICROSYSTEM PRODUCTS

Research institutions and industry have made relentless effort in the past 2 decades in developing and producing smaller and better microelectromechanical (MEM) devices and components. Many of these miniaturized devices, such as silicon gear trains and tongs, were reported by Mehregany et al. [1988]. Following is a list, with brief descriptions, of the MEM devices and components that were produced in recent years. The list, of course, is becoming longer all the time as micromanufacturing technology advances.

1. *Microgears.* Figure 1.9a shows a gear made to a size that is significantly smaller than an ant's head, whereas Figure 1.9b shows a two-level gear made from ceramics. The pitch of the gears is in the order of 100 μm. Both these gears were produced by the LIGA (a German acronym of Lithographie Galvanoformung Abformung) process, as will be described in detail in Chapter 9.

Figure 1.8 | Intelligent inertia sensors used in automobile air bag deployment systems: **(a)** Inertia sensor on chip. **(b)** Packaged sensor on chip.

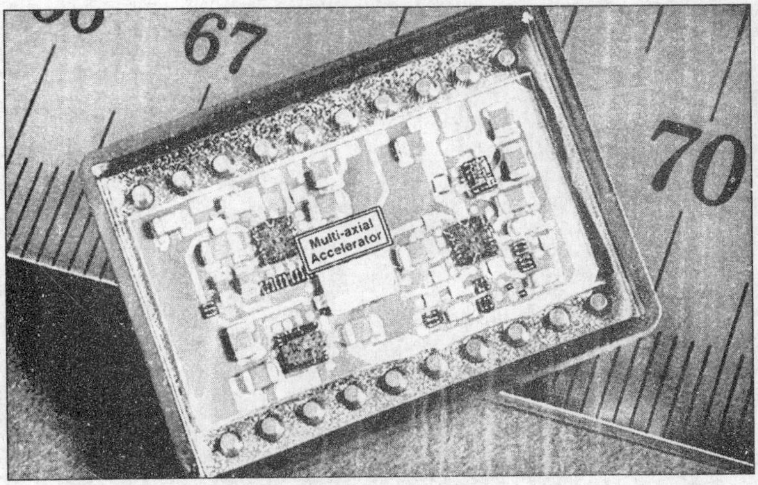

(a)

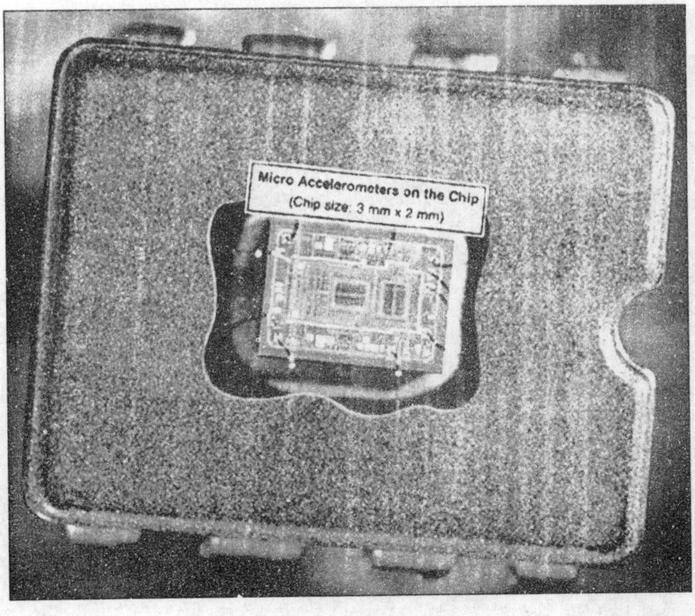

(b)

(**a:** Courtesy of Karlsruhe Nuclear Research Center, Karlsruhe Germany; **b:** courtesy of Analog Devices Inc., Norwood, Massachusetts.)

Figure 1.9 | Microgears produced by the LIGA process: **(a)** Microgear at the tip of an ant's leg. **(b)** Two-level ceramic gear.

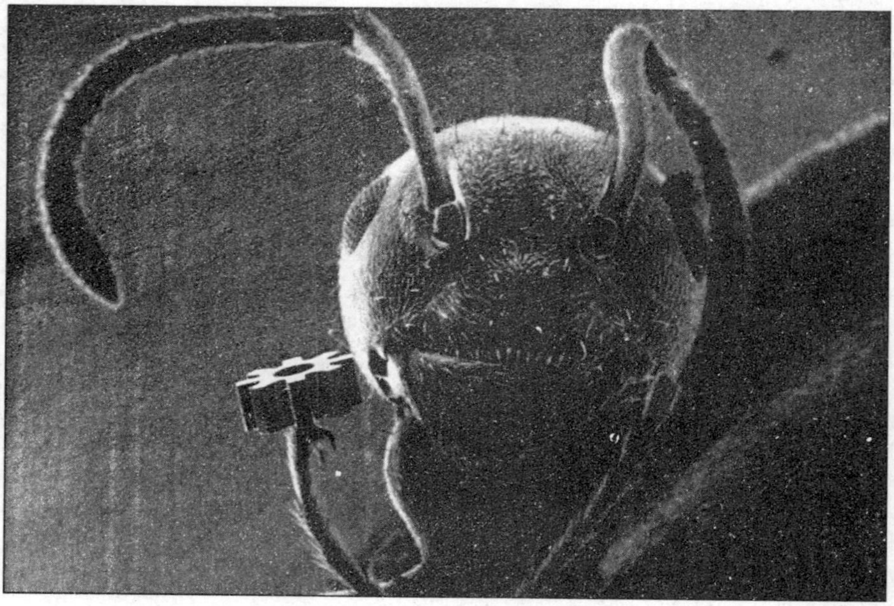

(a)

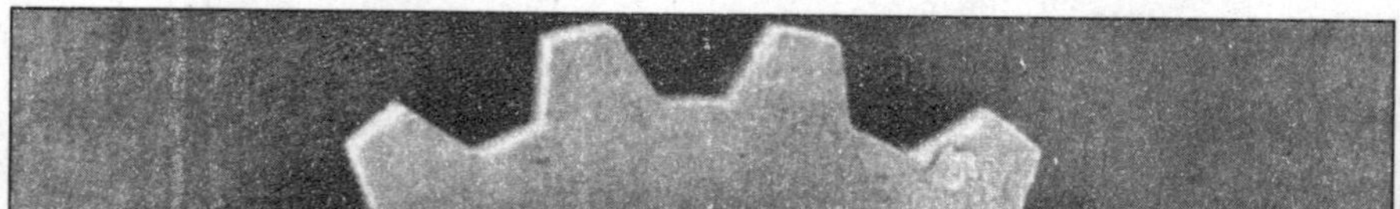

(b)

(**b:** Courtesy of Karlsruhe Nuclear Research Center, Karlsruhe, Germany.)

2. *Micromotors.* Figure 1.10 shows an electrostatic-driven micromotor produced by the LIGA process [Bley 1993]. All three components—the rotcr (the center gear), the stator, and the torque transmission gear—are made of nickel. The toothed rotor, which has a diameter of 700 μm, is engaged to a gear wheel with 250 μm diameter. The latter wheel transmits the torque produced by the motor. The gap between the rotor and the axle and between the rotor and the stator is 4 μm. The height of the unit is 120 μm.

Figure 1.10 I A Micromotor produced by the LIGA process.

(Bley [1993], with permission.)

3. *Microturbines.* A microturbine was produced to generate power [Bley 1993]. As shown in Figure 1.11, the turbine is made of nickel. The rotor has a diameter of 130 μm. A 5-μm gap is provided between the axle and the rotor. The turbine has a height of 150 μm. The entire unit is made of nickel. The maximum rotational speed reaches 150,000 revolutions per minute (rpm), with a lifetime up to 100 million rotations.

4. *Micro-optical components.* These components are extensively used for high-speed signal transmission in the telecommunication industry. Here, we show two such micro-optical components in Figures 1.12 and 1.13. In Figure 1.12 is a micro-optical switch made by a silicon-based manufacturing process. These switches are used to regulate incident light from optical fibers (shown as cylinders in the figure) to appropriate receiving optical fibers. Figure 1.13 shows microlenses made of transparent polymers, PMMA. Each lens shown in the left of the figure has a diameter of 150 μm. These arrays of lenses are combined into micro-objectives for endoscopy with an optical resolution down to 3 μm. These lenses can also be used for copiers,

Figure 1.11 | A microturbine.

(Bley [1993], reproduced with permission.)

Figure 1.12 | Lucent MEMS 1 × 2 optical switch.

(Courtesy of Bell Laboratories of Lucent Technologies.)

Figure 1.13 | Arrays of high-aperture microlenses.

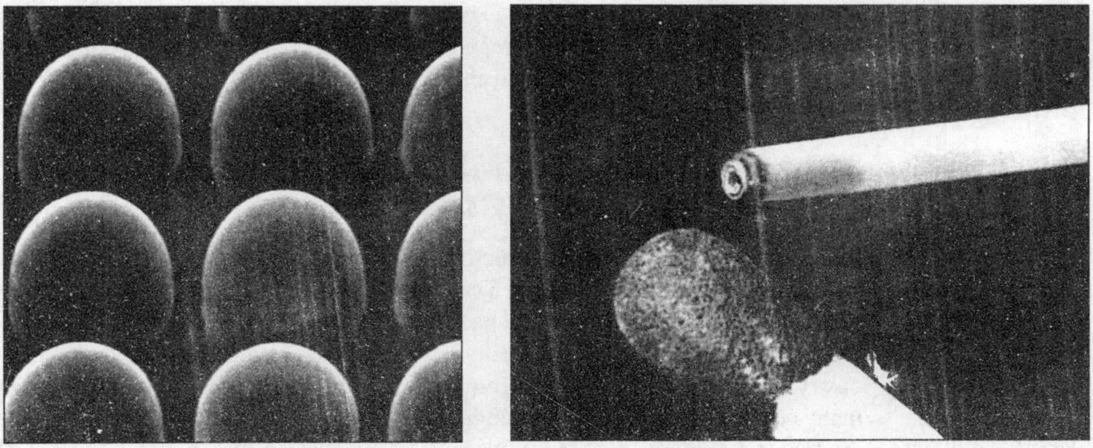

(Reproduced with permission from Bley [1999].)

laser scanners, and printers. At the right is a combination of one such lens with a micro-objective for neurosurgery [Bley 1999].

1.3 | EVOLUTION OF MICROFABRICATION

The size of components of microsystems has been decreasing continuously, with some of the MEMS components made in submicrometer sizes. Fabrication of device components of this minute size is clearly beyond traditional mechanical means such as machining, drilling, milling, forging, casting, and welding. The technologies used to produce these minute components are called *microfabrication technologies*, or *micromachining*. For example, the three-dimensional microstructures can be produced by removing part of the base material by a physical or chemical etching process, whereas thin-film deposition techniques are used to build layers of materials on the base materials. We will learn these microfabrication processes and manufacturing technologies in detail in, respectively, Chapters 8 and 9.

Current micromanufacturing processes involve significant amount of microfabrication processes that were developed for the microelectronics industry in the last 50 years. Microfabrication processing is thus an important part of microsystems design, manufacture, and packaging.

Many would attribute the origin of modern microfabrication to the invention of transistors by W. Schockley, J. Bardeen, and W. H. Brattain in 1947. Indeed, microfabrication technology is inseparable from the integrated circuit (IC) fabrication technology that is used to produce microelectronics. The IC concept first evolved from the production of a monolithic circuit at RCA in 1955 after the invention of transistors. The first IC was produced 3 years later by Jack Kilby of Texas Instruments. It was in the same year that Robert Noyce of Fairchild Semiconductor produced the first planar silicon-based IC. Today, the next generation of IC, i.e., ultra large scale integration (ULSI) can contain 10 million transistors and capacitors on a chip that is smaller than a fingernail.

The term *micromachining* first appeared in public literature in 1982, although some of its key processes were in existence long before. For example, a key fabrication process known as *isotropic etching* of silicon was first used in 1960, and the improved *anisotropic etching* was invented in 1967.

1.4 | MICROSYSTEMS AND MICROELECTRONICS

It is a well-recognized fact that microelectronics is one of the most influential technologies of the twentieth century. The boom of the microelectromechanical systems industry in recent years would not have been possible without the maturity of microelectronics technology. Indeed, many engineers and scientists in today's MEMS industry are veterans of the microelectronics industry, as the two technologies do share many common fabrication technologies. However, overemphasis on the similarity of the two technologies is not only inaccurate, but it can also seriously hinder further advances of microsystems development. We will notice that there are significant differences in the design and packaging of microsystems from that of integrated circuits and microelectronics. It is essential that engineers recognize these differences and develop the necessary methodologies and technologies accordingly. Table 1.1 summarizes the similarities and differences between the two technologies.

Table 1.1 | Comparison of Microelectronics and Microsystems

Microelectronics	Microsystems (silicon-based)
Uses single crystal silicon die, silicon compounds, and plastic	Uses single-crystal silicon die and a few other materials, such as GaAs, quartz, polymers, and metals
Transmits electricity for specific electrical functions	Performs a great variety of specific biological, chemical, electromechanical, and optical functions
Stationary structures	May involve moving components
Primarily 2-D structures	Complex 3-D structures
Complex patterns with high density over substrates	Simpler patterns over substrates
Fewer components in assembly	Many components to be assembled
IC die is completely protected from contacting media	Sensor die is interfaced with contacting media
Mature IC design methodology	Lack of engineering design methodology and standards
Large number of electrical feedthroughs and leads	Fewer electrical feedthroughs and leads
Industrial standards available	No industrial standards to follow
Mass production	Batch production or on customer-needs basis
Fabrication techniques are proved and well documented	Many microelectronics fabrication techniques can be used for production
Manufacturing techniques are proved and well documented	Distinct manufacturing techniques
Packaging technology is relatively well established	Packaging technology is at the infant stage

We may observe from the table that there are indeed sufficient differences between the two technologies. Some of the more significant differences between these two technologies are:

1. Microsystems involve more different materials than microelectronics. Other than the common material of silicon, there are other materials such as quartz and GaAs used as substrates in microsystems. Polymers and metallic materials are common in microsystems produced by LIGA processes. Packaging materials for microsystems include glasses, plastic, and metals, which are excluded in microelectronics.
2. Microsystems are designed to perform a greater variety of functions than microelectronics. The latter are limited to specific electrical functions only.
3. Many microsystems involves moving parts such as microvalves, pumps, and gears. Many require fluid flow through the systems such as biosensors and analytic systems. Micro-optical systems need to provide input/output (I/O) ports for light beams. Microelectronics, on the other hand, does not have any moving component or access for lights or fluids.
4. Integrated circuits are primarily a two-dimensional structure that is confined to the silicon die surface, whereas most microsystems involve complicated geometry in three dimensions. Mechanical engineering design is thus an essential part in the product development of microsystems.
5. The integrated circuits in microelectronics are isolated from the surroundings once they are packaged. The sensing elements and many core elements in microsystems, however, are required to be in contact with working media, which creates many technical problems in design and packaging.
6. Manufacturing and packaging of microelectronics are mature technologies with well-documented industry standards. The production of microsystems is far from that level of maturity. There are generally three distinct manufacturing techniques, as will be described in Chapter 9. Because of the great variety of structural and functional aspects in microsystems, the packaging of these products is indeed in its infant stage at the present time.

The slow advance in the development of microsystems technology is mainly attributed to the complex nature of these systems. As we will learn from Section 1.5, there are many science and engineering disciplines involved in the design, manufacture, and packaging of microsystems.

1.5 | THE MULTIDISCIPLINARY NATURE OF MICROSYSTEM DESIGN AND MANUFACTURE

Despite the fact that micromanufacturing evolved from IC fabrication technologies, there are several other science and engineering disciplines involved in today's microsystems design and manufacturing. Figure 1.14 illustrates the applications of principles of natural science and several engineering disciplines in this process.

Figure 1.14 | Principal science and engineering disciplines involved in microsystem design and manufacture.

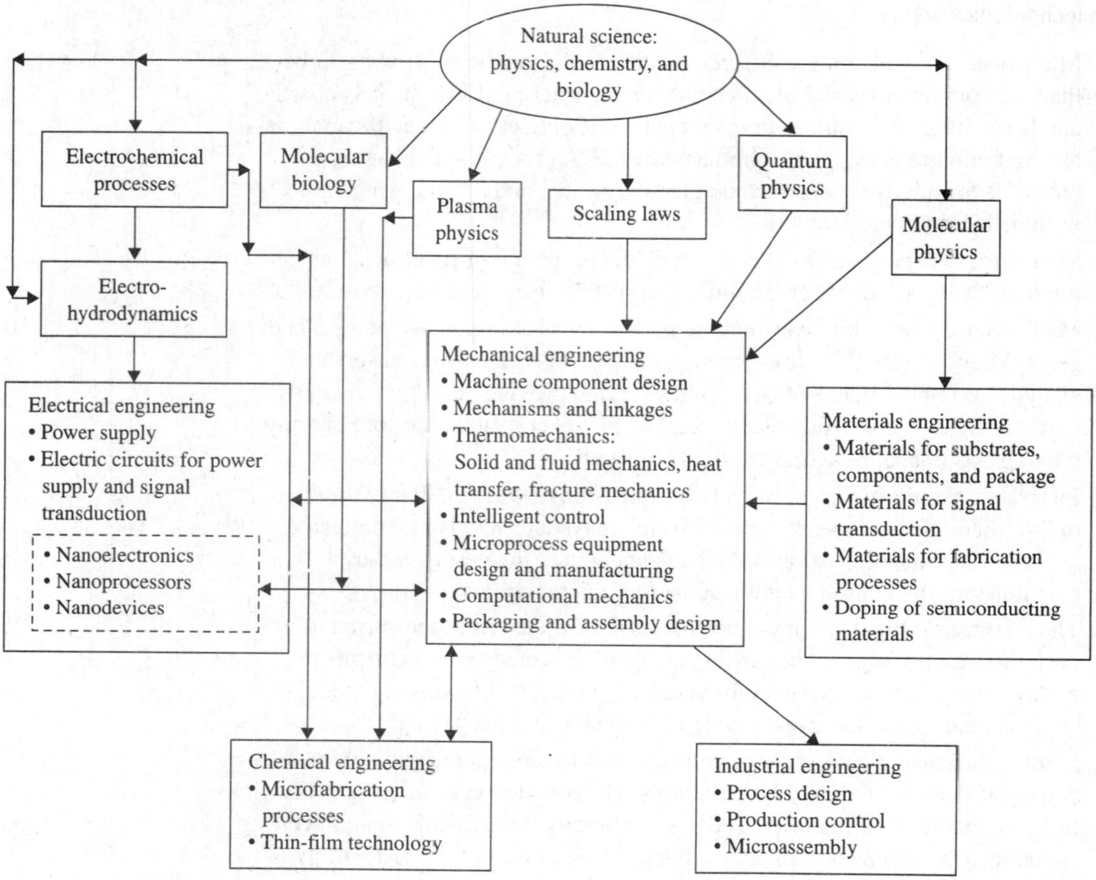

With reference to Figure 1.14, natural science is deeply involved in the following areas:

1. Electrochemistry is widely used in electrolysis to ionize substances in some micromanufacturing processes. Electrochemical processes are also used in the design of chemical sensors. More detail will be given in Chapters 2 and 3.
2. Electrohydrodynamics principles are used as the driving mechanisms in fluid flows in microchannels and conduits, such as those for capillary fluid flow, as will be described in Chapters 5 and 10.
3. Molecular biology is intimately involved in the design and manufacture of biosensors and biomedical equipment, as will be shown in Chapter 2. Much of the basic molecular biology principles are used in nanotechnology to make products such as nanoprocessors and nanodevices.
4. Plasma physics involves the production and supply of ionized gases with high energy. It is required for etching and deposition in many microfabrication

processes. The generation of plasma will be covered in Chapter 3, whereas the application of plasma in microfabrication will be described in Chapter 8.

5. Scaling laws provide engineers with a sense for the scaling down of physical quantities involved in the design of microdevices. We will realize from Chapter 6 that not all physical quantities can be scaled down favorably.
6. Quantum physics is used as the basis for modeling certain physical behaviors of materials and substances in microscales, as will be described in regard to microfluid flow and heat transportation in solids in Chapter 5.
7. Molecular physics provides many useful models in the description of materials at microscales, as well as the alteration of material properties and characteristics used in microsystems, as will be described in Chapters 3 and 7. Molecular dynamics theories are the principal modeling tool for describing mechanical behavior of materials in nanoscale.

Five engineering disciplines are involved in microsystem design, manufacture, and packaging as described below:

1. Mechanical engineering principles are used primarily in the design of microsystem structures and the packaging of the components. These would involve many aspects of design analyses as indicated in the central box in Figure 1.14. Intelligent control of microsystems has not been well developed, but it is an essential part of *micromechatronics systems*, which are defined as intelligent microelectromechanical systems.
2. Electrical engineering involves electrical power supplies and the functional control and signal processing circuit design. For integrated microsystems, e.g., "laboratory-on-a-chip," the IC and microelectronic circuitry that integrates microelectronics and microsystems makes electrical engineering a major factor in the design and manufacturing processes.
3. Chemical engineering is an essential component in microfabrication and micromanufacturing, as will be described in Chapters 8 and 9. Almost all such processes involve chemical reactions. Some of microdevice packaging techniques also rely on special chemical reactions, as will be described in Chapter 11.
4. Materials engineering offers design engineers a selection of available materials that are amenable to microfabrication and manufacturing, as well as packaging. Theories of molecular physics are often used in the design of materials' characteristics, such as doping of semiconducting materials for changing electrical resistivity of the material. Materials engineering plays a key role in the development of chemical, biological, and optical sensors, as will be described in Chapter 2.
5. Industrial engineering relates to the production and assembly of microsystems. Optimum design of the fabrication process and control is essential in microsystem production.

1.6 | MICROSYSTEMS AND MINIATURIZATION

It is fair to say that microsystems, as we have defined them in Section 1.1, are a major step toward the ultimate miniaturization of machines and devices, such as dust-size computers and needle-tip-size robots, that have fascinated many futurists in their seemingly science fiction articles and books. The current microsystem technology, though it is at a relative early stage, has already set the tone for the development of device systems in nanoscale for this new century. [A nanometer (nm) is 10^{-9} meter.] The maturity of nanotechnology will certainly result in the realization of much of the superminiaturized machinery that engineers and scientists have fantasized at the present time.

According to Webster's dictionary, the word *miniature* means a copy on a much-reduced scale. In essence, miniaturization is an art that substantially reduces the size of the original object yet retains the characteristics of the original (and more) in the reduced copy. The need for miniaturization has become more prominent than ever in recent decades, as engineering systems and devices have become more and more complex and sophisticated. The benefits of having smaller components, and hence a device or machine with enhanced capabilities and functionalities, are obvious from engineering perspectives:

- Smaller systems tend to move more quickly than larger systems because of lower inertia of the mass.
- The minute sizes of small devices encounter fewer problems in thermal distortion and vibration because resonant vibration of a system is inversely proportional to the mass. Smaller systems with lower masses have much higher natural frequencies than those expected from most machines and devices in operations.
- In addition to the more accurate performance of smaller systems, their minute size makes them particularly suitable for applications in medicine and surgery and in microelectronic assemblies in which miniaturized tools are necessary.
- Miniaturization is also desirable in satellites and spacecraft engineering to satisfy the prime concerns about high precision and payload size.
- The high accuracy of miniaturized systems in motion and dimensional stability make them particularly suitable for telecommunication systems.

Miniaturization is the only way to have new and competitive engineering systems performing multifunctions with manageable sizes.

The process of miniaturizing engineering systems is an ongoing effort by engineers. We have witnessed many of the results in consumer products with much better performance but much reduced size. Such products include hair dryers, cameras, radios, and telephone sets. We will see from the following examples how these efforts have produced even more staggering results in other fundamentally important engineering products in the last 50 years. There is no doubt that with the rapid development of microsystem technologies, such process will be accelerated at an even more spectacular rate.

Perhaps the foremost example of miniaturization began with the development of integrated circuits in the mid-1950s. As described in Section 1.3, a ULSI chip the size of a small fingernail can contain 10 million transistors. Microprocessors of similar size can now perform over 300 million operations per second. The advances in IC and microprocessor technologies have led to a spectacular level of miniaturization of complex digital electronics computers. Two engineers, J. Presper Eckert and John Mauchly, publicly demonstrated the first general-purpose electronic digital computer called ENIAC (Electronic Numerical Integrator and Calculator) in 1946 at the Moore School of Electrical Engineering at the University of Pennsylvania. The U.S. Army during World War II originally funded this project. As shown in Figure 1.15, the U-shaped computer was 80 ft long by 8.5 ft high and several feet wide. Each of the twenty 10-digit accumulators was 2 ft long. In total, ENIAC had 18,000 vacuum tubes.

Figure 1.15 | The ENIAC digital computer in 1947.

(Used by permission of the University of Pennsylvania's School of Engineering and Applied Science, formerly known as the Moore School of Electrical Engineering.)

At the time of the ENIAC fiftieth anniversary celebration, Dr. J. Van der Spiegel worked with a group of students at the same university to reduce the original ENIAC to the size of a tiny chip with equivalent functional power (shown in Figure 1.16) [Van der Spiegel et al. 1998].

Figure 1.16 | The miniaturized ENIAC.

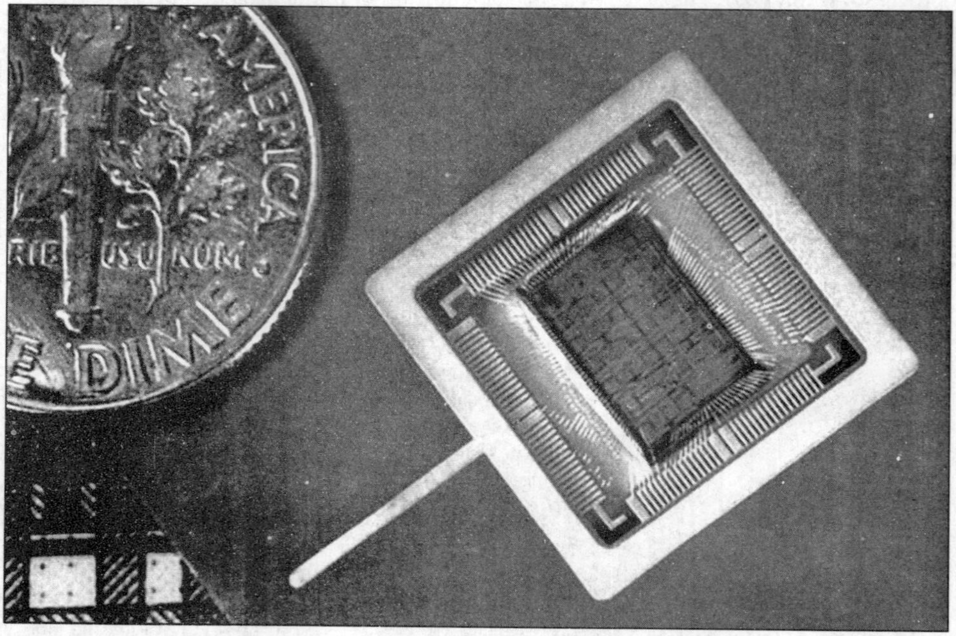

(Used by permission of the University of Pennsylvania's School of Engineering and Applied Science, formerly known as the Moore School of Electrical Engineering.)

In sharp contrast to ENIAC, a typical laptop computer, shown in Figure 1.17, is only about 14 in long × 11 in wide × 2 in thick. In size, ENIAC, estimated at about 4000 ft^3, was more than 5 orders of magnitude bigger than a laptop computer today, but the latter has a computational speed that is 6 orders of magnitudes faster than the ENIAC. This staggering miniaturization took place in just about 50 years!

The microfabrication technology evolved from IC fabrication in the last decade has made machine components in the sizes of micrometers and submicrometers possible. Many of these devices and components have been shown in previous sections. While most sensors and actuating components use silicon as the primary material, other materials such as nickel and ceramics have been used. High-aspect-ratio microstructures (aspect ratio = ratio of the dimensions of the height to those of the surface) have been successfully manufactured by the LIGA process, as shown in Figures 1.9, 10, and 11. This special micromanufacturing process will be described in detail in Chapter 9. The broader use of materials, along with the possibility of producing MEMS with high aspect ratio by the LIGA process have prompted the production of miniaturized machines. Machine tools, such as microlathes and millimeter-size drills have been produced.

Figure 1.17 | A laptop computer as of 1998.

Figure 1.18 shows a miniature autonomous robotic vehicle (MARV) produced by Sandia National Laboratories. The overall size of the vehicle is 1 cubic inch. The intended application of MARV includes microsurgery, miniaturized rovers on planetary missions, and tiny surveillance. The vehicle contains on-board computers, sensors, and control. It can thus perform fully self-contained operations as the mission requires.

Another spectacular micromachine example is the microcars that were produced by Denso Corporation in Japan. Figure 1.19 shows cars the size of rice grains. The smaller car shown in the figure weighs 33 mg. The overall dimensions are 4.785 mm long × 1.730 mm wide × 1.736 mm high. The body was made by nonelectrolysis nickel plating. The nickel film is 30 μm thick. These two cars were fabricated by micro precision machining with semiconductor processes. The electrical motor that drives the cars uses magnetic induction, with a rotor 0.6 mm in diameter. Detailed design methodology for these cars can be found in an article by Teshigahara et al. [1995].

1.7 | APPLICATIONS OF MICROSYSTEMS IN THE AUTOMOTIVE INDUSTRY

MEMS and microsystem products have become increasingly dominant in every aspect of commercial marketplace as the technologies for microfabrication and miniaturization continue to be developed. At the present time, two major commercial

Figure 1.18 | MARV, a miniature autonomous robotic vehicle.

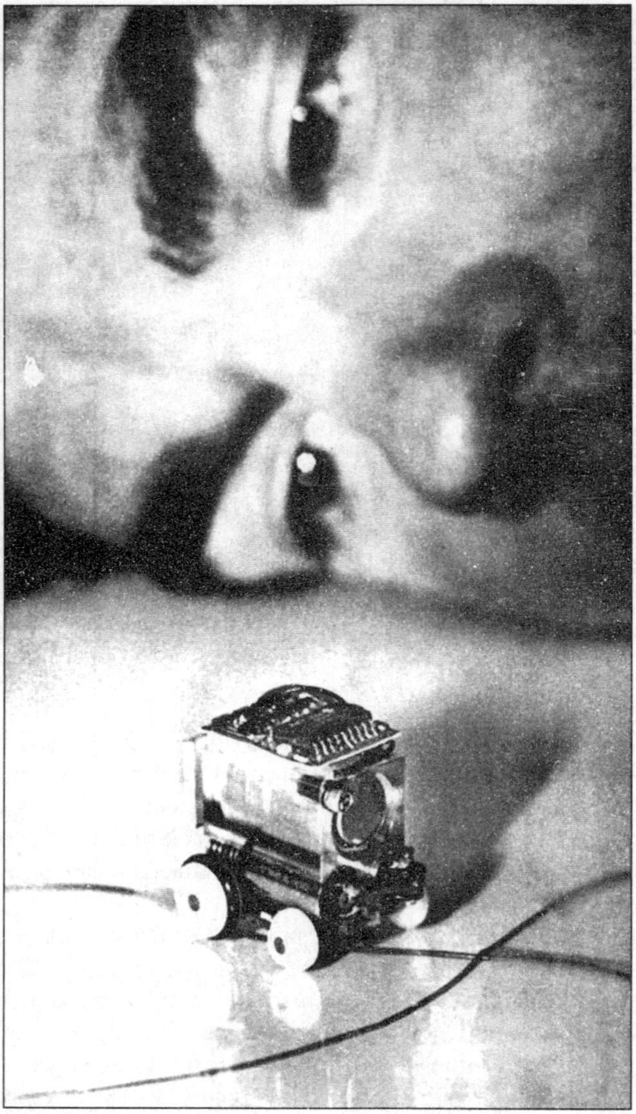

(Courtesy of the Intelligent Systems and Robotics Center, Sandia National Laboratories.)

markets for these products are computer storage systems and automobiles. We will focus our attention on the latter type of products because of its high content of microsensors and actuators.

The automotive industry has been the major user of MEMS technology in the last two decades because of the size of its market. A 1991 report indicated that the industry had a production of 8 million vehicles per year, with 6 million of these in the United States [Sulouff, Jr. 1991]. The primary motivation for adopting MEMS and

Figure 1.19 | Miniature electric cars: **(a)** Compared with rice grains. **(b)** Compared with an American 5-cent coin. **(c)** Compared with the tip of a match stick.

(a)

(b)

(c)

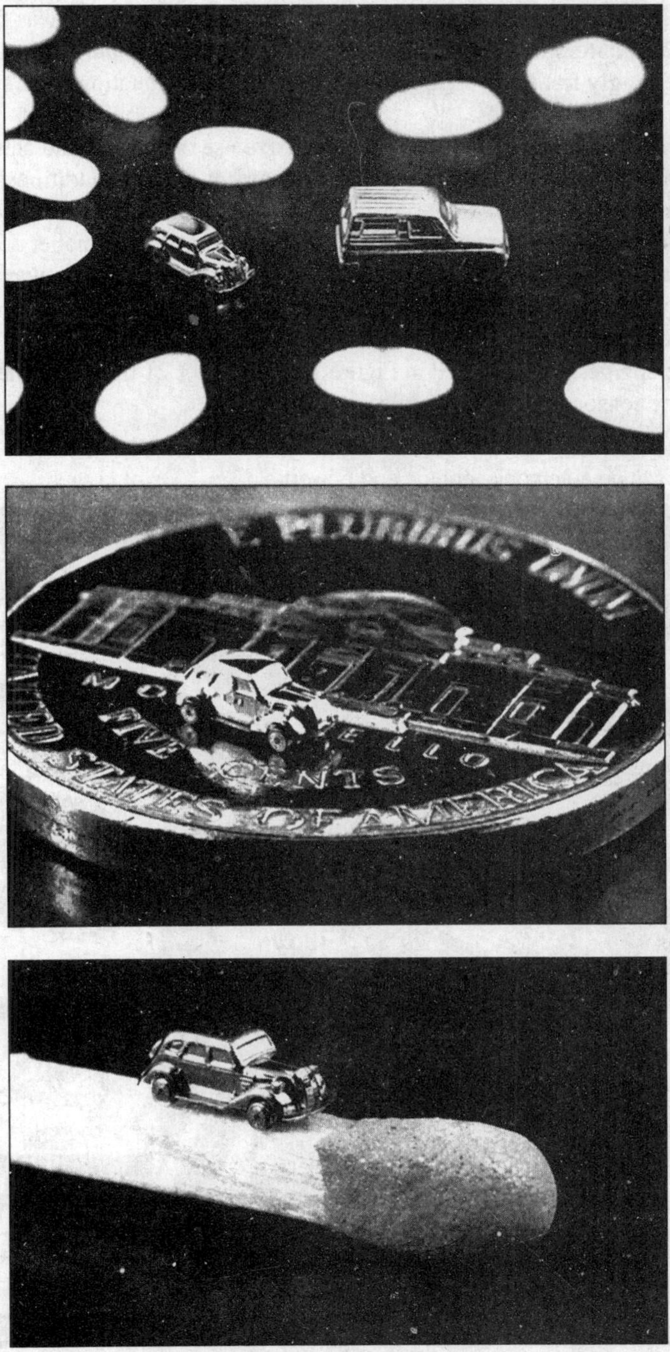

(Courtesy of Denso Research Laboratories, Denso Corporation, Aichi, Japan.)

microsystems in automobiles is to make automobiles safer and more comfortable for the riders and to meet the high fuel efficiencies and low emissions standards required by governments. In all, the widespread use of these products can indeed make the automobile "smarter" for consumers' needs. The term *smart cars* was first introduced in the cover story of a special issue of a trade magazine [*Smart Cars* 1988]. Many of the seemingly fictitious predictions of the intelligent functions of a smart car are in place in today's vehicles.

Smart vehicles are based on the extensive use of sensors and actuators. Various kind of sensors are used to detect the environment or road conditions, and the actuators are used to execute whatever actions are required to deal with these conditions. Microsensors and actuators allow automobile makers to use smaller devices, and thus more of them, to cope with the situation in much more effective ways. Comprehensive summaries on the use of sensors and actuators are available in an early report [Paulsen and Giachino 1989]. Sensors of various kinds are extensively used in automobile engines, as illustrated in Figure 1.20. Figure 1.21 illustrates the application of pressure sensors in an automobile.

Figure 1.20 | Sensors in an automobile engine and powertrain.

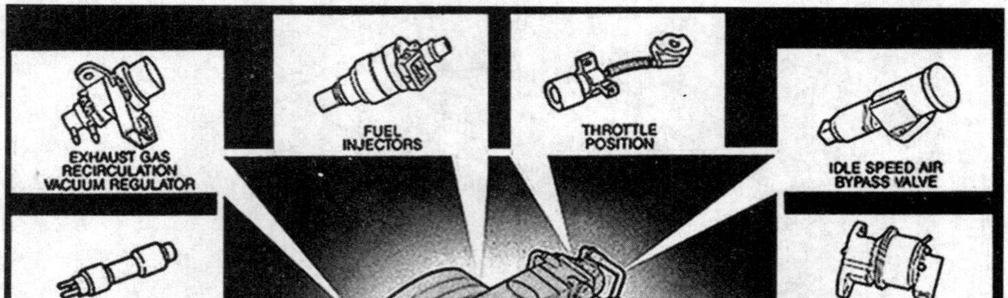

(Paulsen and Giachino [1989], with permission.)

Figure 1.21 | Pressure sensors in automotive applications.

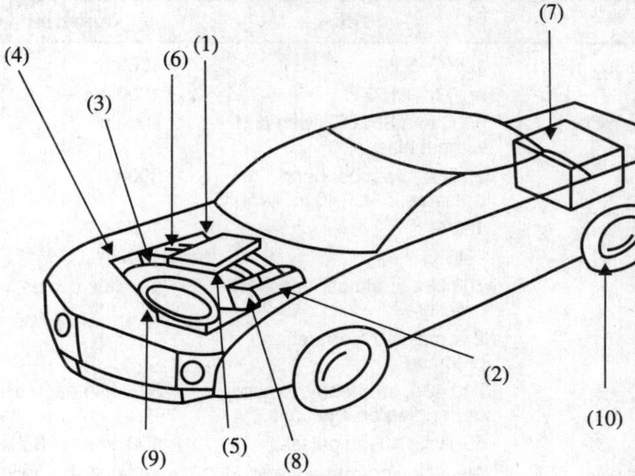

(1) Manifold or temperature manifold absolute pressure sensor
(2) Exhaust gas differential pressure sensor
(3) Fuel rail pressure sensor
(4) Barometric absolute pressure sensor
(5) Combustion sensor
(6) Gasoline direct injection pressure sensor
(7) Fuel tank evaporative fuel pressure sensor
(8) Engine oil sensor
(9) Transmission sensor
(10) Tire pressure sensor

(Chiou [1999], with permission.)

Design and production of automobiles satisfying these expectations presents a major challenge to engineers, as these vehicles are expected to perform in extreme environmental conditions as outlined in several reports [Giachino and Miree 1995, Chiou 1999]. Table 1.2 presents some of these stringent requirements for endurance tests for MEMS components used in the automotive industry.

In addition to the stringent endurance testing conditions for sensors and actuators presented in Table 1.2, they are expected to perform in the extremely harsh environmental conditions that the vehicles function in. Design conditions include a temperature range of -40 to $+125^{\circ}$C and a dynamic loading up to 50g [MacDonald 1990]. Table 1.2 presents some of the endurance testing conditions for pressure sensors [Chiou 1999]. The pressure range for automobiles extends from 100 kPa for a manifold absolute pressure (MAP) sensor to 10 MPa for brake pressure sensors.

Applications of microsystems in automobiles can be categorized into the following four major areas: (1) safety, (2) engine and power train, (3) comfort and convenience, and (4) vehicle diagnostics and health monitoring.

Safety

- Air bag deployment system to protect the driver and passengers from injury in event of serious vehicle collision. The system uses microaccelerometers or micro inertia sensors similar to those illustrated in Figures 1.6 and 1.8.

Table 1.2 | Typical durability requirements.

Tests	Conditions	Duration
High-temperature bias	100°C, 5 V	1000 h
Thermal shock	−40 to +150°C	1000 cycles
Temperature and humidity	85°C and 85% RH with and without bias	1000 h
Pressure, power, and temperature cycles	20 kPa to atmospheric pressure, 5 V, −40 to +150°C	3000 h
Hot storage	150°C	1000 h
Cold storage	−40°C	1000 h
Pressure cycling	700 kPa to atmospheric pressure	2 million cycles
Overpressure	2 × maximum operating pressure	
Vibration	2 to 40*g*, frequency sweep, depending on locations	30+ h in each axis
Shock	50*g* with 10 ms pulses	100 times in 3 planes
Fluid/media compatibility	Air, water, corrosive water, gasoline, methanol, ethanol, diesel fuel, engine oil, nitric and sulfuric acids	Varies with applications

Source: Chiou [1999].

- Antilock braking systems (position sensors).
- Suspension systems (displacement, position and pressure sensors, and microvalves).
- Object avoidance (pressure and displacement sensors).
- Navigation (microgyroscope).

Engine and Power-Train Various sensors have been used in the engines of modern automobiles, as illustrated in Figure 1.20 [Paulsen and Giachino 1989]. A few of these sensors are:

- Manifold control with pressure sensors.
- Airflow control
- Exhaust gas analysis and control (see Figure 1.21)
- Crankshaft positioning
- Fuel pump pressure and fuel injection control
- Transmission force and pressure control
- Engine knock detection for higher power output [Kaneyasu et al. 1995]

The manifold absolute pressure (MAP) sensor was one of the first microsensors adopted by the automotive industry, in the early 1980s. It measures MAP along with the engine speed in rpm to determine the ignition advance. This ignition timing, with optimum air/fuel ratio can optimize the power performance of the engine with low emissions. Early designs of MAP sensors were called SCAP (an abbreviation of silicon capacitive absolute pressure) sensors. A typical SCAP is shown in Figure 1.22

[Paulsen and Giachino 1989]. The manifold gas enters the sensor at the tubular intake shown at the top of Figure 1.21a and b. The intake gas exerts a pressure on a silicon diaphragm, which acts as one of the two electrodes in the chamber as shown in Figure 1.22c. The deflection of the diaphragm from the applied pressure results in the change in the gap between the two electrodes. The applied pressure can be correlated to the change of the capacitance of the electrodes in the capacitor. Appropriate bridge circuits are used to convert the change of the capacitance signal output to electric voltages. Figure 1.22d shows the relative size of the SCAP unit.

Figure 1.22 | Silicon capacitive manifold absolute pressure sensor.

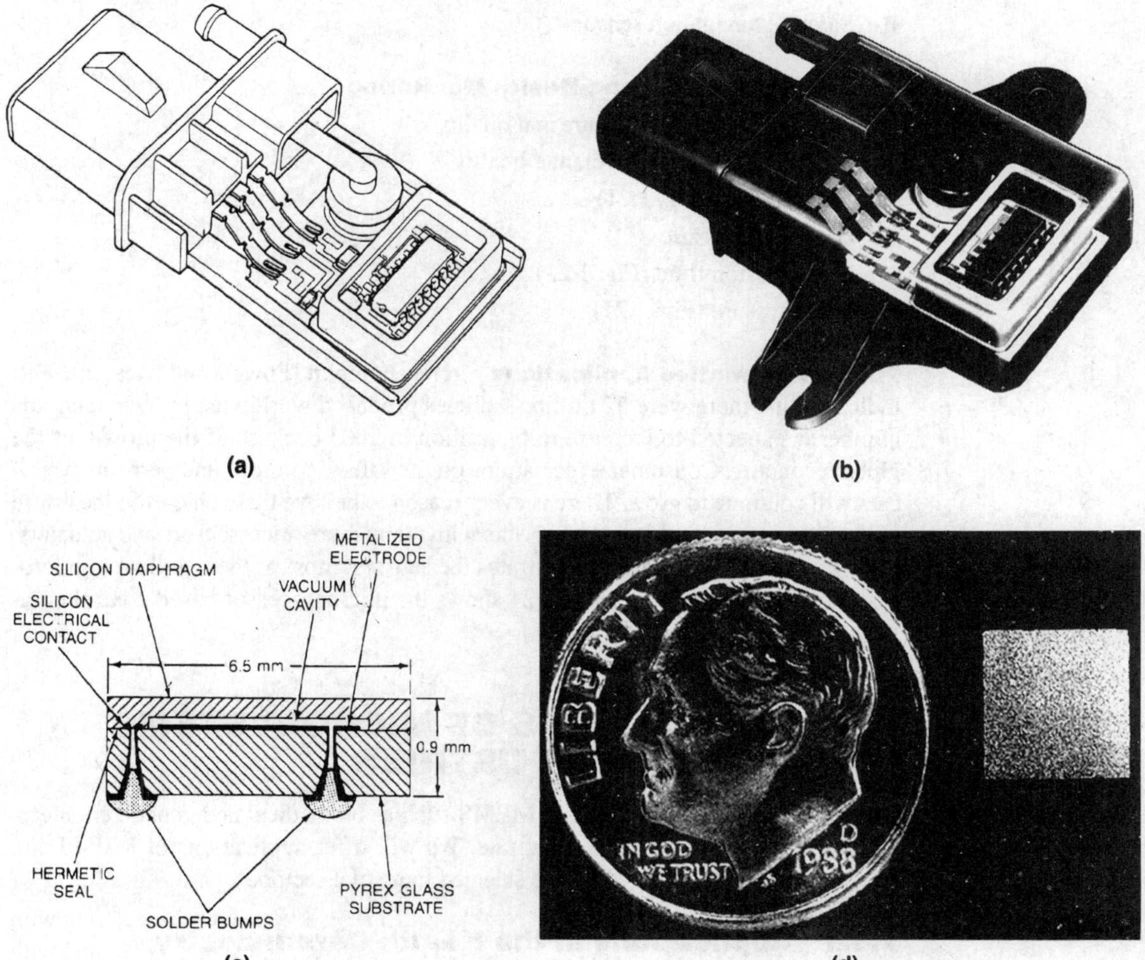

(Courtesy of Ford Motor Company.)

The capacitive pressure sensors have two major shortcomings: bulky size and nonlinear voltage output with respect to the input pressure. Consequently, piezoresistive transducer (PRT) pressure sensors that provide a dc voltage output proportional to the applied pressure have gradually replaced them. The use of piezoresistives as the signal transducer has significantly reduced the size of the sensors.

Comfort and Convenience

- Seat control (displacement sensors and microvalves)
- Rider's comfort (sensors for air quality, airflow, temperature, and humidity controls)
- Security (remote status monitoring and access control sensors)
- Sensors for defogging of windshields
- Satellite navigation sensors

Vehicle Diagnostics and Health Monitoring

- Engine coolant temperature and quality
- Engine oil pressure, level, and quality
- Tire pressure (Fig. 1.21)
- Brake oil pressure
- Transmission fluid (Fig. 1.21)
- Fuel pressure (Fig. 1.21)

Future Automotive Applications A recent report [Powers and Nicastri 1999] indicated that there were 52 million vehicles produced worldwide in 1996, and this number is expected to increase to 65 million in 2005 because of the growth of the global economy. Consumer expectations on the safety, comfort, and performance of cars will continue to grow. There is every reason to believe that vehicles in the future will contain even more microprocessors with many more microsensors and actuators, to be truly smart. Figure 1.23 illustrates the sensors; most of them will be in microscale, in future vehicles. Figure 1.24 shows the microprocessor-based control components in smart cars of the future.

1.8 | APPLICATIONS OF MICROSYSTEMS IN OTHER INDUSTRIES

Other commercial applications of MEMS such as biomedical and genetic engineering are emerging at an astounding rate. We will offer applications of MEMS and microsystems in the following five selected industrial sectors.

1.8.1 Applications in the Health Care Industry

- Disposable blood pressure transducer (DPT). Lifetime 24 to 72 h; annual production 17 million units per year, with a unit price less than $10.

Figure 1.23 | Sensors for future vehicles.

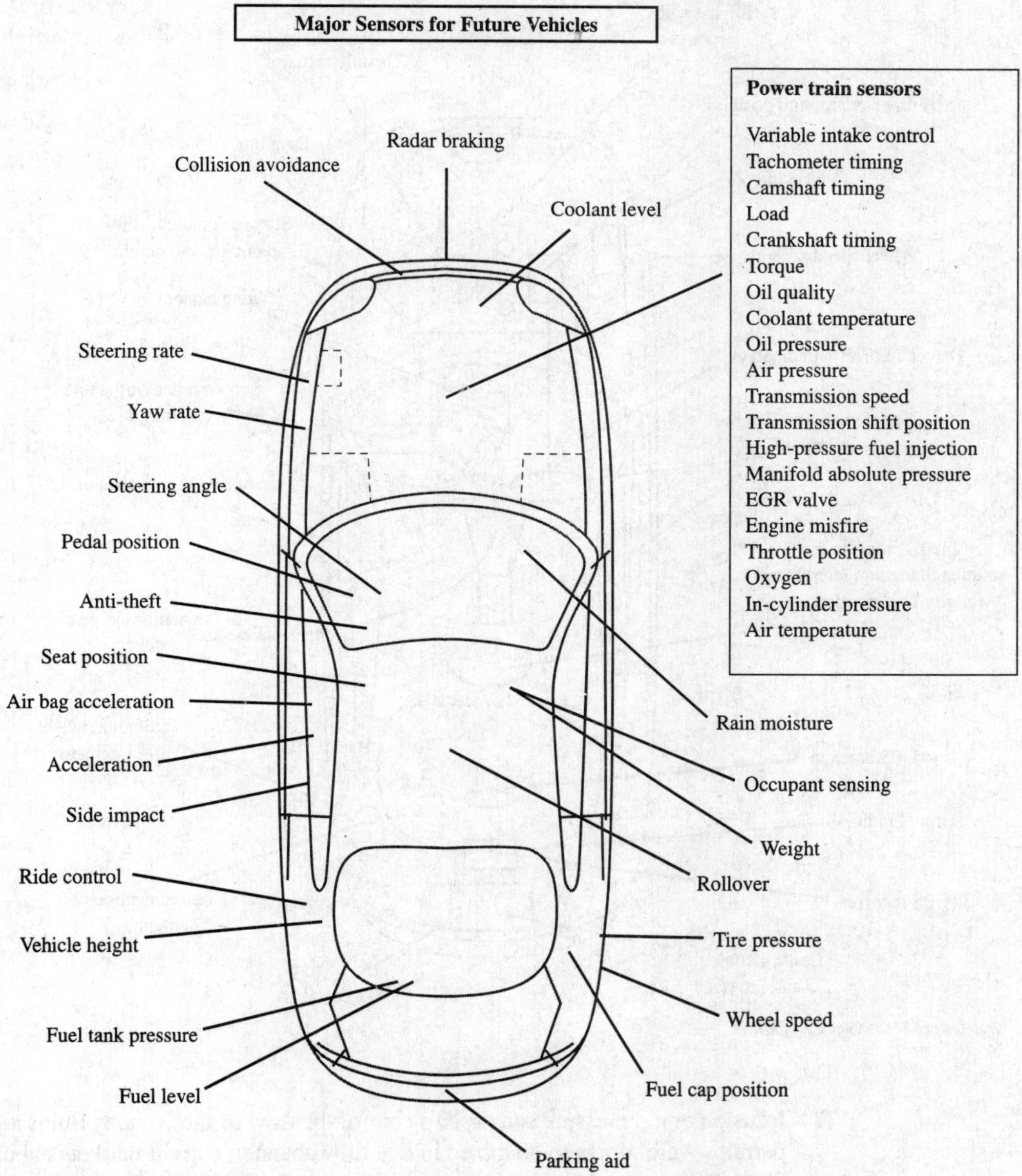

(Courtesy of Ford Motor Company.)

- Intrauterine pressure sensor (IUP). To monitor pressure during child delivery; current market is about 1 million units per year.
- Angioplasty pressure sensor. To monitor the pressure inside the balloon once it is inside the blood vessel; current market is about 500,000 units per year.

Figure 1.24 | Microprocessor control components for future vehicles.

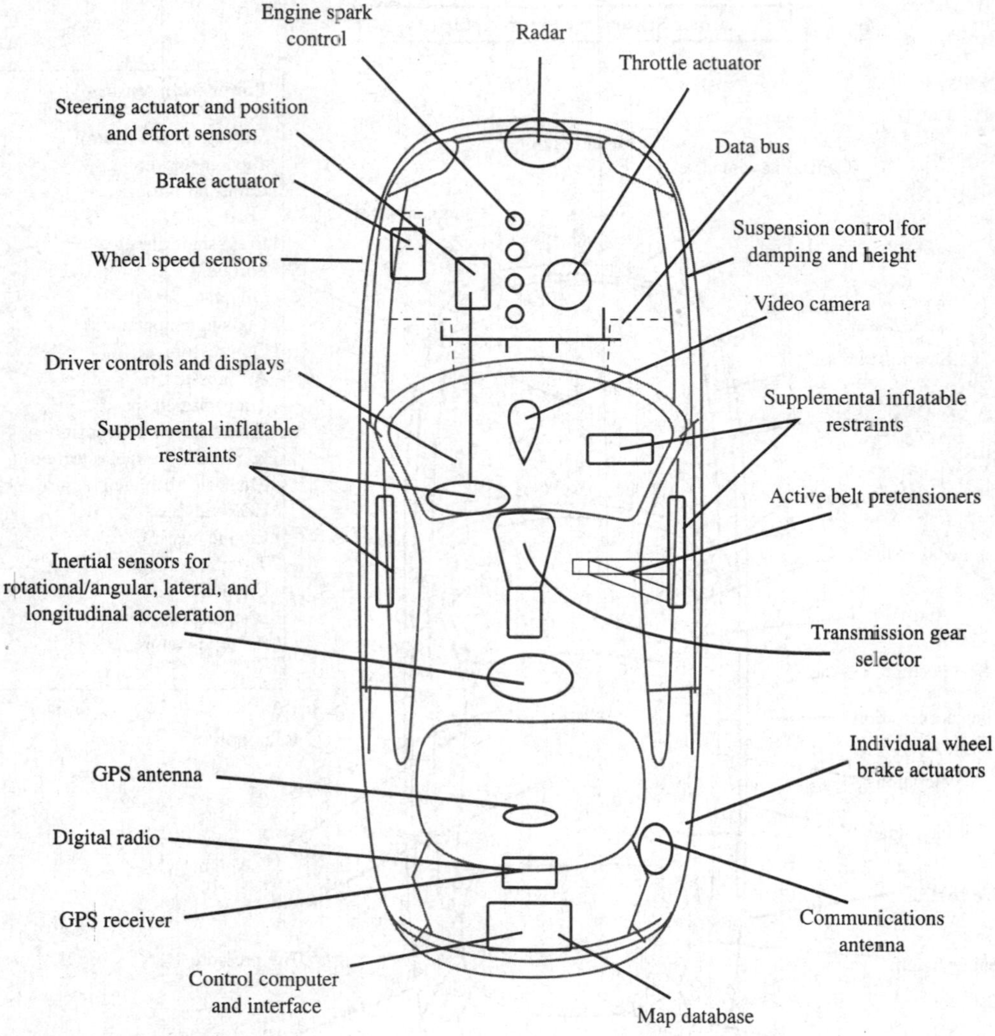

(Courtesy of Ford Motor Company.)

- Infusion pump pressure sensors. To control the flow of intravenous fluids and permit several drugs to be mixed in one flow channel; current market is about 200,000 units per year.
- Other products:

  Diagnostic and analytical systems, such as capillary electrophoresis systems (Chapter 10)

  Human care support systems

Catheter tip pressure sensors
Sphygmomanometers
Respirators
Lung capacity meters
Barometric correction of instrumentation
Medical process monitoring (e.g., drug production by growth of bacteria)
Kidney dialysis equipment

1.8.2 Applications in the Aerospace Industry

- Cockpit instrumentation
  Pressure sensors for oil, fuel, transmission, and hydraulic systems
  Airspeed measurement
  Altimeters
- Safety devices, e.g., ejection seat controls
- Wind tunnel instrumentation (e.g., shear stress sensors)
- Sensors for fuel efficiency and safety
- Microgyroscopes for navigation and stability control
- Microsatellites

A comprehensive summary on possible uses of MEMS and microsystems in space hardware in the near term (less than 10 years) is presented below [Helvajian and Janson 1999]:

- *Command and control systems* with MEMtronics for ultraradiation and temperature-insensitive digital logic and on-chip thermal switches for latch-up and reset
- *Inertial guidance systems* with microgyroscopes, accelerometers, and fiber-optic gyros
- *Attitude determination and control systems* with micro sun and Earth sensors, magnetometers, and thrusters
- *Power systems* with MEMtronic blocking diodes, switches for active solar cell array reconfiguration, and electric generators
- *Propulsion systems* with micro pressure sensors, chemical sensors for leak detection, arrays of single-shot thrusters, continuous microthrusters and pulsed microthrusters
- *Thermal control systems* with micro heat pipes, radiators, and thermal switches
- *Communications and radar systems* with very high bandwidth, low-resistance radio-frequency switches, micromirrors and optics for laser communications, and micro variable capacitors, inductors, and oscillators.

- *Space environment sensors* with micromagnetometers and gravity-gradient monitors (nano-*g* accelerometers).

Considerable effort, as well as progress, has been made in recent years in the development of micro propulsion systems that include micro propellants, nozzles, jet engines, and thrusters [Janson et al. 1999]. Principal supporting mechanical engineering subjects include micro fluid modeling, microcombustion, and rocket science.

1.8.3 Applications in Industrial Products

- Manufacturing process sensors: process pressure transmitters (200,000 units per year)
- Sensors for:

  Hydraulic systems
  Paint spray
  Agricultural sprays
  Refrigeration systems
  Heating, ventilation, and air conditioning systems
  Water level controls
  Telephone cable leak detection

1.8.4 Applications in Consumer Products

- Scuba diving watches and computers
- Bicycle computers
- Fitness gear using hydraulics
- Washers with water-level controls
- Sport shoes with automatic cushioning control
- Digital tire pressure gages
- Vacuum cleaning with automatic adjustment of brush beaters
- Smart toys

1.8.5 Applications in Telecommunications

- Optical switching and fiber-optic couplings
- RF switches
- Tunable resonators

1.9 | MARKETS FOR MICROSYSTEMS

As new MEMS and microsystem products have become available, the market for these products has been expanding rapidly. Tables 1.3 and 1.4 show the world market projections to 2000.

Table 1.3 | Worldwide silicon-based microsensor market

Year	Units, 1000	Revenues, million $	Ave. unit price, $	Revenue growth rate, %
1989	3026	570.50	188.53	—
1990	5741	744.60	129.70	30.5
1991	6844	851.70	124.44	14.4
1992	7760	925.40	119.25	8.6
1993	8816	977.10	110.83	5.6
1994	10836	1116.20	103.00	14.2
1995	13980	1316.30	94.16	17.9
1996	18127	1564.40	86.30	18.9
1997	23514	1857.60	79.00	18.7
1998	30355	2199.80	72.47	18.4
1999	38792	2593.80	66.86	17.9

Principal reference: Frost & Sullivan market intelligence, 1993.

We observe from Table 1.3 that compound annual growth rate since 1992 is 15.9 percent. Unit prices continue to drop (unit price for sensors for automobile is $4 to $10). Pressure sensor market share dropped to 80 percent in total MEMS products. The same share will be 71 percent by 1999.

Table 1.4 | World market projection for MEMS and sensors (in millions of dollars)

	1995	1997	1999	2000
Si sensors in all applications	1316[1]	1858	2549	
Si sensors in automobile	414	621	884	
Si sensors & Si microsystems				3,665
Si-based MEMS[2]	2700			11,900[3]
All sensors	8500			13,100
All industries and scientific instruments	5900			82,000
			Grand-Total:	110,665

Note: 1. United States: 54.4%; Europe: 24.4%; Pacific Rim: 21.2%.
2. Pressure, gas, chemical and rate sensors, accelerometers, valves and actuators, nozzles, displays, microrelays.
3. $5.5 billion in sensors; $2.3 billion in microstructures; $2.6 billion in microsystems; $1.5 billion in microactuators (40% for automobiles).

Principal reference: *Fundamentals of Microfabrication,* Marc Madou, CRC Press, 1997.

PROBLEMS

Part 1. Multiple Choice

1. The largest market share of MEMS products currently belongs to (a) microfluidics, (b) microsensors, (c) microaccelerometers.
2. MEMS components range in size from (a) 1 μm to 1 mm, (b) 1 nm to 1 μm, (c) 1 mm to 1 cm.
3. One nanometer is (a) 10^{-6} m, (b) 10^{-9} m, (c) 10^{-12} m.
4. When we say a device is in mesoscale, we mean the device has a size in the range of (a) 1 μm to 1 mm, (b) 1 nm to 1 μm, (c) 1 mm to 1 cm.
5. The origin of microsystems can be traced back to the invention of (a) transistors, (b) integrated circuits, (c) silicon piezoresistors.
6. A modern integrated circuit may contain (a) 100,000, (b) 1,000,000, (c) 10,000,000 transistors and capacitors.
7. Miniaturization of computers was possible mainly because of (a) better storage systems, (b) replacing vacuum tubes with transistors, (c) the invention of integrated circuits.
8. In general, a microsystem consists of (a) one, (b) two, (c) three components.
9. The microsensor that is commonly used in air bag deployment systems in automobile is a (a) pressure sensor, (b) inertia sensor, (c) chemical sensor.
10. "Laboratory-on-a-chip" means (a) performing experiments on a chip, (b) integration of microsensors and actuators on a chip, (c) integration of microsystems and microelectronics on a chip.
11. The origin of modern microfabrication technology is (a) the invention of transistors, (b) the invention of integrated circuits, (c) the invention of micromachining.
12. The very first significant miniaturization occurred with (a) integrated circuits, (b) laptop computers, (c) mobile telephones.
13. The term *micromachining* first appeared in public in (a) the 1970s, (b) the 1980s, (c) the 1990s.
14. The term *LIGA* refers to (a) a process for micromanufacturing, (b) a microfabrication process, (c) a material treatment process.
15. A typical single ULSI chip may contain (a) one, (b) 10, (c) 100 million transistors.
16. The development of integrated circuits began in (a) the 1960s, (b) the 1970s, (c) the 1950s.
17. The first digital computer, ENIAC, was developed in (a) the 1960s, (b) the 1950s, (c) the 1940s.
18. The *aspect ratio* of a microsystem component is defined as the ratio of (a) the dimensions in the height to those of the surface, (b) the dimensions of the surface to those of the height, (c) the dimensions in width to those of the length.
19. Market value of microsystems is intimately related to (a) volume demand, (b) special features, (c) performance of the products.

20. The most challenging issue facing microsystems technology is (a) the small size of the products, (b) the lack of practical applications, (c) its multidisciplinary nature.

Part 2. Short Questions

1. Give three examples of the objects that you personally recognize to be of the size of approximately 1 millimeter.
2. Explain the difference between MEMS and microsystems.
3. What are most obvious distinctions between microsystems and microelectronics technologies?
4. Why cannot microelectronics technology be adopted in the design and packaging of MEMS and microsystems products?
5. Why cannot traditional manufacturing technologies such as mechanical milling, drilling, and welding be used to produce microsystems?
6. Give at least four distinct advantages of miniaturization of machines and devices.
7. Give two examples of miniaturization of consumer products that you have observed and appreciated.
8. Ask your doctor, or conduct research on your own to find at least one example of application of MEMS in biomedicine that offers significant advantage over the traditional methods.
9. Conduct research on your own to configure an air bag deployment system in an automobile.
10. Describe the proper role that you can play in this multidisciplinary microsystem technology with your academic and professional background.

2 Chapter

Working Principles of Microsystems

CHAPTER OUTLINE

2.1 | INTRODUCTION

As we learned in Chapter 1, the rapid advance in microfabrication technologies has enabled many new MEMS products to emerge in the marketplace. Here, we will learn the working principles of some typical MEMS devices. The reader will soon realize that the design and manufacture of MEMS and microsystem products indeed involves the application of multidisciplinary science and engineering principles, as described in Section 1.5. Learning the working principles of these devices thus offers engineers insights into the design and manufacture of new MEMS products. However, we will present only the essential information that is necessary to describe these working principles. Detailed descriptions of the processes involved in the fabrication and manufacturing of these devices can be found from two principal sources [Madou 1997, Kovacs 1998]. Descriptions of product geometry as well as the fabrication techniques used in producing many microdevices is available in an excellent survey [Trimmer 1997]. Readers should also be aware that some of the microdevices presented in this chapter were chosen for their clever ideas. Many of these devices were not successful in the marketplace.

2.2 | MICROSENSORS

Microsensors are the most widely used MEM devices today. According to Madou [1997], a sensor is a device that converts one form of energy into another and provides the user with a usable energy output in response to a specific measurable input. One such example is to convert the energy that is required to deflect the thin diaphragm in a pressure sensor into an electrical energy (signal) output. A sensor system that includes the sensing element and its associated signal processing hardware is illustrated in Figure 1.1. A smart sensor unit would include automatic calibration, interference signal reduction, compensation for parasitic effects, offset correction, and self-test. All these intelligent functions make this sensor unit an intelligent microsystem, illustrated in Figure 1.7.

Microsensors are used to measure many physical quantities, as described in Chapter 1. Generally, they cover 10 different measured domains [White 1987]; their principal applications are presented below. There are many microsensors available to perform various functions in a variety of industries. Detailed descriptions of many of these sensors can be found in [Madou 1997 and Kovacs 1998]. Here, we will bring to the reader's attention the working principles of some of these sensors.

2.2.1 Acoustic Wave Sensors

The principal application of an acoustic wave sensor is to measure chemical compositions in a gas. These sensors generate acoustic waves by converting mechanical energy to electrical. Acoustic wave devices are also used to actuate fluid flow in microfluidic systems. Actuation energy for this type of sensor is provided by two principal mechanisms: piezoelectric and magnetostrictive. However, the former mechanism is a more popular method for generating acoustic waves. Piezoelectric is a common means for transducing mechanical to electrical energies and vice versa. We will learn more about piezoelectric crystals in Section 2.3.3, and also in Chapter 7.

2.2.2 Biomedical Sensors and Biosensors

Many predict that the biomedical industry will be the next major user of microsystems after the automotive industry. The term *BioMEMS* has been used extensively by the MEMS industry and academe in recent years. It encompasses (1) biosensors, (2) bioinstruments and surgery tools, and (3) systems for biotesting and analysis for quick, accurate, and low-cost testing of biological substances. BioMEMS present a great challenge to engineers, as the design and manufacture of this type of sensor and instrument require the knowledge and experience in molecular biology as well as physical chemistry, in addition to engineering.

Major technical issues involved in the application of MEMS in biomedicine are:

1. Functionality for biomedical operations
2. Adaptivity to existing instruments and equipment
3. Compatibility with biological systems of patients
4. Controllability, mobility, and easy navigation for operations such as those required in a laparoscopy surgery
5. Fabrication of MEMS structures with a high aspect ratio, defined as the ratio of the dimensions in the depth of the structure to the dimensions of the surface.

Microsensors constitute the most fundamental element in any bioMEMS product. There are generally two types of sensors used in biomedicine: (1) *biomedical sensors* and (2) *biosensors*. Biomedical sensors are used to detect biological substances, whereas biosensors may be broadly defined as any measuring devices that contain a biological element [Buerk 1993]. These sensors usually involve biological molecules such as antibodies or enzymes, which interact with analytes that are to be detected.

Biomedical Sensors Biomedical sensors can be classified as biomedical instruments that are used to measure biological substances as well as for medical diagnosis purposes. These sensors can analyze biological samples in quick and accurate ways. These miniaturized biomedical sensors have many advantages over the traditional instruments. They require typically a minute amount of samples and can perform analyses much faster with virtually no dead volume. There are many different types of biomedical sensors in the marketplace.

Electrochemical sensors work on the principle that certain biological substances, such as glucose in human blood, can release certain elements by chemical reaction. These elements can alter the electricity flow pattern in the sensor, which can be readily detected. An example of such a sensor is illustrated as follows [Kovacs 1998].

In Figure 2.1, a small sample of blood is introduced to a sensor with a polyvinyl alcohol solution. Two electrodes are present in the sensor: a platinum film electrode and a thin Ag/AgCl film (the reference electrode). The following chemical reaction takes place between the glucose in the blood sample and the oxygen in the polyvinyl alcohol solution:

$$\text{Glucose} + O_2 \rightarrow \text{gluconolactone} + H_2O_2$$

The H_2O_2 produced by this chemical reaction is electrolyzed by applying a potential to the platinum electrode, with production of positive hydrogen ions, which will flow toward this electrode. The amount of glucose concentration in the blood sample can thus be measured by measuring the current flow between the electrodes.

Figure 2.1 | A biomedical sensor for measuring glucose concentration.

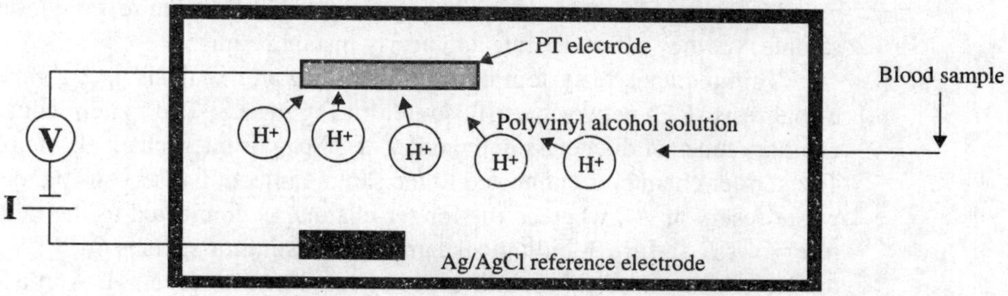

Biosensors Biosensors work on the principle of the interaction of the analytes that need to be detected with biologically derived biomolecules, such as enzymes of certain forms, antibodies, and other forms of protein. These biomolecules, when attached to the sensing elements, can alter the output signals of the sensors when they interact with the analyte. Figure 2.2 illustrates how these sensors are made to function. Proper selection of biomolecules for sensing elements (chemical, optical, etc., as indicated in the right box in the figure) can be used for the detection of specific analyte. In-depth description of biosensors is available in a specialized book by Buerk [1993].

Figure 2.2 | Schematic of biosensors.

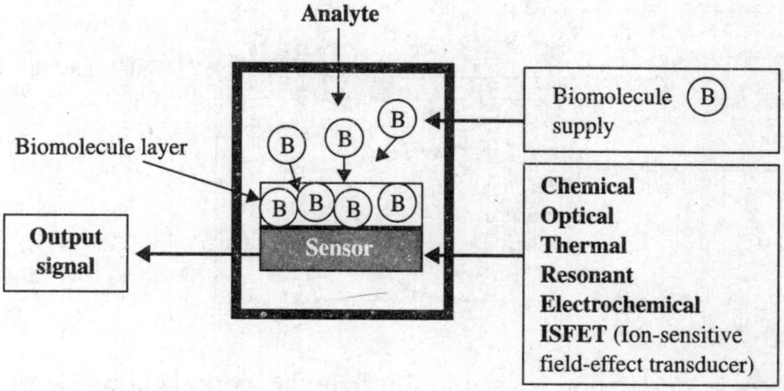

Biotesting and Analytical Systems These systems separate various species in biological samples. Analytes include various biological substances and human genomes. Operation of these systems involves the passage of minute samples in the order of nanoliters in capillary tubes or microchannels driven by electrohydrodynamic means such as *electro-osmosis* and *electrophoresis*. Electrohydrodynamics involves the driving of an ionized fluid by the application of electric fields. This form of fluid flow will be described in Chapter 3. Isolation and separation of species in the capillary tubes or microchannels are achieved by the inherent difference of electro-osmotic mobility of these species. Optical means are used to identify these species after the separation. Often, these minute electrohydrodynamic systems are built with microelectronic circuits for signal transduction, conditioning, and processing. Hundreds and thousands of these capillary tubes can be constructed on a single chip for parallel testing. Microbioanalytical systems typically require minute amounts of sample, yet they offer accurate and nearly instant results.

A simple analyte system used in biotesting and analysis uses a capillary electrophoresis (CE) network, as illustrated in Figure 2.3. The system consists of two capillary tubes of diameters in the order of 30 μm or microchannels of similar sizes. The shorter channel is connected to the sample injection reservoir A and the analyte waste reservoir A', whereas the longer channel is connected to the buffer solvent reservoirs B and B'. A biological sample consisting of species S_1, S_2, S_3, . . . , with distinct electro-osmotic mobilities, is injected into reservoir A. Application of an electric field between the terminals at reservoir A and A' prompts the flow of the injected samples from A to A'. A congregation of the sample forms at the intersection of the two channels because of higher resistance to the flow at that location. A high-voltage electric field is then switched to the terminals B and B'. This electric field can drive the congregated sample in the buffer solvent to flow from reservoir B to B'. The species in the sample can separate in this portion of the flow because of their inherent differences in electro-osmotic mobility.

Figure 2.3 | Schematic diagram of a capillary electrophoresis system.

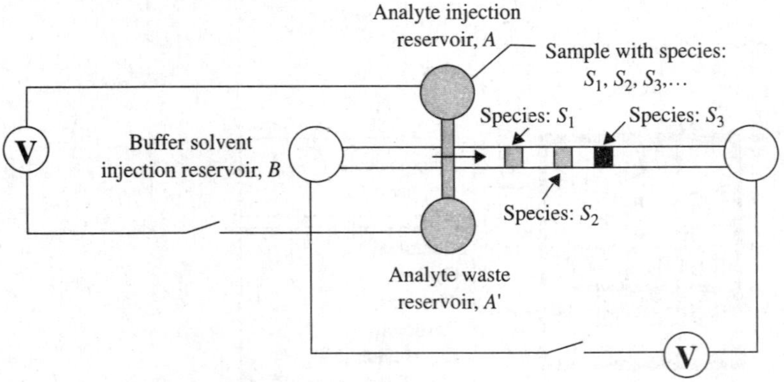

The above description is used to illustrate the working principle of capillary electrophoresis analytical systems. There are obviously other forms of CE systems, as well as other means of biotesting and analytic devices, as described in [Kovacs 1998].

2.2.3 Chemical Sensors

Chemical sensors are used to sense particular chemical compounds, such as various gas species. The working principle of this type of sensor actually is very simple. As we may observe from our day-to-day lives, many materials are sensitive to chemical attacks. For example, most metals are vulnerable to oxidation when exposed to air for a long time. Significant oxide layer built up over the metal surface can change material properties such as the electrical resistance of the metal. This natural phenomenon illustrates the principle on which many microchemical sensors are designed and developed. One may assert that oxygen gas can, in principle, be sensed by measuring the change of electrical resistance in a metallic material as a result of the chemical reaction of oxidation. The presence of oxygen as detected by a chemical sensor, of course, needs to be much more rapid than that of natural oxidation of a metal, and the physical sizes of the samples are on the microscale.

Materials' sensitivity to specific chemicals is indeed used as the basic principle for many chemical sensors. Following are three typical cases worth mentioning [Kovacs 1998].

1. *Chemiresistor sensors.* In Figure 2.4, organic polymers are used with embedded metal inserts. These polymers can cause changes in the electric conductivity of metal when it is exposed to certain gases. For an example, a special polymer called *phthalocyanine* is used with copper to sense ammonia (NH_3) and nitrogen dioxide (NO_2) gases.
2. *Chemicapacitor sensors.* Also in Figure 2.4, some polymers can be used as the dielectric material in a capacitor. The exposure of these polymers to certain gases can alter the dielectric constant of the material, which in turn changes the capacitance between the metal electrodes. An example is to use polyphenyl-acetylene (PPA) to sense gas species such as CO, CO_2, N_2, and CH_4.

Figure 2.4 | Working principle of chemical sensors.

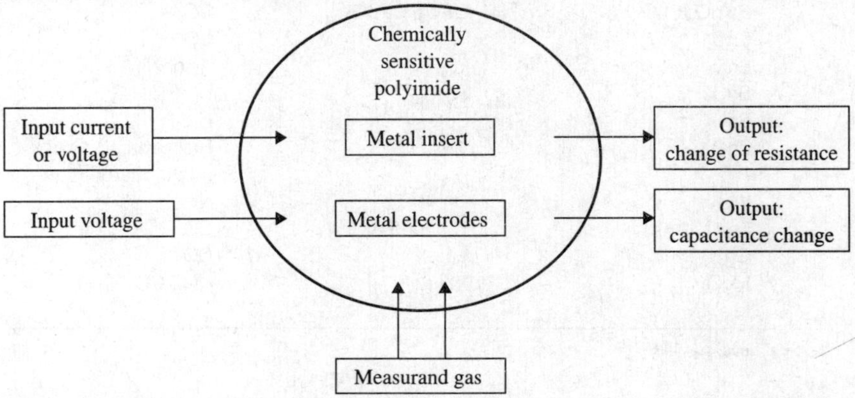

3. *Chemimechanical sensors.* There are certain materials, e.g., polymers, that change shape when they are exposed to chemicals (including moisture). One

may detect such chemicals by measuring the change of the dimensions of the material. An example of such sensor is a moisture sensor using pyraline PI-2722.

4. *Metal oxide gas sensors.* This type of sensor works on a principle similar to that of chemiresistor sensors. Several semiconducting metals, such as SnO_2, change their electric resistance after absorbing certain gases. The process is faster when heat is applied to enhance the reactivity of the measurand gases and the transduction semiconducting metals. Figure 2.5 illustrates a microsensor based on the semiconducting material SnO_2 [Kovacs 1998].

Figure 2.5 | A typical metal oxide gas sensor.

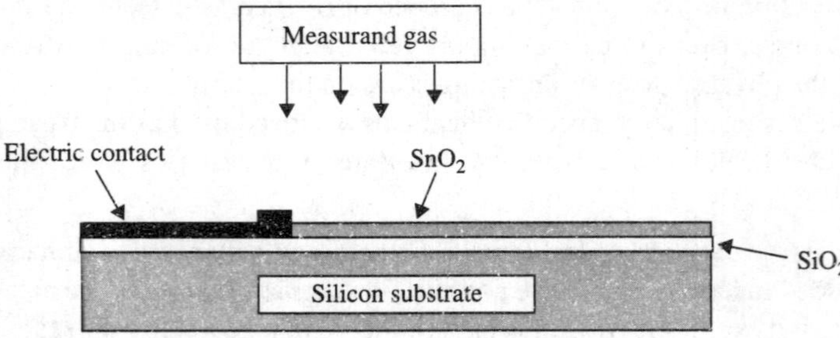

(After Kovacs [1998].)

Better results are obtained if metallic catalysts are deposited on the surface of the sensor. Such deposition can speed up the reactions and hence increase the sensitivity of the sensor. Table 2.1 is a list of metal oxide sensors available for sensing various gases [Kovacs 1998].

Table 2.1 | Available metal oxide gas sensors

Semiconducting Metals	Catalyst Additives	Gas to be Detected
$BaTiO_3$/CuO	La_2O_3, $CaCO_3$	CO_2
SnO_2	Pt + Sb	CO
SnO_2	Pt	Alcohols
SnO_2	Sb_2O_3	H_2, O_2, H_2S
SnO_2	CuO	H_2S
ZnO	V, Mo	Halogenated hydrocarbons
WO_3	Pt	NH_3
Fe_2O_3	Ti-doped Au	CO
Ga_2O_3	Au	CO
MoO_3	None	NO_2, CO
In_2O_3	None	O_3

Source: Kovacs [1998].

2.2.4 Optical Sensors

The principles of interaction between the photons in light and the electrons in the solids that receive the light have been well developed in quantum physics. Devices that can convert optical signals into electronic output have been developed and utilized in many consumer products such as television. Micro-optical sensors have been developed to sense the intensity of light. Solid-state materials that provide strong photon-electron interactions are used as the sensing materials. Figure 2.6 illustrates the four fundamental optical sensing devices [Kovacs 1998].

Figure 2.6 | Optical sensing devices.

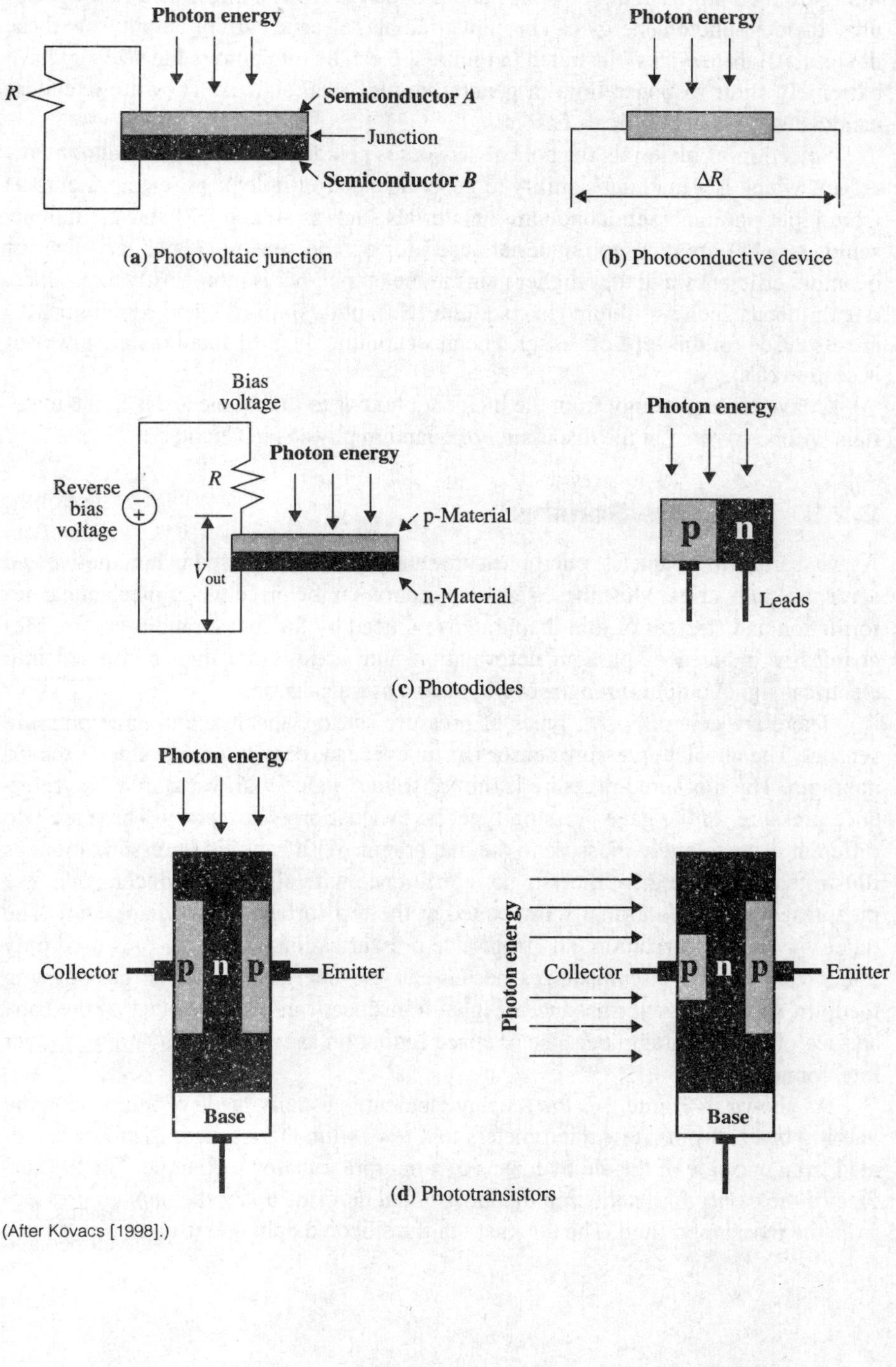

(After Kovacs [1998].)

The photovaltaic junction in Figure 2.6a can produce an electric potential when the more transparent substrate of semiconductor *A* is subjected to incident photon energy. The produced voltage can be measured from the change of electrical resistance in the circuit by an electrical bridge circuit. Figure 2.6b illustrates a special material that changes its electrical resistance when it is exposed to light. The photodiodes in Figure 2.6c are made of p- and n-doped semiconductor layers. The phototransistors in Figure 2.6d are made up of p-, n-, and p-doped layers. As illustrated in these figures, incident photon energy can be converted into electric current output from these devices. All the devices illustrated in Figure 2.6 can be miniaturized in size and have extremely short response time in generating electrical signals. They are excellent candidates for micro-optical sensors.

Selection of materials for optical sensors is principally based on *quantum efficiency,* which is a material's ability to generate electron-hole pairs (electron output) from input photons. Semiconducting materials such as silicon (Si) and gallium arsenide (GaAs) are common materials used for optical sensors. GaAs has superior quantum efficiency and thus higher gains in the output, but is more costly to produce. Alkali metals such as lithium (Li), sodium (Na), potassium (K), and rubidium (Rb) are also used for this type of sensor. The most commonly used alkali metal, however, is cesium (Cs).

Conversion of energy from the incident photons to that of electrons in the materials will be covered in the discussion of quantum physics in Chapter 3.

2.2.5 Pressure Sensors

As we learned in Chapter 1, micropressure sensors are widely used in automotive and aerospace industries. Most these sensors function on the principle of mechanical deformation and stresses of thin diaphragms induced by the measurand pressure. Mechanically induced diaphragm deformation and stresses are then converted into electrical signal output through several means of transduction.

There are generally two types of pressure sensor: absolute and gage pressure sensors. The absolute pressure sensor has an evacuated cavity on one side of the diaphragm. The measured pressure is the "absolute" value with vacuum as the reference pressure. In the gage pressure type, no evacuation is necessary. There are two different ways to apply pressure to the diaphragm. With back side pressurization, as illustrated in Figure 2.7a, there is no interference with signal transducer, such as a piezoresistor, that is normally implanted at the top surface of the diaphragm. The other way of pressurization, i.e., front-side pressurization, Figure 2.7b, is used only under very special circumstances because of the interference of the pressurizing medium with the signal transducer. Signal transducers are rarely placed on the back surface of the diaphragm because of space limitation as well as awkward access for interconnects.

As shown in Figure 2.7, the sensing element is usually made of thin silicon die varying in size from a few micrometers to a few millimeters square. A cavity is created from one side of the die by means of a microfabrication technique. The top surface of the cavity forms the thin diaphragm that deforms under the applied pressure from the measurand fluid. The thickness of the silicon diaphragm usually varies from a few micrometers to tens of micrometers. A constraint base made of metal (called a *header*) or ceramic (Pyrex glass is a common material) supports the silicon die. The deformation of the diaphragm by the applied pressure is transduced into electrical signals by various transduction techniques, as will be described later in this section. The assembly of the sensing elements as shown in Figure 2.7, together with the signal transduction element is then packaged into a robust casing made of metal, ceramic, or plastic with proper passivation of the die.

Figure 2.7 | Cross sections of micro pressure sensors.

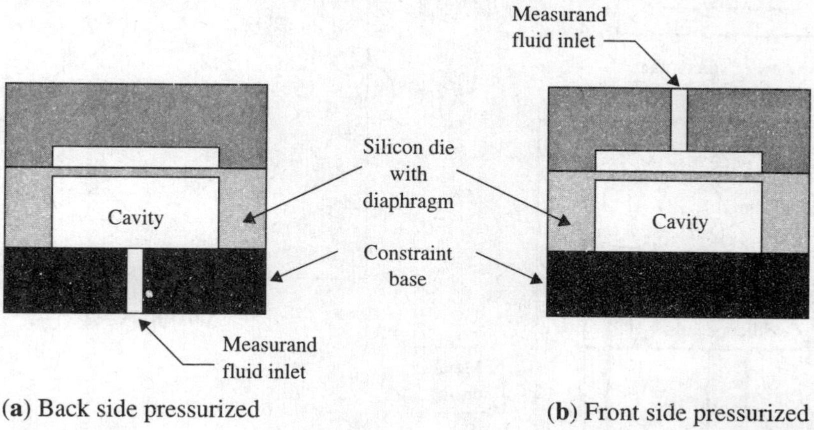

Figure 2.8 schematically illustrates a packaged pressure sensor. The top view of the silicon die shows four piezoresistors (R_1, R_2, R_3, and R_4) implanted beneath the surface of the silicon die. These piezoresistors convert the stresses induced in the silicon diaphragm by the applied pressure into a change of electrical resistance, which is then converted into voltage output by a Wheatstone bridge circuit as shown in the figure. The piezoresistors are essentially miniaturized semiconductor strain gages, which can produce the change of electrical resistance induced by mechanical stresses. In the case illustrated in Figure 2.8, the resistors R_1 and R_3 are elongated the stresses induced by the applied pressure. Such elongation causes an increase of electrical resistance in these resistors, whereas the resistors R_2 and R_4 experience the opposite resistance change. These changes of resistance as induced by the applied measurand pressure are measured from the Wheatstone bridge in the dynamic deflection operation mode as

$$V_o = V_{\text{in}}\left(\frac{R_1}{R_1 + R_4} - \frac{R_3}{R_2 + R_3}\right) \quad [2.1]$$

where V_o and V_{in} are respectively measured voltage and supplied voltage to the Wheatstone bridge.

We will learn the detail characteristics of piezoresistive materials in Chapter 7. Thin wire bonds are used to transmit the voltage change from the piezoresistors bridge through the two metal pads and the interconnect pins as shown in the cross-section view of the sketch. A packaged pressure sensor of this type is shown in the left of Figure 1.2. We will deal with the packaging of these sensors in detail in

Figure 2.8 | A typical micro pressure sensor assembly.

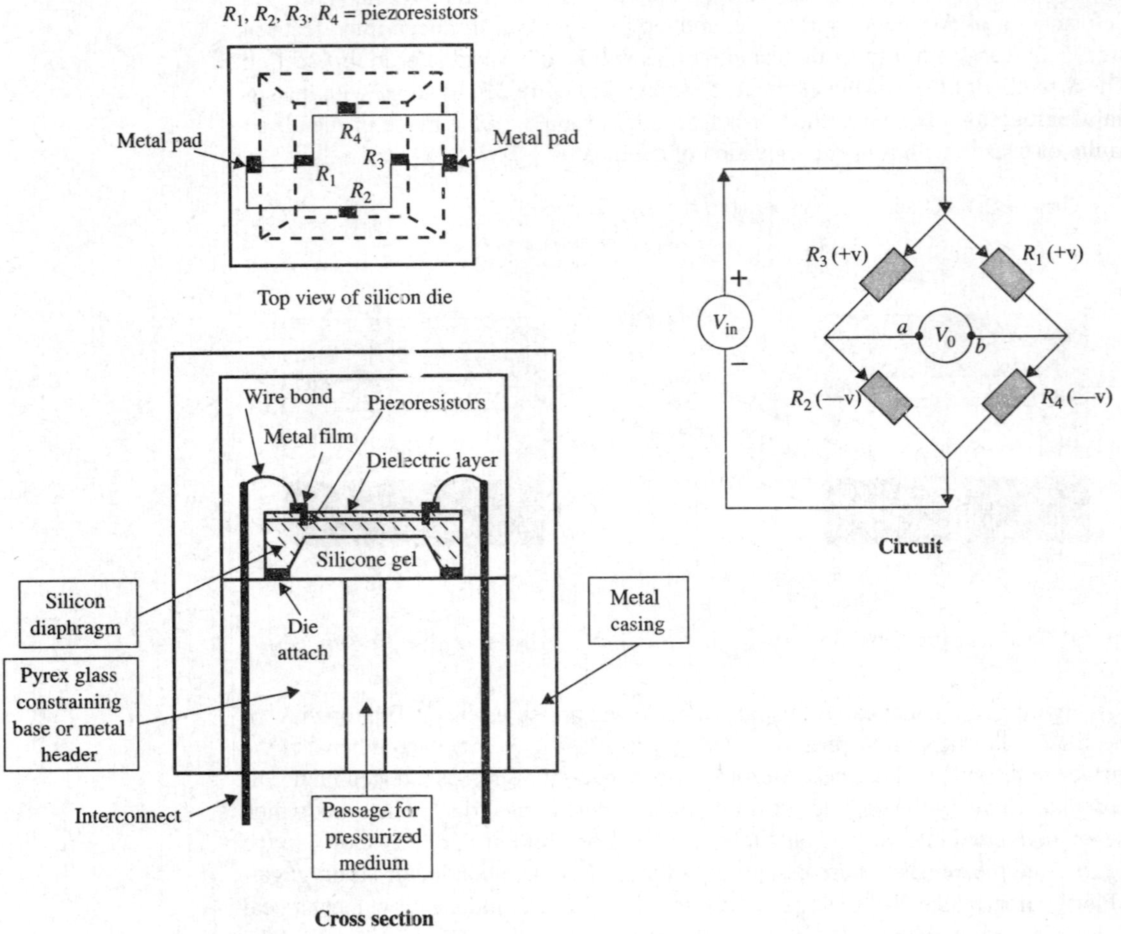

Chapter 11. Micro pressure sensors with piezoresistors have high gains and exhibit a good linear relationship between the in-plane stresses and the resistance change output. However, it has a major drawback: it is temperature sensitive.

Signal transduction in micro pressure sensors can vary, depending on the sensitivity and accuracy required from the sensor. There are several other signal transduction methods developed for these sensors. We will illustrate two of them here.

Figure 2.9 illustrates a micro pressure sensing unit utilizing capacitance change for pressure measurement. Two electrodes made of thin metal films are attached to the bottom of the top cover and the top of the diaphragm. Any deformation of the diaphragm due to the applied pressure will narrow the gap between the two electrodes, leading to a change of capacitance across the electrodes. This method has the advantage of being relatively independent of the operating temperature. The capacitance C

in a parallel plate capacitor can be related to the gap d between the plates by the expression

$$C = \epsilon_r \epsilon_0 \frac{A}{d} \qquad [2.2]$$

in which ϵ_r is the relative permittivity of the dielectric medium and ϵ_0 is the permittivity of free space (vacuum); $\epsilon_0 = 8.85$ pF/m (pF = picofarad = 10^{-12} farad).

Figure 2.9 | Micro pressure sensors using capacitance signal transduction.

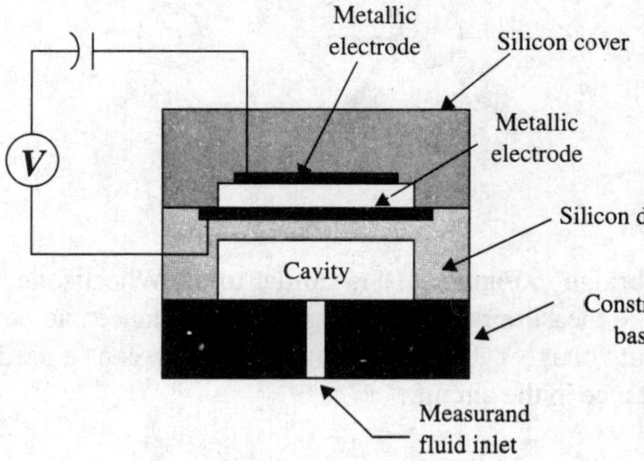

The relative permittivity for several substances is given in Table 2.2.

Table 2.2 | Relative permittivity, ϵ_r of selected dielectric materials

Materials	Relative permittivity
Air	1.0
Paper	2–3.5
Porcelain	6–7
Mica	3–7
Transformer oil	4.5
Water	80
Silicon	12
Pyrex	4.7

EXAMPLE 2.1

Determine the capacitance of a parallel-plate capacitor. The two plates have identical dimensions of $L = W = 1000$ μm with a gap $d = 2$ μm. Air is the dielectric medium between the two plates.

■ Solution

From Table 2.2, we have the relative permittivity $\epsilon_r = 1.0$. From the value $\epsilon_0 = 8.85$ pF/m, and Equation (2.2), the capacitance can be calculated to be 4.43 pF.

Capacitors are common transducers as well as actuators in microsystems. The capacitance variation in a capacitor can be measured by simple circuits such as that illustrated in Figure 2.10 [Bradley et al. 1991].

Figure 2.10 | A typical bridge for capacitance measurements.

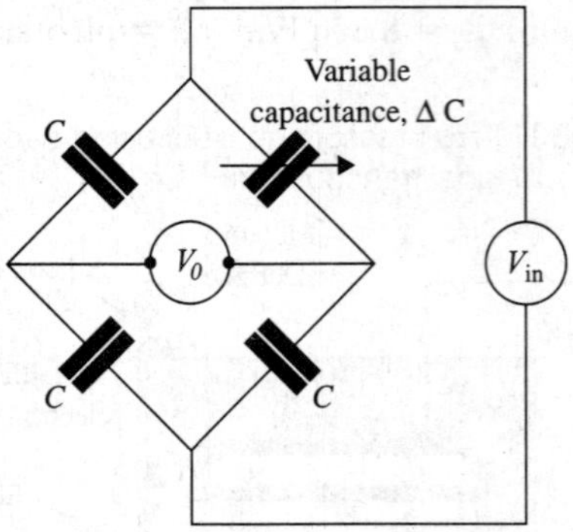

The electrical bridge in Figure 2.10 is similar to the Wheatstone bridge in Figure 2.8 for resistance measurements. The variable capacitance can be measured by measuring the output voltage V_o. The following relationship can be used to determine the variable capacitance in the circuit:

$$V_o = \frac{\Delta C}{2(2C + \Delta C)} V_{in} \qquad [2.3]$$

where ΔC is the capacitance change in the capacitor in the micro pressure sensor and C is the capacitance of the other capacitors in the bridge. The bridge is subjected to a constant supply voltage V_{in}.

Figure 2.11 illustrates a manifold absolute pressure (MAP) sensor for an automobile [Chiou 1999]. This sensor uses a capacitor as signal transducer. The detailed arrangement of the silicon die in the pressure chamber is similar to that shown in Figure 1.22c.

Figure 2.11 | A manifold absolute pressure sensor with capacitor transducer.

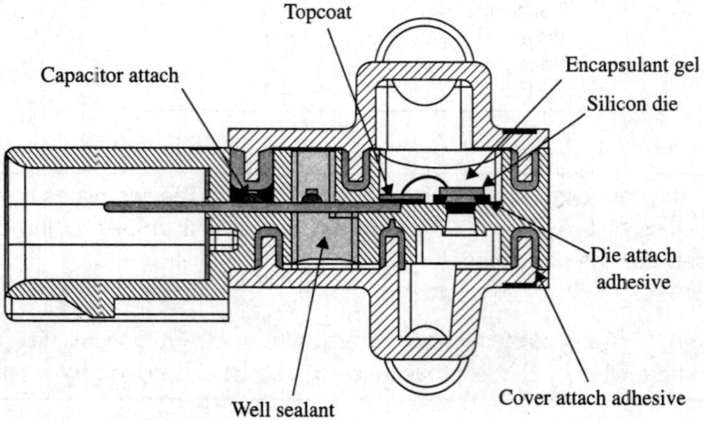

(Courtesy of Motorola Corporation.)

Micro pressure sensors with capacitance signal transduction are not nearly as sensitive to the applied pressure as those with piezoresistors. However, capacitance transducers are not nearly as sensitive to operating temperature as the piezoresistors. On the other hand, one should be aware of a major shortcoming of capacitance transducers: The voltage output from a capacitance bridge circuit is nonlinear to the change of capacitance due to pressure variation. This nonlinear input/output relationship is indicated in Equation (2.3). The following example will illustrate such nonlinear relationship.

EXAMPLE 2.2

Determine the voltage output of the capacitance bridge circuit illustrated in Figure 2.10 with variation of the gap between two flat-plate electrodes as described in Example 2.1.

■ Solution

We may use Equations (2.2) and (2.3) to establish the following table for voltage output:

Gap *d* between electrodes, μm	Capacitance *C*, pF	Change of capacitance Δ*C*, pF	Voltage ratio V_o/V_{in}
2.00	4.430	0	0
1.75	5.063	0.633	0.033
1.50	5.910	1.477	0.071
1.00	8.860	4.430	0.167
0.75	11.813	7.383	0.227
0.50	17.720	13.290	0.300

Figure 2.12a illustrates the construction of a micro pressure sensor using a vibrating beam for signal transduction [Petersen et al. 1991]. A thin n-type silicon beam is installed across a shallow cavity at the top surface of the silicon die. A p-type electrode is diffused at the surface of that cavity under the beam. The p- and n-type silicon layers are doped with, respectively boron and phosphorus, as will be described in detail in Chapter 3. Both p- and n-type silicon are electrically conductive. The beam is made to vibrate at its resonant frequency by applying an ac signal to the diffused electrode in the beam before the application of the pressure to the diaphragm. The stress induced in the diaphragm (and the die) will be transmitted to the vibrating beam. The induced stress along the beam causes a shift of the resonant frequency of the beam. The shift of the resonant frequency of the beam can be correlated to the induced stress and thus to the pressure applied to the silicon diaphragm. Formulas for calculating the corresponding resonant frequency shift in a vibrating beam will be presented in Chapter 4.

This type of signal transduction is insensitive to temperature and provides excellent linear output signals. The disadvantage of such sensors is the cost to fabricate them.

2.2.6 Thermal Sensors

Thermocouples are the most common transducer used to sense heat. They operate on the principle of *electromotive force* (emf) produced at the open ends of two dissimilar

Figure 2.12 | Pressure sensors using a vibrating beam signal transducer.

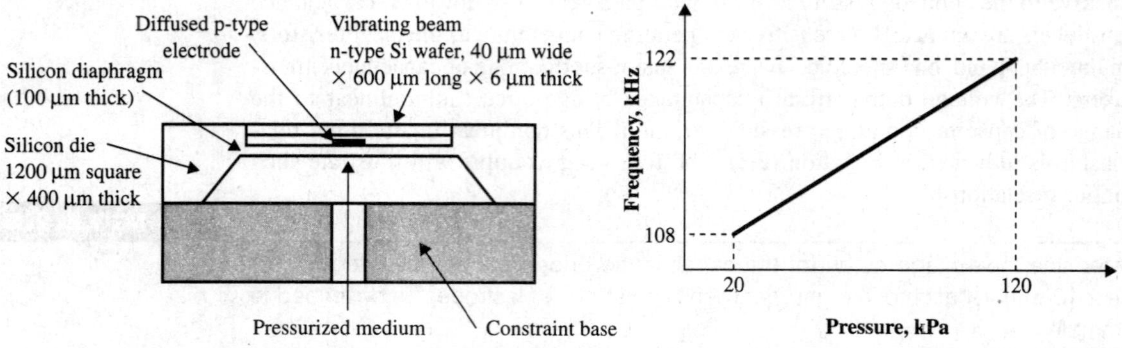

(**a**) Diagram of the pressure-sensing element (**b**) Reported sensitivity

metallic wires when the junction of the wires (called the bead) is heated (see Figure 2.13a). The temperature rise at the junction due to heating can be correlated to the magnitude of the produced emf, or voltage. These wires and the junction can be made very small in size. By introducing an additional junction in the thermocouple circuit, as shown in Figure 2.13b, and exposing that junction to a different temperature than the other, one would induce a temperature gradient in the circuit itself. This arrangement of thermocouples with both hot and cold junctions can produce the *Seebeck effect,* discovered by T. J. Seebeck in 1821. The voltage generated by the thermocouple can be evaluated by $V = \beta \Delta T$ in which β is the Seebeck coefficient and ΔT is the temperature difference between the hot and cold junctions. In practice, the cold junction temperature is maintained constant, e.g., at 0°C, by dipping that junction in ice water. The coefficient β depends on the thermocouple wire materials and the range of temperature measurements. Seebeck coefficients for common thermocouples are given in Table 2.3.

Figure 2.13 | Schematics of thermocouples.

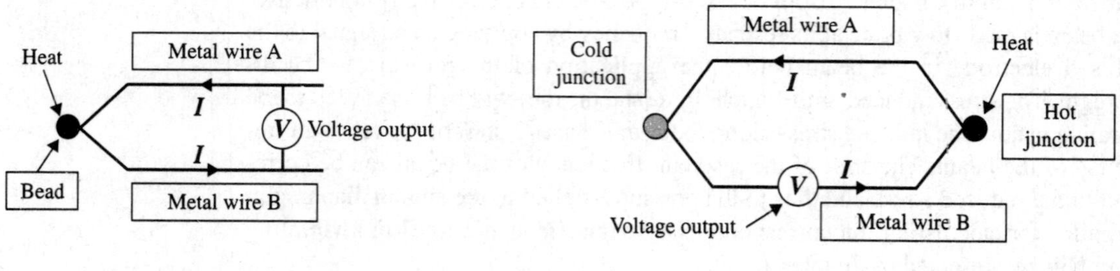

(**a**) A thermocouple (**b**) A dual-junction thermocouple

One serious drawback of thermocouples for micro thermal transducers is that the output of thermocouples decreases as the size of the wires and the beads is reduced. Thermocouples alone are thus not ideal for microthermal sensors.

Table 2.3 | Seebeck coefficients for common thermocouples

Type	Wire materials	Seebeck coefficient, μV/°C	Range, °C	Range, mV
E	Chromel/constantan	58.70 at 0°C	−270 to 1000	−9.84 to 76.36
J	Iron/constantan	50.37 at 0°C	−210 to 1200	−8.10 to 69.54
K	Chromel/alumel	39.48 at 0°C	−270 to 1372	−6.55 to 54.87
R	Platinum (10%)–Rh/Pt	10.19 at 600°C	−50 to 1768	−0.24 to 18.70
T	Copper/constantan	38.74 at 0°C	−270 to 400	−6.26 to 20.87
S	Pt (13%)–Rh/Pt	11.35 at 600°C	−50 to 1768	−0.23 to 21.11

Source: *CRC Handbook of Mechanical Engineering* [1998].

A microthermopile is a more realistic solution for miniaturized heat sensing. Thermopiles operate with both hot and cold junctions, but they are arranged with thermocouples in parallel and voltage output in series. This arrangement is illustrated in Figure 2.14. Materials for thermopile wires are the same as those used in thermocouples—copper/constantan (type T), chromel/alumel (type K), etc.—as shown in Table 2.3.

Figure 2.14 | Schematic arrangement for a thermopile.

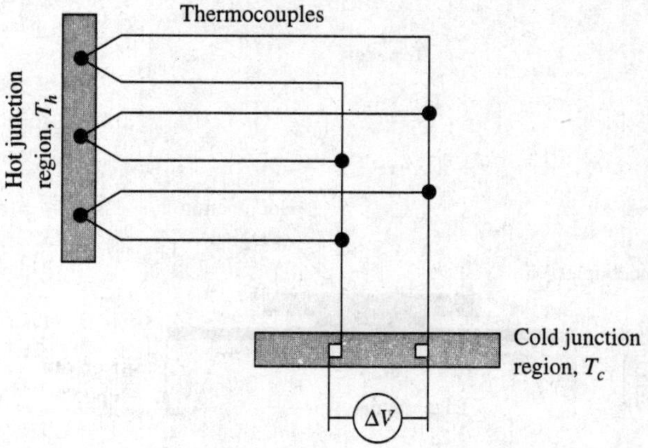

The voltage output from a thermopile can be obtained by the following expression:

$$\Delta V = N\beta\Delta T \qquad [2.4]$$

where

N = number of thermocouple pairs in the thermopile
β = thermoelectric power (or Seebeck coefficient) of the two thermocouple materials, V/K (from manufacturer data or Table 2.3)
ΔT = temperature difference across the thermocouples, K

Choi and Wise [1986] produced the microthermopile represented in Figure 2.15. A total of 32 polysilicon-gold thermocouples were used in the thermopile. The overall dimension of the silicon chip on which the thermopile was built is 3.6 mm × 3.6 mm × 20 μm thick. A typical output signal of 100 mV was obtained from a 500 K blackbody radiation source of Q_{in} = 0.29 mW/cm^2 with a response time of about 50 ms.

Figure 2.15 | Schematic of a microthermopile

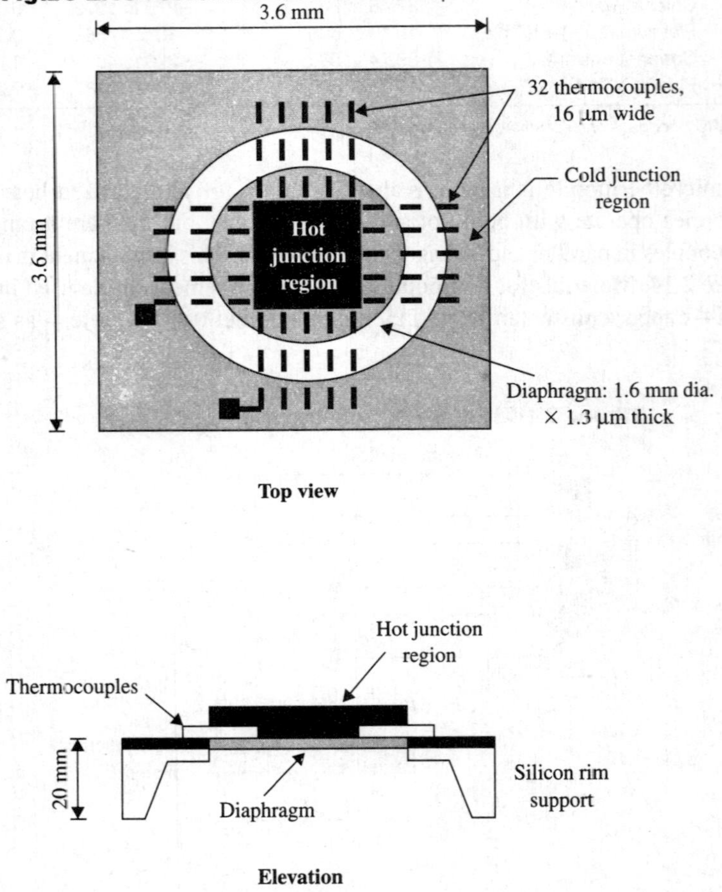

A pressure sensor similar to that shown in Figures 2.9 and 2.11 could be used to detect temperature of a working medium. In such cases, the deformation of the silicon diaphragm is induced by thermal sources instead of pressures as in a pressure sensors. The thermal forces that cause the deflection of the diaphragms can be related to the capacitance of the deflected electrodes. A more sensitive means, however, is to use a thermal pile, or a thermocouple, as the sensing element.

2.3 | MICROACTUATION

Webster's dictionary [Merriam 1995] defines an actuator as "a mechanical device for moving or controlling something." The actuator is a very important part of a microsystem that involves motion. Four principal means are commonly used for actuating motions of microdevices: (1) thermal forces, (2) shape memory alloys, (3) piezoelectric crystals, and (4) electrostatic forces. Electromagnetic actuation is widely used in devices and machines at macroscales. It, however, is rarely used in microdevices because of the unfavorable miniaturization scaling laws, as will be discussed in Chapter 6. We will briefly describe the working principle of each of these four actuation methods in this section.

An actuator is designed to deliver a desired motion when it is driven by a power source. Actuators can be as simple as an electrical relay switch or as complex as an electric motor. The driving power for actuators varies, depending on the specific applications. An on/off switch in an electric circuit can be activated by the deflection of a bimetallic strip as a result of resistance heating of the strip by electric current. On the other hand, most electrical actuators, such as motors and solenoid devices, are driven by electromagnetic induction, governed by Faraday's law.

For devices at micro- or mesoscales, there is little room for the electrical conducting coils required for electromagnetic induction. Consequently, other kinds of driving power sources have to be developed. We will introduce three commonly used sources for microactuation.

2.3.1 Actuation Using Thermal Forces

Bimetallic strips are actuators based on thermal forces. These strips are made by bonding two materials with distinct thermal expansion coefficients. The strip will bend when is heated or cooled from the initial reference temperature because of incompatible thermal expansions of the materials that are bonded together. It will return to its initial reference shape once the applied thermal force is removed. The same principle has been used to produce several microactuators, such as microclamps or valves. In these cases, one of the strips is used as a resistance heater. The other strip could be made from a common microstructural material such as silicon or polysilicon [Riethmuller et al. 1987].

The behavior of thermally actuated bimetallic strips is illustrated in Figure 2.16. The two constituent materials have coefficients of thermal expansion, α_1 and α_2, respectively, with $\alpha_1 > \alpha_2$. The beam made of the bimetallic strips will deform from its original straight shape to a bent shape shown in the right of the figure when it is heated by external sources. The beam is expected to return to its original shape after the removal of the heat.

Figure 2.16 | Thermal actuation of dissimilar materials.

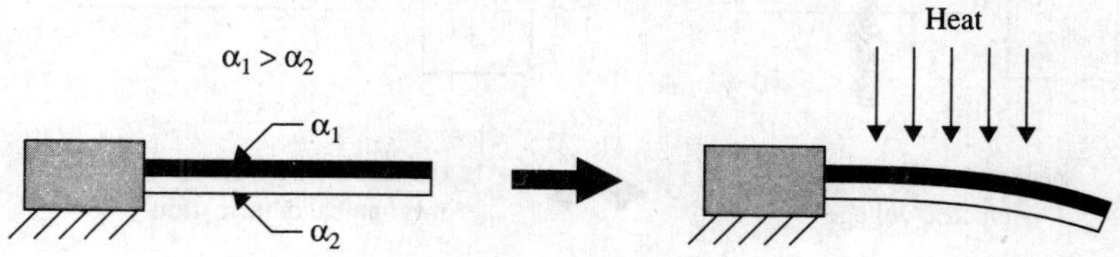

2.3.2 Actuation Using Shape-Memory Alloys

Microactuation can be produced more accurately and effectively by using shape memory alloys (SMA) such as Nitinolor, or TiNi alloys. These alloys tend to return to their original shape at a preset temperature. To illustrate the working principle of a microactuator using SMA, let us refer to Figure 2.17. An SMA strip originally in a bent shape at a designed preset temperature T is attached to a silicon cantilever beam. The beam is set straight at room temperature. However, heating the beam with the attached SMA strip to the temperature T would prompt the strip's "memory" to return to its original bent shape. The deformation of the SMA strip causes the attached silicon beam to deform with the strip, and microactuation of the beam is thus achieved. This type of actuation has been used extensively in micro rotary actuators, microjoints and robots, and microsprings [Gabriel et al. 1988].

Figure 2.17 | Microactuation using shape memory alloys.

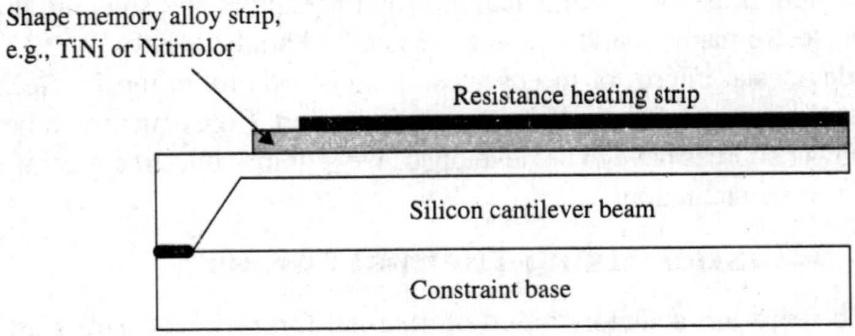

2.3.3 Actuation Using Piezoelectric Crystals

Certain crystals, such as quartz, that exist in nature deform with the application of an electric voltage. The reverse is also valid; i.e., an electric voltage can be generated across the crystal when an applied force deforms the crystal. This phenomenon is illustrated in Figure 2.18.

Figure 2.18 | The piezoelectric effect.

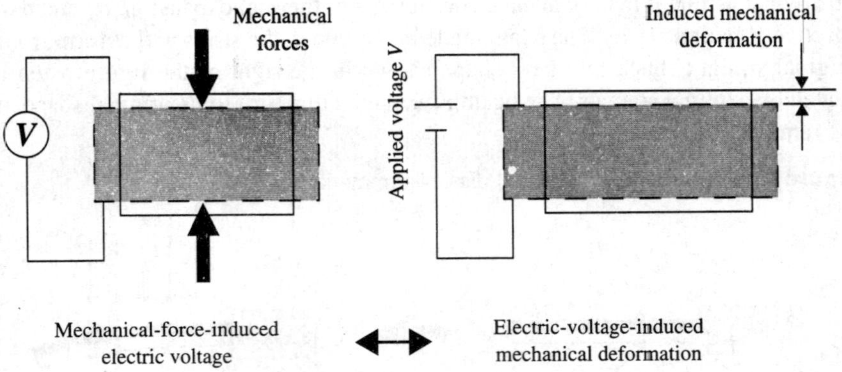

We may attach such a crystal to a flexible silicon beam in a microactuator, as shown in Figure 2.19. An applied voltage across the piezoelectric crystal prompts a deformation of the crystal, which can in turn bend the attached silicon cantilever beam. Piezoelectric actuation is used in a micropositioning mechanism and microclamp reported in [Higuchi et al. 1990].

Figure 2.19 | Actuator using a piezoelectric crystal.

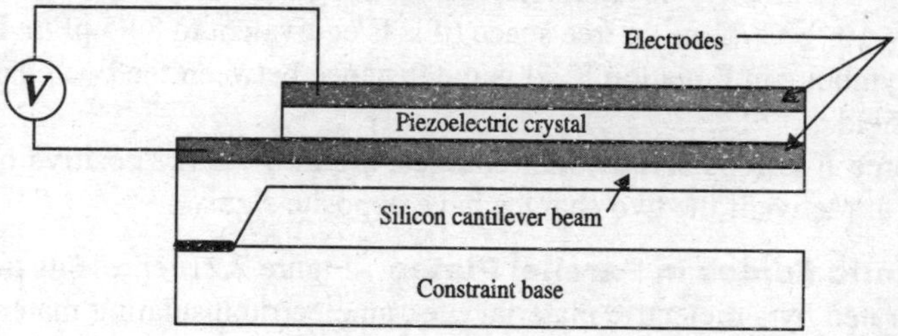

Piezoelectric crystals are essential materials for microactuators. More information on the crystal structures and the mathematical formulas for the determination of electromechanical characteristics of common piezoelectric crystals will be presented in Chapter 7.

2.3.4 Actuation Using Electrostatic Forces

Electrostatic forces are used as the driving forces for many actuators. Accurate assessment of electrostatic forces is an essential part of the design of many micromotors and actuators. A review of the fundamentals of electrostatics is thus necessary in order to use this important power source effectively in the design of microactuators.

Coulomb's Law Electrostatic force F is defined as the electrical force of repulsion or attraction induced by an electric field E. As we have learned from physics, an electric field E exists in a field carrying positive and negative electric charges. Charles Augustin Coulomb (1736–1806) discovered this phenomenon and postulated the mathematical formula for determining the magnitude of the force F between two charged particles.

Figure 2.20 | Two particles in an electric field.

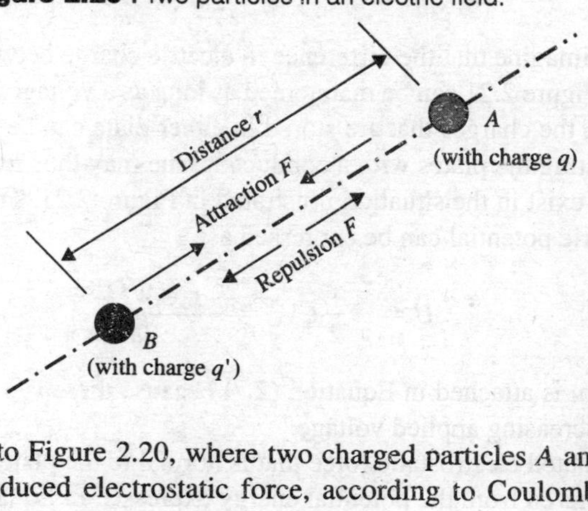

With reference to Figure 2.20, where two charged particles A and B are in an electric field, the induced electrostatic force, according to Coulomb, can be expressed as:

$$F = \frac{1}{4\pi\epsilon} \frac{qq'}{r^2} \qquad [2.5]$$

in which ϵ = permittivity of the material separating the two particles. We will have $\epsilon_0 = 8.85 \times 10^{-12}$ $C^2/N\text{-}m^2$ in free space (this is equivalent to 8.85 pF/m in a capacitor). The symbol r in Equation [2.5] is the distance between the two charged particles in the field.

The force F is repulsive if both charges, q and q', carry positive or negative charges, or attractive if the two charges have opposite signs.

Electrostatic Forces in Parallel Plates Figure 2.21 represents two charged plates separated by a dielectric material (i.e., an electric insulating material) with a gap d. The plates become electrically charged when an electromotive force (emf), or voltage, is applied to the plates. This action will induce capacitance in the charged plates, which can be expressed as

$$C = \epsilon_r \epsilon_0 \frac{A}{d} = \epsilon_r \epsilon_0 \frac{WL}{d} \qquad [2.6]$$

where A is the area of the plates and ϵ_r is the relative permittivity. The relative permittivity ϵ_r in Equation (2.6) for common dielectric materials is presented in Table 2.2.

Figure 2.21 | Electric potential in two parallel plates.

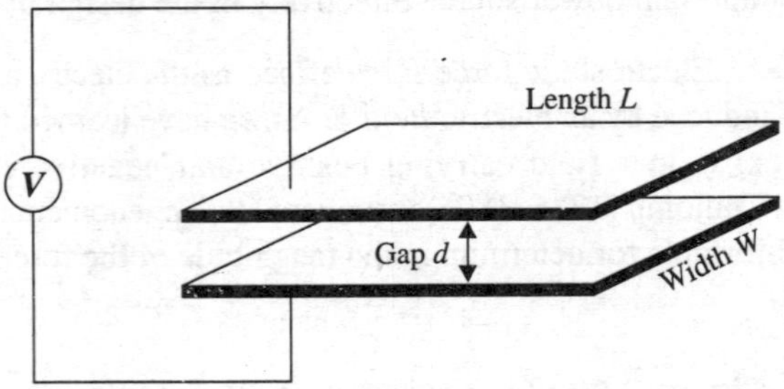

One may imagine that the difference in electric charge between the top and bottom plates in Figure 2.21 can be maintained as long as a voltage is applied to the system. However, the charges that are stored in either plate can be discharged instantly by short circuiting the plates with a conductor. One may thus realize that an electric potential does exist in the situation illustrated in Figure 2.21. The energy associated with this electric potential can be expressed as:

$$U = -\frac{1}{2} CV^2 = -\frac{\epsilon_r \epsilon_0 WLV^2}{2d} \qquad [2.7]$$

A negative sign is attached in Equation (2.7) because there is a loss of the potential energy with increasing applied voltage.

The associated electrostatic force that is normal to the plates (in the d direction) can thus be derived from the potential energy expressed in Equation (2.7) as:

$$F_d = -\frac{\partial U}{\partial d} = -\frac{1}{2} \frac{\epsilon_r \epsilon_0 WLV^2}{d^2} \qquad [2.8]$$

EXAMPLE 2.3

A parallel capacitor is made of two square plates with the dimensions $L = W =$ 1000 μm (or 1 mm), as shown in Figure 2.22. Determine the normal electrostatic force if the gap between these two plates is $d = 2$ μm. The plates are separated by static air.

Figure 2.22 | Normal electrostatic force in a parallel-plate capacitor.

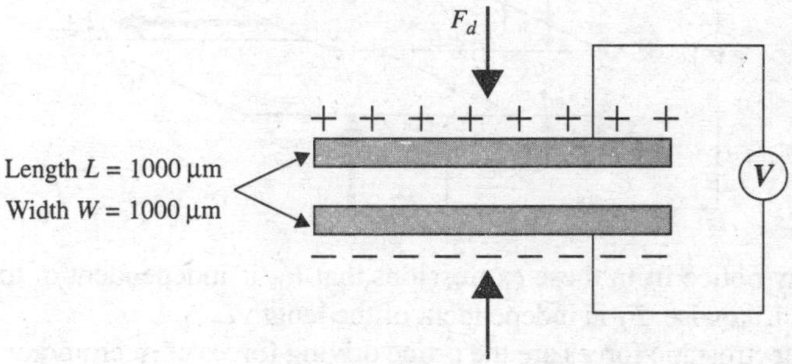

■ Solution

The normal electrostatic force exerted on the plates can be calculated from Equation (2.8) with $\epsilon_r = 1.0$ for air as the dielectric material and $\epsilon_0 = 8.85$ pF/m or 8.85×10^{-12} $C^2/N\text{-}m^2$.

$$F_d = \frac{1.0(8.85 \times 10^{-12})(1000 \times 10^{-6})(1000 \times 10^{-6})V^2}{2(2 \times 10^{-6})^2}$$

$$= -1.106 \times 10^{-6} V^2 \quad \text{newtons (N)}$$

Thus, an 11-mN (millinewton) force is generated with 100 V applied to the plates.

Equation (2.8) also can be used to derive expressions for electrostatic forces in the width W and length L directions. These forces are induced with partial alignment of the plates in the respective directions [Trimmer and Gabriel 1987]. The following general form of Equation (2.8) can be used to derive these forces:

$$F_i = -\frac{\partial U}{\partial x_i} \qquad [2.9]$$

where the index i is the direction in which misalignment occurs, i.e., the width direction in W or the length direction in L.

Thus, by referring to the designation of forces indicated in Figure 2.23, we can write expressions for the two forces in the two directions:

$$F_W = \frac{1}{2}\frac{\epsilon_r \epsilon_0 L V^2}{d} \qquad [2.10]$$

in the width direction, and

$$F_L = \frac{1}{2}\frac{\epsilon_r \epsilon_0 W V^2}{d} \qquad [2.11]$$

in the length direction.

Figure 2.23 | Electrostatic forces on parallel plates.

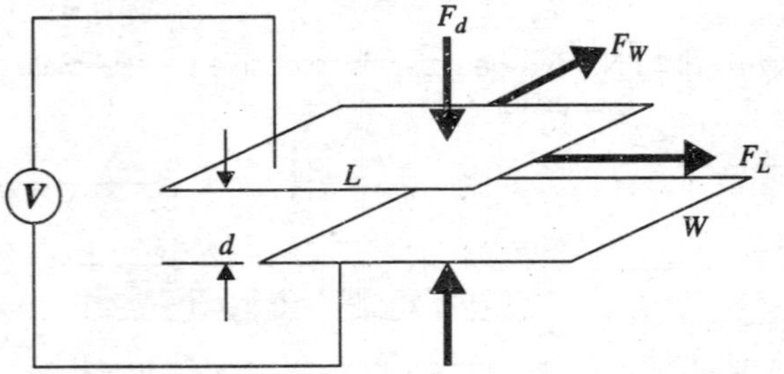

One may notice from these expressions that F_W is independent of the width dimension, W. Likewise, F_L is independent of the length L.

These electrostatic forces are the prime driving forces of micromotors, as will be demonstrated in the microdevices described in Section 2.4. One drawback of electrostatic actuation is that the force that is generated by this method usually is low in magnitude. Its application is thus primarily limited to actuators for optical switches such as that shown in Figure 1.12, and microgrippers and tweezers.

2.4 | MEMS WITH MICROACTUATORS

We will present a few microdevices that function on the principles of microactuation described in the foregoing section.

2.4.1 Microgrippers

The electrostatic forces generated in parallel charged plates can be used as the driving forces for gripping objects, as illustrated in Figure 2.24. As the figure shows, the required gripping forces in a gripper can be provided either by normal forces (Fig. 2.24a) or by the in-plane forces from pairs of misaligned plates (Fig. 2.24b).

Figure 2.24 | Gripping forces in a microgripper.

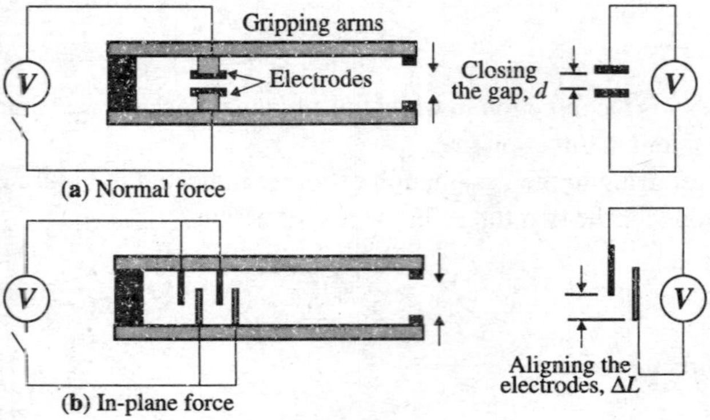

The arrangement that uses normal gripping forces from parallel plates, Figure 2.24a, appears to be simple in practice. A major disadvantage of this arrangement, however, is the excessive space that the electrodes occupy in a microgripper. Consequently, it is rarely used. The other arrangement, with multiple pairs of misaligned plates, is commonly used in microdevices. This arrangement is frequently referred to as the *comb drive*. Comb drive is used in the construction of the microgripper illustrated in Figure 2.25 by Kim et al. [1991]. The gripping action at the tip of the gripper is initiated by applying a voltage across the plates attached to the drive arms and

Figure 2.25 | Schematic diagram of a microgripper.

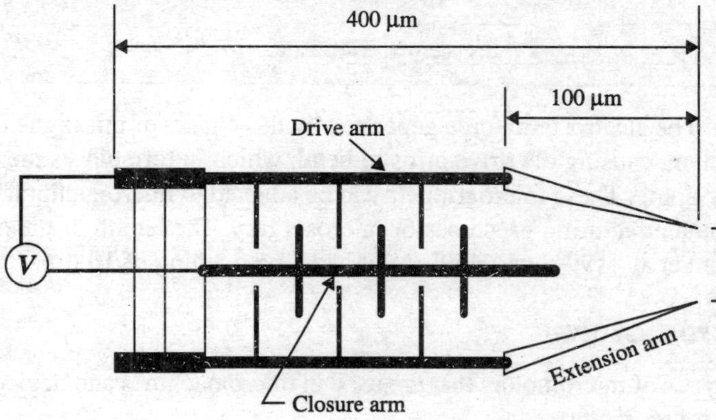

EXAMPLE 2.4

For the comb-driven actuator in Figure 2.26, determine the voltage supply required to pull the moving electrode 10 μm from the unstretched position of the spring. The spring constant k is 0.05 N/m. The comb drive is operated in air. The gap d between the electrodes and the width W of the electrodes are 2 μm and 5 μm respectively.

Figure 2.26 | Schematic of a comb-drive actuator.

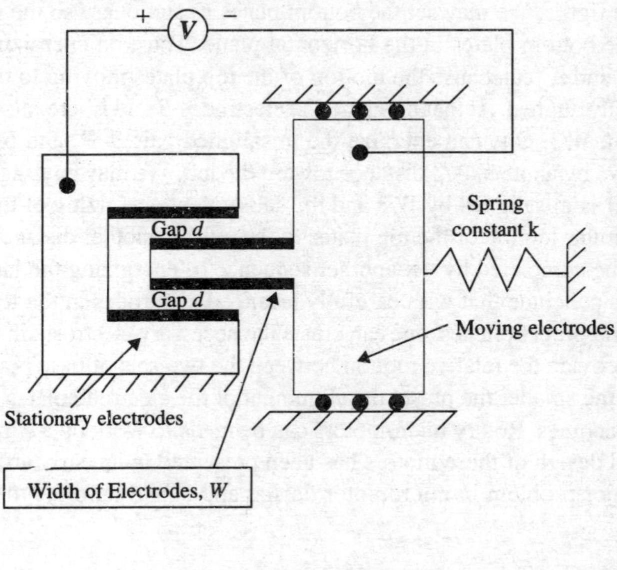

Solution

The required traveling distance of the moving electrodes is $\delta = 10 \times 10^{-6}$ m, which is equivalent to a spring force $F = k\delta = 0.05$ (N/m) $\times 10 \times 10^{-6}$ (m) $= 0.5 \times 10^{-6}$ N.

There are two sets of electrodes in the system; each set needs to generate $0.5F$, or 0.25×10^{-6} N. By using Equation (2.11) with $\epsilon_r = 1.0$, $\epsilon_0 = 8.85 \times 10^{-12}$ C/N-m^2; $W = 5 \times 10^{-6}$ m, and $d = 2 \times 10^{-6}$ m, we obtain the following expression for the required voltage supply:

$$0.25 \times 10^{-6} = \frac{1}{2} \frac{1(1 \times 8.85 \times 10^{-12})(5 \times 10^{-6})}{2 \times 10^{-6}} V^2$$

which gives the required voltage to be $V = 150.33$ V.

the closure arm. The electrostatic force generated by these pairs of misaligned plates tends to align them, causing the drive arms to bend, which in turn closes the extension arms for gripping. These microgrippers can be adapted to micromanipulators or robots in micromanufacturing processes or microsurgery. The length of the gripper produced by Kim et al. [1991] was 400 μm. It had a tip opening of 10 μm.

2.4.2 Micromotors

There are two types of micromotors that are used in micromachines and devices: linear motors and rotary motors.

The actuation forces for micromotors are primarily electrostatic forces. The sliding force generated in pairs of electrically energized misaligned plates, such as illustrated in Figure 2.23, prompts the required relative motion in a linear motor. Figure 2.27 illustrates the working principle of the linear motion between two sets of parallel base plates. Each of the two sets of base plates contains a number of electrodes made of electric conducting plates. All these electrodes have a length W. The bottom base plate has an electrode pitch of W, whereas the top base plate has a slightly different pitch, say $W + W/3$. The two sets of base plates are initially misaligned by $W/3$, as shown in the figure. We may set the bottom plates as stationary so the top plates can slide over the bottom plates in the horizontal plane. Thus, on energizing the pair of electrodes A and A' can cause the motion of the top plates moving to the left until A and A' are fully aligned. At that moment, the electrodes B and B' are misaligned by the same amount, $W/3$. One can energize the misaligned pair B–B' and prompt the top plates to move by another $W/3$ distance toward the left. We may envisage that by then the C–C' pair is misaligned by $W/3$ and the subsequent energizing of that pair would produce a similar motion of the top plates to the left by another distance of $W/3$. The motion will be completed by yet another sequence of energizing the last pair, D–D'. We may thus conclude that with carefully arranged electrodes in the top and bottom base plates and proper pitches, one can create the necessary electrostatic forces that are required to provide the relative motion between the two sets of base plates. It is readily seen that the smaller the preset misalignment of the electrode plates, the smoother the motion becomes. Rotary micromotors can be made to work by a similar principle.

Detailed design of these motors has been presented in an early article [Trimmer 1997]. A major problem in micromotor design and construction is the bearings for

Figure 2.27 | Working principle of electrostatic micromotors.

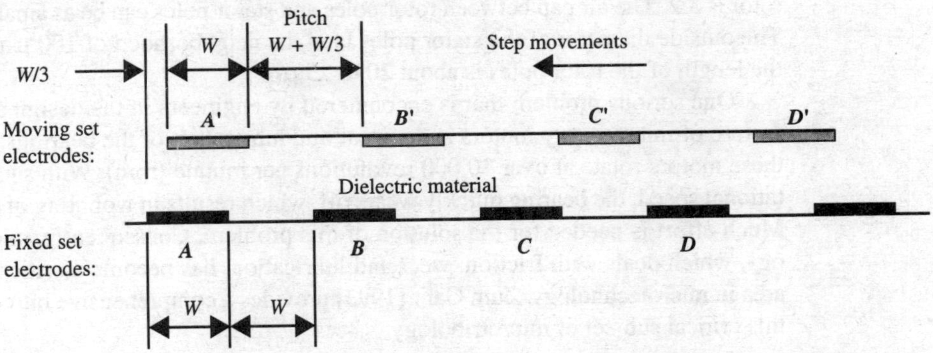

the rotors. Electric levitation principles have been used for this purpose [Kumar and Cho 1991].

Micromotors built on the principles of electrostatic forces are described in detail by Fan et al. [1988], Mehregany et al. [1990], and Gabriel et al. [1988]. Fabrication techniques used to produce these motors are discussed by Lober and Howe [1988].

Rotary motors driven by electrostatic forces can be constructed in a similar way. Figure 2.28 shows a top view of an electrostatically driven micromotor. As can be seen from the figure, electrodes are installed in the outer surface of the rotor poles and the inner surface of the stator poles. As in the case of linear motors, pitches of electrodes in rotor poles and stator poles are mismatched in such a way that they will generate an electrostatic driving force due to misalignment of the energized pairs of

Figure 2.28 | Schematic of a micro rotary motor.

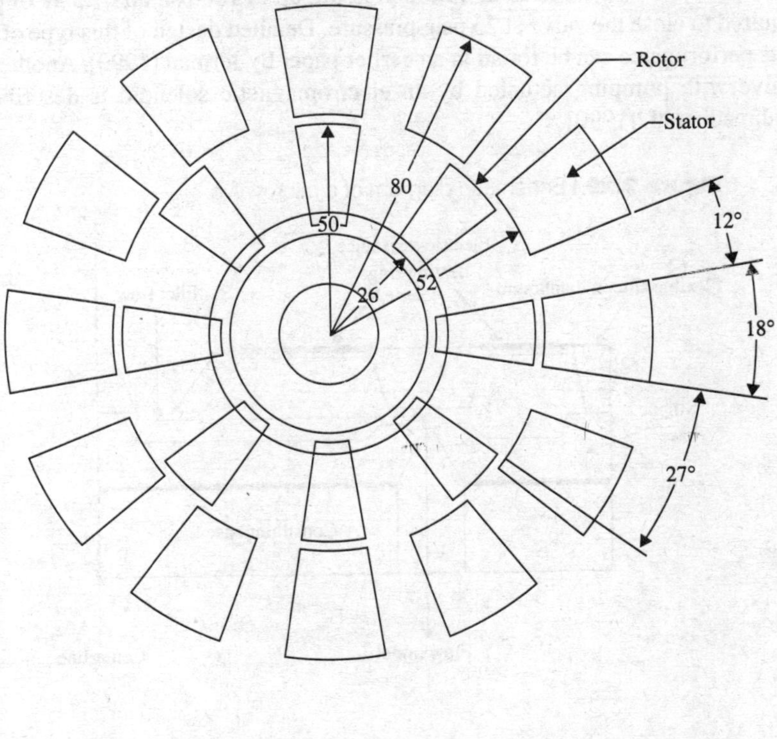

electrodes. The reader will notice that the ratio of poles in the stator to those in the rotor is 3:2. The air gap between rotor poles and stator poles can be as small as 2 μm. The outside diameter of the stator poles is in the neighborhood of 100 μm, whereas the length of the rotor poles is about 20 to 25 μm.

One serious problem that is encountered by engineers in the design and manufacture of micro rotary motors is the wear and lubrication of the bearings. Typically these motors rotate at over 10,000 revolutions per minute (rpm). With such high rotational speed, the bearing quickly wears off, which results in wobbling of the rotors. Much effort is needed for the solution of this problem. Consequently, micro tribology, which deals with friction, wear and lubrication, has become a critical research area in microtechnology. Zum Gahr [1993] provides a comprehensive introduction of this critical subject of microtribology.

2.4.3 Microvalves

Microvalves are primarily used in industrial systems that require precision control of gas flow for manufacturing processes, or in biomedical applications such as in controlling the blood flow in an artery. A growing market for microvalves is in the pharmaceutical industry, where these valves are used as a principal component in microfluidic systems for precision analysis and separation of constituents. Microvalves operate on the principles of microactuation. Jerman [1991] reported one of the earlier microvalve designs. As illustrated in Figure 2.29, the heating of the two electrical resistor rings attached to the top diaphragm can cause a downward movement to close the passage of flow. Removal of heat from the diaphragm opens the valve again to allow the fluid to flow. In Jerman's design, the diaphragm is 2.5 mm in diameter and is 10 μm thick. The heating rings are made of aluminum 5 μm thick. The valve has a capacity of 300 cm^3/min at a fluid pressure up to 100 psi, and 1.5 W of power is required to close the valve at 25 psig pressure. Detailed design of this type of valve and its performance can be found in an earlier paper by Jerman [1990]. Another type of valve with pumping actuated by an electromagnetic solenoid is described by Pourahmadi et al. [1990].

Figure 2.29 | Schematic diagram of a microvalve.

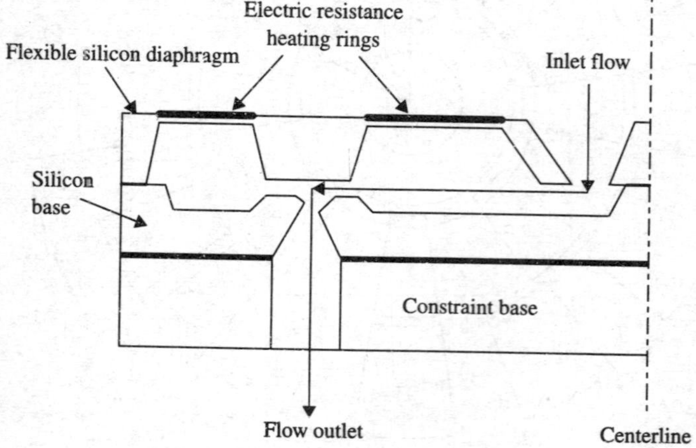

A rather simple microvalve design uses a thermal actuation principle [Henning et al. 1997, Henning 1998]. The cross-section of this type of valve is schematically shown in Figure 2.30. This design is used to control the flow rate from a normally open valve (as shown) to a fully closed state. The downward bending of the silicon diaphragm regulates the amount of valve opening. Bending of the diaphragm is activated by heat supplied to a special liquid in the sealed compartment above the diaphragm. The heat source in this case is the electric resistance foils attached at the top of the device.

Figure 2.30 | A thermally actuated microvalve.

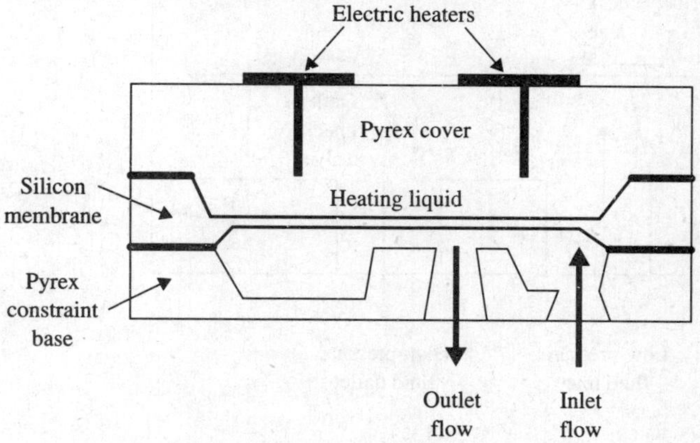

2.4.4 Micropumps

A simple micropump can be constructed by using the electrostatic actuation of a diaphragm as illustrated in Figure 2.31. The deformable silicon diaphragm forms one electrode of a capacitor. It can be actuated and deformed toward the top electrode by applying a voltage across the electrodes. The upward motion of the diaphragm increases the volume of the pumping chamber and hence reduces the pressure in the chamber. This reduction of pressure causes the inlet check valve to open to allow inflow of fluid. The subsequent cutoff of the applied voltage to the electrode prompts the diaphragm to return to its initial position, which causes a reduction of the volume in the pumping chamber. This reduction of volume increases the pressure of the entrapped fluid in the chamber. The outlet check valve opens when the entrapped fluid pressure reaches a designed value, and fluid is released. A pumping action can thus be accomplished. Zengerle [1992] reported the design of this device. The pump in Figure 2.31 has a square shape with a diaphragm 4 mm $\times$ 4 mm $\times$ 25 μm thick. The gap between the diaphragm and the electrode is 4 μm. The actuation frequency is 1 to 100 Hz. At 25 Hz, a pumping rate of 70 μL/min is achieved.

Another type of micropump, called a *piezopump* [Madou 1997] is built on the principle of producing wave motion in the flexible wall of minute tubes in which the fluid flows. Piezoelectric materials coated outside the tube wall generate the wave motion. The wave motion of the tube wall exerts forces on the contained fluid for the required motion. We will present a more detailed description of this type of pumping actuation in Chapter 5.

Figure 2.31 | Schematic diagram of a micropump.

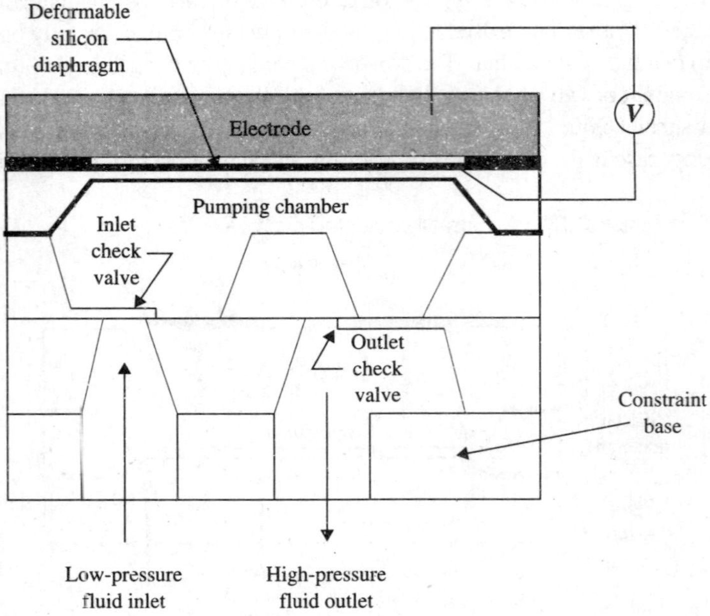

2.5 | MICROACCELEROMETERS

An accelerometer is an instrument that measures the acceleration (or deceleration) of a moving solid. Microaccelerometers are used to detect the associated dynamic forces in a mechanical system in motion. These accelerometers are widely used in the automotive industry as described in Chapter 1. For example, acceleration sensors in the $\pm 2g$ range are used in a car's suspension system and antilock braking system (ABS), whereas $\pm 50g$ range acceleration sensors are used to actuate air bags for driver and passenger safety in event of collision with another vehicle or obstacles. The notation g represents the gravitational acceleration, with a numerical value of 32 ft/s^2 or 9.81 m/s^2. We present microaccelerometer in a separate section because this type of device is often classified as an *inertia sensor,* yet it contains actuation elements.

Most accelerometers are built on the principles of mechanical vibration, as will be described in detail in Chapter 4. Principal components of an accelerometer are a *mass* supported by *springs*. The mass is often attached to a *dashpot* that provides the necessary damping effect. The spring and the dashpot are in turn attached to a casing, as illustrated in Figure 2.32.

In the case of micro accelerometers, significantly different arrangements are necessary because of the very limited space available in microdevices. A minute silicon beam with an attached mass (often called a *seismic mass*) constitutes a spring–mass system, and the air in the surrounding space is used to produce the damping effect. The structure that supports the mass acts as the *spring*. A typical microaccelerometer is illustrated in Figure 2.33; in it, the mass is attached to a cantilever beam or plate, which is used as a spring. A piezoresistor is implanted on the

Figure 2.32 | Typical arrangement of an accelerometer.

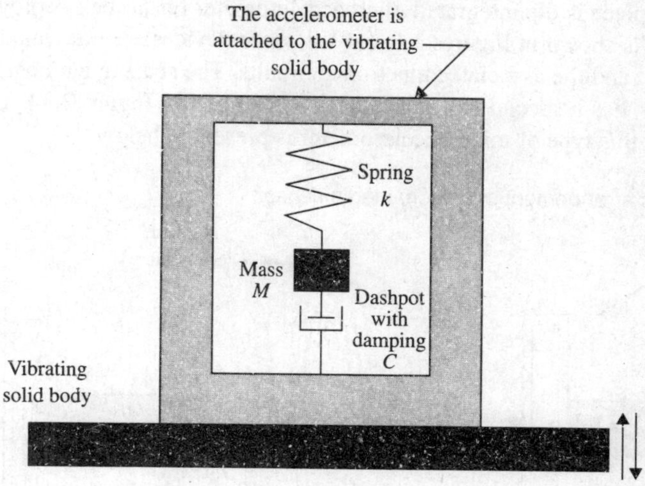

beam or plate to measure the deformation of the attached mass, from which the amplitudes and thus the acceleration of the vibrating mass can be correlated. A mathematical formula that relates the vibrating mass to the acceleration of the casing is available in Chapter 4. Since acceleration (or deceleration) is related to the driving dynamic force that causes the vibration of the solid body to which the casing is attached, accurate measurement of acceleration can thus enable engineers to measure the applied dynamic force. It is not surprising to find that microaccelerometers are widely used as a trigger to activate airbags in automobiles in an event of collision, and also to sense the excessive vibration of the chassis of a vehicle from its suspension system.

Figure 2.33 | Schematic structure of a microaccelerometer.

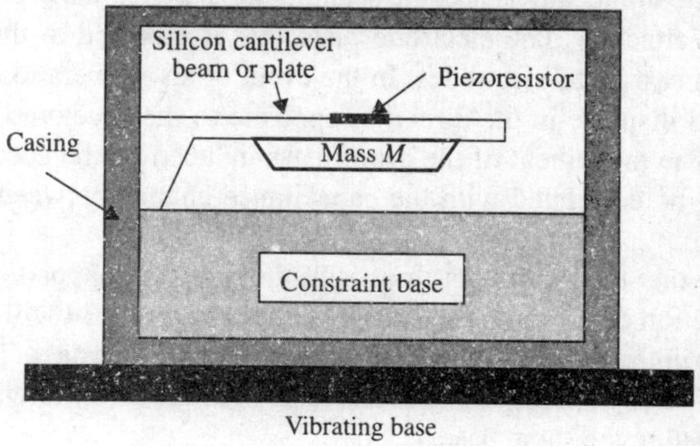

There are many different types of accelerometers available commercially. Signal transducers used in microaccelerometers include piezoelectric, piezoresistive, capacitive, and resonant members [Madou 1997]. Principles of signal conversion of these schemes will be described in Chapter 11.

Perhaps the most widely used microaccelerometer (or inertia sensor) in the current marketplace is the integrated microaccelerometer for air bag deployment in automobiles. As shown in Figures 1.6 and 1.8, these devices are integrated with signal transduction and the associated electronic circuits. The sensing element, i.e., the accelerometer, has a special configuration as illustrated in Figure 2.34. The working principle of this type of microaccelerometer is presented below.

Figure 2.34 | Schematic arrangement of a micro inertia sensor.

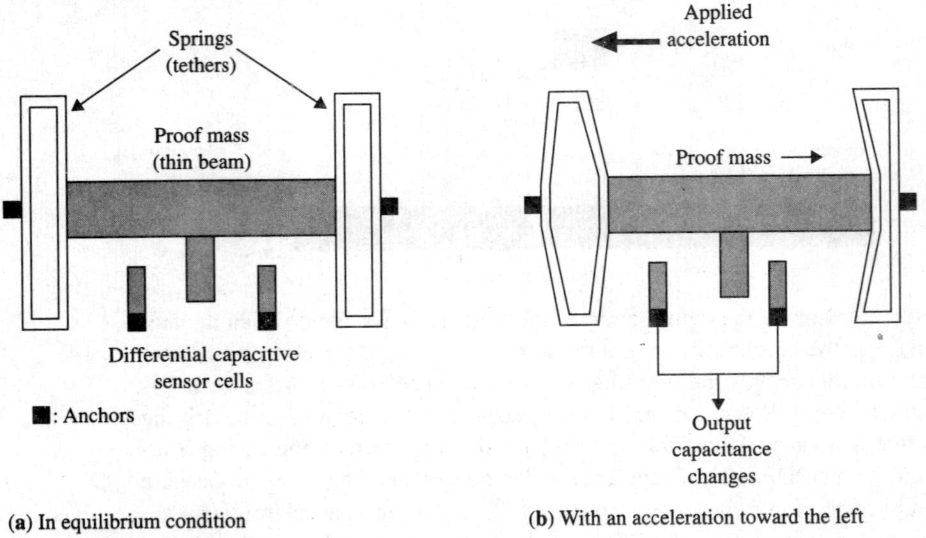

(Doscher [1999].)

With reference to Figure 2.34a, a thin beam is attached to two tethers at both ends. The tethers are made of elastic material and are anchored at one side as shown in the figure. The thin beam acts as the seismic mass (called the *proof mass*) with an electrode plate attached. The electrode plate that is attached to the proof mass is placed between two fixed electrodes. In the event of an acceleration of the unit, the proof mass will displace in the direction opposite to the acceleration, as shown in Figure 2.34b. The movement of the proof mass induced by the acceleration (or deceleration) can be correlated with the capacitance change between the pair of the electrodes.

We realize that the proof mass moves in the direction opposite to the acceleration or deceleration of the unit. The arrangement in Figure 2.34 will measure the acceleration only in the direction along the length of the proof mass. The arrangement in Figure 2.35 is designed for the measurements in both x and y directions when both units are attached to the same base.

A more compact arrangement is illustrated in Figure 2.36, in which the proof mass is replaced by a square plate that can displace in both x and y directions. Electromechanical design principles are presented in references [Doscher 1999, Chau et al. 1995].

Figure 2.35 | Schematic diagram of a dual-axial-motion sensor.

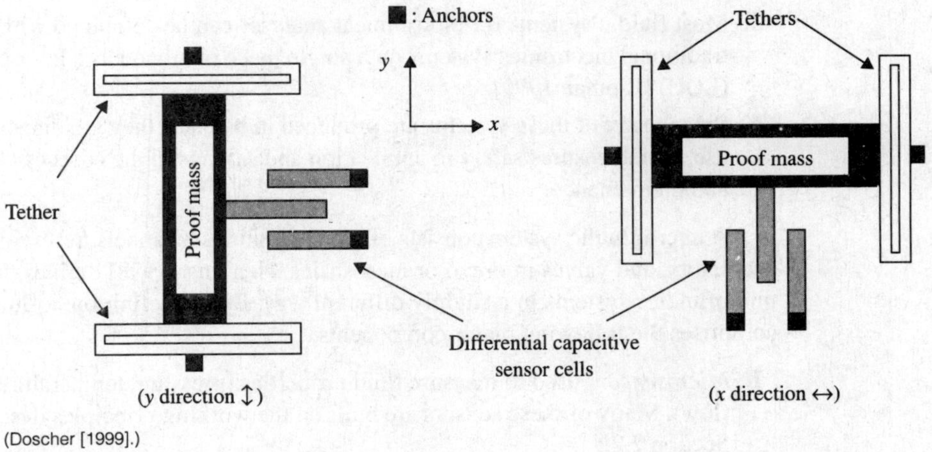

(Doscher [1999].)

Figure 2.36 | Schematic diagram of a compact dual-axial-motion sensor.

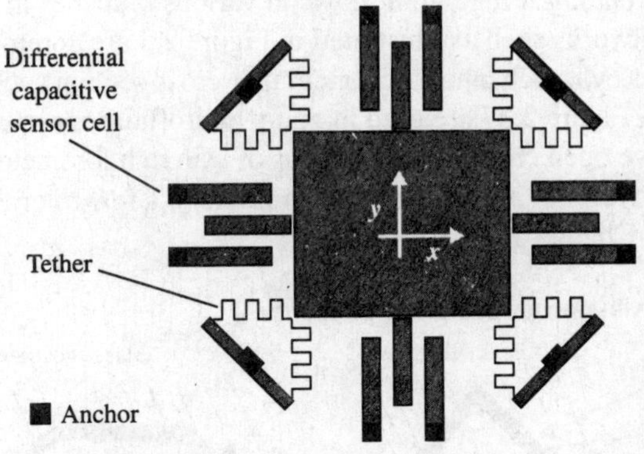

(Courtesy of Analog Devices Inc., Norwood, Massachusetts.)

2.6 | MICROFLUIDICS

Microfluidic systems are widely used in biomedical precision manufacturing processes and pharmaceutical industries. Principal applications of microfluidics systems are for chemical analysis, biological and chemical sensing, drug delivery, molecular separation such as DNA analysis, amplification, sequencing or synthesis of nucleic acids, and for environmental monitoring [Kovacs 1998]. Microfluidics is also an essential part of precision control systems for automotive, aerospace, and machine tool industries. The principal advantages of micro fluidic systems are:

1. The ability to work with small samples, which leads to significantly smaller and less expensive biological and chemical analyses.

2. Most microfluidic systems offer better performance with reduced power consumption.
3. Most fluidic systems for biotechnical analyses can be combined with traditional electronics systems on a single piece of silicon as a lab-on-a-chip (LOC) [Lipman 1999].
4. Since many of these systems are produced in batches, they are disposable after use, which ensures safety in application and savings in the costs of cleaning and maintenance.

A micro fluidic system consists of nozzles, pumps, channels, reservoirs, mixers, oscillators, and valves in micro or mesoscales. Henning [1998] defines the scope of microfluidics systems in a slightly different way. By his definition, a fluidic system comprises the following major components:

1. *Microsensors* used to measure fluid properties (pressure, temperature, and flow). Many of these sensors are built on the working principles described in Section 2.2.
2. *Actuators* used to alter the state of fluids. Microvalves are built on the principles described in Section 2.4.3. Micropumps and compressors operating on similar principle are described in Section 2.4.4.
3. *Distribution channels* regulating flows in various branches in the system. Capillary networks such as illustrated in Figure 2.3 are common in microfluidics. Microchannels of noncircular cross section such as those illustrated in Figure 2.37 are used in many microfluidics systems. They typically have open cross-sectional areas of square micrometers. The channels direct fluids flows of a few hundred nanoliters to a few microliters. As will be described in Chapter 10,

Figure 2.37 | Microchannels with noncircular cross sections.

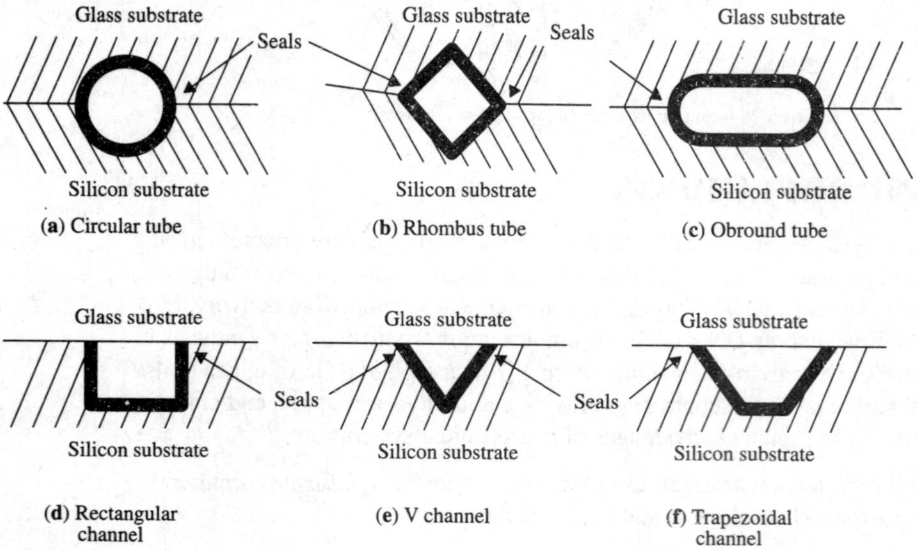

microchannels of noncircular cross sections are usually produced by chemical etching in open channels. The two open channels are bonded to provide the closed conduits. These channels can be produced in lengths of less than a millimeter.

Electrohydrodynamic forces, provided by electro-osmosis or electrophoresis in bio testing and analytical systems (see Section 2.2.2), are used extensively to drive the minute fluid samples through these microchannels. The working principles of electrohydrodynamics will be presented in Chapter 3.

4. *Systems integration* includes integrating the microsensors, valves, and pumps through the microchannels links. This integration also involves the required electrical systems that provide electrohydrodynamic forces, the circuits for transducing and processing the electronic signals, and control of the microfluid flow in the system.

Microfluidic systems can be built with a variety of materials, such as quartz and glasses, plastics and polymers, ceramics, semiconductors, and metals. Working principles of microvalves and pumps have been presented in Sections 2.4.3 and 2.4.4, respectively. The design of these systems requires special considerations, as we will learn from the scaling laws in Chapter 6. Surface-to-mass ratio changes drastically when the systems are scaled down. For example, in the well-known capillary effect on liquid flow through minute tubes or channels, surface tension and viscosity become two major governing factors in the design of the flow system. Electrohydrodynamic pumping is an effective way of moving fluids in microchannels, as will be described in Section 3.8.2. The well-known Navier-Stokes equations presented in Section 5.5 for fluid dynamics can no longer be used in predicting the dynamics of fluid flow in microsystems [Pfahler et al. 1990]. Modified theoretical formulation of microchannel flows will be presented in microfluid dynamics in Chapter 5, and a design case will be presented in Chapter 10.

PROBLEMS

Part 1. Multiple Choice

1. The fundamental working principle of sensors is (1) to convert one form of energy to another form, (2) to convert signals, (3) to convert signs.
2. Acoustic sensors are used to detect (1) sound, (2) temperature, (3) chemical compositions.
3. Medical diagnosis uses (1) biosensors, (2) biomedical sensors, (3) both these sensors.
4. Biomedical sensors and biosensors are (1) the same thing, (2) different things, (3) neither of the above.
5. Biosensors require (1) biomolecules, (2) electrochemical compounds, (3) chemical compounds to work.
6. Chemical sensors work on the principle of (1) interaction of chemical and electrical properties of materials, (2) chemical and biological interaction, (3) mechanical and electrical interaction.
7. Any material that has a change of electrical properties after being exposed to particular gases can be used as a (1) chemical sensor, (2) biosensor, (3) thermal sensor.
8. Optical sensors work on the principle of (1) input heat generated by light, (2) input photon energy by light, (3) impact of electrons on solid surface.
9. Pressure sensors work on the principle of (1) deflecting a thin diaphragm, (2) heating of a thin diaphragm, (3) magnetizing a thin diaphragm by the pressurized medium.
10. The deflection of the thin diaphragm in micropressure sensors is measured by (1) mechanical means, (2) optical means, (3) electrical means.
11. Thermal sensors work on the principle of (1) thermal mechanics, (2) thermometers, (3) thermal electricity.
12. Thermopiles have (1) one, (2) two, (3) three junctions.
13. It takes a minimum of (1) one, (2) two, (3) three different materials to make thermal actuation work.
14. Shape memory alloys are materials that have (1) memory of their shape at the temperature of fabrication, (2) programmed memory of their original shape, (3) memory of their original properties.
15. Piezoelectric actuation works on the principle of (1) electric heating, (2) mechanical–electrical conversion, (3) electrical–mechanical conversion.
16. As the gap between the electrodes grows smaller, the electrostatic forces for actuation (1) grow stronger, (2) grow weaker, (3) do not change.
17. Electrostatic motors work on the principle of (1) closing gaps, (2) alignment of opposing electrodes, (3) both closing and alignment of opposing electrodes.
18. Microaccelerometers are used to measure (1) the velocity, (2) the position, (3) the dynamic forces associated with a rigid body moving at variable speed.
19. Microfluidics is used extensively in (1) thermomechanical, (2) biomedical, (3) electromechanical analysis.

20. A major problem in microchannel flow is (1) capillary effect, (2) friction effect, (3) pressure distribution.

Part 2. Description Problems

1. Give examples of at least two sensors that match the definition given in Section 2.2.
2. What are the advantages and disadvantages of using (1) piezoresistors and (2) capacitors as signal transducer?
3. Describe the three principal signal transduction methods for micropressure sensors. Provide at least one major advantage and one disadvantage of each of these methods.
4. What are principal applications of microsensors, actuators, and fluidics?
5. Why are electrostatic forces used to run micromotors rather than conventional electromagnetic forces? Explain why this actuation technique is not used in macrodevices and machines.
6. Explain why the change of the state of stress in a silicon diaphragm in a micro pressure sensor results in a change of its natural or resonant frequency.
7. Plot the relationship between the output voltage of a capacitor transducer versus the change in the gap *d*, using the geometry and dimensions of the plate electrodes in Example 2.2. What will happen when $d \to 0$? Observe the shape of the plotted curve.
8. Estimate the voltage output for the microthermopile shown in Figure 2.14 if copper wires are used for the thermocouples with the hot junction temperature at 120°C while the cold junction is maintained at 20°C.
9. Describe the four popular actuation techniques for microdevices. Provide at least one major advantage and one disadvantage of each of these techniques.
10. Calculate the electrostatic forces on the plate electrodes with an applied dc voltage at 70 V. The geometry and dimensions of the plate electrodes are described in Example 2.1. The plates are initially misaligned by 20 percent in both length and width directions. Pyrex glass is used as the dielectric material, so there is no gap change with the applied voltage.
11. Calculate the required voltage in Example 2.4 if an arrangement similar to that in Figure 2.25 is used instead of that illustrated in Figure 2.26. Make whatever assumptions necessary for solving the problem.
12. What information is missing from Example 2.4 if you are asked to design a microgripper similar to that shown in Figure 2.25? How would you proceed to finding the missing information?

3 Chapter

Engineering Science for Microsystem Design and Fabrication

CHAPTER OUTLINE

3.1 | INTRODUCTION

The design of most successful microsystem products requires not only the application of theories and principles of mechanical, electrical, materials, and chemical engineering, but it also involves the theories and principles of physics, chemistry, and biology. While it is not possible for any engineer to develop expertise in all these fields, we will nevertheless need to understand the fundamentals of these scientific disciplines from which many of the design principles and microfabrication techniques are developed. Consequently, what will be covered in this chapter are the science topics that are not often familiar to engineers, yet they are closely related to the design, manufacture, and packaging of microsystems, as presented in the subsequent chapters in the book. For brevity, each of the selected science topics will be presented as an overview. Readers are encouraged to look for in-depth descriptions of these topics in the cited references.

3.2 | ATOMIC STRUCTURE OF MATTER

Atoms are the bases of all substances that are known to exist in the universe. Everything on our Earth is made from 96 stable and 12 unstable elements. Each element has a different atomic structure. The basic structure of an atom involves a *nucleus* and the orbiting *electrons*. The nucleus consists of *protons* and *neutrons*. *The difference in atomic structures results in different properties of the elements*. Figure 3.1 illustrates the simplest atomic structure, that of a hydrogen atom. As can be seen from the figure, the nucleus is located at the core of the atom, whereas the lone electron revolves in an *orbit* that is away from the nucleus.

Figure 3.1 | Structure of a hydrogen atom.

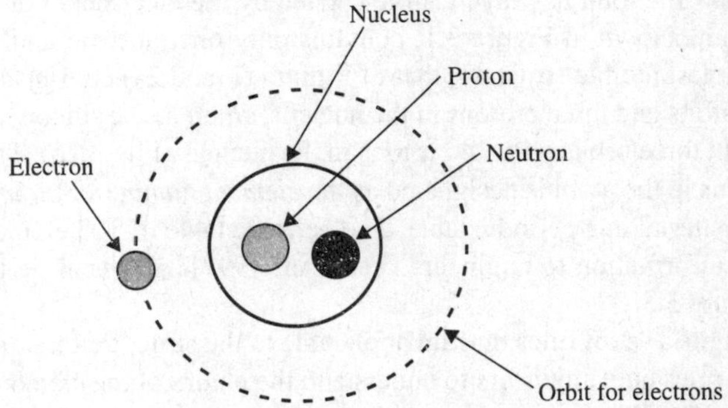

In general, atoms contain more than one electron. Electrons in atoms can exist in more than one orbit, as illustrated in Figure 3.2. They can be moved from one orbit to another orbit with a change of energy that keeps the atom in its natural state. Materials with highly mobile electrons are called *conductors*, whereas those with "immobile" electrons are called insulators or *dielectric* materials.

The outer orbit of atoms has a diameter varying from 2 to 3 $\times$ 10^{-8} cm [or 2 to 3 Å, 1 Å = 10^{-4} μm], which is about 1000 times greater than that of the nucleus. The core of an atom is the *nucleus*, which contains *neutrons* and *protons* (N and P, as shown in Figure 3.2). Neutrons in the atom carry no electric charge. On the other hand, the protons in the nucleus are positive electrical charge carrier whereas the electrons carry negative charges. The respective masses for protons and electrons are 1.67 $\times$ 10^{-24} g and 9.11 $\times$ 10^{-28} g. In a neutral state, the total numbers of protons and

Figure 3.2 | Atomic structures of lithium and silicon.

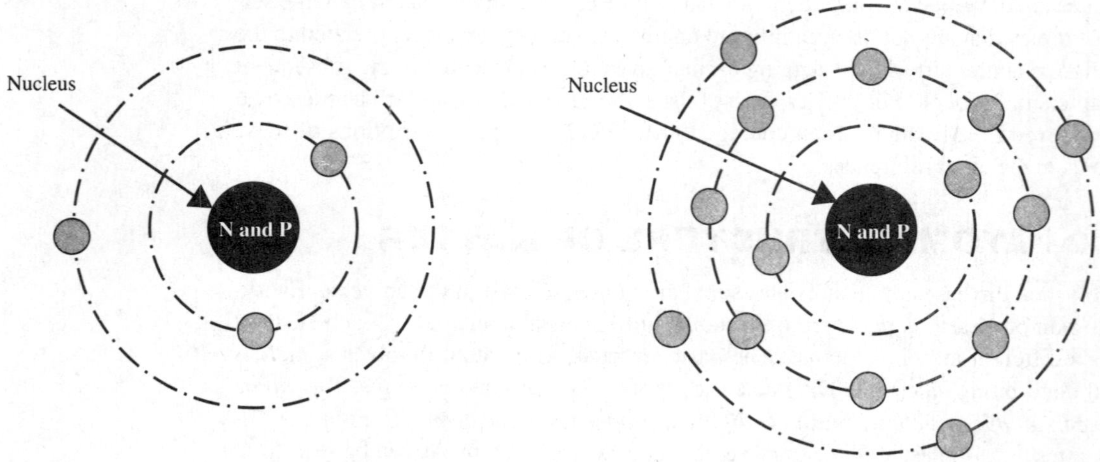

(**a**) Lithium (Li)

(**b**) Silicon (Si)

electrons in an atom are equal to each other. Therefore, the total positive charge in neutrons equals the total negative charge carried by the electrons. For instance, the hydrogen atom, shown in Figure 3.1, contains only one electron and one neutron. The next simplest atomic structure, that of lithium (Fig. 3.2a) consists of three electrons in two orbits and three protons in the nucleus, whereas the silicon atom contains 14 electrons in three orbits with 14 protons in the nucleus (Fig. 3.2b). The total number of electrons in the atom is designated as the *element number* or *atomic number* of the specific material in a periodic table. The periodic table of 103 elements provides much useful information to engineers [Van Zant 1997]. A typical periodic table is shown in Figure 3.3.

Following is a set of rules that are applicable to the structure of atoms [Van Zant 1997]. These rules help engineers to understand the nature of the elements of matters existing on the Earth.

1. In the periodic table, such as the one shown in Figure 3.3, each element contains a specific number of protons, and no two elements have the same number of protons.
2. Elements with the same number of electrons at the outer orbit have similar properties.
3. Elements are stable with eight electrons in the outer orbit.
4. Atoms seek to combine with other atoms to create the stable condition of a full outer orbit, i.e., to have *eight* electrons in their outer orbit.

3.3 | IONS AND IONIZATION

An *ion* is an electrically charged atom or molecule. A negative ion is an atom that contains more electrons than that in its neutral state. A positive ion, on the other hand, is an atom that contains fewer electrons than is necessary to maintain the neutral state.

Figure 3.3 I A periodic table of elements.

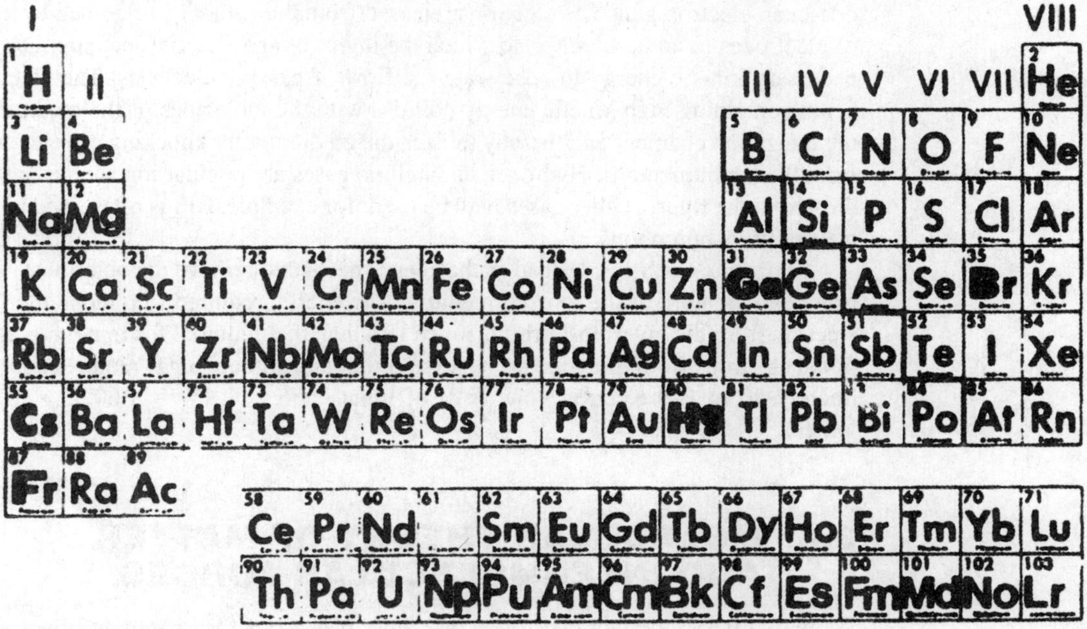

(After Van Zant [1997].)

Ionization is the process of producing ions. Two common methods are available for the production of discrete ions or ion beams: (1) by electrolysis processes and (2) by electron beams. We will describe the electrolysis process in Section 3.8. In either case, external energy is required to initiate and maintain the ionization process. The production of ions from certain substances (gases are common substances to be ionized) by electron beams can be illustrated by the process shown in Figure 3.4.

Figure 3.4 I Schematic diagram of ionization by electron beams.

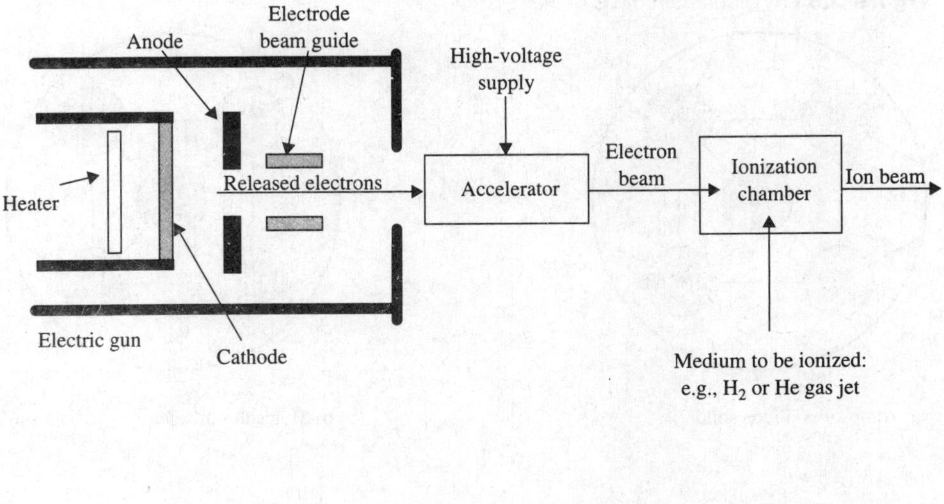

With reference to Figure 3.4, electron beams are generated by heating the cathode in an electron gun. The electrons released from the cathode are guided by a set of electrodes to an accelerator, in which the high-voltage electric field supplies the necessary kinetic energy to accelerate the flow of passing electrons. The electron beam containing high kinetic energy collides with the molecules of the medium in the ionization chamber and thereby ionizes the medium after knocking out electrons from the medium atoms. Hydrogen and helium gases are popular ion sources, as indicated in the figure. Other gases may be used; for example, BF_3 is often used to extract positive boron ions.

Ionization energy is defined as the energy needed to remove the outermost electron from an atom of the ionized medium. The energy required to remove the first electron from the outermost orbit is much less than that required for removing additional electrons from the same orbit. An optimal ionization of a gas requires approximately 50 to 100 electron volts (eV) of input energy (1 eV $= 1.6022 \times 10^{-19}$ joules).

3.4 | MOLECULAR THEORY OF MATTER AND INTERMOLECULAR FORCES

We may observe the fact that force, or energy of various forms, can deform a substance, whether it is a solid, a liquid, or a gas. These deformed substances can restore, either completely or partially, to their original shapes once the applied force or energy is removed. This phenomenon leads us to hypothesize that all matter is made up of particles that are interconnected by bonds that can be compressed or stretched by the application of force or energy. These particles are referred to as the *molecules* of a substance. Molecules are made of atoms, such as the silicon atoms in a silicon wafer, illustrated in Figure 3.5a. Molecules can also involve atoms of chemical compounds such as the hydrogen and oxygen atoms in the case of water molecules, illustrated in Figure 3.5b.

Figure 3.5 | Molecular structure of matters.

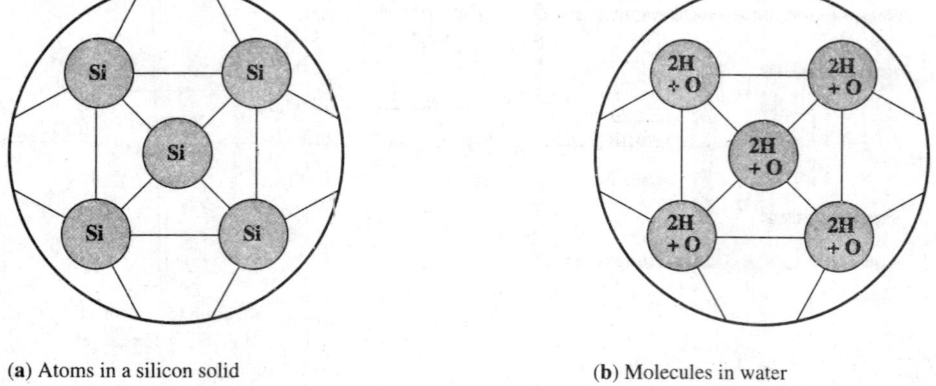

(a) Atoms in a silicon solid

(b) Molecules in water

In either case, as illustrated in Figure 3.5, the atoms are situated in *lattices* that separate them from their neighboring atoms or molecules. The bonding forces are called *atomic cohesive forces* or *intermolecular forces*. These forces in general are called *van der Waals forces*. Van der Waals forces are a major factor in the serious problem of *stiction* of thin films in surface micromachining, as will be described in Chapter 9.

The intermolecular forces are electrostatic in nature. Consequently, the distance between molecules, d, plays an important role in determining the magnitude of these forces. Being electrostatic in nature, these forces are based on Coulomb's law of *attraction* between unlike charges and *repulsion* between like charges as described in Chapter 2. Let us first consider a situation in which a pair of molecules in a natural state are placed a distance d_0 apart. The variation of the force between the molecules is qualitatively illustrated in Figure 3.6.

Figure 3.6 | Variation of intermolecular forces with separation.

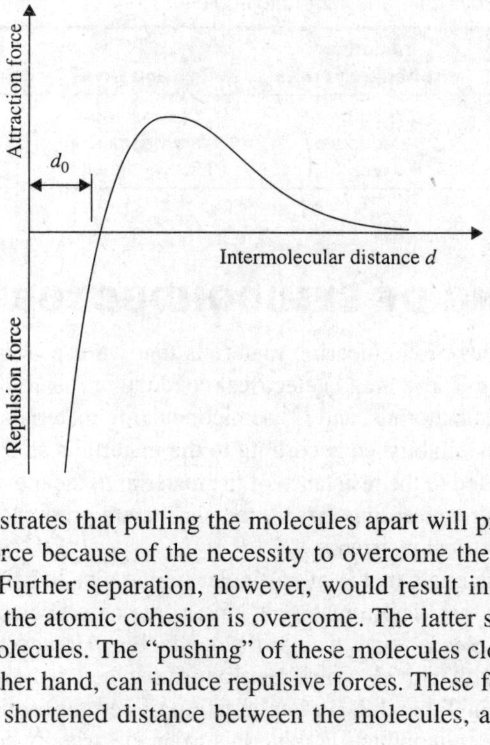

Figure 3.6 illustrates that pulling the molecules apart will prompt the increase of the attraction force because of the necessity to overcome the inherent cohesion of the molecules. Further separation, however, would result in rapid decrease of the attraction once the atomic cohesion is overcome. The latter state is particularly true for gaseous molecules. The "pushing" of these molecules closer than their natural state, on the other hand, can induce repulsive forces. These forces increase dramatically with the shortened distance between the molecules, as illustrated in the figure.

In spite of our perception that intermolecular forces are of electrostatic nature, the magnitude of these forces does not always follow Coulomb's law as presented in Chapter 2. For atoms with electric charges, e.g., ions, in the molecules, the magnitude of these forces can vary from $1/d$ to $1/d^6$, in which d is the separation distance between such atoms in the molecules. Brown and LeMay, Jr. [1981] presented a summary of these forces, given in Table 3.1.

Table 3.1 | Intermolecular forces for molecules with charge-carrying atoms.

Type of interaction	Force or energy dependence on distance	Examples
Ion–ion	$1/d$	Na^+Cl^-, $Mg^{2+}O^{2-}$
Ion–dipole*	$1/d^2$	Na^+—H_2O
Ion–induced dipole	$1/d^4$	K^+—SF_6
Dipole–dipole	$1/d^6$	HCl—HCl; H_2O—H_2O

*A dipole is a system of two equal and opposite charges placed a very short distance apart.

Like electrons in atoms, molecules are not stationary even in their natural state. The atoms in the molecules vibrate from the bonds that keep them together. Thus, all molecules are associated with kinetic energies due to this vibration. Table 3.2 outlines the physical behavior of substances in different states. We will describe these energies at the atomic level in Section 3.9 on quantum physics.

Table 3.2 | Physical behavior of molecules

Substances	Atomic cohesive forces	Vibration level	Level of kinetic energy
Solids	Strong	Least vigorous	Small
Liquids	Moderate	Very vigorous	Moderate
Gases	Weak	Most vigorous	Large

3.5 | DOPING OF SEMICONDUCTORS

There are three types of engineering materials that we use frequently for electromechanical systems. These are (1) electrical conducting materials, (2) electrical insulation or dielectric materials, and (3) semiconducting materials. The classification of these materials is established according to the material's ability to conduct electricity, which is related to the resistance of the material to the movement of electrons. Common materials representing these three classes are presented in Table 3.3 with their respective electrical resistance.

The class of materials that is of particular importance to MEMS and microsystems is *semiconductors*. These materials have some natural electrical conductivity but cannot conduct electricity as well as the conductors. However, they can be made to be a conductor, with their electrical resistivities reduced to the order of 10^{-3} Ω-cm from those shown in Table 3.3, by implantation of certain foreign impurities. The process of turning semiconducting materials to be electrically conducting is called *doping*. By virtue of the foreign impurity added to the semiconducting base material and specific doping patterns, one may control both the intensity and path of electric current flow through the semiconductor material. Microtransistors and microcircuits produced in integrated circuits are formed by using this type of doping process. In MEMS and microsystems, doping of semiconducting materials such as silicon

Table 3.3 | Typical electrical resistivity of insulators, semiconductors, and conductors

Materials	Approximate electrical resistivity ρ, Ω-cm	Classification
Silver (Ag)	10^{-6}	Conductors
Copper (Cu)	$10^{-5.8}$	
Aluminum (Al)	$10^{-5.5}$	
Platinum (Pt)	10^{-5}	
Germanium (GE)	$10^{1.5}$	Semiconductors
Silicon (Si)	$10^{4.5}$	
Gallium arsenide (GaAs)	$10^{8.0}$	
Gallium phosphide (GaP)	$10^{6.5}$	
Oxide	10^{9}	Insulators
Glass	$10^{10.5}$	
Nickel (pure)	10^{13}	
Diamond	10^{14}	
Quartz (fused)	10^{18}	

substrates can also alter the material's resistance to chemical or physical etching, which is a common technique in microfabrication. Doping of silicon is thus often used as a barrier to etching as an "etching stop."

Doping is an essential process to produce p–n junctions in microelectronics. Doping of semiconductors can be achieved by altering the number of electrons in their atoms by implanting foreign atoms with different numbers of electrons. These foreign materials are called *dopants*. The atom that gives up electrons is referred to as the *donor*. The atom that receives extra electrons obvious becomes a negatively charged atom.

We will illustrate the principle of how a semiconductor material such as silicon can be doped to become a conductor. Recall that silicon has four electrons in its outer orbit (see Figure 3.2b). If a material such as boron with three electrons in its outer orbit is doped into the silicon, the combined material has one electron in deficit, and a *hole* for an electron is created. A p-type semiconductor is thus created. Figure 3.7a illustrates such a doping process.

Negatively charged silicon, i.e. n-type silicon, can be created by a doping process that is similar but uses arsenic or phosphorus as the dopant. Both these doping materials have five electrons in their outer orbit. The doping of silicon with these materials will result in an extra electron in the combined material, as illustrated in Figure 3.7b.

The imbalance of electrons in a doped semiconductor facilitates the flow of electrons and thereby increases the conductivity of the material. The degree of the increased conductivity can be related to the reduction of electrical resistivity in the material. Figure 3.8 presents the change of resistivity of silicon with various doses of boron for p-type doping and phosphorus for n-type doping. One may readily see from this figure that the heavier the dose in the doping, the less the resistivity and thus the more electrically conductive the silicon becomes.

Figure 3.7 | Doping of silicon

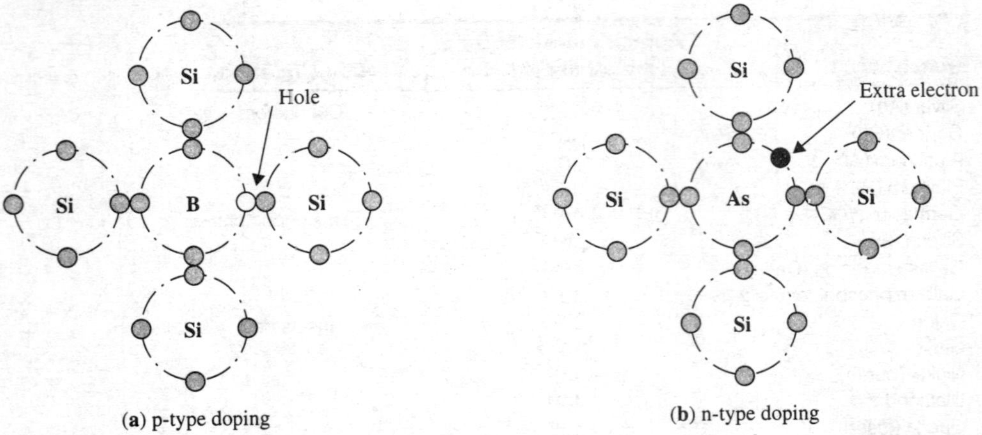

Doping of semiconductors can be achieved by either the diffusion process or ion implantation as described in detail in Chapter 8.

3.6 | THE DIFFUSION PROCESS

The diffusion process is the introduction of a controlled amount of foreign material into selected regions of another material. It is a phenomenon that is frequently present in our day-to-day life. The spread of a drop of dark ink in a pot of clear water is one example of the diffusion process. In such a case, not only can we readily observe the rapid spread of the dark ink in the pot, but also sense the dilution of the dark ink in the mixed liquid with elapsed time. The oxidation of metal in a natural environment is another example for a gas-solid diffusion process. In general, diffusion can take place with liquids to solids, gases to solids, and liquids to liquids as illustrated by the above examples.

Diffusion is a common process that is used in doping silicon wafers with foreign substances as described in the foregoing section. In microfabrication, diffusion is used to oxidize a wafer surface, depositing desired thin films of different materials on the base substrates and building up epitaxial layers over single-crystal substrates in IC fabrication. It plays a major role in the popular *chemical vapor deposition* (CVD) process. Many micromixers and microfluidic devices require the use of diffusion processes. Here, we will present the basic mathematical model that is used to represent this very important physical-chemical process.

Fick's law is used as the basis for mathematical modeling of diffusion processes. It states that the concentration of a liquid A in a liquid B with distinct concentration (Figure 3.9) is proportional to the difference of the concentrations of the two liquids, but is inversely proportional to the distance over which the diffusion effect takes place.

Thus, with reference to Figure 3.9 and with the assumption that $C_1 > C_2$, Fick's law can be expressed in a mathematical form as:

Figure 3.8 | Electric resistivity of silicon versus doses of dopants. (After Van Zant [1997]).

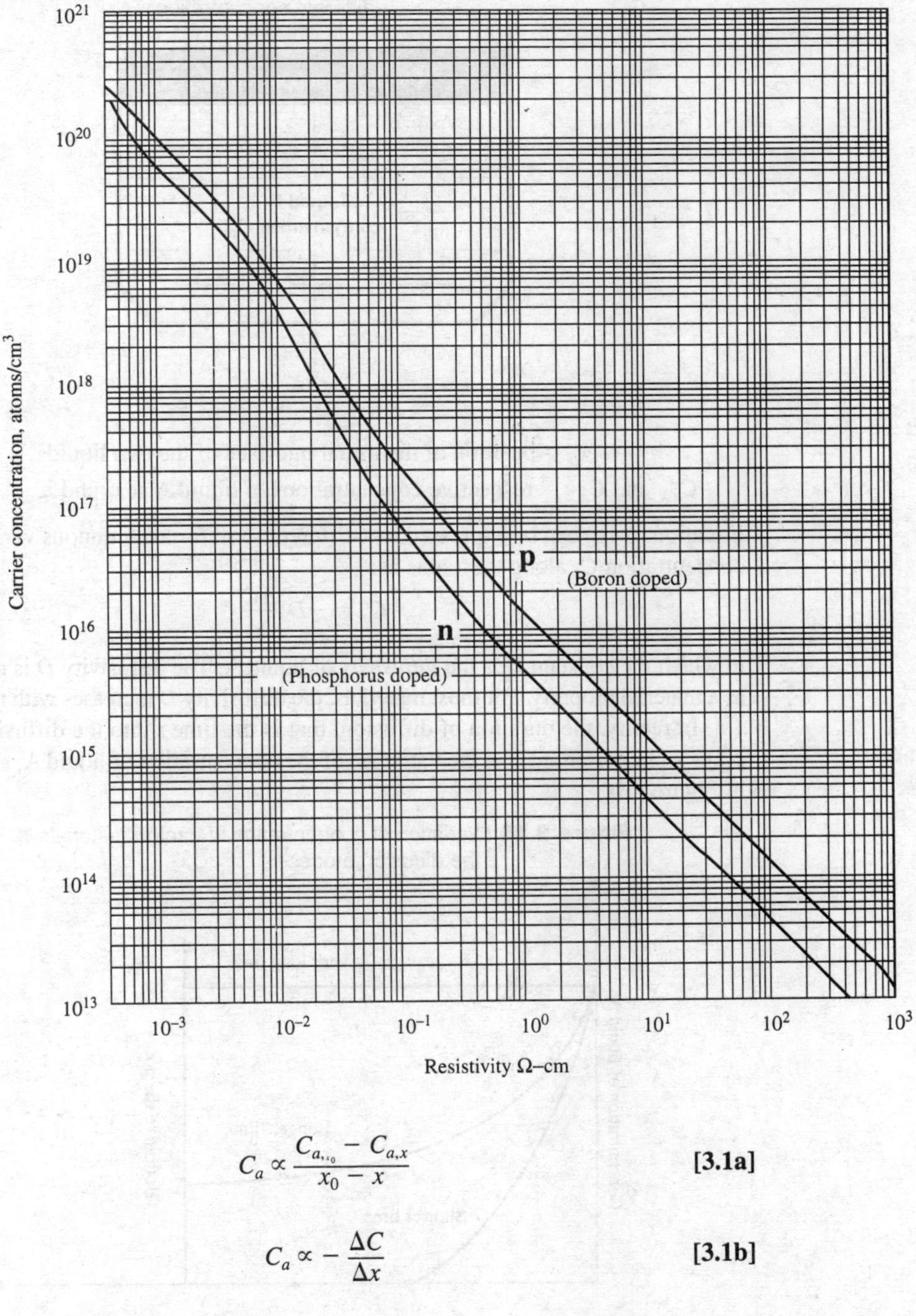

$$C_a \propto \frac{C_{a,x_0} - C_{a,x}}{x_0 - x} \qquad \textbf{[3.1a]}$$

or

$$C_a \propto -\frac{\Delta C}{\Delta x} \qquad \textbf{[3.1b]}$$

where

C_a = the concentration of liquid A at a distance x away from the initial contacting surface per unit area and time t

Figure 3.9 | Diffusion of liquids of different concentration.

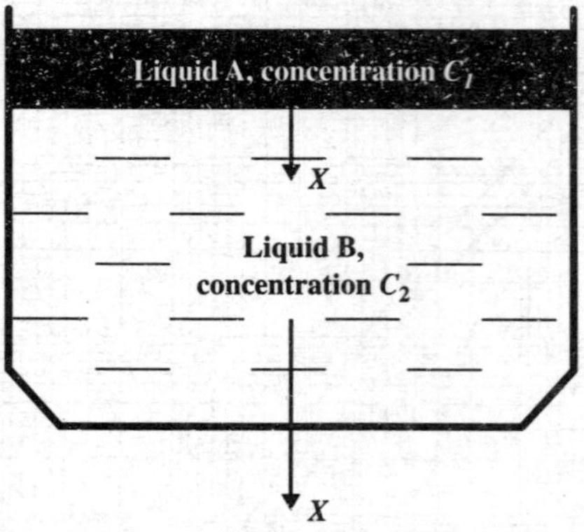

x_0 = position of the initial interface of the two liquids

C_{a,x_0} and $C_{a\,x}$ = respective concentrations of liquid A at x_0 and x.

Equation (3.1b) can be expressed in a different form for a continuous variation of the concentration C_a along the x axis as:

$$C_a = -D\frac{\Delta C}{\Delta x} \qquad [3.2]$$

in which the constant D is the *diffusivity* of liquid A. The diffusivity D is often treated as a material property. For most materials, the diffusivity D increases with temperature.

In reality, the duration of diffusion, that is the time t into the diffusion process, plays an important role in the variation of the concentration of liquid A, as illustrated in Figure 3.10.

Figure 3.10 | Variation of concentration of foreign materials in the diffusion process.

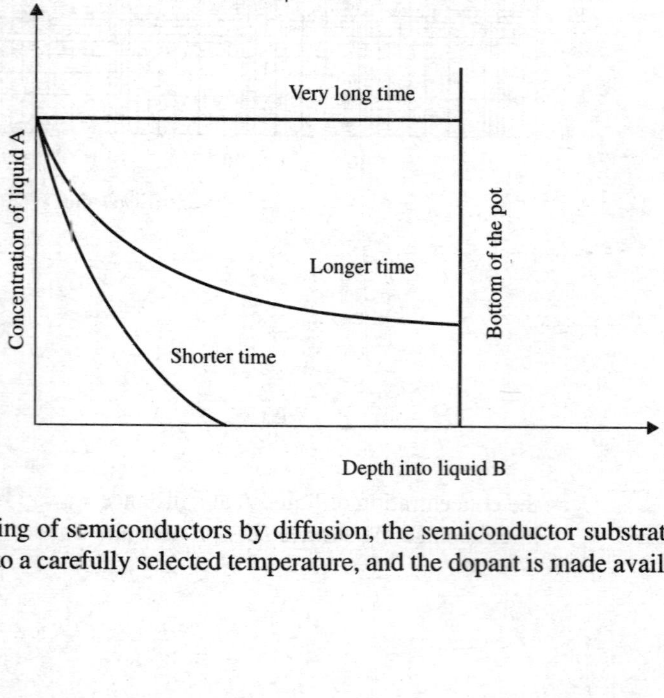

In doping of semiconductors by diffusion, the semiconductor substrates usually are heated to a carefully selected temperature, and the dopant is made available at the surface of the substrate. A mask made of a material that is resistant to the diffusion of the dopant covers the substrate surface during the doping process. The opening made on the mask allows the dopant to be diffused into the substrate surface, and thereby controls the region to be doped. The dopant can diffuse into the substrate until a maximum concentration is reached. This maximum concentration of dopant through diffusion is called *solid solubility*. Figure 3.11 shows solid solubility of common

Figure 3.11 | Solid solubility of selected materials.

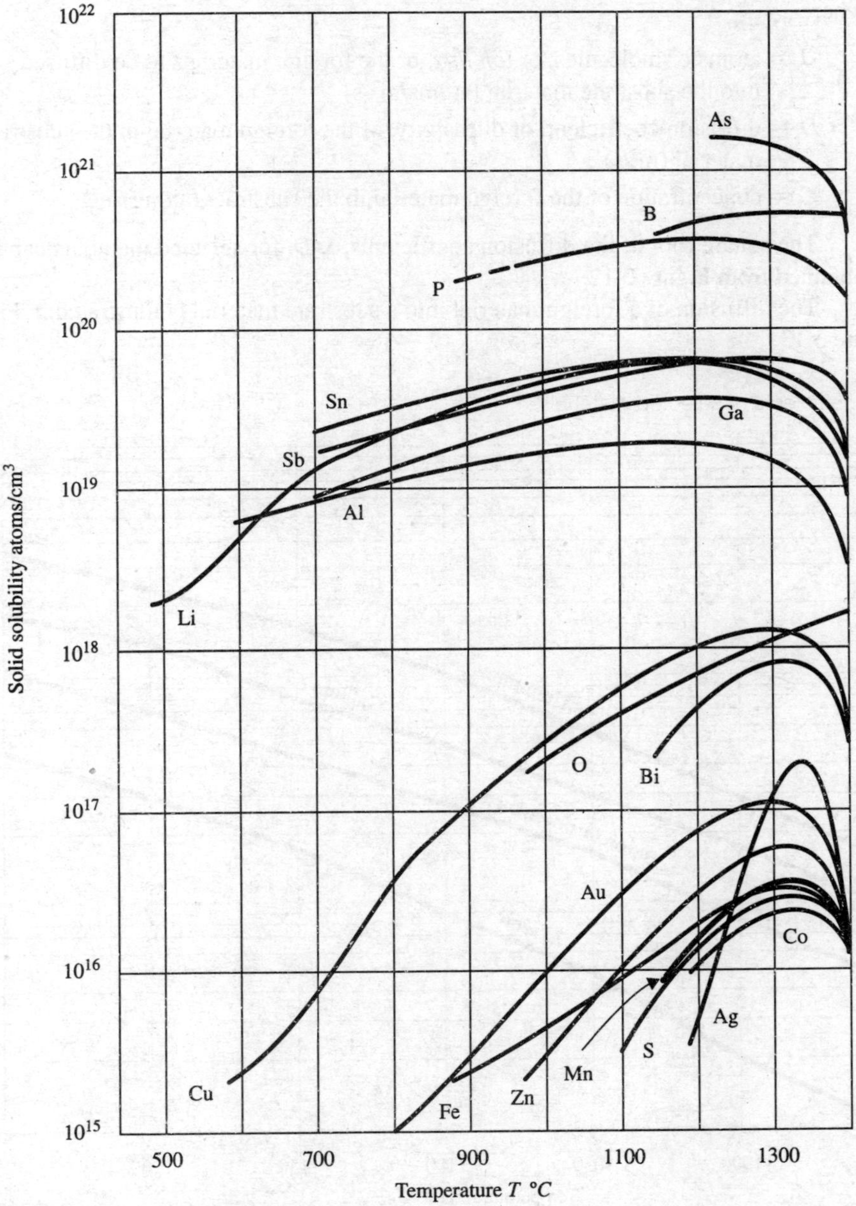

(After Van Zant [1997].)

materials that are used as the foreign substances implanted in silicon substrates in microelectronics and microsystem fabrication. One may readily observe that an optimum temperature of the substrate surface does exist for maximum solubility of each material.

A similar expression to Equation (3.2) for one-dimensional solid-to-solid diffusion can be derived as [Kovacs 1998]:

$$J = -D\frac{\partial C}{\partial x} \tag{3.3}$$

where

J = atoms or molecules, or *ion flux,* of the foreign materials to be diffused into the substrate material (atoms/m^2-s)

D = diffusion coefficient, or diffusivity, of the foreign material in the substrate material (m^2/s)

C = concentration of the foreign material in the substrate (atoms/m^3)

The square root of the diffusion coefficients, $\sqrt{D}$, for selected materials can be obtained from Figure 3.12.

The diffusion of a foreign material into a substrate material is illustrated in Figure 3.13.

Figure 3.12 | Diffusion coefficients of selected materials.

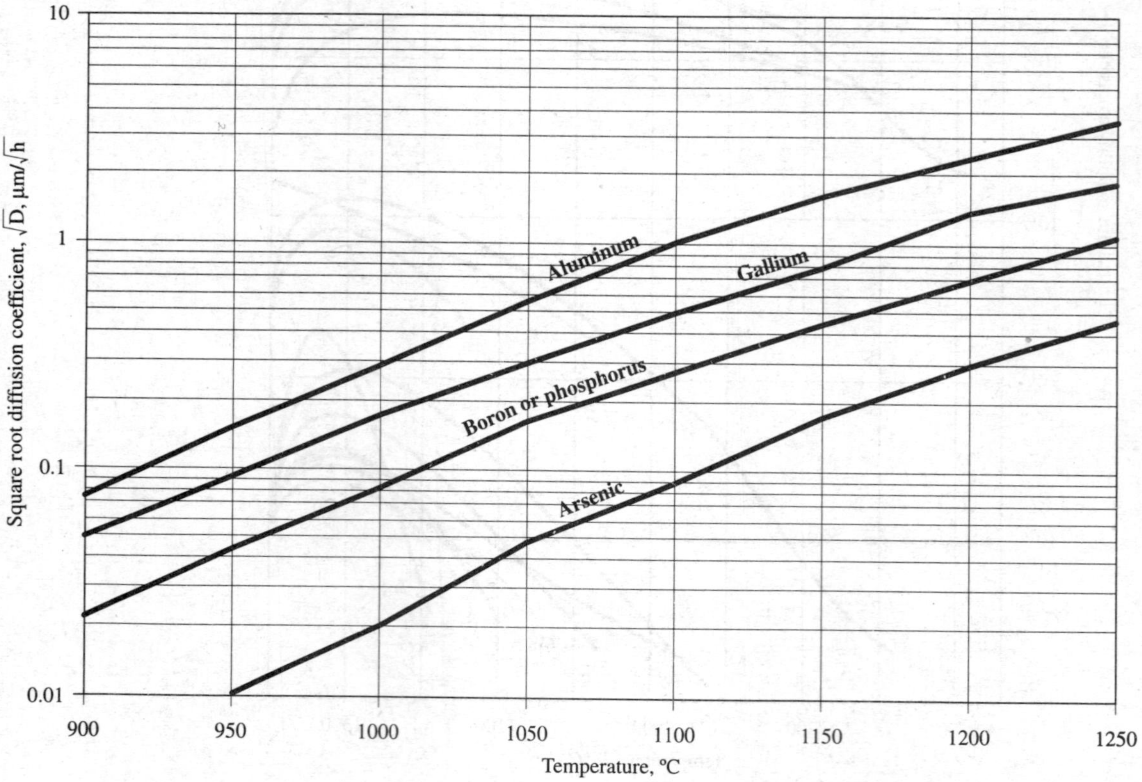

Figure 3.13 | Diffusion of a solid material into a substrate.

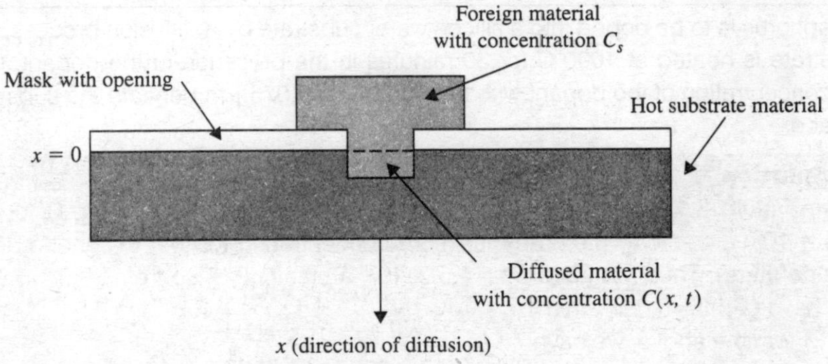

The concentration C in Equation (3.3), $C(x, t)$, at a given depth in the substrate x and at time t for a solid-to-solid diffusion in Figure 3.13 can be determined by solving the diffusion equation derived from Fick's law in the following form:

$$\frac{\partial C(x, t)}{\partial t} = D \frac{\partial^2 C(x, t)}{\partial x^2} \quad [3.4]$$

The solution of Equation (3.4) can be shown to take the following form with a localized diffusion zone in the substrate:

$$C(x, t) = C_s \operatorname{erfc}\left(\frac{x}{2\sqrt{Dt}}\right) \quad [3.5]$$

where $C(x, t)$ = the concentration of the foreign material at a depth x into the substrate at time t, and C_s = the solid solubility at the diffusion temperature (Fig. 3.11).

The solution in Equation (3.5) is expressed in terms of the complementary error function erfc (X), which is defined as 1 − erf (X) with erf (X) being the *error function*. Values of erf (X) can be found in mathematical tables, such as those in Abramowitz and Stegun [1964]. Table 3.4 provides approximate values of erf (X) and therefore erfc (X).

Table 3.4 | Error functions with selected variables

X	erf (*X*)	*X*	erf (*X*)	*X*	erf (*X*)	*X*	erf (*X*)
0.0	0.0						
0.05	0.0564	0.55	0.5633	1.05	0.8624	1.55	0.9716
0.10	0.1125	0.60	0.6039	1.10	0.8802	1.60	0.9763
0.15	0.1680	0.65	0.6420	1.15	0.8961	1.65	0.9804
0.20	0.2227	0.70	0.6778	1.20	0.9103	1.70	0.9838
0.25	0.2763	0.75	0.7112	1.25	0.9229	1.75	0.9867
0.30	0.3286	0.80	0.7421	1.30	0.9340	1.80	0.9891
0.35	0.3794	0.85	0.7707	1.35	0.9438	1.85	0.9911
0.40	0.4284	0.90	0.7969	1.40	0.9523	1.90	0.9923
0.45	0.4755	0.95	0.8209	1.45	0.9597	1.95	0.9942
0.50	0.5205	1.00	0.8427	1.50	0.9661	2.00	0.9953

EXAMPLE 3.1

Phosphorus is to be doped into a silicon wafer substrate by a diffusion process. The substrate is heated at 1000°C for 30 minutes in the presence of the dopant. Find the concentration of the dopant with the depth $x = 0.075$ μm beneath the substrate surface.

Solution

From Figure 3.11, we get the solid solubility of phosphorus at 1000°C to be 4.5×10^{20} atoms/cm³ (i.e., the maximum concentration for that material at this temperature). Thus, we have $C_s = 4.5 \times 10^{20}$ atoms/cm³. Also from Figure 3.12, we have $(D)^{1/2} = 0.085\ \mu m/(h)^{1/2}$.

From the above, we have:

$$2\sqrt{Dt} = 2\sqrt{(0.085)^2 \times 30/60} = 2\sqrt{0.003612} = 2 \times 0.0601 = 0.1202\ \mu m$$

Therefore the concentration of phosphorus at a depth x after 30 minutes into diffusion can be calculated from Equation (3.5):

$$C(x, 0.5) = C_s \operatorname{erfc}\left(\frac{x}{2\sqrt{Dt}}\right) = 4.5 \times 10^{20} \operatorname{erfc}\left(\frac{x}{0.1202}\right)$$

By using Table 3.4, we can estimate the concentration of phosphorus at a depth of 0.075 μm to be:

$$C(0.075, 0.5) = 4.5 \times 10^{20} \operatorname{erfc}\left(\frac{0.075}{0.1202}\right) = 4.5 \times 10^{20} \operatorname{erfc}(0.624)$$

$$= 9 \times 10^{20}[1 - \operatorname{erf}(0.624)] = 1.7 \times 10^{20} \text{ atoms/cm}^3$$

3.7 | PLASMA PHYSICS

Plasma is a gas that carries electrical charges. It contains approximately equal numbers of electrons and positively charged ions. So, as a whole, plasma is a mixture of neutral ionized gas. Plasma is a key ingredient in microfabrication, as it contains a large number of positive ions with extremely high kinetic energy that can be used to perform the following essential functions in micro fabrication:

1. Assist in depositing foreign materials in a base material such as a silicon substrate
2. Assist in penetrating desirable foreign substances into a base material to facilitate the implantation of the foreign materials
3. Remove a portion of base material by knocking out the atoms from that material

One advantage of using plasma in microfabrication is the relative ease of manipulating its flow by electrostatic forces or magnetic fields. Plasma-assisted etching and sputtering and plasma-enhanced vapor deposition are popular and effective techniques in microfabrication, as will be described in detail in Chapter 8.

Figure 3.4 showed how ions can be generated for the production of plasma. In most cases, free electrons are first produced by an electron gun, as illustrated in Figure 3.4. These electrons travel at a high speed after passing through an accelerator. The fast electrons enter the chamber as illustrated in Figure 3.4, where they knock more electrons from the ionized medium, e.g., a carrier gas of H_2, He, or BF_3. According to Ruska [1987], the following activities happen in a plasma generator, as illustrated in Figure 3.14: (1) ionization, (2) dissociation, (3) excitation, and (4) recombination.

Figure 3.14 | Schematic of a plasma generator.

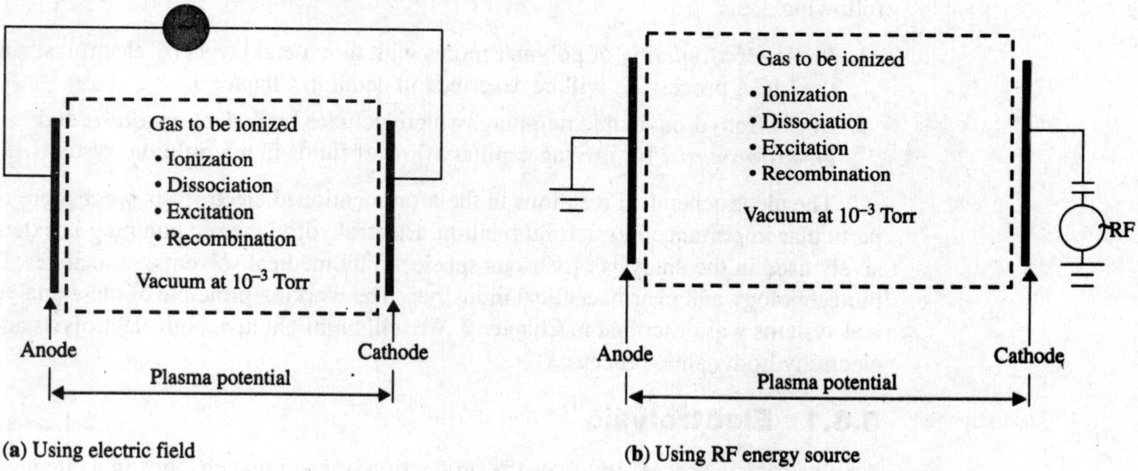

Ionization In this process, an electron is knocked loose from the atom of the medium M, resulting in a positive charged molecule, or ion:

$$e^- + M \rightarrow M^+ + 2e^- \quad [3.6]$$

Dissociation A molecule of the medium M_2 breaks down into smaller fragments, with or without ionization:

$$e^- + M_2 \rightarrow M + M + e^-$$
$$\rightarrow M^+ + M + 2e^- \quad [3.7]$$

Excitation The molecules hold together, but absorb energy from the fast electrons present in the chamber, to be raised to an excited electronic state (M*):

$$e^- + M \rightarrow M^* + e^- \quad [3.8]$$

Recombination

$$e^- + M^+ \rightarrow M \text{ (or } M^*) \quad [3.9]$$

Since electrons are the sources of ionization that sustain the plasma, continuous supply of electrons must be maintained. Consequently, energy sources such as high voltage electric field or radio-frequency (RF) sources must be applied to the generator at all times. Such energy is necessary to allow the cathode in the generator to release electrons on a continuous basis.

3.8 | ELECTROCHEMISTRY

Electrochemistry, which is the study of chemical reactions caused by the passage of an electric current, is an important subject in microsystem design and fabrication. Electrochemical reactions are used in many engineering processes. Principal applications of electrochemistry in microfabrication and microsystem design are in the following areas:

1. In the *electroplating* of polymer molds with thin metal layers by electrolysis in the LIGA process, as will be described in detail in Chapter 9.
2. In electrohydrodynamic pumping, which includes both *electrophoresis* and *electro-osmosis;* for driving capillary flow of fluids in microfluidic systems.

The electrochemical reactions in the aforementioned electrolysis process are of particular importance in microfabrication. Electrohydrodynamic pumping is extensively used in the analysis of various species in biomedical solvents, or analytes, in biotechnology and pharmaceutical industries. The working principle of these analytical systems was described in Chapter 2. We will highlight here both electrolysis and electrohydrodynamic processes.

3.8.1 Electrolysis

An electrolysis process involves the production of chemical changes in a chemical compound or solution by oppositely charged constituents (or ions) moving in opposite directions under an electric potential difference. As we have learned from the foregoing section, a solution that carries ions is electrically conductive. A solution that conducts electric current is called an *electrolyte* and the vessel that holds the electrolyte is called an *electrolytic cell.* Electrolytes may be liquid solutions or fused salts, or ionically conducting solids.

Passing electric current through the fluid in the electrolytic cell can produce electrically charged ions in the electrolyte. Since electric current is the flow of electrons, the free electrons in the current can alter the atomic structures in the fluid molecules. This can lead to atoms with unbalanced electrons, and free ions with electric charges are thus produced in the solution.

In addition to the electrolytic cell, a pair of submerged electrodes is required in order to provide an electric potential in an electrolysis process. A simple example of electrolysis is the decomposition of molten sodium chloride (NaCl) in an electrolytic cell, as illustrated in Figure 3.15.

Electrodes are placed at the two ends of the electrolytic cell. An electric potential is established after the electrodes are connected to a dc source. This source, which usually is a battery, acts as an electron pump that pushes electrons into one electrode, making it a *cathode*, and pulling electrons from the other electrode to make it an

Figure 3.15 | An electrolysis process.

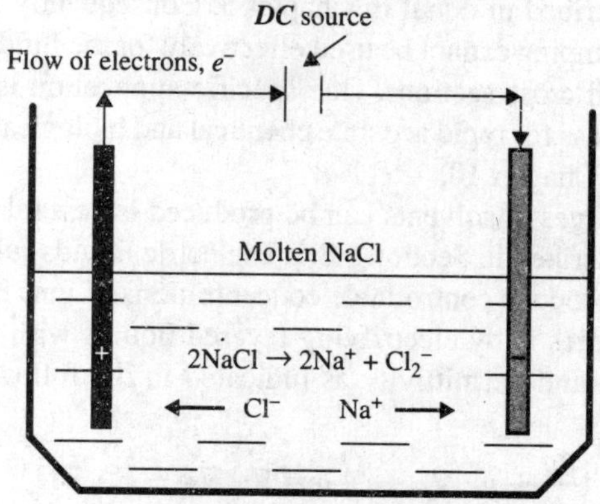

anode. The passing of the electric current decomposes the molten salt into positively charged Na ions and negatively charged Cl ions following the chemical reaction:

$$2NaCl \rightarrow 2Na^+ + Cl_2^- \qquad [3.10]$$

The Na^+ ions from the decomposed molten salt pick up electrons from the cathode and result in a decrease of electrons at that electrode. As the Na^+ ions in the vicinity of the cathode are depleted, additional Na^+ ions migrate in. In a similar fashion, there is a simultaneous movement of Cl^- ions toward the anode. The salt (Na) and the chlorine (Cl) are thus collected at the respective cathode and anode throughout the electrolysis process.

Electrolysis is a very useful technique in separating and extracting chemical compounds. The value of this process for microsystem design and fabrication, however, lies in the fact that an ionized fluid may be set in motion by an electric potential field. The motion of ions in electrolytes under an electric field leads to the very important effect known as *electrohydrodynamics* of microfluid flow, as will be described in Section 3.8.2.

3.8.2 Electrohydrodynamics

Electrohydrodynamics (EHD) deals with the motion of fluids driven by an electric field applied to the fluids. There are two principal applications of EHD in microsystems: (1) electro-osmotic pumping and (2) electrophoretic pumping. These unique pumping techniques are used to move chemical and biological fluids in channels with extremely small cross sections, ranging from square micrometers to square millimeters at flow rates in the order of cubic micrometers per second. Microfluidics are widely used in the pharmaceutical industry and in biochemistry analyses using extremely small sample quantities in the order of a few hundred nanoliters (1 nanoliter (nL) = 10^{-9} liter).

Electrohydrodynamic pumping involves no moving mechanical parts such as rotating impellers. Often, it is the only effective way to move fluids in extremely small channels because of the capillary effect. This effect in fluid flow in small conduits is principally due to the surface tension and the van der Waals forces in fluid molecules, as will be described in detail in Chapter 5. Consequently, conventional volumetric mechanical pumping cannot be used effectively for the fluid movement through these extremely small cross sections. One specific application is the capillary electrophoresis (CE) process for rapid accurate chemical and biological analysis, as will be described in detail in Chapter 10.

Free electric charges in solvents can be produced in several ways, for example, by electrolytes as described in Section 3.8.1. Dielectric liquids subjected to very high electric voltage can produce controllable concentrations of ions in the fluid. Another way to generate charges is by electrifying layered liquids with spatial gradients of electric conductivity and permittivity as indicated in the following equation [Bart et al. 1990]:

$$\frac{\partial \rho}{\partial t} + v \cdot \nabla \rho + \frac{\Omega}{\epsilon}\rho = \mathbf{E} \cdot \nabla \epsilon - \mathbf{E} \cdot \nabla \Omega \qquad [3.11]$$

where

ρ = free charge density
$\mathbf{E}$ = electric field
ϵ = permittivity of the fluids
Ω = ohmic conductivity of the fluids
$\nabla\rho$, $\nabla\epsilon$, $\nabla\Omega$ = gradients of free charge density, permittivity, and ohmic conductivity, respectively, in the fluids

Figure 3.16 illustrates the generation of ionic solution in an electro-osmotic pumping in a conduit [Bart et al. 1990]. The solution consists of two materials with different electric resistivities, Ω_1 and Ω_2, and permittivities, ϵ_1 and ϵ_2. A material interface is formed after the application of an electrical field across the conduit walls. Free charges are produced by the gradient of electric conductivity, which is represented by the difference in the resistivities and of the permittivities between the two materials. The signs of the surface charges in the material layers are determined by the permittivities of the materials. For instance, if material layer 1 were close to being an insulator, then the surface charge induced would be opposite to the sign of the electrode array.

Figure 3.16 | EHD pumping of a solvent.

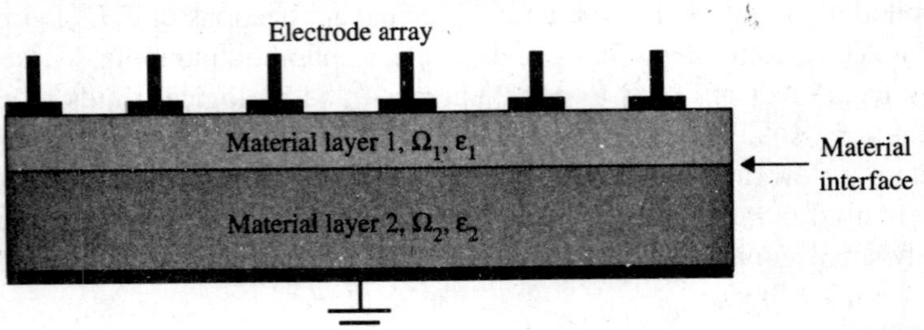

(Bart et al. [1990].)

Once free charges are generated in the fluid, pumping of the fluid along the longitudinal direction can be accomplished by applying electric field to the electrodes in the array along the length of the channel, as shown in Figure 3.16. Figure 3.17 illustrates the principle of electro-osmotic pumping of homogeneous fluids and Figure 3.18 illustrates the principle of electrophoretic pumping of heterogeneous fluids.

Figure 3.17 I Illustration of electro-osmotic pumping.

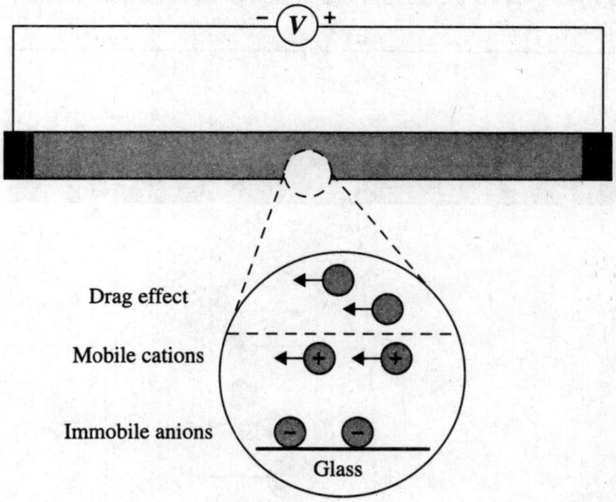

(After Kovacs [1998].)

Electro-osmotic Pumping Electro-osmotic pumping is used to move electrically neutral fluids through channels of extremely small cross sections. The condition is that the walls of the conduit or channel must have attached, immobile charges. A glass wall such as shown in Figure 3.17 can produce such immobile charges along the surface if it is coated either with ionizable materials, e.g., deprotonated silanol groups, or with strongly absorbed charged species that are present in the fluid [Manz et al. 1994]. In such cases, electro-osmotic motion of fluid in microchannels of capillary tubes can be produced in any electrolyte fluid, as described in Section 3.8.1. Because the gradient of the concentration of electric charges decreases toward the center of the conduit, a dual layer of fluid with varying concentration of charges is formed, as shown in Figure 3.17. The charges in the double layer can be moved with the applied electric charges along the longitudinal direction. The momentum of the moving charges can drive the solvent through the flow channel. A unique feature of electro-osmotic flow is that a uniform velocity profile of the moving fluid across the cross section of the channel of the tube is obtained.

Electrophoretic Pumping This method of pumping is often used to separate minute foreign particles or species from the bulk fluid. Microfluidic systems with electrophoresis are widely used in biomedical and pharmaceutical industries. The movement of ions of the particles in a heterogeneous medium is prompted by an applied high-voltage electric field, as illustrated in Figure 3.18. The ions with different charges move in opposite directions along the channel in the separation process. When the flow is fully developed, the ions in the stream can automatically separate themselves by their inherent electro-osmotic mobility under the influence of the applied electric field. It is a highly desirable situation in chemical and biological analyses, in which the separation of various minute species is often a difficult task.

Figure 3.18 | Electrophoretic pumping of ions in a fluid.

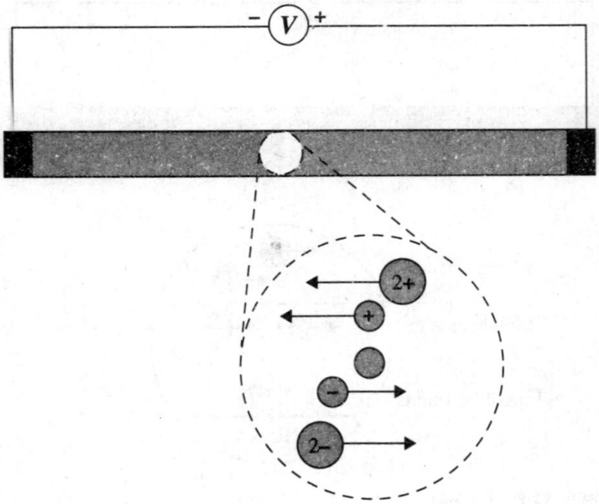

(After Kovacs [1998], Manz et al. [1994].)

The rate of flow in both electro-osmotic and electrophoretic pumping is linearly proportional to the applied electric field. These pumping methods involve no moving mechanical parts. It is especially suitable for miniaturized systems. Only a small sample of analyte is required in most cases with this type of pumping in microfluidic systems.

The velocity of the moving free charges in electrophoretic pumping can be determined by the following expression [Kovacs 1998]:

$$v_i = \frac{z_i q}{6\pi r_i \mu} E \qquad \textbf{[3.12]}$$

where

v_i = velocity of an ion i of charge z_i in a homogeneous fluid
E = applied electric field in volts
r_i = radius of ion i
μ = the viscosity of the fluid
q = charge on the electron = 1.6022×10^{-19} C.

It is obvious from Equation (3.12) that the velocity of the free ions and thus the fluid flow is linearly proportional to the applied electric field in both electrohydrodynamic-pumping techniques. We will revisit these pumping systems in Section 5.6 in Chapter 5 and Section 10.7.2 in Chapter 10.

3.9 | QUANTUM PHYSICS

As MEMS and microsystems continue to be miniaturized in physical size to submicrometer scales, we can expect a more pronounced influence of atomic effects on the design and fabrication processes. One example is the effect of the van der Waals forces on the mechanical design of these systems. The physical laws and theories that were derived at macroscale are no longer valid without significant modifications to accommodate the atomic effects.

In addition to the van der Waals forces, another critical atomic effect that can affect the engineering design of microsystems is the means of transportation of energy, i.e. heat, in solids in submicrometer scale. Such effects can be better illustrated by quantum physics.

A *quantum* represents the smallest amount of energy that any system can gain or lose. It occurs at the atomic scale. Max Planck (1858–1947) postulated the quantization of energy around 1900 in his formulation of thermal radiation. This concept led to the development of a fundamentally important topic in physics—quantum physics. Planck postulated that atoms in a substance oscillate back and forth from their neutral positions in a lattice (see Figure 3.5) when they are energized. These oscillations radiate energy that can be determined by the following simple formula:

$$E = nh\nu \qquad \textbf{[3.13]}$$

where

n = integer number = 1, 2, 3, . . .
h = Planck's constant = 6.626076×10^{-34} J-s
ν = frequency of oscillation, s^{-1}.

The unit of energy associated with the oscillation of atoms indicated in Equation (3.13) is called a *phonon*, which is transmitted through a substance in the form of acoustic waves at extremely high frequency, in the order of 10^{13} Hz.

Albert Einstein proposed his photon theory in 1905, in which he postulated that light behaves as if energy is concentrated into localized bundles (a stream of particles) called *photons*. The energy associate with a single photon is

$$E = h\nu \qquad \textbf{[3.14]}$$

in which ν is the frequency of the light.

With Einstein's photon theory, we are able to explain many phenomena in optics that the classical wave theory failed to support. One such example is the photoelectric effect, as described in Chapter 2.

One important concept for us to appreciate is that both phonons and photons are *energy carriers*. Neither of them has mass, despite the fact that they are treated as "particles" in quantum physics. A familiar phenomenon, the light emitted from heating of a metal by electricity demonstrates this. The higher the electric energy that is pumped into the metal, the brighter the emitted light becomes. If light is carried by photons, as postulated by Einstein's photon theory, a brighter light should carry more photon energy. Thus, one should not be surprised that one can feel the emitted photons as the heat from a glowing metal heater, but not in the form of forces, as photons do not have mass.

Because both phonons and photons are energy carriers, we can expect these carriers to play major roles in heat transfer in substances that have dimensions near the atomic scale. Section 5.10 on micro heat conduction in Chapter 5 will present a formulation for a quantitative assessment of heat transportation in media from a quantum physics perspective.

PROBLEMS

Part 1. Multiple Choice

1. Everything on our Earth is made from (1) 86, (2) 96, (3) 106 stable elements, and each element has a different atomic structure.
2. The core of an atom is a (1) neutron, (2) nucleus, (3) electron.
3. Elements have different properties because they have different (1) atomic structures, (2) chemical compositions, (3) physical compositions.
4. Elements that have similar properties when they have the same number of (1) electrons, (2) protons, (3) nuclei in the outer orbit of their respective atomic structures.
5. A nucleus contains (1) neutrons and protons, (2) electrons and protons, (3) neutrons and electrons.
6. Protons carry (1) positive, (2) negative, (3) no charge.
7. Electrons carry (1) positive, (2) negative, (3) no charge.
8. Neutrons carry (1) positive, (2) negative, (3) no charge.
9. The outer orbit of atoms has a diameter that is (1) 100, (2) 1000, (3) 10,000 times of that of a nucleus.
10. A periodic table consists of (1) 96, (2) 103, (3) 108 elements.
11. Silicon atoms contain (1) 8, (2) 10, (3) 14 electrons.
12. Silicon atoms have (1) 4, (2) 6, (3) 8 electrons in their outer orbit.
13. An ion carries (1) electric charge, (2) magnetic charge, (3) electrostatic charge.
14. Ionization energy is the energy required to remove (1) neutrons, (2) protons, (3) electrons from the outer orbit of an atom.
15. Molecules are made of bounded (1) electrons, (2) atoms, (3) nuclei.
16. The forces that bind the atoms in a molecule are called (1) intermolecular forces, (2) electroatomic forces, (3) interatomic forces.
17. The intermolecular forces are (1) van der Waals, (2) electrostatic, (3) electromagnetic in nature.
18. The intermolecular forces, in general, are (1) proportional, (2) equal to, (3) inversely proportional to the distances between molecules.
19. The physical behavior of solid molecules is typically (1) strong in kinetic energy and atomic cohesive forces, (2) weak in kinetic energy and atomic cohesive forces, (3) weak in kinetic energy but strong in atomic cohesive forces.
20. Positive silicon can be produced by doping with (1) boron atoms, (2) phosphorus atoms, (3) either kind of atom.
21. Negative silicon can be produced by doping with (1) boron atoms, (2) phosphorus atoms, (3) either kind of atom.
22. Silicon is a semiconducting material. It can be made more electrically conductive by (1) a doping process, (2) a diffusion process, (3) an electric implantation process.
23. N-type silicon is (1) less, (2) more, (3) about equally conductive as p-type silicon when is doped with same dose of dopant.
24. Diffusion is a good way to (1) coat, (2) implant, (3) remove foreign materials in silicon substrates.

25. Diffusion analysis is based on (1) Fourier's law, (2) Fick's law, (3) Hooke's law.
26. Plasma is a gas that (1) does, (2) does not, (3) may carry electric charges.
27. To maintain a plasma, one needs to keep supplying (1) high temperature, (2) high pressure, (3) high electrical field to the plasma chamber.
28. Electrochemistry involves (1) chemical reactions, (2) ionization, (3) decomposition of any substance caused by an electric current.
29. Electrolysis involves the production of (1) chemicals, (2) chemical changes, (3) ionization in a substance by the application of an electric potential.
30. Electrolysis uses (1) an ac, (2) a dc, (3) either an ac or a dc power supply.
31. An electrolyte is (1) an electrode, (2) the container, (3) the solution that conducts electric current in an electrolysis process.
32. An anode is the (1) positive, (2) negative, (3) neutral electrode.
33. A cathode is the (1) positive, (2) negative, (3) neutral electrode.
34. Electrohydrodynamics deals with (1) dissolution, (2) motion, (3) solidification of a fluid under an applied electric field.
35. The principal use of electrohydrodynamics in microsystems is to (1) conduct electrolysis of minute chemicals, (2) move minute amounts of fluid, (3) detect minute amounts of fluid.
36. Electro-osmotic pumping is used to move minute amounts of (1) homogeneous, (2) heterogeneous, (3) any fluid in capillary passages.
37. Electropheretic pumping is used to move minute amounts of (1) homogeneous, (2) heterogeneous, (3) any fluid in capillary passages.
38. Quantum physics is used to describe (1) physical movement of atoms, (2) energy transport, (3) collisions of atoms in MEMS and nanosystems.
39. A quantum represents the smallest amount of (1) mass, (2) volume, (3) energy that any system can gain or lose.
40. Photons have the mass equal to (1) an electron, (2) a neutron, (3) zero.

Part 2. Descriptive Problems

1. Express the size and weight of a hydrogen and a silicon atom in nanometers (nm), where 1 nm = 10^{-9} meter.
2. What do the "atomic numbers" and "group numbers" in a periodic table mean?
3. What is the desirable level of electric resistivity to make a semiconductor electrically conducting?
4. Explain the physical meaning of the negative sign attached to Fick's law in Equations (3.2) and (3.3).
5. Determine the optimum temperatures of silicon substrates for which doping of arsenic, phosphorus, and boron are to be carried out by a diffusion process. Also, what is the solubility of these dopants at these optimal temperatures?
6. Plot the distributions of the concentration of phosphorus atoms as a function of depth in the silicon substrate in Example 3.1 at the times 30, 60, 90, 120, 150, and 180 minutes.
7. Determine the time required to dope boron into the silicon substrate so that the resistivity of the doped silicon at the depth of 2 μm is 10^{-3} Ω-cm.

8. Describe how ions are produced in an electrolysis process.
9. Describe the principles of electrophoresis and electro-osmosis. Where are these processes used in MEMS and microsystems?
10. Explain the capillary effect in microfluid flow, and why conventional mechanical pumping cannot move fluids in small channels with capillary effect.
11. Describe the role of quantum physics in the design of MEMS and microsystems.

4 Chapter

Engineering Mechanics for Microsystem Design

CHAPTER OUTLINE

4.1 | INTRODUCTION

As we learned in Chapter 2, most microsystems are made of three-dimensional structures that often involve heat transmission as well as solid/fluid interactions. Moreover, many of the components in microsystems are essentially machine components in microscale, such as gears, springs, bearings, linkages, and mechanisms. It is thus necessary to use mechanical engineering and machine design principles in dealing with the design and packaging of microsystems.

In this and the subsequent chapters, we will learn how mechanical engineering principles in solid and fluid mechanics and heat transfer, derived from theories of continua, can be used in the design of various components of micro- and mesoscales in microsystems. We will also discuss the limits of using these phenomenological models derived for continua for components at the small scale of single-digit micrometers. In these circumstances, many of the science principles presented in Chapter 3 will be used as the basis for modifications to suit small structures in the order of submicrometers.

Engineering mechanics, which involves both solid and fluid mechanics, is the base for mechanical design of microsystems. Proper functioning of microsystems and their structural integrity, as well as reliable system packaging, require the application of engineering mechanics principles. The field of engineering mechanics is a specialized area of engineering science. It is not possible for us to deal with the entire subject in this chapter. What we will present in sequel is just "scratching the surface" with a few selected topics that are useful in the design and packaging of microdevices and systems.

Mechanics, by a traditional definition, is a branch of engineering science that studies the relationship between the applied *forces* and the resulting *motions*. The word *motion* in microsystems can involve either rigid body motion, as in a microaccelerometer, or the deformation of solids, as in deformable diaphragms in micro pressure sensors. The principles of dynamics and mechanical vibration are frequently applied in the design and operation of microaccelerometers and pressure sensors using beams vibrating at resonant frequencies for high sensitivities in output. Solid mechanics principles are also extensively used in the design of system packaging.

Fluid mechanics, on the other hand, is involved in the design of microvalves and microfluidics. Often, the classical theories and formulations that are derived for idealized loading and geometry are not sufficient for the analyses required in the design of MEMS and microsystems. In such cases, numerical techniques such as computational fluid dynamics (CFD) and the finite element method are used for refined solutions. Commercial CFD and finite element codes are commonly used by industry in the design of these products.

We will also learn how the principles of heat transfer derived for solids in macroscale can be used in assessing the temperature field in MEMS structure. Again, substantial modification of these principles for components at sub-micrometer scale will be outlined. The application of fluid mechanics and heat transfer in microsystems design will be presented in Chapter 5.

Units for Solid Mechanics Problems It is prudent for engineers to be consistent in adopting a set of units for stress analyses of microsystems. The following International System (SI) units are recommended for such applications.

The unit *newton* (N) is used for forces. A force of 1 N is defined as "a force required to give 1 kilogram (kg) mass an acceleration of 1 meter per square second (m/s^2). Since 1 kg of mass has a weight of 9.81 N, we can readily recognize that 1 kg force equals 9.81 Ns. The reader will realize that the gravitational acceleration, $g = 9.81\ m/s^2$, is used in the above formulation of the force unit. Recommended units for thermophysical quantities used in microsystem design, and conversion between SI units and the Imperial System can be found on the inside front cover of this text.

The unit *meter* (m) is recommended for length. Thus, 1 μm = 10^{-6} m, and 1 millimeter (mm) = 1000 μm. The unit *kilogram* (kg) is recommended for the mass in all computations.

The *pascal* (Pa) is used for pressures and stresses; 1 Pa is equivalent to 1 N/m^2, from which 1 MPa = 10^6 Pa = $10^6\ N/m^2$.

4.2 | STATIC BENDING OF THIN PLATES

In Chapter 2, we learned that many micro pressure sensors work on the principle of converting strain in a deformed thin silicon diaphragm, induced by applied pressure, to the desired form of electronics output. For most cases, these diaphragms, either in circular, square, or rectangular shape, can be treated as thin plates subjected to lateral bending by uniformly applied pressure.

Several closed-form solutions of plate bending can be found in a useful reference [Timoshenko and Woinowsky-Krieger 1959]. The governing differential equation for the deflection of a rectangular plate subject to lateral bending can be expressed as:

$$\left(\frac{\partial^2}{\partial x^2} + \frac{\partial^2}{\partial y^2}\right)\left(\frac{\partial^2 w}{\partial x^2} + \frac{\partial^2 w}{\partial y^2}\right) = \frac{p}{D} \tag{4.1}$$

in which $w = w(x, y)$ is the lateral deflection of a flat plate due to the uniformly distributed applied pressure p. The x–y plane defines the plate as shown in Figure 4.1.

Figure 4.1 | Bending of a rectangular plate.

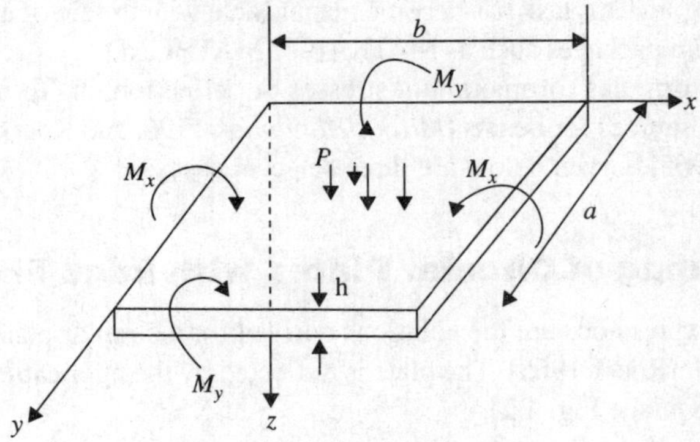

The parameter D is the flexural rigidity of the plate, which can be expressed as:

$$D = \frac{Eh^3}{12(1 - \nu^2)} \tag{4.2}$$

where E and ν are Young's modulus and Poisson's ratio, respectively, of the plate material and h is the thickness of the plate as shown in Figure 4.1.

The components of bending moments about the x and y axes (see Fig. 4.1) and the bending stresses can be computed from the deflection $w(x, y)$, obtained from the solution of Equation (4.1) as shown below:

The bending moments:

$$M_x = -D\left(\frac{\partial^2 w}{\partial x^2} + \nu \frac{\partial^2 w}{\partial y^2}\right) \tag{4.3a}$$

$$M_y = -D\left(\frac{\partial^2 w}{\partial y^2} + \nu \frac{\partial^2 w}{\partial x^2}\right) \tag{4.3b}$$

$$M_{xy} = D(1 - \nu)\frac{\partial^2 w}{\partial x \partial y} \tag{4.3c}$$

Bending stresses:

$$(\sigma_{xx})_{\max} = \frac{6(M_x)_{\max}}{h^2} \tag{4.4a}$$

$$(\sigma_{yy})_{\max} = \frac{6(M_y)_{\max}}{h^2} \tag{4.4b}$$

$$(\sigma_{xy})_{\max} = \frac{6(M_{xy})_{\max}}{h^2} \tag{4.4c}$$

The solution of the plate deflection $w(x, y)$ obtained from Equation (4.1) with appropriate boundary conditions is used to obtain the bending moments from Equation (4.3). The maximum bending stresses can thus be computed from Equation (4.4) with respective maximum bending moments. The solution of Equation (4.1) used to require quite a tedious effort. This task has become manageable with the aid of a personal computer and software packages such as MATLAB or MATHCAD.

Simplified formulas for maximum stresses or deflection due to bending can be found in several sources for beams [*Mark's Handbook* 1996, and Roark 1965] and for plates [Roark 1965]. Given below are three selected cases.

4.2.1 Bending of Circular Plates with Edge Fixed

The following expressions are the solutions derived for a circular plate with a radius a and thickness h [Roark 1965]. The plate is deflected by the application of a uniform pressure loading p (see Fig. 4.2).

Figure 4.2 | A circular plate subjected to uniform pressure loading.

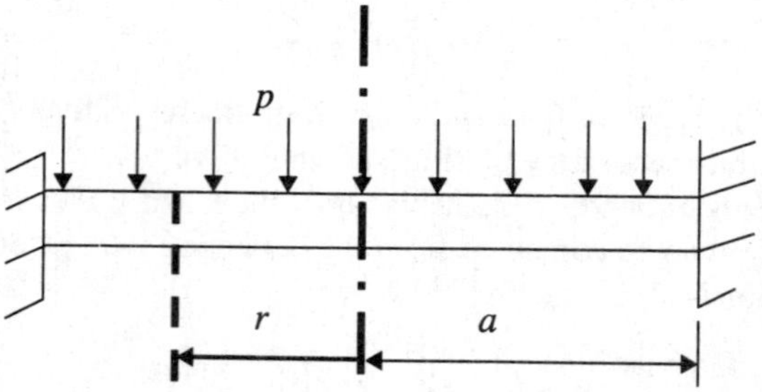

The maximum radial stress is

$$(\sigma_{rr})_{\max} = \frac{3W}{4\pi h^2} \qquad [4.5a]$$

at the edge, and the maximum hoop (tangential) stress is

$$(\sigma_{\theta\theta})_{\max} = \frac{3\nu W}{4\pi h^2} \qquad [4.5b]$$

at the edge.

Both these stresses at the center of the plate become

$$\sigma_{rr} = \sigma_{\theta\theta} = \frac{3\nu W}{8\pi h^2} \qquad [4.6]$$

The maximum deflection occurs at the center with the value:

$$w_{\max} = -\frac{3W(m^2 - 1)a^2}{16\pi E m^2 h^3} \qquad [4.7]$$

where $W = (\pi a^2)p$ and $m = 1/\nu$. The negative sign attached to the deflection in Equation (4.7) indicates its downward direction.

EXAMPLE 4.1

Determine the minimum thickness of the circular diaphragm of a micro pressure sensor made of silicon as illustrated in Figure 4.3. The diaphragm has a diameter of 600 μm and its edge is rigidly fixed to the silicon die. The diaphragm is designed to withstand a pressure of 20 MPa without exceeding the plastic yielding strength of 7000 MPa. The silicon diaphragm has a Young's modulus E = 190,000 MPa and a Poisson's ratio $\nu = 0.25$.

Figure 4.3 | A circular diaphragm for a pressure sensor.

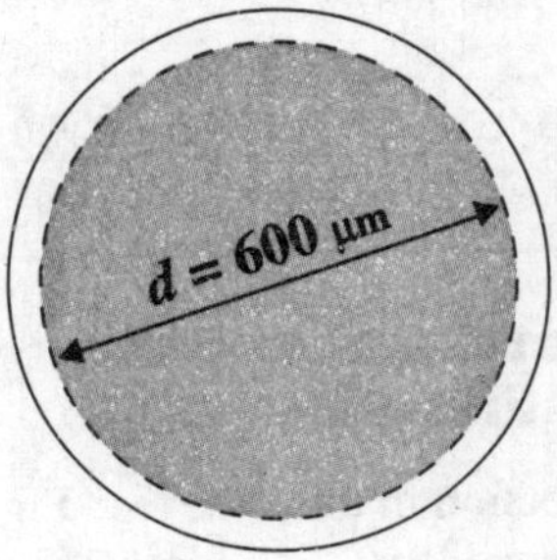

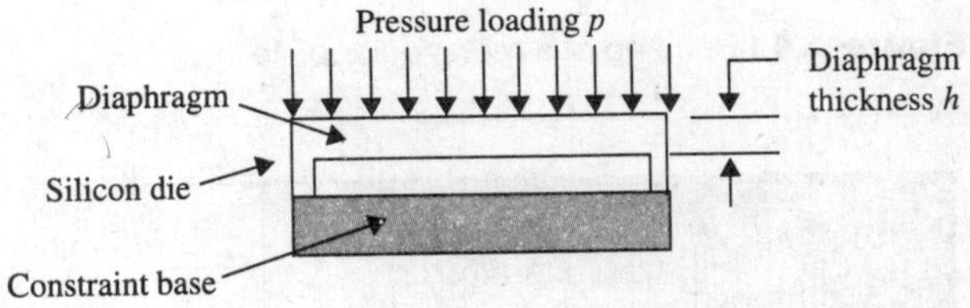

Solution

From Equation (4.5a), we have the thickness of a thin plate expressed as

$$h = \sqrt{\frac{3W}{4\pi(\sigma_{rr})_{max}}}$$

or, from Equation (4.5b),

$$h = \sqrt{\frac{3\nu W}{4\pi(\sigma_{\theta\theta})_{max}}}$$

Since we will design the diaphragm so that its maximum stress is kept below the plastic yielding strength of the material, or $(\sigma_{rr})_{max} \leq \sigma_y = 7000$ MPa and $(\sigma_{\theta\theta})_{max} \leq \sigma_y = 7000$ MPa, and also because Poisson's ratio $\nu = 0.25 \leq 1.0$, it is obvious that the first of the two expressions shown above for thickness h should be used.

The equivalent load W in the expression can be evaluated as:

$$W = (\pi a^2)p = 3.14 \times (300 \times 10^{-6})^2 \times (20 \times 10^6) = 5.652 \text{ N}$$

From Equation (4.5a),

$$h = \sqrt{\frac{3 \times 5.652}{4 \times 3.14 \times (7000 \times 10^6)}} = 13.887 \times 10^{-6}\text{ m}$$

or the minimum diaphragm thickness $h = 13.887\ \mu\text{m}$.

The maximum deflection of the diaphragm, w_{max}, can be determined by using Equation (4.7):

$$w_{max} = -\frac{3W(m^2-1)a^2}{16\pi E m^2 h^3}$$

With $m = 1/\nu = 1/0.25 = 4.0$,

$$w_{max} = -\frac{3 \times 5.652(16-1) \times (300 \times 10^{-6})^2}{16 \times 3.14 \times (190{,}000 \times 10^6) \times 16 \times (13.887 \times 10^{-6})^3} = -55.97\ \mu\text{m}$$

4.2.2 Bending of Rectangular Plates with All Edges Fixed

A closed-form solution for deflection in this case (see Fig. 4.4) is available in Timoshenko and Woinowsky-Krieger [1959]. A simplified solution can be found in Roark [1965].

Figure 4.4 | Bending of a rectangular plate.

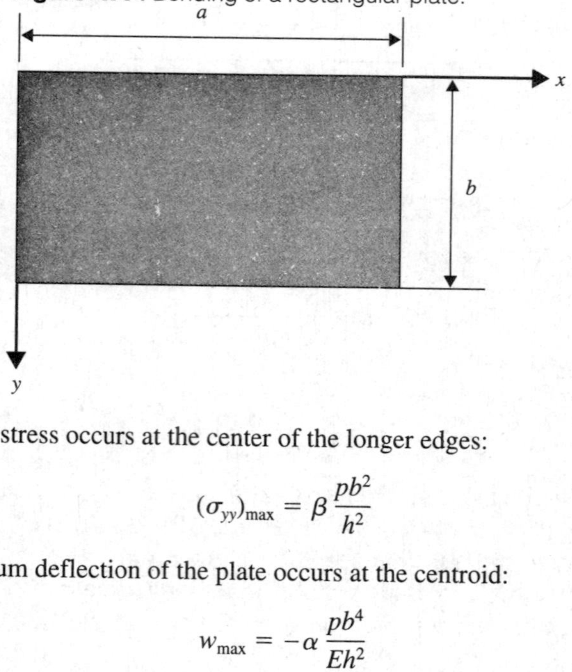

The maximum stress occurs at the center of the longer edges:

$$(\sigma_{yy})_{max} = \beta \frac{pb^2}{h^2} \quad \textbf{[4.8]}$$

and the maximum deflection of the plate occurs at the centroid:

$$w_{max} = -\alpha \frac{pb^4}{Eh^2} \quad \textbf{[4.9]}$$

The coefficients α and β in Equations (4.8) and (4.9) can be determined from Table 4.1.

Table 4.1 | Coefficients for maximum stress and deflection in a rectangular plate

a/b	1	1.2	1.4	1.6	1.8	2.0	∞
α	0.0138	0.0188	0.0226	0.0251	0.0267	0.0277	0.0284
β	0.3078	0.3834	0.4356	0.4680	0.4872	0.4974	0.5000

EXAMPLE 4.2

A rectangular diaphragm has dimensions $a = 752$ μm and $b = 376$ μm, as illustrated in Figure 4.4. These edge dimensions will result in the same plane area as the circular diaphragm in Example 4.1. The thickness of the diaphragm, the applied pressure, and material properties remain identical to those in Example 4.1. Determine the maximum stress and deflection in the diaphragm.

Solution

We have the edge ratio $a/b = 2.0$, which leads to $\alpha = 0.0277$ and $\beta = 0.4974$ from Table 4.1. The diaphragm thickness $h = 13.887$ μm and the applied pressure $p = 20$ MPa are used in determining the maximum stress and deflection.

Thus, we may obtain the maximum stress from Equation (4.8) as

$$(\sigma_{yy})_{max} = \beta \frac{pb^2}{h^2} = 0.4974 \frac{(20 \times 10^6)(376 \times 10^{-6})^2}{(13.887 \times 10^{-6})^2} = 7292.8 \times 10^6 \text{ Pa}$$

or $(\sigma_{yy})_{max} = 7292.8$ MPa, which is greater than the plastic yield strength of silicon of $\sigma_y = 7000$ MPa—a situation that can be interpreted as "unsafe."

The maximum deflection of the diaphragm can be determined from Equation (4.9):

$$w_{max} = -\alpha \frac{pb^4}{Eh^3} = -\alpha \frac{pb}{E}\left(\frac{b}{h}\right)^3 = -\frac{0.0277 \times (20 \times 10^6) \times 376 \times 10^{-6}}{190000 \times 10^6}$$

$$\left(\frac{376 \times 10^{-6}}{13.887 \times 10^{-6}}\right)^3 = -21.76 \times 10^{-6} \text{ m}$$

The maximum deflection at the center of the diaphragm is $w_{max} = 21.76$ μm, which is less than that in the circular diaphragm.

4.2.3 Bending of Square Plates with All Edges Fixed

Square diaphragms are common in micro pressure sensors (see Fig. 2.7). This geometry is favored because it allows easy slicing (or "dicing") of the silicon sensing elements from standard sized wafers.

A typical square plate is illustrated in Figure 4.5. By letting the two edges of the plate be equal to each other, i.e., $a = b$ in Equations (4.8) and (4.9), we can get the following expressions for the maximum stress and deflection for a square plate.

Figure 4.5 | Bending of a square plate.

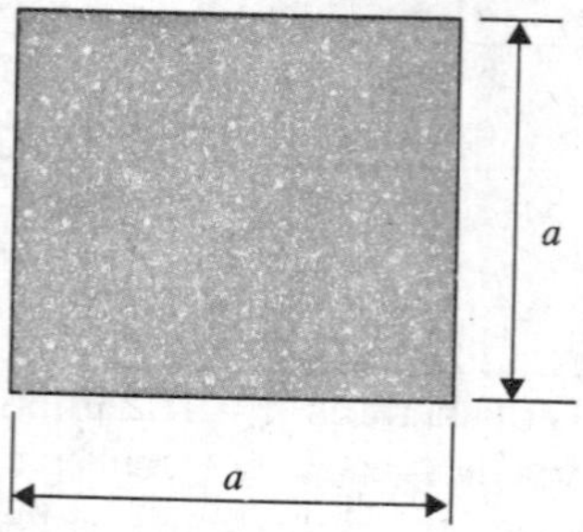

The maximum stress occurs at the center of each edge:

$$\sigma_{max} = \frac{0.308pa^2}{h^2} \tag{4.10}$$

and the maximum deflection of the plate occurs at the center of the plate:

$$w_{max} = -\frac{0.0138pa^4}{Eh^3} \tag{4.11}$$

The stress at the center of the plate can be derived to be:

$$\sigma = \frac{6p(m+1)a^2}{47mh^2} \tag{4.12}$$

and the strain at the center is:

$$\epsilon = \frac{1-\nu}{E}\sigma \tag{4.13}$$

EXAMPLE 4.3

A square silicon diaphragm with 532-μm edge length is subjected to the same pressure loading of $p = 20$ MPa as in Examples 4.1 and 4.2. The diaphragm has the same thickness, 13.887 μm. All material properties are identical to those given in Example 4.1. Determine the maximum stress and deflection in the diaphragm under the applied pressure.

■ Solution

The maximum stress in the diaphragm can be determined by using Equation (4.10) with $a = 532 \times 10^{-6}$ m, $h = 13.887 \times 10^{-6}$ m, and $p = 20 \times 10^6$ P:

$$\sigma_{max} = \frac{0.308pa^2}{h^2} = \frac{0.308 \times (20 \times 10^6)(532 \times 10^{-6})^2}{(13.887 \times 10^{-6})^2} = 9040 \times 10^6 \text{ Pa}$$

or $\sigma_{max} = 9040$ MPa, which is much greater than the yield strength of the silicon die of $\sigma_y = 7000$ MPa.

The maximum deflection of the diaphragm can be determined from Equation (4.11):

$$W_{max} = -\frac{0.0138pa^4}{Eh^3} = -\frac{0.0138pa}{E}\left(\frac{a}{h}\right)^3$$

$$= -\frac{0.0138(20 \times 10^6) \times (532 \times 10^{-6})}{190{,}000 \times 10^6}\left(\frac{532 \times 10^{-6}}{13.887 \times 10^{-6}}\right)^3 = -43 \times 10^{-6}\ \text{m}$$

or $W_{max} = 43\ \mu\text{m}$.

We may summarize the results obtained from Examples 4.1 to 4.3 in the following way, based on the same diaphragm area, thickness, and applied pressure.

Geometry of diaphragm	Maximum stress, MPa	Maximum deflection, μm
Circular	7000	55.97
Rectangular (a/b = 2.0)	7293	21.76
Square	9040	43.00

It is conceivable from the above summary that the circular diaphragm results in the most favorable situation. The deflection is excessive in this case, but a mechanical stopper can be installed in the die to limit such deflection. The square diaphragm appears to be the least favored geometry from both the strength and deflection points of view. However, because of the ease of wafer dicing and the symmetry of the geometry, it is still a popular geometry in the pressure sensor industry.

EXAMPLE 4.4

Determine the deflection and maximum stress in a square diaphragm used in a micro pressure sensor as illustrated in Figure 4.6a. The geometry and dimensions of the diaphragm are shown in Figure 4.6b. The expected maximum applied pressure loading to the micro pressure sensor is p = 70 MPa. The silicon diaphragm has the following material properties: Young's modulus E = 190,000 MPa and Poisson's ratio $\nu = 0.25$.

Solution

In reality, the entire top (front) surface of the silicon diaphragm is subjected to the applied pressure. However, we will assume that only the central portion of the diaphragm with an area of $783 \times 783\ \mu\text{m}^2$ is deflectable. The active load-carrying portion of the silicon diaphragm thus has the edge length a = 783 μm or 783×10^{-6} m, and a thickness h = 266 μm. The maximum stress in the diaphragm can be computed from Equation (4.10) to be:

$$\sigma_{max} = \frac{0.308 \times (70 \times 10^6) \times (783 \times 10^{-6})^2}{(266 \times 10^{-6})^2} = 186.81\ \text{MPa}$$

at the midpoint of each edge, and

$$W_{max} = -\frac{0.0138 \times 70(783 \times 10^{-6})^4}{190{,}000(266 \times 10^{-6})^3} = -10{,}153 \times 10^{-11}\ \text{m}$$

One may readily see from the computed results that the maximum stress in the diaphragm is much below the yield strength of the material, which is 7000 MPa. The maximum deflection of 0.1015 μm is small because of the small area that is subjected to the pressure.

Figure 4.6 | Stress and deformation in a silicon diaphragm.

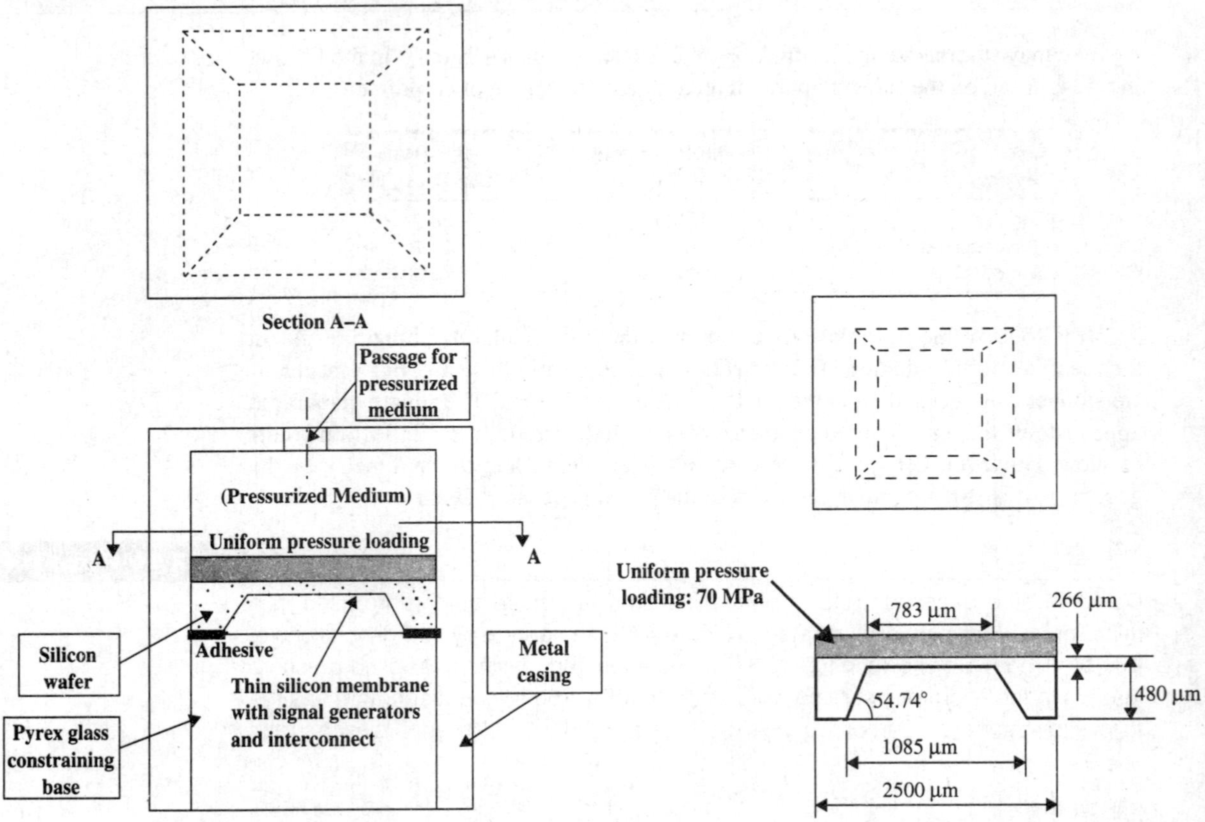

(a) A front-side pressure sensor

(b) Dimensions of the silicon diaphragm

4.3 | MECHANICAL VIBRATION

The theory of mechanical vibration is the basis for microaccelerometer design. Applications of these devices, such as the inertia sensors used in the air bag deployment systems in automobiles, are presented in Chapter 1, and the working principles are described in Chapter 2. We will review the fundamental principles of mechanical vibration and its application in microaccelerometer design in this section.

4.3.1 General Formulation

The simplest mechanical vibration system is the mass–spring system as illustrated in Figure 4.7a. The mass that is hung from the spring with a spring constant k vibrates from its initial equilibrium position because of the application of a small instantaneous disturbance to the mass. The displacement of the mass at a given time t, $X(t)$, can be obtained from the solution of the following equation of motion derived from Newton's second law:

$$m\frac{d^2X(t)}{dt^2} + kX(t) = 0 \qquad [4.14]$$

The general solution of Equation (4.14) has the form

$$X(t) = C_1 \cos(\omega t) + C_2 \sin(\omega t) \qquad [4.15]$$

in which the circular frequency of the vibrating mass is

$$\omega = (k/m)^{1/2} \qquad [4.16]$$

and C_1 and C_2 are arbitrary constants to be determined by appropriate initial conditions.

The frequency of the vibrating mass is

$$f = \frac{\omega}{2\pi}$$

The circular frequency ω in Equation (4.16) is often referred to as the *natural frequency* of the system—a very important quantity in assessing the resonant vibration of solid structures including microdevices. It has units of radians per second, or rad/s.

Figure 4.7 | Simple mechanical vibration systems.

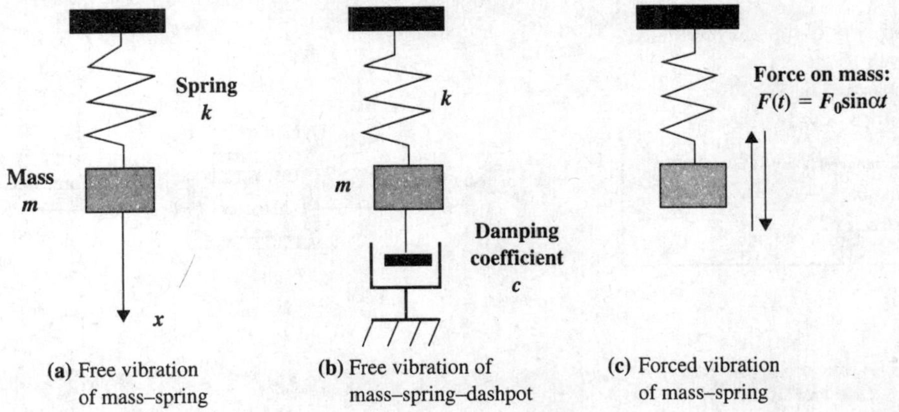

EXAMPLE 4.5

Determine the amplitude and frequency of vibration of a 10-mg mass suspended from a spring with a spring constant $k = 6 \times 10^{-5}$ N/m. The vibration of the mass is initiated by a small "pull" of the mass downward by an amount $\delta_{st} = 5$ μm.

Solution

The description of the physical situation in this example matches the one that Equation (4.14) represents, with the initial displacement of

$$X(0) = \delta_{st} = 5 \times 10^{-6} \text{ m}$$

and the initial velocity of

$$\dot{X}(0) = 0$$

The solution in Equation (4.15) is applicable for the problem, with $C_2 = 0$ and $C_1 = \delta_{st} = 5 \times 10^{-6}$ m from the above conditions.

The amplitude of the vibrating mass at any given time t is thus:

$$X(t) = 5 \times 10^{-6} \cos \omega t \qquad [4.17]$$

in which the circular frequency ω is determined from Equation (4.16) as

$$\omega = \sqrt{\frac{k}{m}} = \sqrt{\frac{6 \times 10^{-5}}{10^{-5}}} = 2.45 \text{ rad/s}$$

and the corresponding frequency is

$$f = \frac{\omega}{2\pi} = \frac{2.45}{2 \times 3.14} = 0.39 \text{ cycles/s (cps)}$$

The natural frequency ω_n of the spring–mass system is same as the circular frequency, or $\omega_n = 2.45$ rad/s.

EXAMPLE 4.6

Find the same answers as in Example 4.5 for a balanced mass–spring system as illustrated in Figure 4.8 with the spring constant $k_1 = k_2$.

Figure 4.8 | A balanced spring–mass system.

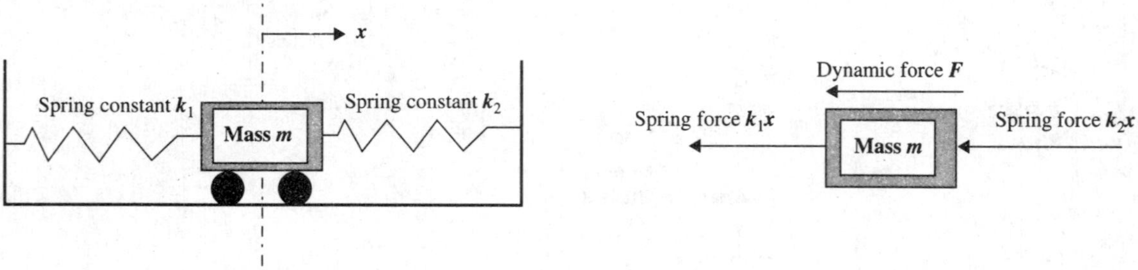

■ Solution

The free-body diagram shown in Figure 4.8 leads to the following equation of motion:

$$m\ddot{X}(t) + (k_1 + k_2)X(t) = 0 \qquad [4.18]$$

The solution of Equation (4.18) is similar to that for Equation (4.14), with $k = k_1 + k_2$.

Hence, the amplitude of the vibrating mass remain the same as in Equation (4.17), but the circular frequency, ω, and thus the natural frequency, becomes:

$$\omega_n = \sqrt{\frac{k_1 + k_2}{m}} = \sqrt{\frac{(6 + 6) \times 10^{-5}}{10^{-5}}} = 3.464 \text{ rad/s}$$

One will realize that the solution in Equation (4.15) for free vibration of a mass-spring system will be oscillatory with constant amplitude. The oscillation of the mass about its initial equilibrium position will extend indefinitely with time, which obviously is not realistic.

Now, if we introduce a dashpot into the system as shown in Figure 4.7b, the dashpot will induce a resistance to the motion of the vibrating mass, or a *damping* effect, that results in the reduction in the amplitude.

Let us assume that the dashpot has a damping coefficient c that will generate a retarding damping force proportional to the velocity of the vibrating mass. The equation of motion in Equation (4.14) is modified to give:

$$m\frac{d^2X(t)}{dt^2} + c\frac{dX(t)}{dt} + kX(t) = 0 \qquad \textbf{[4.19]}$$

The instantaneous position of the mass, $X(t)$, in Equation (4.19) will take one of three forms, depending on the magnitude of the *damping parameter* defined as $\lambda = c/2m$.

Case 1. $\lambda^2 - \omega^2 > 0$, an overdamping situation:

$$X(t) = e^{-\lambda t}\left(C_1 e^{t\sqrt{\lambda^2-\omega^2}} + C_2 e^{-t\sqrt{\lambda^2-\omega^2}}\right) \qquad \textbf{[4.20a]}$$

where C_1 and C_2 are arbitrary constants and ω is given in Equation (4.16).

Figure 4.9 illustrates the possible solutions of Equation (4.20a). We can envisage from these diagrams that the amplitude of vibration of the mass drops rapidly in this case. Overdamping is thus desirable in the design of machines and devices, including microsystems, that are vulnerable to excessive vibration. The design engineer should select a proper damper in the design of such microdevices.

Figure 4.9 | The overdamping situation.

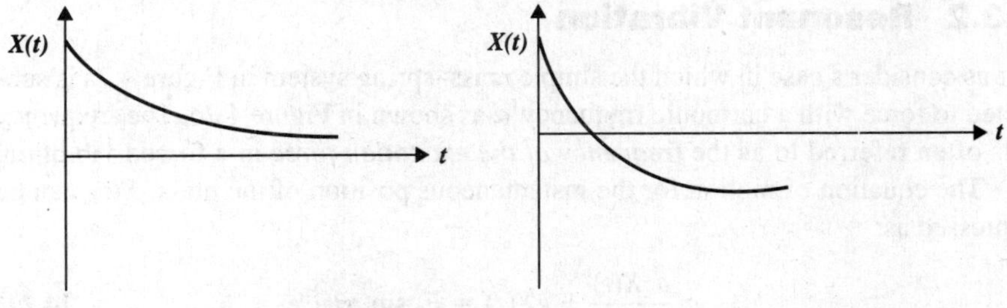

Case 2. $\lambda^2 - \omega^2 = 0$, a critical damping situation:

$$X(t) = e^{-\lambda t}(C_1 + C_2 t) \qquad \textbf{[4.20b]}$$

The above solution is illustrated in Figure 4.10. These representations indicate that the amplitude of vibration decreases initially, followed by a slight increase before the eventual decay. It is not as desirable a situation as the overdamping case.

Case 3. $\lambda^2 - \omega^2 < 0$, an underdamping situation:

$$X(t) = e^{-\lambda t}\left(C_1 \cos\sqrt{\omega^2 - \lambda^2}\, t + C_2 \sin\sqrt{\omega^2 - \lambda^2}\, t\right) \qquad \textbf{[4.20c]}$$

Figure 4.10 | The critical damping situation.

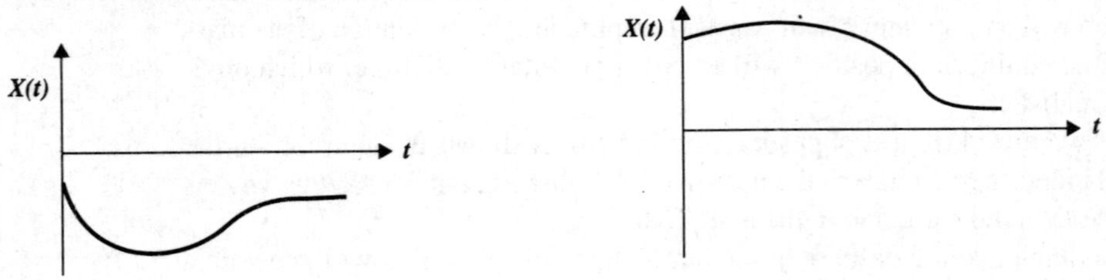

Figure 4.11 illustrates underdamping of a system. We realize from the graphical representation in Figure 4.11 that, although the amplitude of vibration decays continuously, the mass remains in oscillatory motion for a long time. This situation obviously is the least desirable as far as machine design is concerned.

Figure 4.11 | The underdamping situation.

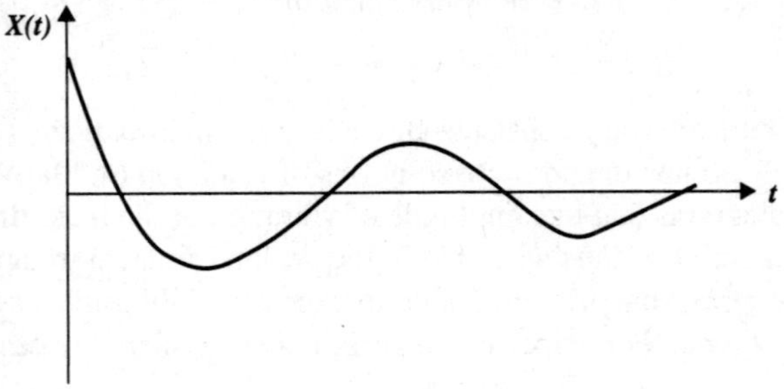

4.3.2 Resonant Vibration

Let us consider a case in which the simple mass–spring system in Figure 4.7a is subjected to force with a harmonic frequency α as shown in Figure 4.7c. The frequency α is often referred to as the *frequency of the excitation force* in a forced vibration.

The equation of motion for the instantaneous position of the mass, $X(t)$, can be expressed as:

$$m\frac{d^2X(t)}{dt^2} + kX(t) = F_0 \sin \alpha t \qquad [4.21]$$

in which F_0 is the maximum amplitude of the applied force.

Solving Equation (4.21) for $X(t)$ gives

$$X(t) = \frac{F_0}{\omega(\omega^2 - \alpha^2)}(-\alpha \sin \omega t + \omega \sin \alpha t) \qquad [4.22]$$

It is seen from Equation (4.22) that $X(t) \to 0/0$; i.e., $X(t)$ is indeterminate when $\alpha = \omega$. However, upon applying L'Hôpital's rule, the following solution is obtained for this special case:

$$X(t) = \frac{F_0}{2\omega^2} \sin \omega t - \frac{F_0}{2\omega} t \cos \omega t \qquad \textbf{[4.23]}$$

One may observe that $X(t) \to \infty$ (i.e., a very large amplitude of vibration) in a very short time when $\alpha = \omega$, as illustrated in Figure 4.12. This phenomenon is referred to as the *resonant vibration* of a mass–spring system.

Figure 4.12 | Rapid increase of amplitude in resonant vibration.

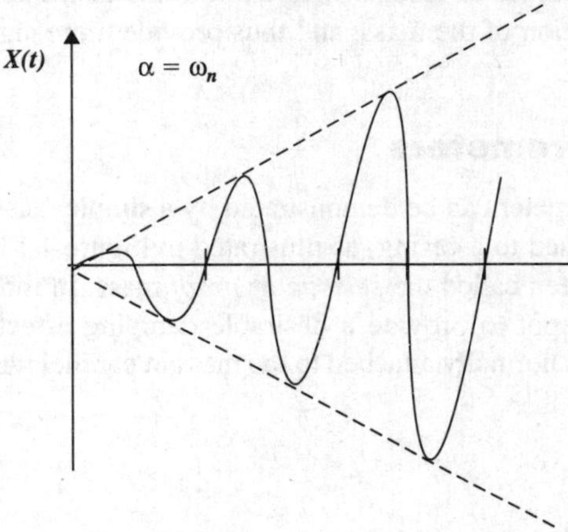

Although the equation for resonant vibration is derived for a simple mass–spring system, *resonant vibration* occurs in other forms of structures whenever the condition of the applied force frequency α equals the natural frequency ω of the structure. For microdevices of complex geometry, there are in theory an infinite number of *modes* in which resonant vibration can take place. Each of these modes is associated with a distinct deformed geometry of the device during the resonant vibration. These multimode resonant vibrations of structures can be attributed to the infinite number of natural frequencies inherent in the structural system.

The method used to determine the natural frequencies associated with various modes of vibration is similar to that used for the simple spring–mass system. The only difference is that the term ω_n is used to represent the resonant vibration of a structure of complex geometry:

$$\omega_n = \sqrt{\frac{K}{M}} \qquad \textbf{[4.24]}$$

The frequency ω_n in Equation (4.24) is called the *natural frequency* of a structure at the nth mode, where n is the mode number, with $n = 1, 2, \ldots$. The analysis involving the determination of ω_n ($n = 1, 2, \ldots$) is referred to as the *modal analysis* of a structure. In the case of modal analysis of a structure such as a microdevice, the stiffness constant K and the mass M in Equation (4.24) are replaced by the respective stiffness matrix $[K]$ and the mass M is replaced by the mass matrix $[M]$ of the structure. These matrices are obtainable from a finite element analysis. Resonant vibration in the structure occurs whenever the frequency of the applied source of vibration α equals any of the natural frequencies of the structure, ω_n $(n = 1, 2, \ldots)$.

The consequences of resonant vibration of a structure can be catastrophic. Thus, engineering design of structures generally attempts to avoid such occurrences by raising the natural frequencies of the structure so that all conceivable frequencies of vibration induced by the applied excitation forces will not reach even the lowest of all modes of the natural frequencies. However, exceptions are made in the case of accelerometer design, where a close-to-resonant vibration of the instrument can result in larger amplitude of vibration of the mass, and thus provide more significant and sensitive output signals.

4.3.3 Microaccelerometers

The principle of an accelerometer can be demonstrated by a simple mass attached to a spring that in turn is attached to a casing, as illustrated in Figure 4.13a. The mass used in accelerometers is often called the *seismic* or *proof* mass. In most cases, the system also includes a dashpot to provide a desirable damping effect. A dashpot with damping coefficient c is normally attached to the mass in parallel with the spring (Fig. 4.13b).

Figure 4.13 | Schematic configurations of accelerometers.

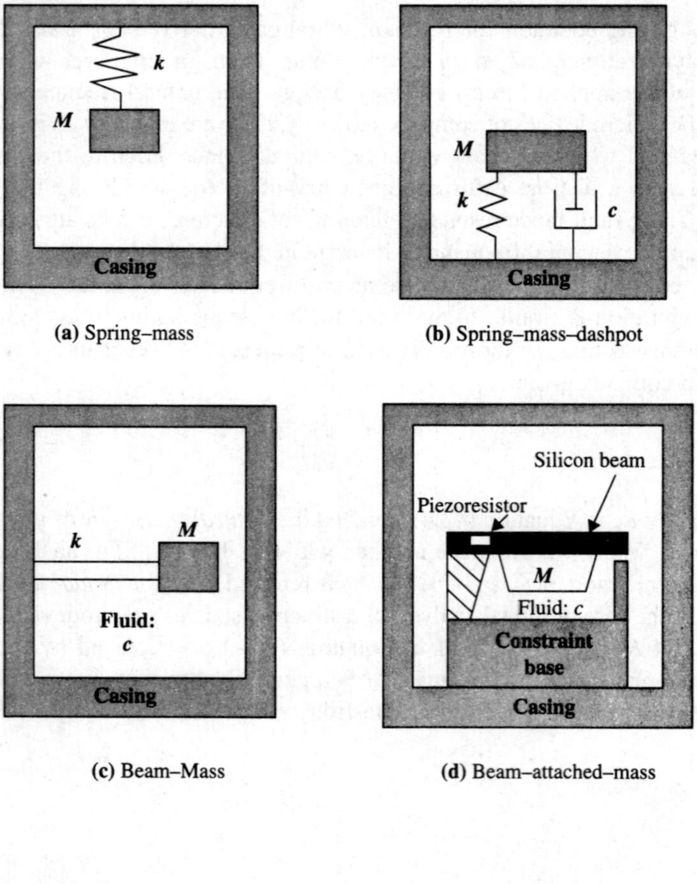

(a) Spring–mass (b) Spring–mass–dashpot

(c) Beam–Mass (d) Beam–attached–mass

For accelerometers built in miniature, the coil spring and dashpot in Figure 4.13a and b occupy too much space. Alternative arrangements of components are those shown in Figure 4.13c [Petersen 1982, Putty and Chang 1989] and Figure 13d [Barth 1990]. The basic configuration of microaccelerometers includes a casing that contains either a cantilever beam and an attached mass as illustrated in Figure 4.13c, or a hanging mass from a thin beam or diaphragm as shown in Figure 4.13d. In a beam–mass or diaphragm–mass system, the elastic beam or diaphragm replaces the spring and the entrapped air or fluid provides the damping effect.

An alternative arrangement for a more compact microaccelerometer is illustrated in Figure 2.34, where the beam mass is attached to two tether springs made of thin silicon frames. The beam–mass moves in the longitudinal direction when it is subjected to an acceleration or deceleration in the same direction.

If the casing of an accelerometer is attached to a vibrating machine or device, when such a machine or device is subject to dynamic or impact loads, the mass (called *seismic mass*) will vibrate. The acceleration of the vibrating mass in the accelerometer can be measured from its instantaneous position by using a potentiometer or comparable means. The theoretical correlation of the amplitude of vibration of the mass and the acceleration of the casing can be found in many textbooks on mechanical vibration.

4.3.4 Design Theory of Accelerometers

We will present in this subsection the theoretical formulation on the amplitude of vibration of the seismic mass in an accelerometer. This formulation can also be used as the basis of the design of microaccelerometers.

Figure 4.14a illustrates a typical accelerometer that consists of a seismic mass supported by a spring and a dashpot. The casing of the vibration system is attached to a vibrating machine with an amplitude of vibration, $x(t)$, that can be described by a harmonic motion, expressed mathematically as

$$x(t) = X \sin \omega t \qquad [4.25]$$

where X is the maximum amplitude of vibration of the base, t is time, and ω is the circular frequency of vibration of the base.

Figure 4.14 | A typical accelerometer.

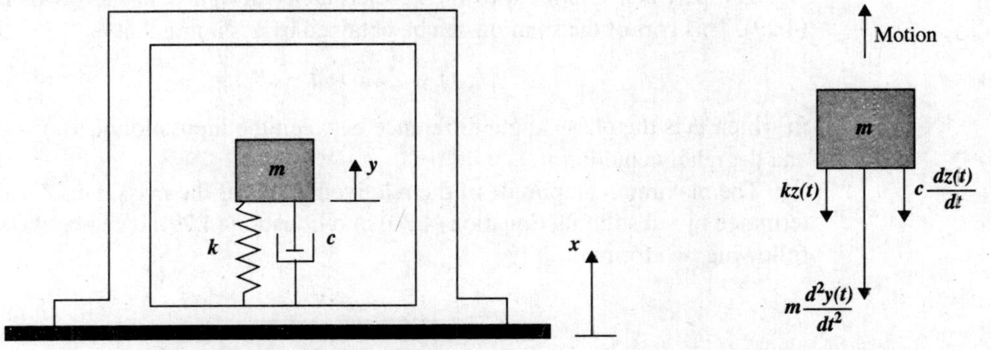

(a) Schematic of an accelerometer

(b) Forces on seismic mass

If we designate $y(t)$ to be the amplitude of vibration of the mass m from its initial equilibrium position, then the relative, or net motion, of the mass m with reference to the base can be expressed as:

$$z(t) = y(t) - x(t) \tag{4.26}$$

from which we have

$$\dot{z}(t) = \dot{y}(t) - \dot{x}(t) \quad \text{and} \quad \ddot{z}(t) = \ddot{y}(t) - \ddot{x}(t)$$

in which

$$\dot{z}(t) = \frac{dz(t)}{dt} \quad \text{and} \quad \ddot{z}(t) = \frac{d^2z(t)}{dt^2}$$

and so on.

The equation of motion of the seismic mass can be derived from Newton's law of dynamic force equilibrium, as illustrated in Figure 4.14b, as:

$$-kz(t) - c\dot{z}(t) - m\ddot{y}(t) = 0 \tag{4.27}$$

Substituting the relation $\ddot{y}(t) = \ddot{z}(t) + \ddot{x}(t)$ into Equation (4.27) leads to:

$$m\ddot{z}(t) + c\dot{z}(t) + kz(t) = -m\ddot{x}(t) \tag{4.28}$$

Since the attached machine vibrates with an amplitude $x(t) = X \sin \omega t$, as in Equation (4.25), Equation (4.28) takes the form:

$$m\ddot{z}(t) + c\dot{z}(t) + kz(t) = mX\omega^2 \sin \omega t \tag{4.29}$$

Equation (4.29) is a second-order nonhomogeneous differential equation, and its solution consists of two parts, i.e., the complementary solution (CS) and the particular solution (PS).

The CS of Equation (4.29) can be obtained from the homogeneous portion of that equation:

$$m\ddot{z}(t) + c\dot{z}(t) + kz(t) = 0$$

which is similar to Equation (4.19) with the form of solution presented in Equations (4.20a, b, and c). In any of these three cases for the CS, the amplitude of vibration $z(t)$ is illustrated in Figures 4.9, 4.10, and 4.11.

The part that is critical to the accelerometer design is in the PS of Equation (4.29). This part of the solution can be obtained by assuming that

$$z(t) = Z \sin(\omega t - \phi) \tag{4.30}$$

in which ϕ is the phase angle difference between the input motion $x(t) = X \sin \omega t$ and the relative motion $z(t)$.

The maximum amplitude of the relative motion of the mass, i.e., Z, can be determined by substituting Equation (4.30) into Equation (4.29). It can be shown in the following two forms:

$$Z = \frac{\omega^2 X}{\sqrt{\left(\frac{k}{m} - \omega^2\right)^2 + \left(\frac{\omega c}{m}\right)^2}} \tag{4.31a}$$

and

$$\phi = \tan^{-1} \frac{\frac{\omega c}{m}}{\frac{k}{m} - \omega^2} \qquad \textbf{[4.31b]}$$

Alternatively, the above solution can be expressed as:

$$Z = \frac{\omega^2 X}{\omega_n^2 \sqrt{\left[1 - \left(\frac{\omega}{\omega_n}\right)^2\right]^2 + \left[2h \frac{\omega}{\omega_n}\right]^2}} \qquad \textbf{[4.32a]}$$

and

$$\phi = \tan^{-1} \frac{2h\left(\frac{\omega}{\omega_n}\right)}{1 - \left(\frac{\omega}{\omega_n}\right)^2} \qquad \textbf{[4.32b]}$$

where

$$\omega_n = \sqrt{\frac{k}{m}}$$

is the natural frequency of the undamped free vibration of the accelerometer and

$$h = \frac{c}{2m\omega_n} = \frac{c}{c_c}$$

is the ratio of the damping coefficients of the damping medium in the micro-accelerometer to its critical damping, with $c_c = 2m\omega_n$ [see the case in Equation (4.20b)].

One may readily observe that, when the system approaches resonant vibration, i.e., $\omega \rightarrow \omega_n$, the amplitude of vibration $Z \rightarrow X/(2h)$ from Equation (4.32a). Since h is related to the damping effect, a free vibration with $h = 0$ will lead to an "infinitely large" amplitude of vibration of the mass. Hence the selection of the damping parameter h is critical in accelerometer design.

The effect of damping on the amplitude of vibration of the mass is qualitatively illustrated in Figure 4.15. Curves included in the figure were drawn from Equation (4.32a). One may observe from this illustration that when the frequency of vibration of the system ω is much greater than the natural frequency of the accelerometer, i.e., $\omega \gg \omega_n$, the maximum relative amplitude Z is approximately equal to the maximum amplitude of vibration of the seismic mass, or $Z = X$. Precise graphs of Figure 4.15, with Z/X versus ω/ω_n as well as $\omega_n Z/\ddot{X}$, can be found in several mechanical vibration textbooks. These graphs serve a useful purpose in the design of accelerometers. On the other hand, if the frequency of vibration of the machine, ω, is much smaller than the natural frequency of the microaccelerometer, i.e., $\omega \ll \omega_n$, we will observe the following relationship [Dove and Adams 1964]:

$$Z \approx -\frac{a_{\text{base, max}}}{\omega_n^2} \qquad \textbf{[4.33]}$$

where $a_{\text{base, max}}$ is the maximum acceleration of the machine on which the accelerometer is attached.

Figure 4.15 | Amplitude of vibration of an accelerometer.

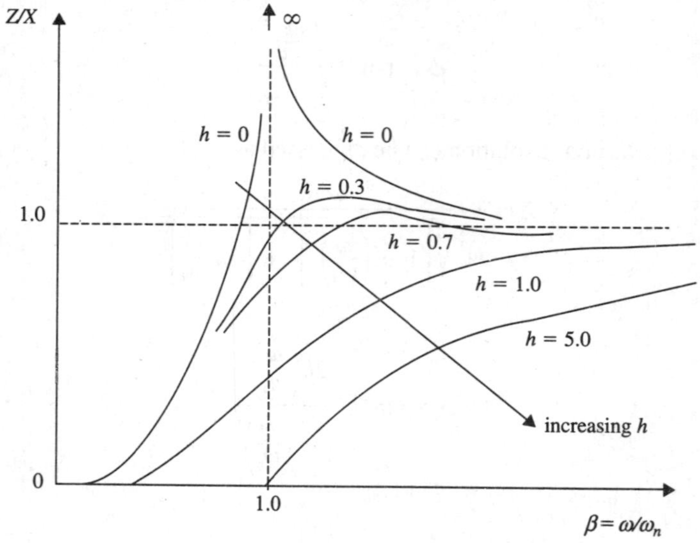

EXAMPLE 4.6

Derive a formula for estimating the natural frequency of a micro accelerometer involving a beam-mass as illustrated in Figure 4.13c with negligible damping effect.

■ Solution

We may idealize the configuration in Figure 4.13c in a simple cantilever beam subjected to an equivalent concentrated load $W = Mg$ at the free end as shown in Figure 4.16. It is obvious that W is the weight of the mass M and g is gravitational acceleration.

Figure 4.16 | A beam spring.

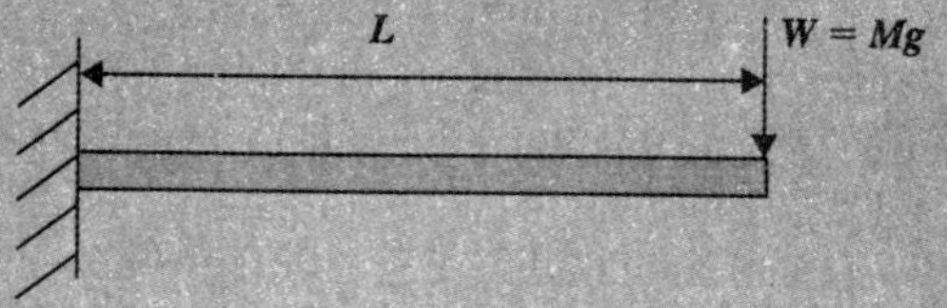

One will recall that the maximum deflection of the beam, $\delta = WL^3/(3EI)$ in which E and I are the respective Young's modulus of the beam material and the area moment of inertia of the beam cross section.

Since the spring constant of a material k is defined as load/deflection, we have for the equivalent spring constant for the beam:

$$k = \frac{W}{\delta} = \frac{3EI}{L^3} \qquad [4.34]$$

The above equivalent spring constant leads to the approximate natural frequency ω_n of the microaccelerometer:

$$\omega_n = \sqrt{\frac{k}{M}} = \sqrt{\frac{3EI}{ML^3}} \quad \textbf{[4.35]}$$

in which M is the mass of the seismic mass attached to the beam and the mass of the beam is neglected.

The reader is reminded that the above example is good only as an approximation, as many microaccelerometers are constructed with suspended mass from a cantilever plate instead of a beam. The equivalent spring constant derived from the simple beam theory obviously is not applicable to these cases. A silicon-based microaccelerometer with a proof mass attached to a cantilever plate was constructed [Bryzek et al. 1992]. It had a diaphragm (i.e., the cantilever plate) of 3.4 mm × 3.4 mm × 1.25 mm thick. The displacement of the mass M and hence the acceleration of the vibrating diaphragm is correlated to the associated strains in the diaphragm by the piezoresistors implanted at the root of the cantilever plate by a diffusion process. There were two "overrange" stops installed in the device to prevent possible damage of the sensor from mishandling. Because of the complex geometry of the accelerometer, a finite element method was used to determine the natural frequencies of the device.

EXAMPLE 4.7

Determine the equivalent spring constant k and the natural frequency ω_n of a cantilever beam element in a microaccelerometer as illustrated in Figure 4.17. The beam is made of silicon with a Young's modulus of 190,000 MPa.

Figure 4.17 | A beam spring and a seismic mass.

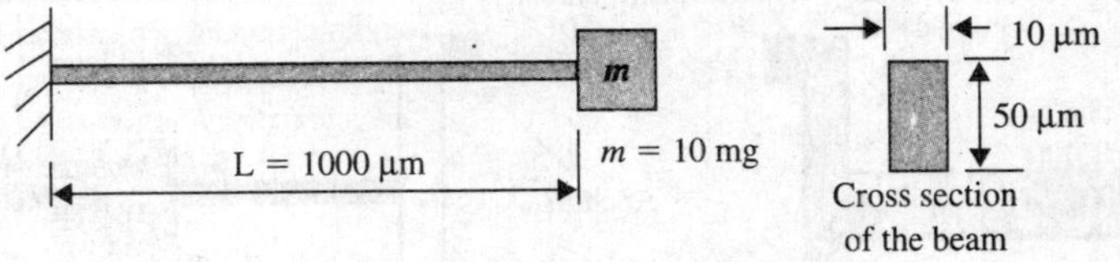

Solution

The moment of inertia of the beam cross section is:

$$I = \frac{(10 \times 10^{-6})(50 \times 10^{-6})^3}{12} = 0.1042 \times 10^{-18}\ \mathrm{m}^4$$

The spring constant k of the beam can be determined from Equation (4.34) as:

$$k = \frac{3(190{,}000 \times 10^6)(0.1042 \times 10^{-18})}{(1000 \times 10^{-6})^3} = 59.39\ \mathrm{N/m}$$

The natural frequency ω_n of the beam spring–mass system is calculated by using Equation (4.24) as:

$$\omega_n = \sqrt{\frac{k}{m}} = \sqrt{\frac{59.39}{10^{-5}}} = 2437\ \mathrm{rad/s}$$

The unit kilogram is used for the mass m in the above calculation.

Other types of beam springs can be used for microaccelerometer design. The reader will readily derive the equivalent spring constants for the centrally loaded beams illustrated in Figure 4.18a to be $k = (48EI)/L^3$ for the simply supported beam spring and $k = (192EI)/L^3$ for the fixed-end beam spring.

Figure 4.18 | Beam springs for microaccelerometers.

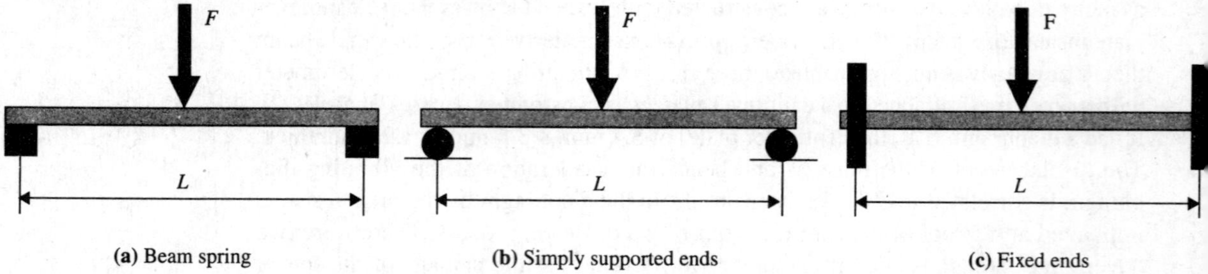

EXAMPLE 4.8

Determine the natural frequency of a "force-balanced" microaccelerometer similar to the one illustrated in Figure 2.34. An idealized form of the structure is illustrated in Figure 4.19. Dimensions of the beam springs are indicated in Figure 4.20.

Figure 4.19 | Basic structure of a force-balanced microaccelerometer.

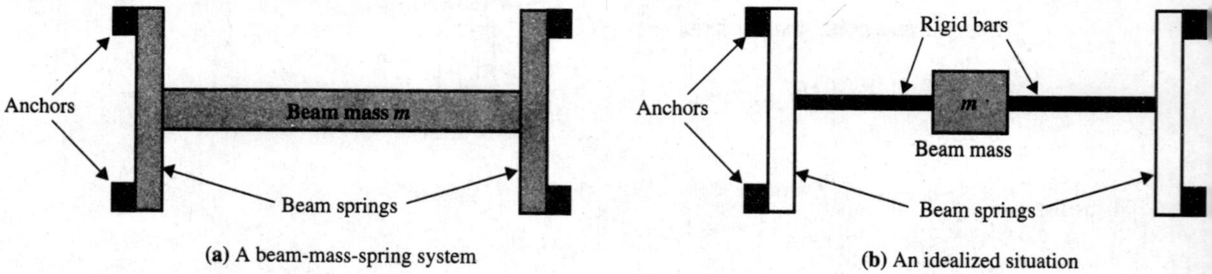

Figure 4.20 | Dimensions of a force-balanced microaccelerometer.

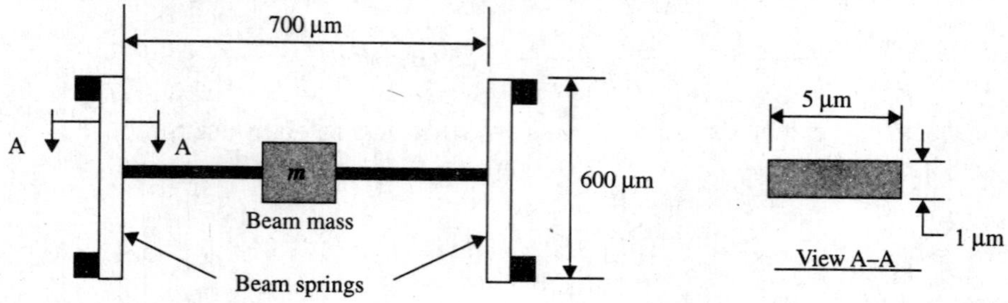

The entire structure is assumed to be made of silicon with Young's modulus, $E = 190{,}000$ MPa. The proof mass, $m = 3 \times 10^{-6}$ kg, is assumed to be concentrated at the centroid of the moving beam as illustrated in Figure 4.19b.

■ Solution

The area moment of inertia of the beam spring I is first computed as:

$$I = \frac{(1 \times 10^{-6})(5 \times 10^{-6})^3}{12} = 10.42 \times 10^{-24}\ \text{m}^4$$

■ Case 1. Beams are simply supported at the ends, Figure 4.18b

The equivalent spring constant k is:

$$k = \frac{48EI}{L^3} = \frac{48(190{,}000 \times 10^6)(10.42 \times 10^{-24})}{(600 \times 10^{-6})^3} = 0.44\ \text{N/m}$$

The natural frequency of the arrangement in Figure 4.20 is similar to a mass vibrating between two springs as described in Example 4.6 and Figure 4.8. We thus have:

$$\omega_n = \sqrt{\frac{2k}{m}} = \sqrt{\frac{2 \times 0.44}{3 \times 10^{-6}}} = 542\ \text{rad/s}$$

■ Case 2. Beams are rigidly fixed at the ends, Figure 4.18c

The equivalent spring constants k takes a different form:

$$k = \frac{192EI}{L^3} = \frac{192(190{,}000 \times 10^6)(10.42 \times 10^{-24})}{(600 \times 10^{-6})^3} = 1.76\ \text{N/m}$$

The corresponding natural frequency ω_n is:

$$\omega_n = \sqrt{\frac{2k}{m}} = \sqrt{\frac{2 \times 1.76}{3 \times 10^{-6}}} = 1083\ \text{rad/s}$$

EXAMPLE 4.9

Determine the displacement from its neutral equilibrium position of the mass of the microaccelerometer in Example 4.8 one (1) ms after the deceleration from its initial velocity of 50 km/h to a stand still.

■ Solution

We have derived the equivalent spring constant $k_{eq} = 1.76$ N/m from Case 2 of Example 4.8. Thus, by using this equivalent spring constant and the situation simulated in Figure 4.8, we have the equation of motion given in Equation (4.14) in a new form:

$$m\frac{d^2X(t)}{dt^2} + 2k_{eq}X(t) = 0$$

or

$$\frac{d^2X(t)}{dt^2} + \omega^2 X(t) = 0$$

in which

$$\omega = \sqrt{\frac{2k_{eq}}{m}} = \sqrt{\frac{2 \times 1.76}{3 \times 10^{-6}}} = 1083 \text{ rad/s}$$

The two initial conditions for the above differential equation are:

$$X(t)|_{t=0} = 0$$

for the initial displacement and

$$\left.\frac{dX(t)}{dt}\right|_{t=0} = 50 \text{ km/h} = 13.8888 \text{ m/s}$$

for the initial velocity.

The general solution of the differential equation is similar to that shown in Equation (4.15); i.e.,

$$X(t) = C_1 \cos(\omega t) + C_2 \sin(\omega t)$$

The arbitrary constants C_1 and C_2 can be determined by the above two initial conditions to be $C_1 = 0$ and $C_2 = 0.01282$, which lead to the following expression for the instantaneous position of the centroid of the beam mass:

$$X(t) = 0.01282 \sin(1083.2t)$$

The position of the beam mass at $t = 1$ ms $= 10^{-3}$ s is obtained from the above expression to be:

$$X(10^{-3}) = 0.01282 \sin(1083.2 \times 10^{-3}) = 2.4 \times 10^{-4} \text{ m} \quad \text{or} \quad 0.24 \text{ mm}$$

from its initial equilibrium position.

4.3.5 Damping Coefficients

The damping coefficient c in Equation (4.19) has a significant effect on the physical behavior of a mechanical vibration system. Figure 4.15 illustrates that the amplitude of the vibrating seismic mass is sensitive to the h factor, which is related to the damping coefficient of the surrounding medium.

Damping is a form of resistance induced by the friction between the surface of the vibrating mass and the surrounding fluid that can be either liquid or gas. In microaccelerometer design, damping can occur in two distinct ways:

1. Squeeze-film damping for those microaccelerometers involving the compression of the surrounding damping fluid by the vibrating mass, such as the case illustrated in Figure 4.13
2. Shear resistance for the force-balanced type of microaccelerometers depicted in Figures 1.6, 2.34, and 4.19

In either case, the damping coefficient c can be determined from the following simple relationship:

$$F_D(t) = cV(t) \tag{4.36}$$

in which F_D is the resistance force to the moving mass, c is the damping coefficient, and $V(t)$ is the velocity of the moving mass.

We will derive the expressions for estimating the damping coefficients in the design of micro accelerometers.

Damping Coefficient in a Squeeze Film The following derivations are based on the information provided in two sources [Newell 1968 and Starr 1990]. The system illustrated in Figure 4.21 represents a vibrating strip with length $2L$ and width $2W$ that squeezes the damping fluid in a narrow gap $H(t)$. If y(t) is the instantaneous position of the strip, then the velocity of the moving strip is expressed as $\dot{y}(t)$.

Figure 4.21 | Squeeze-film damping.

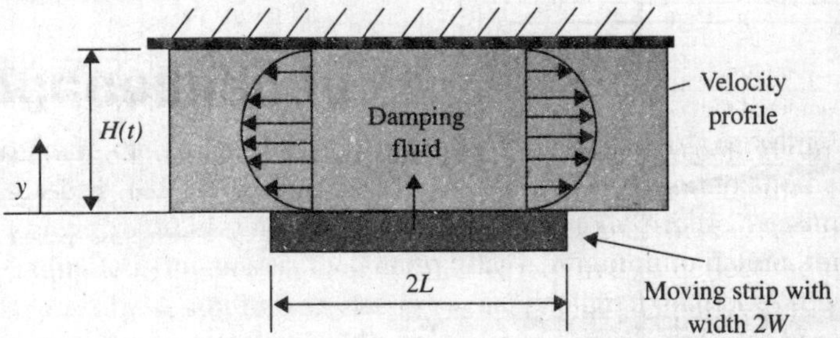

For noncompressible damping fluid media, the following expression for the resistance force F_D is obtained:

$$F_D = 16f\left(\frac{W}{L}\right)W^3L\left(\frac{dy(t)}{dt}\right)H_0^3 \tag{4.37}$$

where H_0 is the nominal thickness of the fluid film.

The derivative of $y(t)$ in Equation (4.37) represents the velocity of the moving strip. Thus the squeeze damping coefficient c can be obtained by combining Equation (4.36) with the relationship in Equation (4.37):

$$c = 16f\left(\frac{W}{L}\right)W^3LH_0^3 \tag{4.38}$$

Numerical values of the function $f(W/L)$ in Equation (4.38) versus W/L are given in Table 4.2. Clearly, the damping coefficient in noncompressible squeeze films is independent of the fluid properties.

Table 4.2 | Geometric function for squeeze-film damping

$\frac{W}{L}$	$f\left(\frac{W}{L}\right)$	$\frac{W}{L}$	$f\left(\frac{W}{L}\right)$
0	1.00	0.6	0.60
0.1	0.92	0.7	0.55
0.2	0.85	0.8	0.50
0.3	0.78	0.9	0.45
0.4	0.72	1.0	0.41
0.5	0.60		

EXAMPLE 4.10

Estimate the damping coefficient of a microaccelerometer using a cantilever beam spring as illustrated in Figure 4.22.

Figure 4.22 | A microaccelerometer with liquid damping fluid.

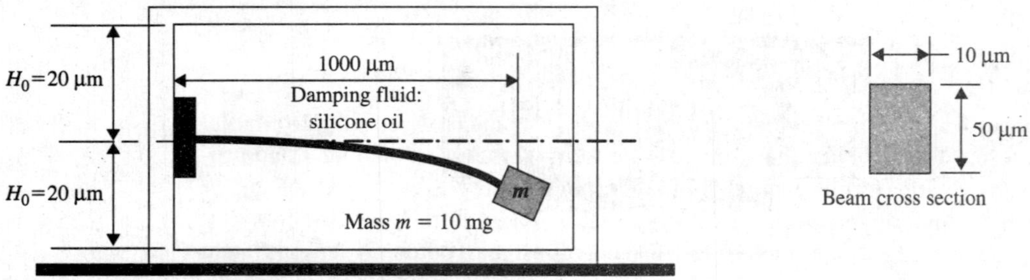

Solution

The damping coefficient for the case in Figure 4.22 can be determined by using Equation (4.38) with $L \approx 500 \times 10^{-6}$ m and $W = 5 \times 10^{-6}$ m. The function $f(W/L)$ is 0.992 by interpolation from Table 4.2 with $W/L = 0.01$.

The nominal film thickness H_0 can be determined by the average maximum deflection of the beam along the length, or simply by setting the gap $H_0 = 20$ μm as shown in Figure 4.22. The adoption of the latter value leads to a very small value of $c = 8 \times 10^{-33}$ N-s/m.

For a squeeze film made of compressible fluids such as air, a squeeze number S is introduced. This number takes the following form [Starr 1990]:

$$S = \frac{12 \mu \omega L^2}{H_0^2 P_a} \quad [4.39]$$

where

μ = dynamic viscosity of the gas film, N-s/m^2
ω = frequency of the moving strip, rad/s
L = characteristic length, m
P_a = ambient gas pressure, Pa

The damping effect is included in the stiffness enhancement, Δk, of the equivalent spring constant k as:

$$\frac{\Delta k}{k} = \alpha f_k(\varepsilon)\left(\frac{\omega}{\omega_n}\right)hS \quad \textbf{[4.40]}$$

in which the constant $\alpha = 0.8$ for long narrow strips such as beam springs, $h = c/2m\omega_n$, with ω_n the natural frequency of the accelerometer, and ε = the ratio of the displacement of the moving mass to the film thickness.

The function $f_k(\varepsilon)$ can be determined by the following expression:

$$f_k(\varepsilon) = (1 + 3\varepsilon^2 + 3\varepsilon^4/8)/(1 - \varepsilon^2)^3 \quad \textbf{[4.41]}$$

Equation (4.40) is qualitatively plotted against the input frequency ω in Figure 4.23.

Figure 4.23 | Squeeze-film damping coefficient by compressible fluids.

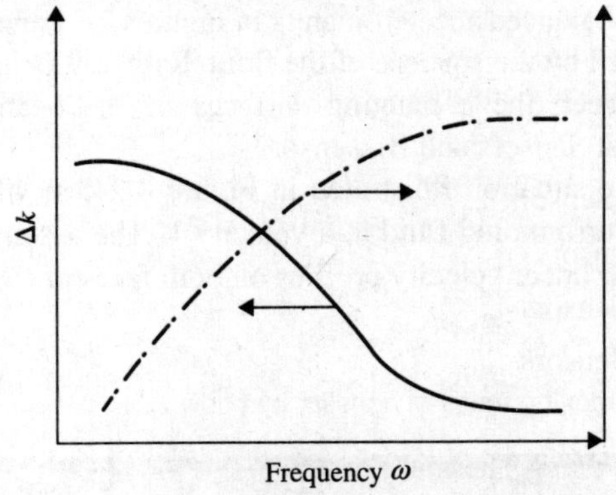

The reader needs to exercise caution because the above expressions are derived from the theory of continuum fluid dynamics based on the assumption of nonslip fluid flow. The validity of using the continuum fluid mechanics theories is limited by the film thickness, or the gap $H(t)$ in Figure 4.21. Depending on the types of gas films, the theory applies only when the gap $H > H_{\text{threshold}}$, in which $H_{\text{threshold}}$ is governed by the mean free path of the gas molecules as will be defined in Chapter 5. For air at 25°C and 1 atm, the mean free path is about 0.09 μm (or 90 nm). Slip flow of gases may take place in film gaps that are less than 100 times the mean free path length. Thus, the above formulation is valid only when the air film thickness $H > 9$ μm.

We notice the appearance of dynamic viscosity of fluid μ in Equation (4.39). Indeed, this fluid property is an important factor in assessing the damping of a vibrating system. Table 4.3 presents the dynamic viscosity of seven selected fluids. The same property for many other fluids can be found in *Marks' Standard Handbook for Mechanical Engineers* [1996].

Table 4.3 | Dynamic viscosity for selected fluids (in 10^{-6} N-s/m^2)

Compressible fluids					
	0°C	**20°C**	**60°C**	**100°C**	**200°C**
Air	17.08	18.75	20.00	22.00	25.45
Helium	18.60	19.41	21.18	22.81	26.72
Nitrogen	16.60	17.48	19.22	20.85	24.64
Noncompressible fluids					
	0°C	**20°C**	**40°C**	**60°C**	**80°C**
Alcohol	1772.52	1199.87	834.07	591.80	432.26
Kerosene	2959.00	1824.23	1283.18	971.96	780.44
Fresh water	1752.89	1001.65	651.65	463.10	351.00
Silicone oil*		740			

*Pfahler et al. 1990.

Microdamping in Shear For microaccelerometers that involve a thin vibrating beam mass (often referred to as the *proof mass*) moving in the surrounding fluid, as in the case of the force-balanced accelerometers in Figures 2.34 and 4.19, the damping effect is not provided by the squeeze of the fluid. Rather, it is induced by the resistance in shear between the contacting surfaces of the beam mass and the surrounding fluid.

Let us consider the situation illustrated in Figure 4.24, in which the moving mass m moves in the surrounding fluid at a velocity V. The assumed nonslip fluid flow condition results in linear velocity profiles on both faces of the beam as shown in the figure.

Figure 4.24 | A moving mass in nonslip fluid flow.

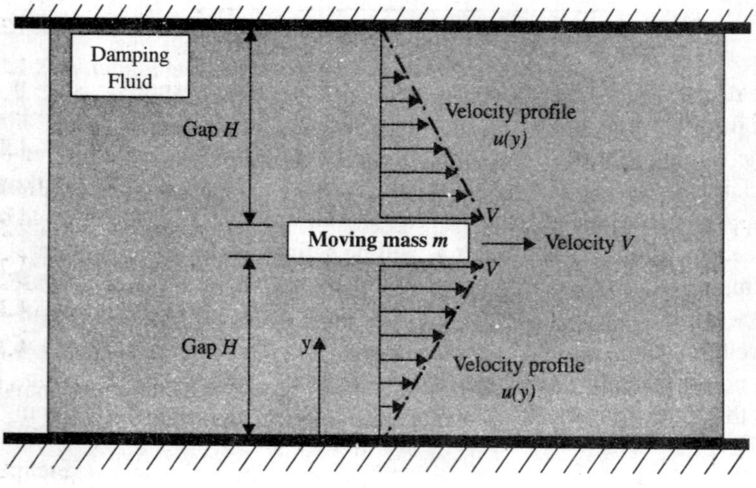

The shear stress τ_s at either the top or bottom face of the beam mass can be expressed as:

$$\tau_s(y) = \mu \frac{du(y)}{dy} \qquad [4.42]$$

where μ = dynamic viscosity of the damping fluid and $u(y)$ = velocity profile in the fluid, with the fluid velocity V at the fluid/solid interfaces. The velocity profile in the present case follows a linear relationship, i.e., $u(y) = Vy/H$ in which H is the width of the gap between the top or bottom surface of the beam and the wall of the enclosure.

We can solve for the shear stress at the contacting surfaces, τ_0, by using Equation (4.42) with the above velocity function:

$$\tau_0 = \frac{\mu V}{H}$$

from which we can determine the equivalent shear forces F_D acting on both the top and bottom faces of the beam mass to be:

$$F_D = \tau_0(2Lb) = \frac{2\mu Lb}{H} V$$

where L and b are the length and width of the beam mass.

The damping coefficient c can thus be determined from Equation (4.36) as:

$$c = \frac{F_D}{V} = \frac{2\mu Lb}{H} \qquad [4.43]$$

EXAMPLE 4.11

Estimate the damping coefficient in a force-balanced micro accelerometer, as illustrated in Figure 4.25, for air and for silicone oil as the damping fluid. Assume that the microaccelerometer operates at 20°C.

Solution

From Figure 4.25, we have the following physical dimensions: width of the beam mass, $b = 5 \times 10^{-6}$ m; length of the beam, $L = 700 \times 10^{-6}$ m, and width of the gap, $H = 10 \times 10^{-6}$ m.

The viscosity for air at 20°C is $\mu_{air} = 18.75 \times 10^{-6}$ N-s/m² from Table 4.3. Thus the damping coefficient c is determined from Equation (4. 43) as:

$$c = \frac{2\mu_{air}\, Lb}{H} = \frac{2(18.75 \times 10^{-6})(700 \times 10^{-6})(5 \times 10^{-6})}{10 \times 10^{-6}} = 1.3125 \times 10^{-8} \text{ N-s/m}$$

For the case involving silicone oil as the damping fluid, we have $\mu_{si} = 740 \times 10^{-6}$ N-s/m² at 20°C from Table 4.3, which leads to the following expression damping coefficient:

$$c = \frac{2\mu_{si}\, Lb}{H} = \frac{2(740 \times 10^{-6})(700 \times 10^{-6})(5 \times 10^{-6})}{10 \times 10^{-6}} = 51.8 \times 10^{-8} \text{ N-s/m}$$

Figure 4.25 | Damping of a force-balanced microaccelerometer.

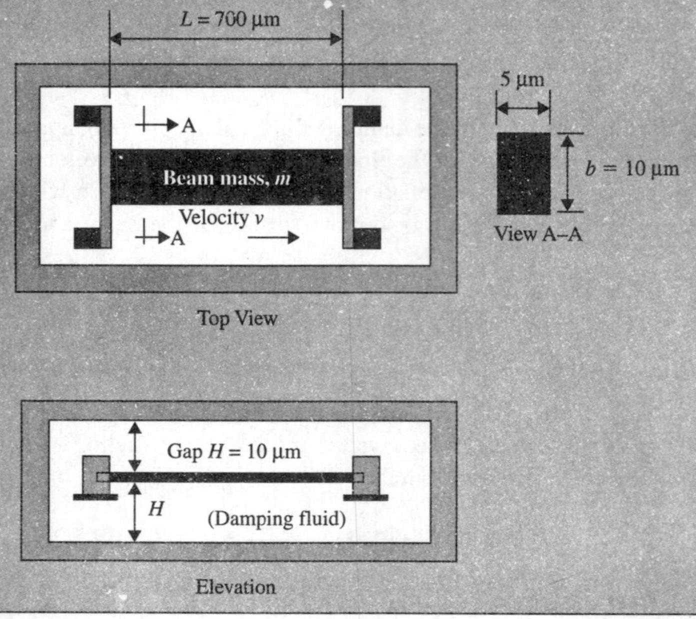

As we have observed from the last two examples, the damping effect in microaccelerometers is indeed very small. This situation leads to a very small value of h in Equation (4.32a). Consequently one may expect a large signal output from micro accelerometer with low value of h as can be observed from Figure 4.15. The low damping coefficients enhance the signal outputs in microaccelerometers.

EXAMPLE 4.12

Two vehicles with respective masses m_1 and m_2 are traveling in opposite directions at velocities V_1 and V_2, as illustrated in Figure 4.26. Each vehicle is equipped with an inertia sensor (or microaccelerometer) built with a cantilever beam as configured in Example 4.7. Estimate the deflection of the proof mass in the sensor in vehicle 1 with mass m_1, and also the strain in the two piezoresistors embedded underneath the top and bottom surfaces of the beam near the support after the two vehicles collide. We have the following information on the microaccelerometer from Example 4.7:

Figure 4.26 | Two vehicles in a course of head-on collision.

The equivalent spring constant k = 59.39 N/m and the natural frequency ω_n = 2437 rad/s. The beam has the properties: Young's modulus, $E = 1.9 \times 10^{11}$ N/m^2; moment of inertia of the cross section, $I = 0.1042 \times 10^{-18}$ m^4; the nominal length of the beam, L = 1000 μm; and the distance from outer surface to the centroid, $C = 25 \times 10^{-6}$ m.

We will have the following conditions for the computations: m_1 = 12,000 kg; m_2 = 8000 kg; $V_1 = V_2$ = 50 km/h.

■ Solution

We postulate that the two vehicles will tangle together after the collision, and the entangled vehicles move at a velocity V as illustrated in Figure 4.27. This final velocity of the entangled vehicles, V, can be obtained by using the principle of conservation of momentum:

$$V = \frac{m_1 V_1 - m_2 V_2}{m_1 + m_2} = \frac{12{,}000 \times 50 - 8000 \times 50}{12{,}000 + 8000} = 10 \text{ km/h}$$

Figure 4.27 | Vehicles after the collision.

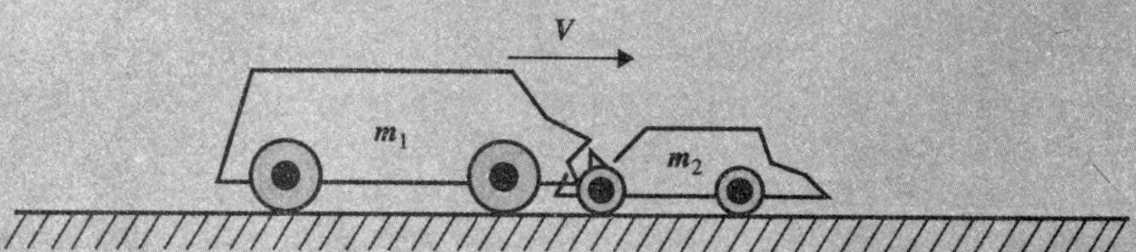

The decelerations of both vehicles can be determined by the following expressions:

$$\ddot{X} = \frac{V - V_1}{\Delta t} \quad \text{for vehicle with } m_1$$

$$\ddot{X} = \frac{V - V_2}{\Delta t} \quad \text{for vehicle with } m_2$$

where Δt is the time required for the deceleration.

Let us assume that it takes 0.5 s for vehicle 1 to decelerate from 50 km/h to 10 km/h after the collision. Thus the time for deceleration of the vehicle m_1 is Δt = 0.5 s in the above expressions. We may thus compute the deceleration of vehicle m_1 to be:

$$\ddot{X} = a_{\text{base}} = \frac{(10 - 50) \times 10^3/3600}{0.5} = -22.22 \text{ m/s}^2$$

We may assert that the frequency of vibration, ω of either vehicle in the case of collision is much smaller than the natural frequency of the accelerometer, ω_n; i.e., $\omega \ll \omega_n$. We may thus justify the use of the expression in Equation (4.33) for the maximum movement of the proof mass, Z:

$$Z \approx -\frac{a_{\text{base}}}{\omega_n^2} = \frac{-22.22}{(2437)^2} = -3.74 \times 10^{-6} \text{ m} \quad \text{or} \quad -3.74 \text{ μm}$$

We thus expect the proof mass to move by the amount of 3.74 μm during the collision, which means the maximum deflection of the cantilever beam is the same amount at the free end.

The corresponding force for such deflection at the free end is:

$$F = \frac{3EIZ}{L^3} = \frac{3(1.9 \times 10^{11})(0.1042 \times 10^{-18})(3.74 \times 10^{-6})}{(1000 \times 10^{-6})^3} = 2.2213 \times 10^{-4}\ \text{N}$$

The equivalent maximum bending moment $M_{max} = FL$, where

$$M_{max} = (2.2213 \times 10^{-4}) \times 10^{-3} = 2.2213 \times 10^{-7}\ \text{N-m}$$

from which we can compute the maximum bending stress σ_{max} near the support to be:

$$\sigma_{max} = \frac{M_{max}\,C}{I} = \frac{(2.2213 \times 10^{-7})(25 \times 10^{-6})}{0.1042 \times 10^{-18}} = 53.30 \times 10^6\ \text{Pa}$$

$$= 53.30\ \text{MPa}$$

The corresponding maximum strains in the piezoresistors in these locations are:

$$\varepsilon_{max} = \frac{\sigma_{max}}{E} = \frac{53.30 \times 10^6}{190 \times 10^9} = 02.81 \times 10^{-4} = 0.0281\%$$

4.3.6 Resonant Microsensors

We have learned from Equation (4.16) that the natural frequency of a simple mass–spring system is related to the spring constant and the vibrating mass. In reality, a structure or a device has geometry that is far more complicated than that of a simple mass–spring system. Yet the device of complex geometry is just as vulnerable to resonant vibration as in a simple mass–spring system. Determination of the natural frequency of a device of complex geometry is related to the device's *stiffness matrix* [**K**], and the *mass matrix* [**M**] as described analogously for simple stiffness K and mass M in Section 4.3.2. The natural frequency of various vibration modes of a device can thus be expressed in a form similar to that for a simple mass–spring system:

$$\omega_n = \sqrt{\frac{[\mathbf{K}]}{[\mathbf{M}]}} \qquad [4.44]$$

Both the stiffness and mass matrices in Equation (4.44) are obtainable from a finite-element analysis, as will be described in Section 10.5 in Chapter 10. The [**K**] matrix is related to both the geometry and material properties of the device. One can thus perceive that any change of the stress in a structure component will result in the change of the component's geometry, which in turn will alter the [**K**] matrix. Consequently, the natural frequency of a device or a component can vary with the change of the states of the stresses in the structure. Shifting of natural frequencies of a device is expected when it is subject to a different state of stress.

As we observe from Figures 4.12 and 4.15, resonant vibrations occur when the casing to which an accelerometer is attached vibrates at a frequency that is approaching the natural frequency of the device; i.e., $\beta = \omega/\omega_n \rightarrow 1.0$. The amplitude of vibration of the proof mass reaches the peak value at this frequency. For a micro-accelerometer, the device provides the most sensitive output at this frequency. Thus, the advantage of using the peak sensitivity of a vibration-measuring device at resonant frequency has been adopted in microsensor design. Howe [1987] developed a theory for the natural frequency of a vibrating beam in Mode 1 subjected to a longitudinal force as

$$\omega_{n,1}^2 = \frac{\int_0^L \frac{EI}{2}\left[\frac{d^2Y_1(x)}{dx^2}\right]^2 dx + \int_0^L \frac{F}{2}\left[\frac{dY_1(x)}{dx}\right]^2 dx}{\int_0^L \frac{1}{2}\rho bh\, Y_1^2(x)\, dx} \qquad [4.45]$$

where

F = applied axial force to a vibrating beam, which can be converted to the normal stress in the beam
ρ = mass density of the beam material
E = Young's modulus
I = area moment of inertia of the beam cross section
b and h = respective width and thickness of the rectangular beam cross section

The function $Y_1(x)$ in Equation (4.45) is the amplitude of the vibrating beam, as illustrated in Figure 4.28. The function $Y_1(x)$ can be obtained from the solution $y(x, t)$ in the form of $y(x, t) = Y_1(x) \exp(i\omega t)$, in which $i = \sqrt{-1}$ and ω is the frequency of the vibrating beam.

Figure 4.28 | A vibrating beam subject to tensile force.

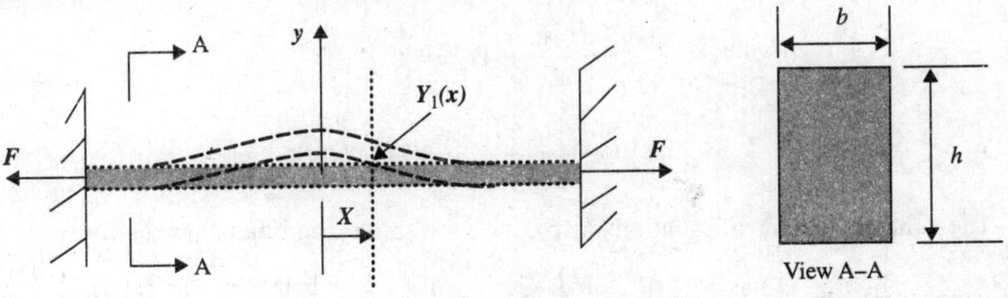

The solution, $y(x, t)$ in a vibrating beam, can be obtained from solving the following partial differential equation:

$$\alpha^2 \frac{\partial^4 y(x,t)}{\partial x^4} + \frac{\partial^2 y(x,t)}{\partial t^2} = 0 \qquad [4.46]$$

in which $\alpha = \sqrt{EI/\gamma}$ with γ to be the mass/unit length of the beam.

The solution of Equation (4.46) obviously requires the specified initial and boundary conditions. Solutions of several cases with clamped end conditions are available for the design of microbridges [Bouwstra and Geijselaers 1991]. The solution of a special case with the following initial and boundary conditions is presented below [Trim 1990].

The situation is for a simply supported beam vibrating in the first mode, as illustrated in Figure 4.29. The following initial and end conditions are used in solving Equation (4.46):

Figure 4.29 | A beam in free vibration.

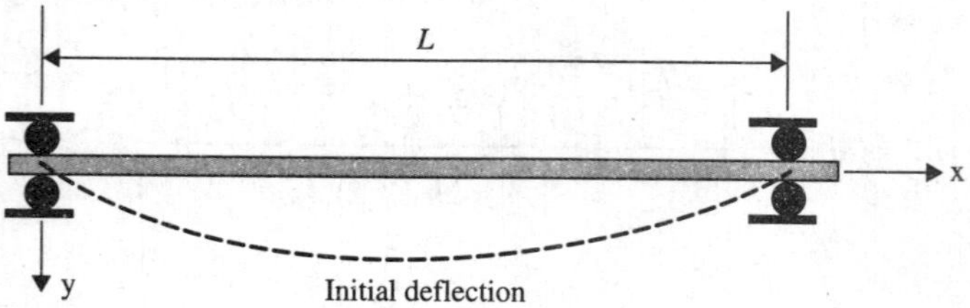

The initial deflection and the velocity of the beam in static equilibrium are used as the initial conditions:

$$y(x, 0) = f(x) = x \sin \frac{\pi x}{L} \qquad 0 \leq x \leq L$$

$$\left. \frac{\partial y(x, t)}{\partial t} \right|_{t=0} = 0 \qquad 0 \leq x \leq L$$

Boundary conditions for a simply-supported beam at $t > 0$ require that

$$y(0, t) = 0$$

$$y(L, t) = 0$$

$$\left. \frac{\partial^2 y(x, t)}{\partial x^2} \right|_{x=0} = 0$$

$$\left. \frac{\partial^2 y(x, t)}{\partial x^2} \right|_{x=L} = 0$$

The solution, $y(x, t)$ in Equation (4.46), with the above conditions has the form:

$$y(x, t) = \frac{L}{2} \sin \frac{\pi x}{L} \cos \frac{\pi^2 \alpha t}{L^2} - \frac{16L}{\pi^2} \sum_{n=1}^{\infty} \frac{n}{(4n^2 - 1)^2} \sin \frac{2n\pi x}{L} \cos \frac{4n^2 \pi^2 \alpha t}{L^2} \qquad \textbf{[4.47]}$$

The number $n = 1, 2, 3, \ldots$ in Equation (4.47) represents the various modes of vibration of the beam. It is customary to take the first mode (i.e., $n = 1$) for the solution in the design process. One may deduce the maximum amplitude of vibration, i.e., the function $Y_1(x)$ for Equation (4.45).

Equation (4.45) has indicated that the natural frequency of a vibrating beam can be changed with a change of the applied axial force (or the normal stress). This principle is used to construct a highly sensitive pressure sensor [Petersen et al. 1991]. In this microdevice, a single-crystal p-type silicon beam 40 μm wide × 6 μm thick × 600 μm long is attached to the middle of the front face cavity of the diaphragm in a pressure sensor by means of fusion bonding. The beam is first excited to resonance by an ac voltage signal applied to the diffused deflection electrode beneath the beam. This takes typically a few millivolts. Electrostatic forces from the ac power source intermittently pull the beam downward at the proper frequency to excite the beam to vibrate at its natural frequency. The resonant vibration of the beam is detected by the piezoresistors diffused into the ends of the beam. A schematic diagram of this type of pressure sensor is shown in Figure 4.30.

Figure 4.30 I Resonant vibrating beam pressure sensor.

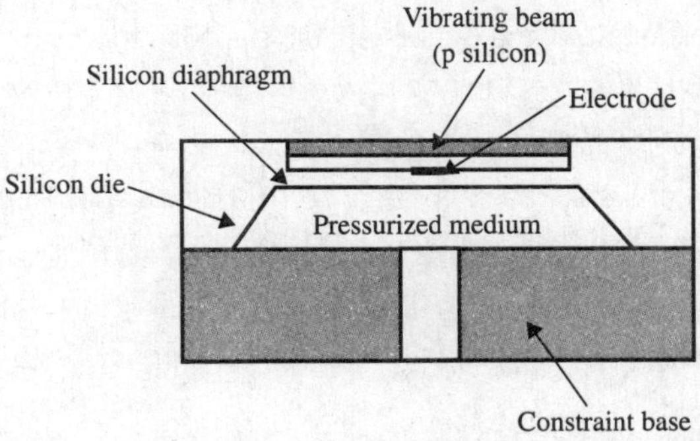

The application of pressure at the back face of the diaphragm would induce tensile stress in the vibrating beam and thus alter its natural frequency. Adjustment of the supply excitation voltage is necessary to get the beam to vibrate at resonance in the new state. The applied pressure can be calibrated to correspond to the shifting of the resonant frequency of the vibrating beam. This method has proved to offer much higher sensitivity than the micro pressure sensors that use other forms of signal transduction such as piezoresistors or capacitors. An overall measurement accuracy of 0.01 percent is achieved. The corresponding sensitivity of the pressure sensor was presented in Figure 2.12.

EXAMPLE 4.13

Determine the shift of natural frequency of a fixed-end beam made of silicon, with geometry and dimensions shown below when subjected to a longitudinal stress at 187 MPa. The dimensions of the beam are width $b = 40 \times 10^{-6}$ m, depth (thickness) $h = 6 \times 10^{-6}$ m, length $L = 600 \times 10^{-6}$ m. The following properties are applicable to the silicon beam:

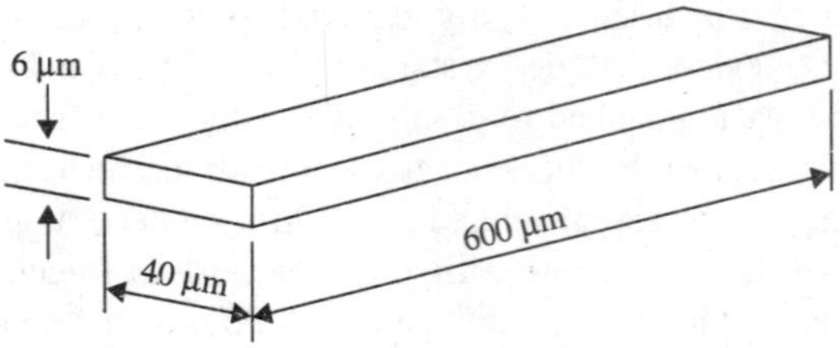

Young's modulus $E = 190{,}000$ MPa

Mass density $\rho = 2300$ kg/m^3

Solution

The parameters that are required for the computation are:

Mass per unit length of the beam, $\gamma = 5.52 \times 10^{-7}$ kg/m

Weight per unit length of the beam, $w = 5.4096 \times 10^{-6}$ N/m

Area moment of inertia $I = 7.2 \times 10^{-22}$ m^4

The partial differential equation in Equation (4.46) is used to compute the amplitude of vibration with the following end and initial conditions:

End conditions:

$$\left.\frac{\partial y(x, t)}{\partial x}\right|_{x=0} = 0 \text{ and } y(x, t)|_{x=0} = 0 \qquad \text{at the end } x = 0$$

$$\left.\frac{\partial y(x, t)}{\partial x}\right|_{x=L} = 0 \text{ and } y(x, t)|_{x=L} = 0 \qquad \text{at the other end } x = L$$

Initial conditions:

$$\left.\frac{\partial y(x, t)}{\partial t}\right|_{t=0} = 0 \qquad \text{for the initial velocity}$$

$$y(x, t)|_{t=0} = f(x) = \frac{w}{24EI}(x - L)^2 x^2 \qquad \text{for initial sag of the beam due to its own weight (from handbooks)}$$

The separation of variables technique is used for the solution of the partial differential equation. The solution, $y(x, t)$ in Equation (4.46) with the above end and initial conditions can be expressed as:

$$y(x, t) = \sum_{n=1}^{\infty} K_n \left[\cosh(\lambda_n x) - \cos(\lambda_n x) - \frac{\cosh(\lambda_n L) - \cos(\lambda_n L)}{\sinh(\lambda_n L) - \sin(\lambda_n L)}[\sinh(\lambda_n x) - \sin(\lambda_n x)]\right] \cos[\alpha(\lambda_n)^2 t]$$

where the coefficient K_n is obtained from the following integrals:

$$K_n = \frac{\int_0^L f(x)X(x)\,dx}{\int_0^L [X(x)]^2\,dx}$$

in which the function $X(x)$ is part of the above general solution of $y(x, t)$, i.e.,

$$X(x) = [\cosh(\lambda_n x) - \cos(\lambda_n x)] - \frac{\cosh(\lambda_n L) - \cos(\lambda_n L)}{\sinh(\lambda_n L) - \sin(\lambda_n L)}[\sinh(\lambda_n x) - \sin(\lambda_n x)]$$

The function $f(x)$ is given as one of the two initial conditions.

The number n in the infinite series solution represents the mode number of vibration. For the problem on hand, we are interested only in the mode 1 vibration, i.e., $n = 1$. The corresponding eigenvalue λ_1 can be evaluated as $4.73/L$. The natural frequency of the beam in mode 1 is:

$$\omega_1 = \lambda_1^2 \sqrt{\frac{EI}{\gamma}} = 9.783 \times 10^5 \text{ rad/s}$$

or

$$\omega_1 = \frac{9.783 \times 10^5}{2\pi} = 1.557 \times 10^5 \text{ Hz}$$

The function $Y_1(x)$ in Equation (4.45) is used to determine the shift of natural frequencies due to applied stress. It can be obtained from the solution of $y(x, t)$ shown above with $n = 1$:

$$Y_1(x) = K_1 X_1(x)$$

in which K_1 and $X_1(x)$ are the respective coefficients K_n and $X(x)$ evaluated at $n = 1$ in the above expressions.

The applied longitudinal force to the beam is $F = A\sigma$, in which A is the cross-sectional area of the beam (240×10^{-12} m^2) and σ is the applied stress at 187 MPa. We can thus compute the shift of the natural frequency of the beam under this stress from Equation (4.45) to be:

$$\omega_{1,s} = 1.932 \times 10^6 \text{ rad/s} \quad \text{or} \quad 3.075 \times 10^5 \text{ Hz}$$

4.4 | THERMOMECHANICS

Many microsystems are either fabricated at high temperatures, such as in fusion bonding or oxidation processes, or are expected to operate at elevated temperatures in the case of engine cylinder pressure sensors in an automotive application. Some transduction components, such as piezoresistors, are highly sensitive to the operating temperature. Thermal effects are thus an important factor in the design and packaging of microsystems.

There are generally three serious effects on micromachines and devices exposed to elevated temperatures:

4.4.1 Thermal Effects on Mechanical Strength of Materials

Most engineering materials exhibit reductions in the stiffness (e.g., Young's modulus) and the yield and ultimate strengths with increasing temperature, as illustrated in

Figure 4.31. These reductions are more drastic with plastics and polymers. Fortunately, materials used for many core elements of microsensors and actuators, including silicon, quartz, and Pyrex glass, are relatively insensitive to the temperature. Significant changes of material properties are not expected in these device components at elevated temperatures. Also from Figure 4.31, we observe that thermophysical properties of most materials increase with the temperature. Again, these changes are more pronounced in packaging materials such as adhesives, sealing, and die-protection materials used in most microsystems. Table 4.4 presents temperature-dependent specific heat and coefficient of thermal expansion of silicon.

Figure 4.31 | Variation of material properties with temperature.

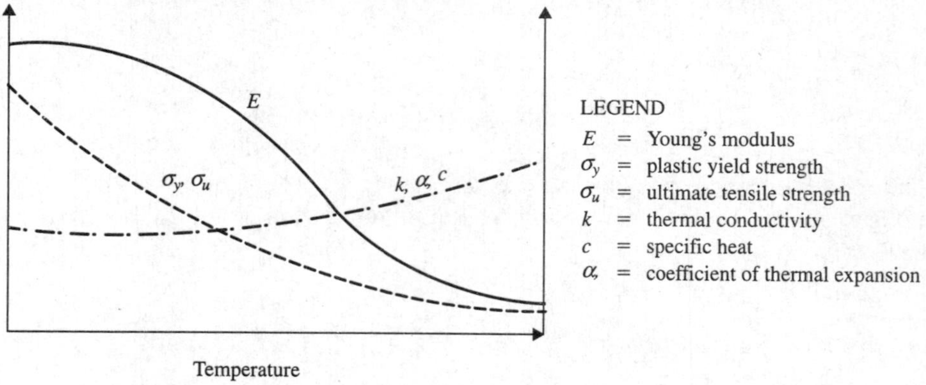

Table 4.4 | Temperature-dependent thermophysical properties of silicon

Temperature, K	Specific heat, J/g-K	Coefficient of thermal expansion, 10^{-6}/K
200	0.557	1.406
220	0.597	1.715
240	0.632	1.986
260	0.665	2.223
280	0.691	2.432
300	0.713	2.616
400	0.785	3.253
500	0.832	3.614
600	0.849	3.842

Source: "Properties of Silicon," EMIS Group, INSPEC, New York, 1988.

4.4.2 Creep Deformation

Creep deformation can occur in materials when the material's temperature exceeds half of the *homologous melting point* of the material (i.e., melting point on the absolute temperature scale). Creep is a form of deformation of the material without being subjected to additional mechanical loads. Creep deformation can develop in some components of microdevices such as adhesives and solder joints over a period of time, as illustrated in Figure 4.32. There are generally three stages of creep deformation: primary, steady-state, and tertiary creep. Prolonged exposure of these materials to elevated operating temperatures can lead to detrimental tertiary creep with catastrophic failure of the device.

Figure 4.32 | Creep deformation of materials at elevated temperature.

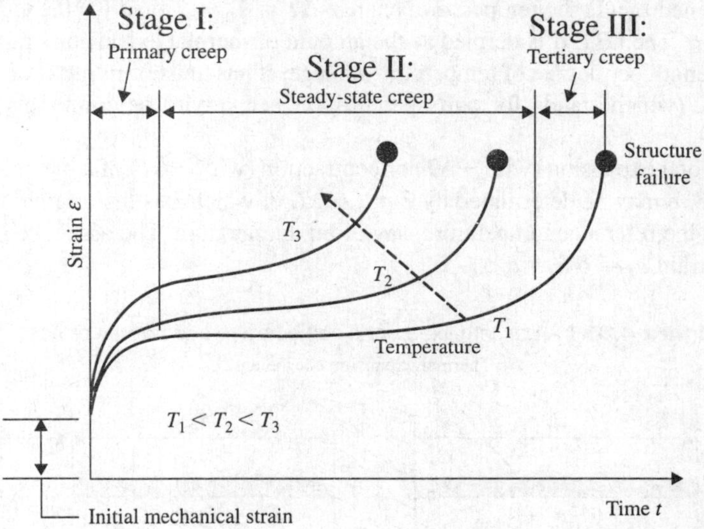

Another important fact about material's creep deformation is the effect of increasing operating temperature. The creep curves become steeper as temperature increases. We can readily see from Figure 4.32 that the distinction between the three stages of creep becomes less distinct at higher temperatures. The structure takes a shorter time to fail in creep at these temperatures than at lower operating temperatures.

Although creep can be a serious design consideration for plastics, polymers and bonding materials in microsystems, most sensing and actuating components made of silicon or its compounds typically have very high melting points, which makes creep less a problem. There is an abundance of publications on the creep behavior of solder alloys relating to microelectronics solder joint reliability [Hsu and Zheng 1993, Hsu et al. 1993].

4.4.3 Thermal Stresses

Thermal stresses are induced in microdevices operating at elevated temperatures either due to mechanical constraints or due to mismatch of the coefficients of thermal expansion (CTE) of the mating components. Thermal stresses can also be induced in devices with little or no mechanical constraint as a result of nonuniform temperature distributions in the structure. Since most microsystems are made of components of different materials such as layered thin films, thermal stresses due to mismatch of

CTE need to be accurately assessed during the design stage, as excessive thermal stresses can cause failure of microdevices. Thermal stress analysis is thus an important part of the design of microsystems.

To many engineers, a well-known physical phenomenon is that materials expand when they are heated up and contract when they are cooled down. The amount of expansion or contraction of a material subjected to the change of thermal environment is determined by (1) the temperature change $\Delta T = T_2 - T_1$ and (2) the CTE of the material, α. The CTE α is defined as the amount of thermal expansion of a material per unit length per degree of temperature change. It has units of in/in/°F, or m/m/°C, or ppm/K (which stands for parts per million per kelvin) as commonly used in industry.

The total expansion (with $+\Delta T$) or contraction (with $-\Delta T$) of a bar as shown in Figure 4.33b may be determined by $\delta = L\alpha\,\Delta T$, in which L is the original length of the bar at the reference temperature, say room temperature. The associated induced thermal strain $\varepsilon_T = \delta/L = \alpha\,\Delta T$.

Figure 4.33 | A bar with fixed ends subjected to a temperature rise.

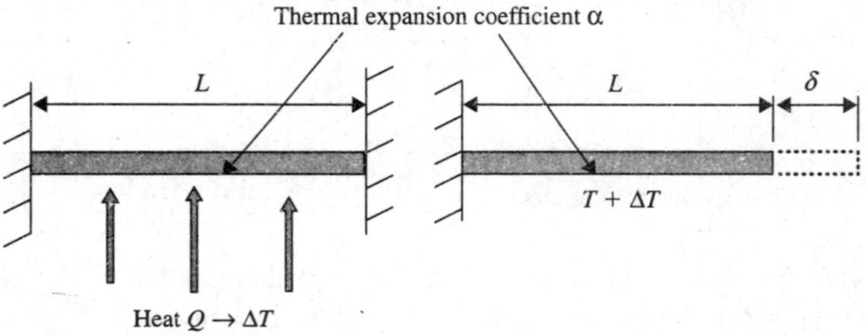

(a) A bar with both ends fixed

(b) Free Thermal expansion with one end constraint removed

Now let us examine the situation in Figure 4.33a further. The bar is fixed at both ends. We can expect a compressive stress induced in the bar with an application of a temperature rise $\Delta T = T_2 - T_1$. The magnitude of the stress is:

$$\sigma_T = E\varepsilon_T = -\alpha E\,\Delta T \quad \textbf{[4.48]}$$

The derivation of the above expression can be carried out as follows: By referring to Figure 4.33b, let us first evaluate the amount of thermal expansion of the bar if one end of the bar were set free. The amount of free thermal expansion of the bar is $\delta = L\,(\alpha\,\Delta T)$. However, that end of the bar is actually fixed; i.e., the end is not released, as shown in Figure 4.33a. This situation can be considered to be equivalent to having an equivalent mechanical force "pushing" back the free-expanded bar with an amount δ. The required "pushing" force would produce an equivalent compressive force, which produces a compressive strain of $\varepsilon_T = -\delta/L = -\alpha\,\Delta T$. The corresponding stress is thus equal to a compressive stress of $\sigma_T = E\varepsilon_T = -\alpha E\,\Delta T$, as shown in Equation (4.48).

The following example will illustrate how thermal stress and deformation due to mismatch of thermal expansion coefficients can be determined. The reader will recognize the application of this case in the design of a microactuator that involves bilayer strips.

Figure 4.34 shows the schematic of a bilayer beam with two strips bonded together. Bilayer strips of this kind are commonly used in microactuators such as microtweezers actuated by thermal means. The fact that the two strips have distinct thermal expansion coefficients can make the strip curve either upward or downward when it is subjected to a temperature rise or drop. This actuated curvature can result in either opening or closing an electric circuit.

Figure 4.34 | A bimaterial strip subject to a temperature change.

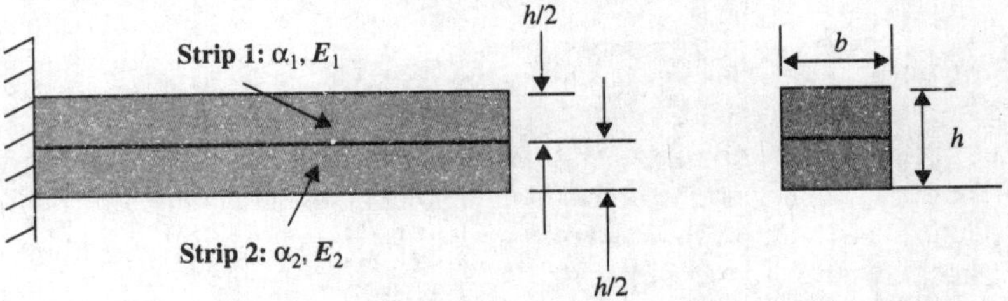

For the case illustrated in Figure 4.34, the interface force F and the resulting curvature of the strip can be determined by the following formulas [Boley and Weiner 1960]:

$$F = \frac{(\alpha_2 - \alpha_1)T}{8} \frac{hb}{\left(\frac{1}{E_1} + \frac{1}{E_2}\right)} \quad [4.49]$$

and the curvature ρ of the deformed strip is

$$\rho = \frac{2h}{3(\alpha_2 - \alpha_1)T} \quad [4.50]$$

The terms E_1 and E_2 in Equation (4.49) are the respective Young's moduli of strip 1 and 2.

EXAMPLE 4.14

A microactuator made of a bilayer strip—an oxidized silicon beam—is illustrated in Figure 4.35a. A resistance heating film is deposited on the top of the oxide layer. Estimate the interfacial force between the Si and SiO_2 layers and the movement of the free end of the strip with a temperature rise $\Delta T = 10°C$. Use the following material properties:

Young's modulus: $E_{SiO2} = E_1 = 385{,}000$ MPa; $E_{Si} = E_2 = 190{,}000$ MPa

CTE: $\alpha_{SiO2} = \alpha_1 = 0.5 \times 10^{-6}/°C$; $\alpha_{Si} = \alpha_2 = 2.33 \times 10^{-6}/°C$

Figure 4.35 | A bilayered strip actuator.

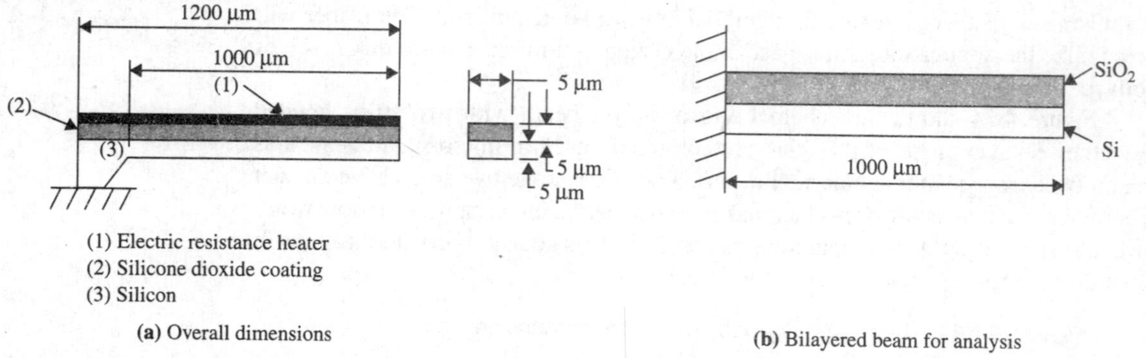

■ **Solution**

We may use Equations (4.49) and (4.50) to calculate the interfacial force and the radius of curvature of the heated strip with $h = 10 \times 10^{-6}$ m and $b = 5 \times 10^{-6}$ m.

Thus, by substituting the above material properties and the dimensions into Equations (4.49) and (4.50), we have the values for the interfacial force, F and the radius of curvature, ρ as follows:

$$F = \frac{(2.33 \times 10^{-6} - 0.5 \times 10^{-6})10}{8} \frac{(10 \times 10^{-6})(5 \times 10^{-6})}{\frac{1}{385{,}000 \times 10^{6}} + \frac{1}{190{,}000 \times 10^{6}}}$$

$$= 14.55 \times 10^{-6} \text{ N}$$

$$\rho = \frac{2(10 \times 10^{-6})}{3(2.33 \times 10^{-6} - 0.5 \times 10^{-6})10} = 0.3643 \text{ m}$$

We realize that the interfacial force is so small that it can be entirely neglected. The radius of curvature of the bent strip, however, needs to be used to assess the corresponding movement of the free end, which can be approximated from the situation illustrated in Figure 4.36.

Figure 4.36 | Movement of the free end of a bilayer strip actuator.

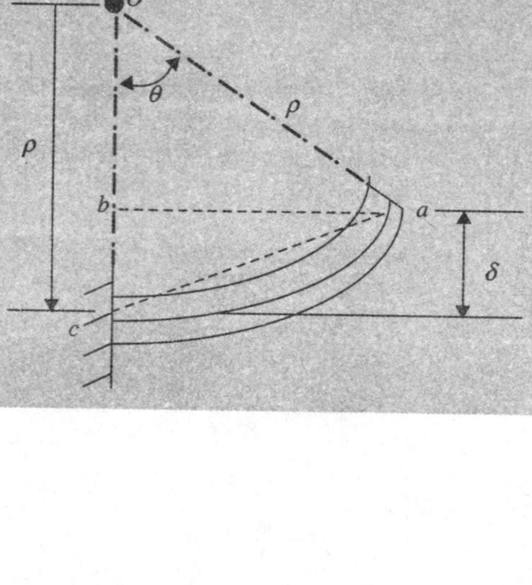

Since the radius of curvature of the arc is $\rho = 0.3643$ m, the corresponding perimeter of a complete circle with this radius is $2\pi\rho = 2.2878$ m. The angle θ that corresponds to the arc can be approximated by:

$$\frac{\theta}{\text{arc }(ac)} \approx \frac{\theta}{\text{line }(ac)} = \frac{\theta}{1000 \times 10^{-6}} = \frac{360}{2.2878}$$

which leads to $\theta = 0.1574°$

Consequently, we may use the relationship derived from the triangle ΔOab to obtain the movement of the free end δ to be:

$$\delta \approx \rho - \rho \cos\theta = 0.3643 - 0.3643 \cos(0.1574°)$$

$$= 1.373 \times 10^{-6}\text{ m} \qquad \text{or } 1.373\ \mu\text{m}$$

Theories of thermomechanics, which include those for thermoelastic-plastic stress analysis of structures of complex geometry and loading and boundary conditions, as well as the associated creep and fracture of these structures due to thermal effects, can be found in a reference book [Hsu 1986]. Finite element formulation for the above thermomechanical analyses is also available in the same book.

As we have learned from Chapter 2, many MEMS components are in the shapes of thin plates and beams. We will outline the closed-form solutions of thermal stress distributions in these structure components in the following sections. Detailed derivation of these formulas is available in several sources, including Boley and Weiner [1966]. Components of stress and the associated strains in a three-dimensional solid under static equilibrium conditions are defined in Figure 4.37. The displacements along the respective x, y, and z directions are designated as $u(x, y, z)$, $v(x, y, z)$ and $w(x, y, z)$.

Figure 4.37 | Designation of stress components in a solid.

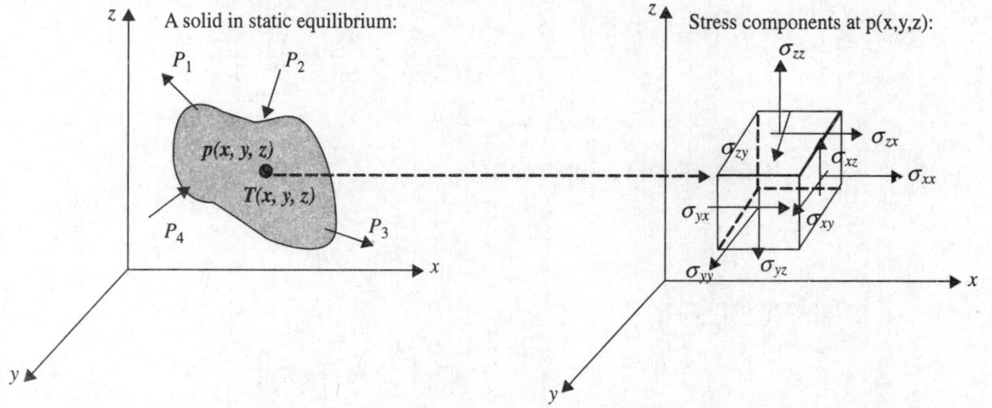

Thermal Stresses in Thin Plates with Temperature Variation through the Thickness. Figure 4.38 illustrates a thin plate of arbitrary shape defined by a cartesian coordinate system. Plane stress is assumed for this thin plate, which means the following stress situations with temperature variation along the thickness direction, or $T = T(z)$:

$$\sigma_{xx}(x, y, z) = f_1(z) \qquad \text{and} \qquad \sigma_{yy}(x, y, z) = f_2(z)$$

$$\sigma_{zz} = \sigma_{xz} = \sigma_{yx} = \sigma_{yz} = 0$$

Figure 4.38 | Thermal stresses in a thin plate.

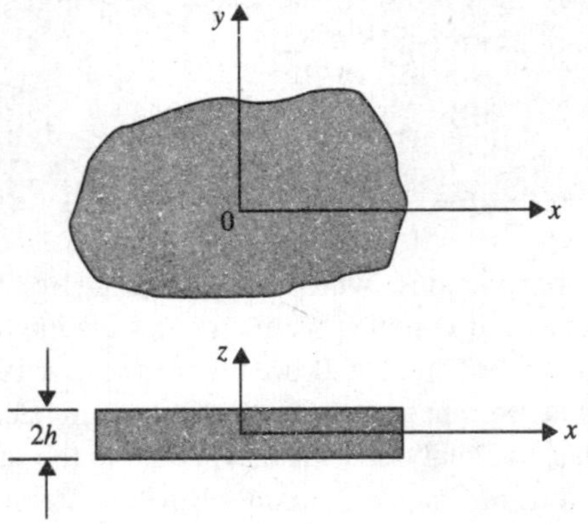

Thermal stresses:

$$\sigma_{xx} = \sigma_{yy} = \frac{1}{1-\nu}\left[-\alpha E T(z) + \frac{1}{2h}N_T + \frac{3z}{2h^3}M_T\right] \tag{4.51}$$

Thermal strains:

$$\varepsilon_{xx} = \varepsilon_{yy} = \frac{1}{E}\left(\frac{N_T}{2h} + \frac{3z}{2h^3}M_T\right) \tag{4.52a}$$

$$\varepsilon_{zz} = -\frac{2\nu}{(1-\nu)E}\left(\frac{N_T}{2h} + \frac{3z}{2h^3}M_T\right) + \frac{1+\nu}{1-\nu}\alpha T(z) \tag{4.52b}$$

$$\varepsilon_{xy} = \varepsilon_{yz} = \varepsilon_{zx} = 0$$

Displacement components:

$$u = \frac{x}{E}\left(\frac{N_T}{2h} + \frac{3z}{2h^3}M_T\right) \tag{4.53a}$$

along the x axis,

$$v = \frac{y}{E}\left(\frac{N_T}{2h} + \frac{3z}{2h^3}M_T\right) \tag{4.53b}$$

along the y axis, and

$$w = -\frac{3\,M_T}{4h^3\,E}(x^2 + y^2) + \frac{1}{(1-\nu)E}\left[(1+\nu)\alpha E \int_0^z T(z)\,dz - \frac{\nu z}{h}N_T - \frac{3\nu z^2}{2h^3}M_T\right] \quad [4.53c]$$

along the z axis.

The normal thermal force N_T and thermal moment M_T are expressed in terms of the temperature function $T(z)$ as follows:

$$N_T = \alpha E \int_{-h}^{h} T(z)\,dz \quad [4.54a]$$

$$M_T = \alpha E \int_{-h}^{h} T(z) z\,dz \quad [4.54b]$$

in which α is the coefficient of thermal expansion, E is the Young's modulus, and ν is the Poisson's ratio of the material.

Thermal Stresses in Beams Due to Temperature Variation through the Depth This situation is illustrated in Figure 4.39. The beam has a cross-sectional area of $A = 2bh$ and the corresponding area moment of inertia $I = 2h^3b/3$.

Figure 4.39 | Thermal stresses in beam.

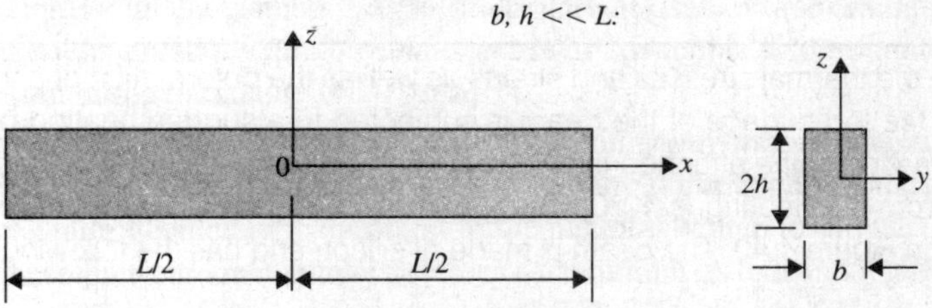

The bending stress $\sigma_{xx}(x, z)$ is the dominant stress component to be concerned with. It can be expressed as:

$$\sigma_{xx}(x, z) = -\alpha E T(z) + \frac{b\,N_T}{A} + \frac{z(b\,M_T)}{I} \quad [4.55]$$

with $\sigma_{zz} = \sigma_{xz} = 0$.

The associated strain components are:

$$\varepsilon_{xx}(x, z) = \frac{1}{E}\left[\frac{b\,N_T}{A} + \frac{z}{I}(b\,M_T)\right] \quad [4.56a]$$

$$\varepsilon_{zz}(x, z) = -\frac{\nu}{E}\left[\frac{b\,N_T}{A} + \frac{z}{I}(b\,M_T)\right] + \left(\frac{1+\nu}{E}\right)\alpha T(z) \quad [4.56b]$$

with $\varepsilon_{xz} = 0$.

The displacement components are:

$$u(x, z) = \frac{x}{E}\left[\frac{b\,N_T}{A} + \frac{z}{I}(b\,M_T)\right] \quad \textbf{[4.57a]}$$

along the x direction, and

$$w(x, z) = -\frac{b\,M_T}{2EI}x^2 - \frac{\nu}{E}\left[\frac{b\,N_T}{A}z + \frac{z^2}{2I}(b\,M_T)\right] + \alpha\left(\frac{1+\nu}{E}\right)\int_0^z T(z)\,dz \quad \textbf{[4.57b]}$$

along the z direction.

The normal force and the bending moment due to thermal force, N_T and M_T, are expressed in Equations (4.54a) and (4.54b), respectively.

The curvature of the bent beam can be determined by the following expression:

$$1/\rho \approx -\frac{b\,M_T}{EI} \quad \textbf{[4.58]}$$

in which ρ is the radius of curvature of the bent beam.

EXAMPLE 4.15

Determine the thermal stresses and strains as well as the deformation of a thin beam 1 μs after the top surface of the beam is subjected to a sudden heating by the resistance heating of the attached thin copper film. The temperature at the top surface resulting from the heating is 40°C. The geometry and dimensions of the beam are illustrated in Figure 4.40. The beam is made of silicon and has the following material properties:

Mass density ρ = 2.3 g/cm^3
Specific heat c = 0.7 J/g-°C
Thermal conductivity k = 1.57 w/cm-°C (or J/cm-°C-s)
Coefficient of thermal expansion α = 2.33 × 10^{-6}/°C
Young's modulus E = 190,000 × 10^{6} N/m^2
Poisson's ratio ν = 0.25

Figure 4.40 | A silicon beam subject to heating at top surface.

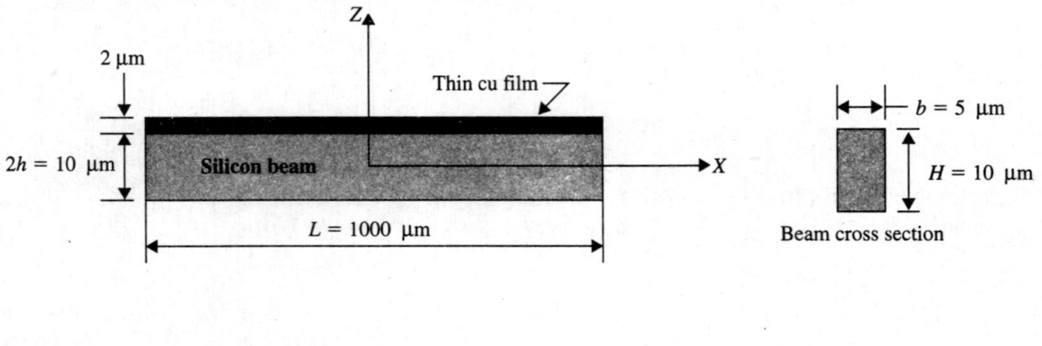

Solution

Thermal stresses, strains, and displacements induced in the beam by the sudden heating of the top surface can be computed by Equations (4.55), (4.56), and (4.57). However, as in almost all thermal stress analyses, one needs to determine the temperature distribution in the structure first. In the present case, we need to obtain the temperature variation across the depth of the beam, i.e., $T(z)$ at a given time t, as required in these equations.

The temperature $T(z, t)$ in the beam can be obtained either from a heat conduction analysis with appropriate boundary conditions, as will be described in Chapter 5, or from the solution of a similar but approximate case involving the heating of a half-infinite space, as illustrated in Figure 4.41.

Figure 4.41 | Temperature in half-space.

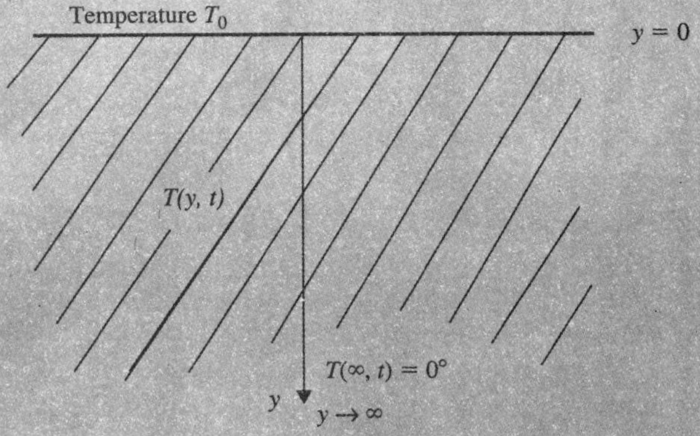

As will be described in Chapter 5, the situation in Figure 4.41 can be mathematically modeled with a partial differential equation and appropriate initial and boundary conditions as follows:

$$\frac{\partial^2 T(y, t)}{\partial y^2} = \frac{1}{\alpha}\frac{\partial T(y, t)}{\partial t} \quad \textbf{[4.59]}$$

The initial condition is $T(y, 0) = 0$, and the boundary conditions are $T(0, t) = T_0$ and $T(\infty, t) = 0$.

The corresponding solution has the form:

$$T(y, t) = T_0 \operatorname{erfc}\left(\frac{y}{2\sqrt{\alpha t}}\right) \quad \textbf{[4.60]}$$

in which α is the thermal diffusivity of the solid.

The corresponding thermal diffusivity of the silicon beam for the present case can be computed as follows:

$$\alpha = \frac{k}{\rho c} = \frac{1.57(\text{J/cm-°C-s})}{2.3(\text{g/cm}^3) \times 0.7(\text{J/g-°C})} = 0.9752 \text{ cm}^2\text{/s}$$

or $\alpha = 97.52 \times 10^{-6}$ m^2/s for silicon.

The function erfc (x) in Equation (4.60) is the complementary error function as described in Chapter 3.

Thus, by substituting the relevant parameters into Equation (4.60), we can obtain the temperature distribution along the depth of the beam at 1 μs after its top surface is heated at 40°C as:

$$T(y, 1\ \mu s) = 40\ \text{erfc}\left(\frac{y}{2\sqrt{(97.52 \times 10^{-6})(10^{-6})}}\right) = 40\ \text{erfc}\left(\frac{10^6 y}{19.75}\right) \qquad \textbf{[a]}$$

The application of the above solution to the beam as illustrated in Figure 4.40 requires a transformation of coordinates from the y coordinate in Figure 4.41 to the z coordinate in Figure 4.40 by letting $z = -y + h$, or $y = h - z$, with $h = 5\ \mu m$.

The solution in Equation (a) after the transformation takes the form:

$$T(z, 1\ \mu s) = 40\ \text{erfc}\left[\frac{10^6 (5 \times 10^{-6} - z)}{19.75}\right] \qquad \textbf{[b]}$$

By substituting the above expression for $T(z)$ into Equation (4.55), we can evaluate the thermal stress at 1 μs, and also the corresponding strains and displacements from Equations (4.56) and (4.57), respectively.

■ Alternative Numerical Solution

The determination of the thermal force N_T in Equation (4.54a) and the bending moment M_T in Equation (4.54b) for the present case will require the integration of the function $T(z, 1\ \mu s)$ in Equation (b), which involves the complementary error function erfc (z). This task is manageable with mathematical solution software such as MathCad or MatLAB. An alternative approach is to derive a polynomial function that can approximate the temperature distribution as expressed in Equation (b). This polynomial function can then be used to obtain the numerical solution for the problem on hand.

We can obtain the numerical solution of the temperature distribution along the depth of the beam by evaluating the complementary error function in Equation (b) by using the numerical values of error functions in Table 3.4. The tabulated value of $T(z)$ at $t = 1\ \mu s$ is given below:

y in Equation (a)	**$z \times 10^{-6}$ m in Equation (b)**	**erfc $\left[\frac{10^6 (5 \times 10^{-6} - z)}{19.75}\right]$**	**$T(z)$, °C at 1 μs**
0	5	0	40.00
1	4	0.0506	37.72
2	3	0.1013	35.44
3	2	0.1519	33.20
4	1	0.2025	30.30
5	0	0.2532	28.81
6	−1	0.3038	26.69
7	−2	0.3544	24.63
8	−3	0.4051	22.64
9	−4	0.4557	20.74
10	−5	0.5063	18.92

We will find that a linear polynomial function in the following form fits the temperature distribution in Equation (b) reasonably well:

$$T(z) = 2.1 \times 10^6 z + 28.8 \qquad \textbf{[c]}$$

where z is the coordinate shown in Figure 4.40 in units of micrometers.

We can thus determine the normal force and the bending moment by using $T(z)$ from Equation (c) in Equations (4.54a) and (4.54b) as follows:

$$N_T = \alpha E \int_{-h}^{h} T(z)\, dz$$
$$= (2.33 \times 10^{-6})(190{,}000 \times 10^6) \int_{-5\times10^{-6}}^{5\times10^{-6}} (2.1 \times 10^6 z + 28.8)\, dz$$
$$= 127.5 \text{ N}$$

$$M_T = \alpha E \int_{-h}^{h} T(z)\, z\, dz$$
$$= (2.33 \times 10^{-6})(190{,}000 \times 10^6) \int_{-5\times10^{-6}}^{5\times10^{-6}} (2.1 \times 10^6 z + 28.8) z\, dz$$
$$= 77.4725 \times 10^{-6} \text{ N-m}$$

At this stage, we need to find both the cross-sectional area of the beam, A, and the area moment of inertia, I, which are

$$A = (5 \times 10^{-6})(10 \times 10^{-6}) = 5 \times 10^{-11} \text{ m}^2$$

$$I = \frac{bH^3}{12} = \frac{1}{12}(5 \times 10^{-6})(10 \times 10^{-6})^3 = 4.167 \times 10^{-22} \text{ m}^4$$

We are now ready to compute the induced thermal stresses, strains and displacements by the temperature distribution in Equation (c):

From Equation (4.55):

$$\sigma_{xx}(z, 1\ \mu\text{s}) = -(2.33 \times 10^{-6})(190{,}000 \times 10^6)(2.1 \times 10^6 z + 28.8)$$
$$+ \frac{(5 \times 10^{-6})127.5}{5 \times 10^{-11}} + \frac{z(5 \times 10^{-6})(77.4725 \times 10^{-6})}{4.167 \times 10^{-22}}$$
$$= -4.427 \times 10^5(2.1 \times 10^6 z + 28.8) + 127.5 \times 10^5$$
$$+ 92.95 \times 10^{10} z \quad \text{Pa}$$

We will have the maximum bending stress $\sigma_{xx,\,max}$ 1 μs after the heating by substituting $z = 5 \times 10^{-6}$ m into the above expression. We will obtain:

$$\sigma_{xx,\,max} = -500 \text{ Pa in compression}$$

The associated thermal strain components can be evaluated from Equations (4.56a) and (4.56b) as follows:

$$\varepsilon_{xx}(z) = \frac{1}{190{,}000 \times 10^6}\left[\frac{(5 \times 10^{-6})127.5}{5 \times 10^{-11}} + \frac{z(5 \times 10^{-6} \times 77.4725 \times 10^{-6})}{4.167 \times 10^{-22}}\right]$$

$$= 67.11 \times 10^{-6}(1 + 0.73 \times 10^5 z)$$

$$\varepsilon_{zz}(z) = 0.25(-67.11 \times 10^{-6})(1 + 0.73 \times 10^5 z)$$

$$+ \frac{1 + 0.25}{190{,}000 \times 10^6}(2.33 \times 10^{-6})(2.1 \times 10^6 z + 28.8)$$

$$= (-16.78 \times 10^{-6} - 1.23z) + (3.22 \times 10^{-11} z + 44.15 \times 10^{-17})$$

The maximum strains occur at the top surface. Numerical values of these strains can be obtained by letting $z = 5 \times 10^{-6}$ m in the above expressions:

$$\varepsilon_{xx,\,max} = \varepsilon_{xx}(5 \times 10^{-6}) = 91.61 \times 10^{-6} = 0.0092\%$$

$$\varepsilon_{zz,\,max} = \varepsilon_{zz}(5 \times 10^{-6}) = -22.93 \times 10^{-6} = -0.0023\%$$

The displacements at the top free corners with $x = \pm 500 \times 10^{-6}$ m and $z = 5 \times 10^{-6}$ m can be computed from Equations (4.57a) and (4.57b):

$$u = \frac{500 \times 10^{-6}}{190{,}000 \times 10^6}\left[\frac{(5 \times 10^{-6})127.5}{5 \times 10^{-11}} + \frac{(5 \times 10^{-6})(5 \times 10^{-6})(77.47 \times 10^{-6})}{4.167 \times 10^{-22}}\right]$$

$$= 0.046\ \mu\text{m}$$

$$w = -\frac{(5 \times 10^{-6})(77.47 \times 10^{-6})(500 \times 10^{-6})^2}{2(190{,}000 \times 10^6)(4.167 \times 10^{-22})}$$

$$- \frac{0.25}{190{,}000 \times 10^6}$$

$$\left[\frac{(5 \times 10^{-6})127.5}{5 \times 10^{-11}}(5 \times 10^{-6}) + \frac{(5 \times 10^{-6})^2(5 \times 10^{-6})(77.47 \times 10^{-6})}{2(4.167 \times 10^{-22})}\right]$$

$$+ \frac{1 + 0.25}{190{,}000 \times 10^6}(2.33 \times 10^{-6}) \int_0^{5 \times 10^{-6}} (2.1 \times 10^6 z + 28.8)z\,dz$$

$$= -0.612\ \mu\text{m}$$

The curvature of the bent beam is estimated from Equation (4.58) to be:

$$\frac{1}{\rho} = -\frac{(5 \times 10^{-6})(77.47 \times 10^{-6})}{(190{,}000 \times 10^6)(4.167 \times 10^{-22})} = -4.892\ \text{m}^{-1}$$

4.5 FRACTURE MECHANICS

Many MEMS devices involve the binding of thin films of distinct materials as illustrated in Chapter 2. These different materials with distinct properties are bonded together by various means, as will be described in Chapter 8 on microfabrication technology and Chapter 11 on the packaging of microsystems. These bonding techniques include chemical and physical depositions, thermal diffusion, soldering, adhesions, etc. Interfaces of bonded structures are likely places where structural failure will take place. Fracture mechanics is a way to assess the structural integrity of these interfaces.

Fracture of bonded interfaces can occur when excessive forces act normal to the interfaces, and/or when shear forces act on the interfaces. Some of the linear elastic fracture mechanics (LEFM) principles can be applied in the design analysis in order to mitigate possible delamination of interfaces in MEMS and microsystems.

LEFM is an engineering discipline on its own. It is not possible to cover this subject in much detail in this section. What we will learn here is the basic concept and the formulation of LEFM that may be used in the microstructure design.

4.5.1 Stress Intensity Factors

Let us consider a small crack existing in a solid subjected to a stress field as illustrated in Figure 4.42. The stress field has a tendency to deform the crack in three distinct modes. These deformation modes, illustrated in Figure 4.43, are Mode I: the opening mode, Mode II: the edge sliding mode, and Mode III: the tearing mode.

Figure 4.42 | A crack in a solid subjected to a stress field.

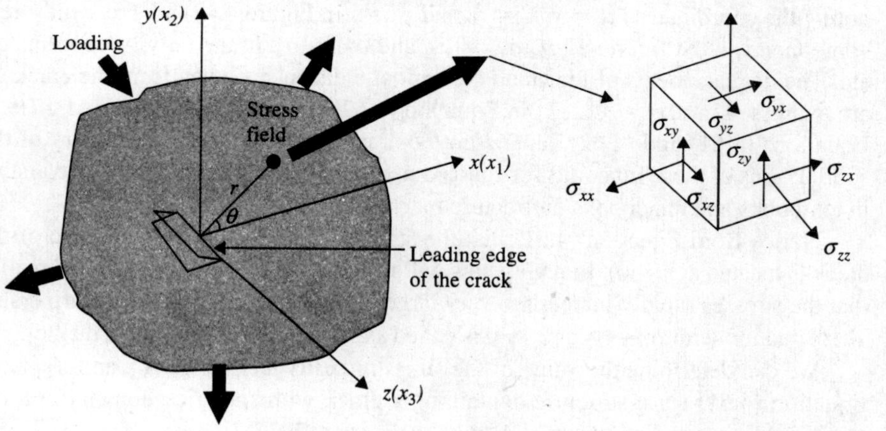

Figure 4.43 | Three modes of fracture.

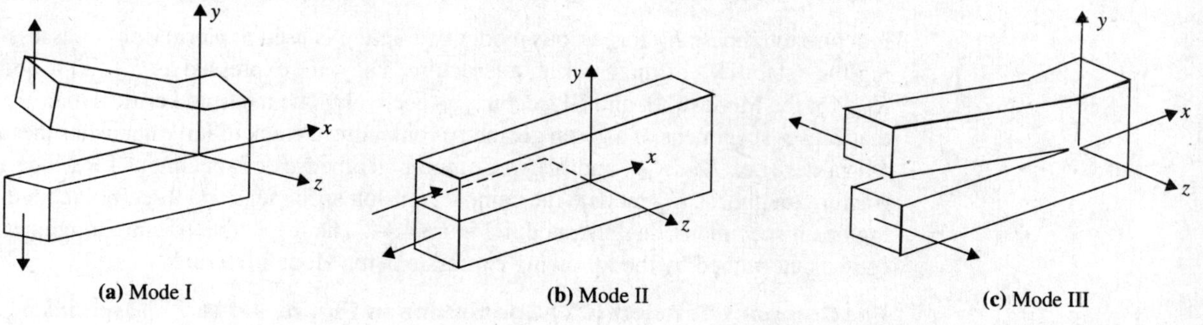

The *opening mode* (Mode I) is associated with local displacements in which the crack surfaces tend to move apart in a direction perpendicular to these surfaces (symmetric with respect to the x–y and z–x planes). The *edge-sliding mode* (Mode II) is characterized by displacements in which the crack surfaces slide over one another and remain perpendicular to the leading edge of the crack (symmetric with respect to the x–y plane and skew-symmetric with respect to the x–z plane). The *tearing mode* (Mode III) is defined by the crack surfaces sliding with respect to one another parallel to the leading edge of the crack (skew-symmetric with respect to the x–y and x–z planes).

The *stress intensity factor K* is introduced in LEFM to assess the stress field surrounding a crack. By referring to the situation in Figure 4.42, the stress and displacement components around a crack can be expressed as:

$$\sigma_{ij} = \frac{K_{\mathrm{I}} \quad \text{or} \quad K_{\mathrm{II}} \quad \text{or} \quad K_{\mathrm{III}}}{\sqrt{r}} f_{ij}(\theta) \qquad \textbf{[4.61]}$$

$$u_i = K_{\mathrm{I}} \quad \text{or} \quad K_{\mathrm{II}} \quad \text{or} \quad K_{\mathrm{III}}\sqrt{r}\, g_i(\theta) \qquad \textbf{[4.62]}$$

where K_{I}, K_{II}, and K_{III} are the stress intensity factors of Mode I, II, and III fracture, respectively.

The stress components σ_{ij} in Equation (4.61), with subscripts $i, j = 1, 2, 3$ denoting the coordinates ($x_1 = x, x_2 = y$, and $x_3 = z$ in Figure 4.42) are the same stress components in the figure. Thus, $\sigma_{11} = \sigma_{xx}$ and $\sigma_{23} = \sigma_{yz}$ in the (x, y, z) coordinates, etc. The displacements of the small element at a distance r away from the crack tip are expressed in u_i ($i = 1, 2, 3$) in Equation (4.62). The functions $f_{ij}(\theta)$ and $g_i(\theta)$ in Equations (4.61) and (4.62) depend on the crack length c and the geometry of the solid. Forms of these functions for solid structures of different geometry are available in textbooks and handbooks on fracture mechanics.

Notice from Equation (4.61) that the stresses $\sigma_{ij} \to \infty$ as $r \to 0$ (the tip of the crack is located at $r = 0$). In reality, this will not happen. However, we should realize that the stresses rapidly increase to very large numbers near the tip of a sharp crack. This situation with $\sigma_{ij} \to \infty$ as $r \to 0$ is called *stress singularity* near the crack tip.

We can determine the value of the *stress intensity factors* K_{I}, K_{II}, and K_{III} from Equation (4.61) for a structure containing a crack with specific geometry and the stress field surrounding the crack using the theory of LEFM.

4.5.2 Fracture Toughness

Fracture toughness K_C for various modes of fracture is used as a criterion for assessing the "stability" of the crack in a structure. They are expressed as $K_{\mathrm{I}C}$, $K_{\mathrm{II}C}$, and $K_{\mathrm{III}C}$ for the Modes I, II, and III fracture, respectively. The measured critical load, P_{cr} that causes specimens of a given geometry containing a crack to fail establishes these K_C values (i.e., $K_{\mathrm{I}C}$, $K_{\mathrm{II}C}$, and $K_{\mathrm{III}C}$). Common geometries of specimens for Mode I fracture toughness $K_{\mathrm{I}C}$ include the compact tension specimens and three-point bending beam specimens, as illustrated in Figure 4.44. The respective fracture toughness can be determined by the following expressions for Mode I fracture:

For Compact Tension (CT) Specimens in Figure 4.44a The specimen is subjected to a tension to open the crack. A notch is machined at midheight in the

Figure 4.44 | Specimens used to determine Mode I fracture toughness.

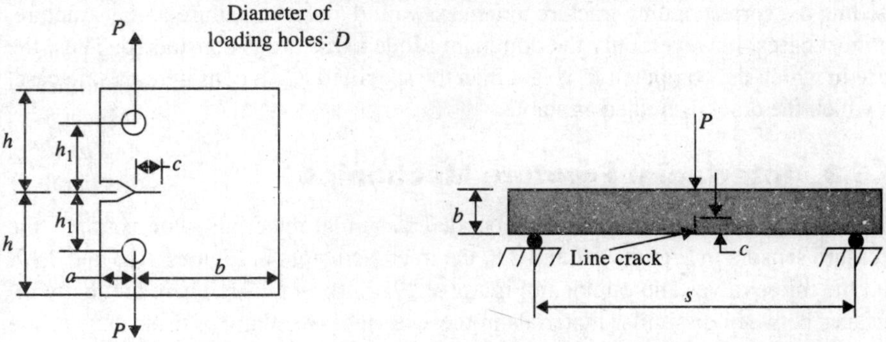

(a) Compact tension specimen (b) Three-point bending specimen

specimen to facilitate the fracture. A minute crack with the length c is created at the tip of the notch by fatigue load of low magnitude. This minute crack initiates its growth and subsequent unstable propagation under the applied loading. Fracture toughness of the CT specimen material can be determined by the following formula:

$$K_{IC} = \sigma_C \sqrt{\pi c}\, F(c/b) \qquad \textbf{[4.63]}$$

in which c is the crack length and $\sigma_C = P_{cr}/A$, with P_{cr} = the applied load P that causes the specimen to fracture with crack propagation. A = cross section of the bulk specimen.

The function $F(c/b)$ can be found in fracture mechanics handbooks. An approximation of $F(c/b) = 1$ is used if such reference is not readily available. This function value is for the case that the specimen is subjected to uniform tensile load.

Compact tension specimens used to determine fracture toughness need to satisfy the following specification for the geometry, with reference to Figure 4.44a:

$$h = 0.6b \qquad h_1 = 0.275b \qquad D = 0.25b \qquad a = 0.25b$$

The thickness of the specimen $T = b/2$

For Three-Point Bending Specimens, Figure 4.44b The expression in Equation (4.63) can be used to determine the fracture toughness of the specimen material, except that the function $F(c/b)$ takes the following form:

$$F(c/b) = 1.09 - 1.735(c/b) + 8.2(c/b)^2 - 14.18(c/b)^3 + 14.57(c/b)^4 \qquad \textbf{[4.64a]}$$

for the case $s/b \leq 4$, and

$$F(c/b) = 1.107 - 2.12(c/b) + 7.71(c/b)^2 - 13.55(c/b)^3 + 14.25(c/b)^4 \qquad \textbf{[4.64b]}$$

for $s/b \leq 8$.

Fracture toughness so determined from the measurements of the critical loading P_{cr} on specific specimens is used as a design criterion similar to the allowable stress used in a conventional machine design. In practice, the engineer will conduct a stress analysis to determine the stress field around a hypothetical crack of certain size and orientation in the structure. Equation (4.61) is used to determine the stress intensity factors K_I, K_{II}, and K_{III}. The stress intensity factors so determined are compared with the corresponding fracture toughness K_{IC}, K_{IIC}, and K_{IIIC}. Computed K values exceeding the corresponding fracture toughness would mean the failure of the structure. In most cases, however, only the dominant Mode I fracture is considered. Thus, the case in which the computed K_I is less than the specified K_{IC} is considered a safe case, in which the crack is treated as stable.

4.5.3 Interfacial Fracture Mechanics

Many MEMS devices are made with bonded dissimilar materials—for example, the pressure sensors in Figures 2.7 and 2.8, the microactuators in Figures 2.16 and 2.17, and the microvalves and pumps in Figures 2.29, 2.30, and 2.31. There are many interfaces between dissimilar materials in these devices. As many of these devices are expected to operate millions of working cycles, the strength of these bonded interfaces has become a serious concern to the design engineer.

The analytical assessment of fracture strength of interfaces is much more complicated than that for single-mode fracture as presented in the foregoing subsections. A major complication is that most of these interfaces are subject to simultaneous Mode I and II (and III in some cases) loading. Mixed-mode fracture analysis is significantly different from what was established for single mode fracture analysis. Here again, we will only outline the general approach to the solution of this type of problem.

Let us refer to the interface of a bonded bimaterial structure as illustrated in Figure 4.45. The two bonded materials have distinct properties such as Young's moduli, E_1 and E_2 and Poisson's ratios ν_1 and ν_2. The interface is subjected to simultaneous loading in both the normal and lateral shearing directions with respective stress fields of σ_{yy} and σ_{xy}. These stress fields make the interface vulnerable to coupled opening mode (Mode I) and the shearing mode (Mode II) fracture.

Figure 4.45 | An interface of a bimaterial structure.

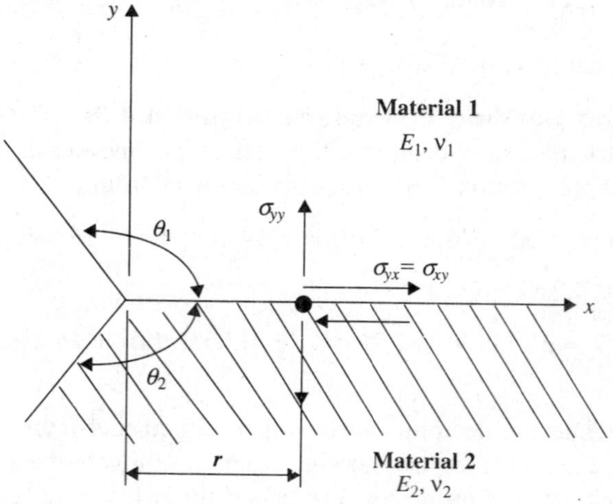

The stresses, σ_{ij} with $i, j = x, y$ near the interfacial crack can be expressed by the following equation, which is similar to Equation (4.61):

$$\sigma_{ij} = \frac{K_{\mathrm{I}} \text{ or } K_{\mathrm{II}}}{r^{\lambda}} + L_{ij} \ln(r) + \text{terms}$$

where K_{I} and K_{II} are the respective stress intensity factors for Mode I and Mode II fracture, λ is a *singularity parameter*, and L_{ij} are constants.

For the region that is close to the tip of the interface (i.e., the distance r is very small), the contribution of the term ln (r) to the stress field can be neglected. We thus have the stresses near the tip of the interface where the stress singularity dominates the stress distribution as:

$$\sigma_{ij} = \frac{K_{\mathrm{I}} \text{ or } K_{\mathrm{II}}}{r^{\lambda}} \quad [4.65]$$

from which we can express the stress components along the interface as:

$$\sigma_{yy} = \frac{K_{\mathrm{I}}}{r^{\lambda_{\mathrm{I}}}} \quad [4.65a]$$

and

$$\sigma_{xy} = \frac{K_{\mathrm{II}}}{r^{\lambda_{\mathrm{II}}}} \quad [4.65b]$$

in which λ_{I} and λ_{II} are, respectively, singularity parameters for the materials in Mode I and Mode II fracture.

The stress intensity factors K_{I} and K_{II} in Equations (4.65a) and (4.65b) can be obtained by linearizing both these equations in logarithm scales as:

$$\ln(\sigma_{yy}) = -\lambda_{\mathrm{I}} \ln(r) + \ln(K_{\mathrm{I}}) \quad [4.66a]$$

$$\ln(\sigma_{xy}) = -\lambda_{\mathrm{II}} \ln(r) + \ln(K_{\mathrm{II}}) \quad [4.66b]$$

Equations (4.66a) and (4.66b) are expressed respectively in graphical form in Figure 4.46a and 4.46b. In Figure 4.46, r_0 represents a short range from the crack tip, in which the stresses vary linearly, whereas r_e is the closest distance away from the crack tip with which numerical values of stresses are available from a finite element analysis.

Figure 4.46 | Graphical representation of stress intensities in Mode I and II.

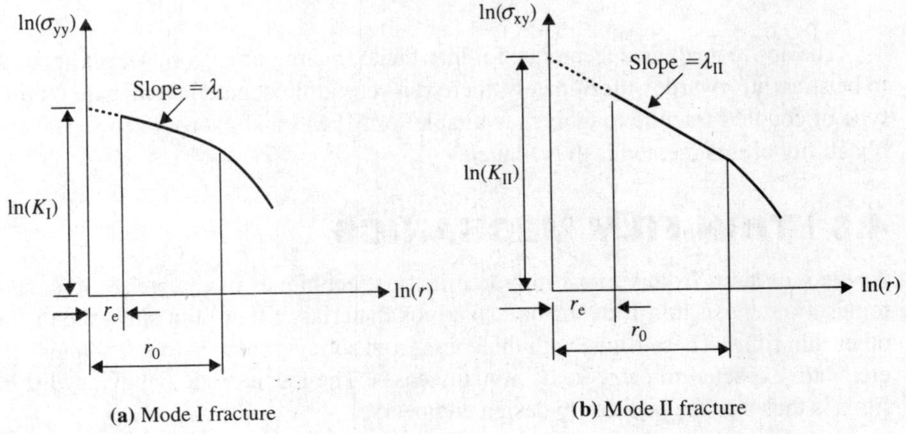

For a typical interfacial fracture analysis, the stress distributions in both materials along the interface are determined either by the classical method or by a finite element analysis. These stress distributions in both mating materials closest to the interface are then plotted in natural logarithm scales as shown in Figure 4.46. The stress intensity factors K_I and K_{II} of the materials near the interface can be obtained from the intercepting points at the y axes in the respective $\ln(\sigma_{yy})$ versus $\ln(r)$ and $\ln(\sigma_{xy})$ versus $\ln(r)$ plots. The slopes of the straight-line portion of the plots represent the singularity parameters λ_I and λ_{II} for both modes of fracture.

The stress intensity factors K_I and K_{II} obtained for the mating materials from the above procedure will enter the following mixed-mode fracture criterion in the form of:

$$\left(\frac{K_I}{K_{IC}}\right)^2 + \left(\frac{K_{II}}{K_{IIC}}\right)^2 = 1 \qquad [4.67]$$

in which the fracture toughness, K_{IC} and K_{IIC}, is determined by experimental methods for the same mating materials. A database for coupled Mode I and II fracture toughness for a limited number of microelectronics materials was reported [Nguyen et al. 1997]. Equation (4.67) can be represented graphically by a quarter-ellipse as illustrated in Figure 4.47. An interface of two materials is considered "safe" if the K_I and K_{II} computed by the above procedure for both materials are located within the elliptical envelope.

Figure 4.47 | Fracture criterion for interfaces.

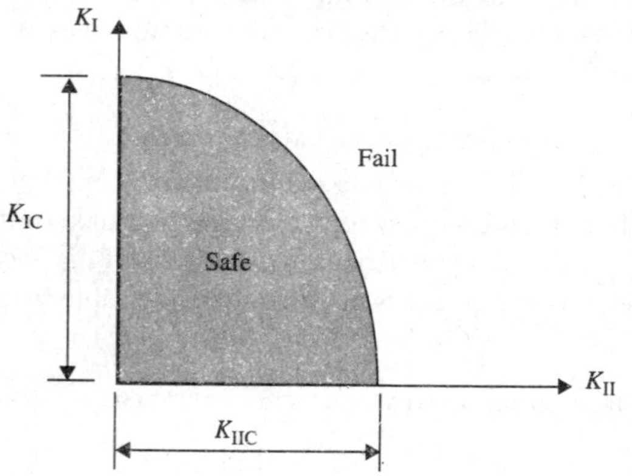

The above analytical technique for interfacial fracture of layer materials appears to be straightforward. Unfortunately, there is a very limited material database on this type of coupled fracture toughness available for MEMS and microsystems. The applicability of this method is thus limited.

4.6 | THIN-FILM MECHANICS

As we will learn from various microfabrication techniques in Chapter 8, it is customary to deposit thin films made of various materials either on a substrate or on other thin films. These films, with thickness from submicrometers to a few micrometers, are expected to carry loads in many cases. The mechanistic behavior of thin films is thus a prime concern to design engineers.

Quantitative assessment of stresses in thin films is not available for design analysis at this time for the following reasons:

1. These films are so thin that the atomic bonding forces (such as the van der Waals forces as described in Chapter 3) become an important factor in the material's strength in that direction. Credible methods for quantifying these forces are not currently available.
2. Almost all deposition techniques used in producing thin films result in inherent *residual stresses* in the film after the fabrication. These stresses, also termed *intrinsic stresses,* are induced either by the difference in coefficients of thermal expansion of the materials in the case of multilayered thin films, or by the reaggregation of crystalline grains of thin-film materials during the deposition process. Residual stresses in thin films are a major problem in MEMS design. A finite element analysis on an oxidation process for a 4-μm thin silicon oxide film over a silicon substrate at 1000°C indicated a compressive residual stress as high as 2000 MPa in the oxide film [Hsu and Sun 1998]. This falls in the range of 10 to 5000 MPa residual stresses in thin films according to Madou [1997]. The same reference indicates that deposition of thin metal films on silicon substrates result in residual stresses that are in the range of 10 to 100 MPa, but in tension. These tensile stresses can cause cracking of the thin metal films.
3. Distribution of the material in the thin films is not likely to be uniform across the area on which they are deposited. The use of the continuum mechanics theory using average material behavior in assessing the stresses in thin films is thus not realistic.

The total stresses σ_T in a thin film over a thick substrate can be expressed as follows [Madou 1997]:

$$\sigma_T = \sigma_{\text{th}} + \sigma_m + \sigma_{\text{int}} \qquad [4.68]$$

where

σ_{th} = thermal stress induced by the operating temperature as described in Section 4.4
σ_m = stresses induced by the applied mechanical loads
σ_{int} = the intrinsic stresses

Numerical evaluation of the first two stress components in Equation (4.68) often requires the use of finite element codes, whereas the assessment of the intrinsic stresses is usually conducted with empirical means developed by relevant measurements as well as by the engineer's own experience.

4.7 | OVERVIEW OF FINITE ELEMENT STRESS ANALYSIS

Almost all MEMS and microsystems involve complex three-dimensional geometry. These devices are fabricated under severe thermal and mechanical conditions and subject to harsh loading during operations. Furthermore, these structures often consist of multilayers of dissimilar thin films bonded together by either physical or chemical means. Credible stress analysis of these complex systems cannot be achieved by conventional methods with closed-form solutions as presented in this chapter. The finite element method is the only viable way to attain credible solutions. Following is a brief introduction of this powerful method. Theories and formulation are deliberately omitted. They can be found in a number of selected references such as Hsu [1986], Hsu [1992], and Desai [1979]. The fundamental principles of deriving the key *element equations* will be presented in Chapter 10. What the reader will learn from the following is the essential input/output information that is required in using commercial finite element codes such as the popular ANSYS® code used by the MEMS industry.

4.7.1 The Principle

In essence, the principle of the finite element method (FEM) is "divide and conquer". The very first step in a finite element analysis is to divide the whole solid structure made of continua into a *finite* number of subdivisions of special shapes (called the *elements*) interconnected at the corners or specific points on the edges of the elements (called the *nodes*). A variety of geometry of elements is usually available from the element library offered by most commercial codes.

Once the structure is subdivided into a great many elements (a process called *discretization*), engineering analysis is performed on these elements instead of the whole structure. Solutions obtained at the element level are assembled to get the corresponding solutions for the whole structure.

Credible results from a finite element analysis (FEA) are attainable with intelligent discretization of the structure. Two principal rules need to be followed:

1. Place denser and smaller elements in the parts of the structure with an abrupt change of geometry where high stress or strain concentrations are expected.
2. Avoid using elements with high aspect ratio, which is defined as the ratio of the longest dimension to the shortest one in the same element. The user is advised to keep this aspect ratio below 10.

Figure 4.48 shows the discretization of the silicon die/diaphragm/constraint base of a micro pressure sensor [Schulze 1998]. Only a quarter of the silicon die is included in the finite element model because of the quarter plane symmetry of the die/diaphragm/constraint base assembly.

4.7.2 Engineering Applications

FEM is used in almost every engineering discipline. Particularly well-developed common applications are in the following three areas, which happen to be necessary for microsystems design.

1. *Stress/strain analysis of solid structures.* FEM will provide solutions in stresses (σ), strains (ε), and displacements (U) for each element in a discretized model. The displacements are the primary unknown quantities to be solved in this type of analysis.

Figure 4.48 | A finite-element model for the silicon die/constraint base assembly of a micro pressure sensor.

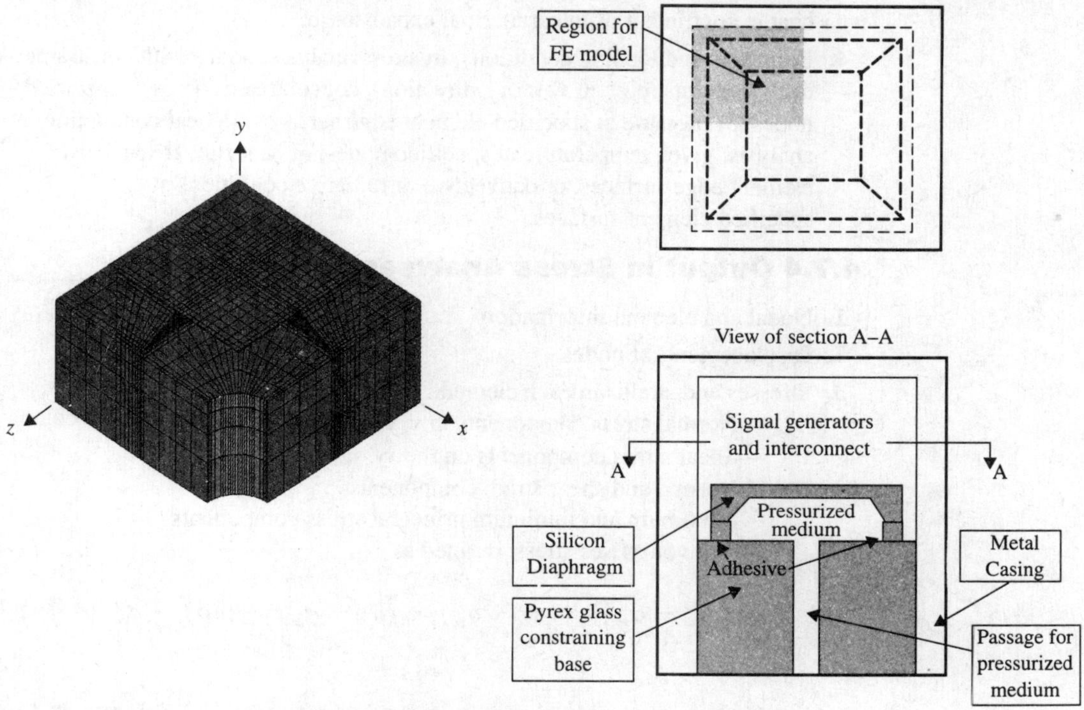

2. *Heat conduction analysis.* It can handle all modes of heat transfer including conduction, convection, and radiation. Various boundary conditions such as heat flux (q) and convective/radiative types can be handled. Expected results include temperature in the elements or at the nodes.
3. *Fluid dynamics.* The FEM codes for fluid dynamics are built on the basis of Navier-Stokes equations, as will be presented in Chapter 5. Numerical solutions will include mass flow, velocities, and driving pressures in selected control volumes of the flowing fluid.

4.7.3 Input Information to FEA

1. General information:
   a. Profile of the structure geometry
   b. Establish the coordinates: x–y for plane; r–z for axisymmetrical; x–y–z for three-dimensional geometry
2. Develop FE mesh (e.g. in Fig. 4.48). For automatic mesh generation, the user usually specifies the desirable density of nodes and elements in specific regions. Information generated for the FE analysis includes: node number, nodal coordinates, nodal conditions (e.g., constraints, applied forces), element number, element description (e.g., involved nodes)
3. Material property input: In stress analysis: Young's modulus E; Poisson ratio ν; shear modulus of elasticity, G; yield strength, σ_y; ultimate tensile strength σ_u.

In heat conduction analysis: mass density ρ; thermal conductivity k; specific heat c; coefficient of linear thermal expansion α.

4. Boundary and loading conditions. In stress analysis: nodes with constrained displacements (e.g., in x, y, or z direction), concentrated forces at specified nodes, or pressure at specified element edge surfaces. In heat conduction analysis: given temperature at specified nodes, or heat flux at specified element edge surfaces, or convective or radiative conditions at specified element surfaces.

4.7.4 Output in Stress Analyses from FEA

1. Nodal and element information
2. Displacements at nodes
3. Stresses and strains in each element:
   —Normal stress components in x, y, and z directions
   —Shear stress components on the xy, xz, and yz planes
   —Normal and shear strain components
   —Maximum and minimum principal stress components
   —The von Mises stress, defined as

$$\bar{\sigma} = \frac{1}{\sqrt{2}}[(\sigma_{xx} - \sigma_{yy})^2 + (\sigma_{xx} - \sigma_{zz})^2 + (\sigma_{yy} - \sigma_{zz})^2 + 6(\sigma_{xy}^2 + \sigma_{yz}^2 + \sigma_{xz}^2)]^{1/2} \tag{4.69}$$

in which σ_{xx}, σ_{yy}, and σ_{zz} are respective stress components along the x, y, and z axes and σ_{xy}, σ_{yz}, and σ_{xz} are the shear stress components in the element as defined in Figure 4.37. The von Mises stress is used as the "representative" stress in a multiaxial stress situation. It is compared with the yield strength σ_y for plastic yielding, and to σ_u for the prediction of the rupture of the structure.

4.7.5 Graphical Output

All modern commercial finite element analysis codes offer graphical output in the form of:

- A solid model of the structure such as shown in Figure 10.25 for a microgripper
- User input discretized solid structure with a finite element mesh such as shown in Figure 4.48 for a micro pressure sensor die
- A deformed or/and undeformed solid model of the structure
- Color-coded zones to indicate the distribution of stresses, strains, and displacements over the selected plane of the structure
- Color-coded zones to indicate other required output quantities such as temperature or pressure in the structure
- Animated movements or deformation of the structure under specified operating conditions

4.7.6 General Remarks

The finite element method (FEM) is widely used by the MEMS and microsystem industry for design analysis and simulation of the systems. Indeed, because of the inherent complex geometry, loading, and boundary conditions of MEMS and microsystems, FEM appears to be the only viable tool for such purposes. However, there are several facts that users need to be aware of. Following are a few specific ones that are relevant to MEMS and microsystems design analyses and simulations:

1. Engineers need to be aware of the fact that FEM offers only approximate solutions to whatever problem is being analyzed. Credible results are obtainable by the intelligent use of the finite element code by the user. As mentioned in Section 4.7.1, the user needs to recognize stress/strain concentrations by allowing denser and smaller elements in the parts of the structure with abrupt change of geometry. It is also necessary to avoid using elements with high aspect ratio for computational accuracy. The limit on aspect ratio makes discretization of thin-film structural components in MEMS and microsystems an insurmountable task. The user's knowledge and experience in selecting the right elements for special thermomechanical functions of the structural components, such as the spring and contact elements from the available element library, will be desirable. Another user-related issue is the proper establishment of the loading and boundary conditions in the discretized model. In many finite element codes, the pressure loading needs to be properly converted to the concentrated forces at the adjacent nodes on the applicable boundary surfaces. This issue often becomes more complicated in transient thermal analyses.
2. One must be sure of the credibility of the input material properties to the analysis. Because of the relatively short history of the MEMS and microsystem industry, most thermophysical properties of the materials used for these products are lacking. We will present some of these material properties in Chapter 7, but these properties are far from being sufficient for proper design analyses. Information on some phenomenological behavior of MEMS structures, such as the interfacial fracture toughness as described in Section 4.5 and the intrinsic stress model in Section 4.6, required for modeling and simulations is not currently available. An ongoing effort is being made by scientists and engineers to generate more complete material databases. The user is thus cautioned to be selective in the input material properties to the finite element analyses.
3. The user should be aware of the fact that unless otherwise specified, most commercial general-purpose finite element codes are developed on the theories of continua. A typical example is the common use of constitutive relations such as the generalized Hooke law for solid mechanics or the Fourier law for heat conduction. The theories on continuum mechanics and heat conduction are time tested for macrosystems. However, they are not valid for MEMS and microsystems components in submicrometer scales. Commercial finite element

codes thus cannot be indiscriminately used for MEMS and microsystems design analyses at submicrometer scales.

4. The reader may have come to the realization from Section 4.6 that it is difficult to predict the mechanistic behavior of MEMS and microsystems, which often consist of components made of layers of thin films, or are bonded to dissimilar materials. The fabrication of these components introduces unavoidable residual and intrinsic stresses in the structure. As we will learn from later chapters, virtually all microfabrication techniques result in such adverse effects in MEMS and microsystems. The incorporation of these effects in the design analysis requires the coupling of the finite element analysis and the microfabrication processes. Some developers in computer-aided design (CAD) software packages have made such an effort in coupling, as will be described in Chapter 10. The reader is cautioned to be aware of this fact when using a commercial finite element code in a design analysis.

Despite the aforementioned shortcomings of the current state of the art of commercially available finite element codes, they are still the only viable means for engineers to acquire a good insight on whatever MEMS and microsystems they are involved with in design and simulation. Engineers are merely reminded of a fact that the power of FEM, like many other tools for craftsmen and professionals, is in the intelligence of the user. The user needs to have sound knowledge of the tool and be aware of its limitations before he or she can maximize the benefits of the tool.

PROBLEMS

Part 1. Multiple Choice

1. In general, mechanical engineering principles derived for continua can be used for MEMS components with the size (1) larger than 1 nanometer, (2) larger than 1 micrometer, (3) larger than 1 picometer.
2. The theory of thin plate bending can be used to assess (1) the deflection only, (2) stresses only, (3) both the deflection and stresses in thin diaphragms of micro pressure sensors.
3. Square diaphragms are the (1) most popular, (2) somewhat popular, (3) least popular geometry for micro pressure sensors.
4. From a mechanics point of view, the most favored diaphragm geometry in micro pressure sensors is (1) circular, (2) square, (3) rectangular.
5. The principal theory used in microaccelerometer design is (1) plate bending, (2) mechanical vibration, (3) strength of materials.
6. The natural frequency of a microdevice is determined by its (1) mass, (2) structure stiffness, (3) mass and structure stiffness.
7. Microdevices in theory contain (1) one, (2) several, (3) an infinite number of natural frequencies.
8. The analysis that attempts to determine several or all natural frequencies of a microdevice is called (1) modal, (2) vibration, (3) model analysis.
9. "Resonant" vibration of a device made of elastic materials occurs when the frequency of the excitation force (1) approaches, (2) equals, (3) exceeds any of the natural frequencies of the device.
10. The dashpot in a mass–spring vibration system serves the purpose of including the (1) damping, (2) acceleration, (3) deceleration effect on the system.
11. The damping effect in most microaccelerometer design is (1) very important, (2) somewhat important, (3) not important.
12. The damping effect by compressible fluids (1) increases, (2) decreases, (3) remains unchanged with increase of the input frequency of the vibrating mass.
13. The movement of the beam mass in force-balanced microaccelerometers is usually measured by (1) piezoresistor, (2) piezoelectric, (3) capacitance changes.
14. A vibrating beam will have its natural frequency (1) increased, (2) decreased, (3) unchanged with increase of longitudinal stress in tension.
15. Thermal stresses can be induced in mechanically constrained microdevice components by (1) uniform temperature rise, (2) nonuniform temperature rise, (3) any temperature rise.
16. Thermal stresses induced in a microdevice component made of dissimilar materials are due to (1) the difference of coefficients of thermal expansion of the materials, (2) the weakness of the bonding interface, (3) the degradation of materials after bonding.

17. Thermal stresses are induced in microdevice components free of mechanical constraints by (1) uniform temperature change, (2) nonuniform temperature change, (3) uniform temperature with time.
18. The creep deformation in a material becomes serious (1) at any temperature, (2) above half the melting point, (3) above half the homologous melting point.
19. The *homologous melting point* of a material is defined as the melting point on the scale of (1) absolute temperature, (2) Celsius temperature, (3) Fahrenheit temperature.
20. The parts of microsystems that are obviously vulnerable to creep failure are (1) solder bonds, (2) epoxy resin bonds, (3) silicone rubber bonds.
21. There are generally (1) two, (2) three, (3) four modes of fracture at the interfaces of microdevices.
22. The most frequently occurring fracture failure mode in microstructures is (1) Mode III, (2) Mode II, (3) Mode I.
23. Interfaces in microdevices are vulnerable to (1) mixed Mode I and II, (2) mixed Mode I and III, (3) mixed Mode II and III failure.
24. Fracture mechanics analysis of interfaces in microstructures requires the distribution of (1) normal stresses, (2) shear stresses, (3) both the normal and shear stresses at the vicinity of the interface.
25. The finite element method is a viable analytical tool for microstructures of (1) simple geometry, (2) complex geometry and loading/boundary conditions, (3) complex loading and boundary conditions.
26. The very first step in a finite element analysis is (1) to find the approximate solution, (2) to set the governing equation and boundary condition, (3) to subdivide the continuum into a number of subdivisions, a process called *discretization.*
27. The primary unknown quantity in a finite element analysis is the quantity that (1) appears in the formulation, (2) the most important quantity, (3) the most desirable quantity to be determined.
28. The primary unknown quantity in a stress analysis by the finite element method is (1) stress, (2) strain, (3) displacement.
29. The constitutive relation in a finite element analysis relates (1) the construction of appropriate formulations, (2) the primary and other essential quantities, (3) the loading and boundary conditions.
30. The von Mises stress represents (1) the stress component following the von Mises principle, (2) the stress for a specific material, (3) stresses in a structure of complex geometry.

Part 2. Computation Problems

1. Use the material properties in Table 7.3 in Chapter 7 to determine the maximum deflection of a circular diaphragm made of aluminum with conditions specified in Example 4.1.

2. If the stress required to produce a measurable signal output in a square diaphragm in a pressure sensor is 350 MPa, what will be the required thickness of the diaphragm? The diaphragm is an integral part of a silicon die that is shaped from a wafer of 100 mm in diameter in the (100) plane (i.e., with a 54.74° angle in the slope from the bottom face into the cavity as illustrated in Fig 4.6). The die has a plane area of 3 mm × 3 mm. A pressurized medium is applied at the front side of the silicon die.
3. Derive the expressions for the equivalent spring constants for the three beam springs illustrated in Figure 4.18.
4. A microdevice component, 5 g in mass, is attached to a fine strip made of silicon, as illustrated in Figure P4.1. The equivalent spring constant of the strip is 18,240 N/m. Both the mass and the strip-spring are made of silicon. The mass is pulled down by 5 μm initially and is released at rest. Determine (a) the natural frequency of the simulated mass-spring system. (b) The maximum amplitude of vibration.

Figure P4.1 | A strip–spring–mass system.

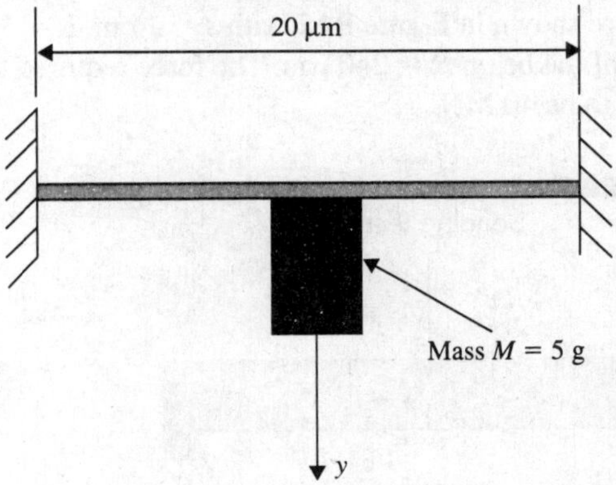

5. Determine the time required to break the strip spring in Problem 4 when a force of $F(t) = 5\cos(1910t)$ N is applied to the mass at time $t > 0$, where t is the time into the vibration. Assume the strip breaks at a deflection of 1 mm. The vibration of the strip–mass system begins when the system is at rest.
6. Determine the thickness of the beam–spring in Example 4.9 if the maximum allowable deflection of the beam is 5 mm.
7. Assess the effect of damping on the amplitude of vibration of a cantilever beam–mass system.
8. Assess the effect of damping on the movement of the beam–mass in Example 4.11.
9. Use a force-balanced microaccelerometer such as that described in Example 4.11 for the situations described in Example 4.12. Determine the maximum movement of the beam mass. The damping effect can be neglected.

10. The microactuator described in Example 4.14 is expected to operate with a temperature rise from 10°C to 50°C. Plot the movements of the free end of the actuator with respect to the range of temperature rise. Use a temperature increment of 10°C.
11. Compute the maximum thermal stresses, thermal strains, and displacements induced in a beam by a temperature variation in the thickness direction, $T(z) = 2 \times 10^6 z + 30$ in degrees Celsius. The geometry and dimensions of the beam are given in Figure 4.40. The beam is made of silicon.
12. For the same conditions and the required answer described in Example 4.15 with the width of the beam being $b = 100$ μm. (*Hint:* Use the formulations for thin plates.)
13. Find the maximum thermal stress, strain, and displacements in a beam subjected to heating at its top surface at time $t = 0.1$ μs, 0.5 μs, and 1.5 μs. The geometry and loading and other conditions are identical to that described in Example 4.15. Plot the results versus time during the period of heating.
14. Compute the fracture toughness for a silicon substrate using a three-point bending specimen as illustrated in Figure 4.44b. Dimensions of the beam specimen are shown in Figure P4.2 with $s = 1$ cm, $b = 5$ mm, $c = 100$ μm. The width of the beam $B = 240$ μm. The force required to break the specimen is assumed to be 40 MN.

Figure P4.2 | Measurements of fracture toughness of silicon in a three-point bending test.

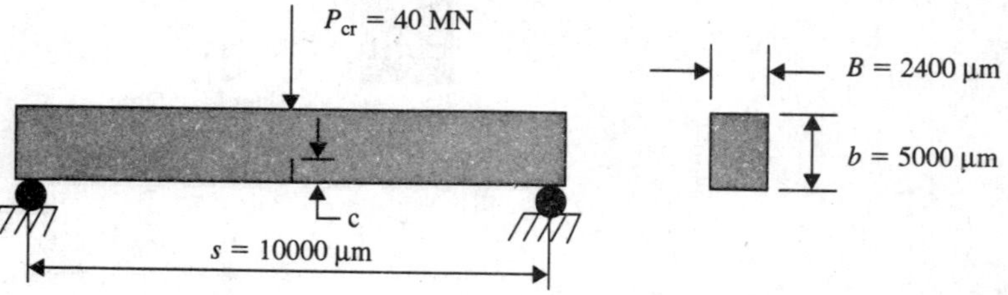

15. What fracture toughness of the silicon specimen would you get if the width B of the specimen is 100 times greater than that used in Problem 14. What would you have observed from the comparison of the results obtained from these two cases?

Chapter 5

Thermofluid Engineering and Microsystem Design

CHAPTER OUTLINE

5.1 | INTRODUCTION

Many MEMS devices and microsystems such as the manifold absolute pressure sensors (MAPS) presented in Chapter 2 are expected to handle hot gases. Many others are expected to operate at elevated temperatures. Still, there are microvalves, pumps and fluidics for biosensors that involve moving fluids in both liquid and gaseous forms. Mechanical design of these systems requires the application of theories and principles on fluid dynamics and heat transfer. Thermofluids engineering principles are also used in microfabrication process design as will be described in Chapter 8. For example, gas flow over hot substrate surfaces need to be properly controlled in many thin-film deposition processes.

In this chapter, we will first review thermofluids engineering principles that are derived from classical continuum theories and how these principles can be used in the design of MEMS and microsystems. This will be followed by highlights on the current developments in gas dynamics and heat conduction in solids at submicrometer and nanoscale. Scientists and engineers are making relentless efforts in this frontier area of research, as the trend of miniaturization of MEMS and microsystems not only continues but accelerates.

5.2 | OVERVIEW OF THE BASICS OF FLUID MECHANICS IN MACRO- AND MESOSCALES

Fluid mechanics studies fluids either in motion (fluid dynamics) or at rest (fluid statics), as well as solid/fluid interactions. For MEMS and microsystems design, fluid dynamics is of primary interest to engineers.

There are two principal types of fluids that microsystems design engineers need to deal with. These are (1) noncompressible fluids, e.g., liquids, and (2) compressible fluids, e.g., gases. Both types of fluids have applications in micro- and nanoscale systems.

Fluids are aggregations of molecules. These molecules are closely spaced in liquids and widely spaced in gases. Unlike solids, fluid molecules are much more widely spaced than the molecule size. Further, these molecules are not fixed in a lattice, but move freely relative to each other. We may thus view fluids as a substance that have volume but no shape. Unlike solids, fluids cannot resist any shear force or shear stress without moving. All fluids have viscosity that causes friction when they are set in motion. Viscosity is a measure of the fluid's resistance to shear when the fluid is in motion. Thus, it is necessary to have a driving pressure to make a fluid flowing in a conduit, a channel, or, as in the case of fluidics, system of conduits.

As mentioned in the foregoing section, theories derived from continuum fluids can be used for studying the flow of fluids in macro- and mesoscales. A *continuum fluid* is viewed as a fluid with its properties varying continuously in the space. For this reason, differential calculus can be applied to analyze this substance.

5.2.1 Viscosity of Fluids

Fluids can be put to motion by even a slight shear force. The induced shear strain in a bulk of fluid as illustrated in Figure 5.1 can be expressed by the change of the right angle at the corners in Figure 5.1a to the angle θ in Figure 5.1b after the deformation in shear.

Figure 5.1 I Shear deformation of a fluid.

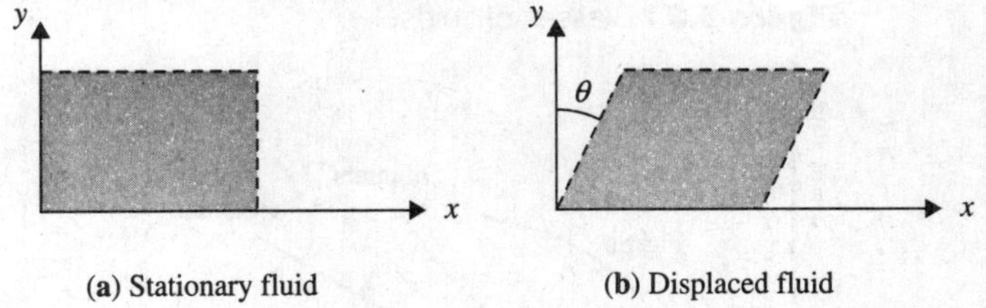

The shear deformation in Figure 5.1 is considered possible by a relative motion of a pair of plates placed at the top and the bottom of the bulk fluid as illustrated in Figure 5.2. We notice that the deformed fluid in Figure 5.2b is produced by the motion of the top plate at a velocity u_0. A *nonslip* boundary condition at the interfaces of the fluid at both the top and bottom plates is assumed in the situation. The relative motion of the top and bottom plates in parallel represents a *shear* force that causes the flow of the fluid. The associate shear stress τ is considered proportional to the rate of the change of the induced shear strain θ. Mathematically, this relationship can be expressed as:

Figure 5.2 | Shear flow of fluid by a moving plate.

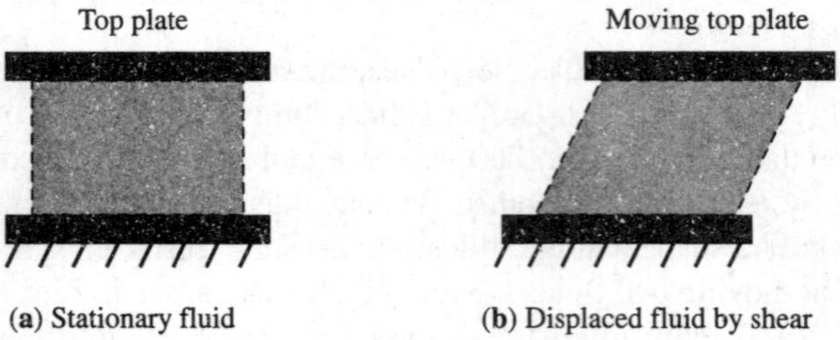

$$\tau \propto \frac{\delta\theta}{\delta t}$$

or in the form of an equation as:

$$\tau = \mu \frac{\delta\theta}{\delta t}$$

For the case with $\delta t \to 0$, we have:

$$\tau = \lim_{\delta t \to 0} \left(\mu \frac{\delta\theta}{\delta t} \right) = \mu \frac{d\theta}{dt} \qquad [5.1]$$

The proportional constant μ is called the *dynamic viscosity,* or *viscosity,* of the fluid. It is a very important property of a fluid. We notice the linear relationship between the shear stress and the shear strain rate exhibited in Equation (5.1). Fluids that exhibit such linear relationship are classified as *newtonian fluids.* Other classifications of fluid are illustrated in Figure 5.3.

Figure 5.3 | Classes of fluids.

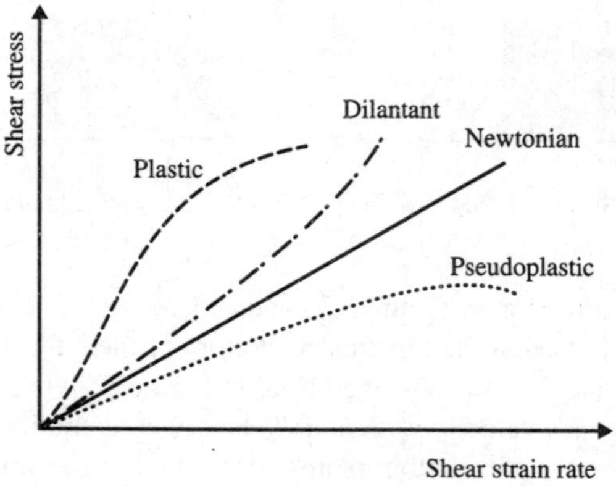

(White [1994].)

The velocity profile of the fluid flow in Figure 5.2 is illustrated in Figure 5.4, in which we perceive a linear variation of the velocity in the *y* direction. One may readily prove that $d\theta/dt$ in Equation (5.1) is equal to $du(y)/dy$, in which $u(y)$ represents the velocity of the fluid at a distance *y* from the stationary state at the bottom plate. The following expression for the shear stress, τ is obtained:

$$\tau = \mu \frac{du(y)}{dy} \qquad [5.2]$$

Figure 5.4 I Velocity profile in a newtonian fluid flow.

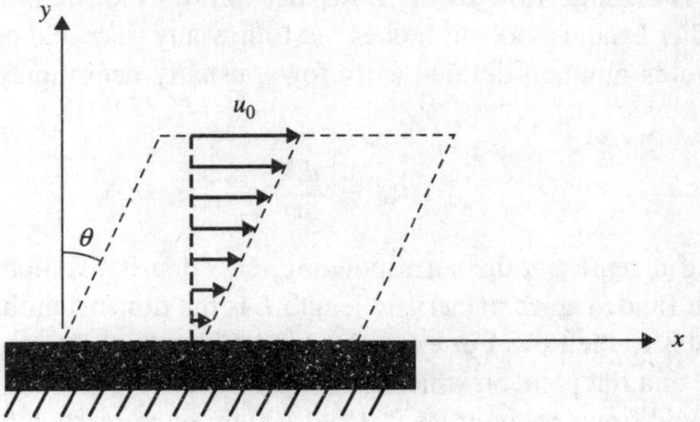

Numerical values of dynamic viscosity in Equation (5.2) for seven selected fluids are presented in Table 4.3 in Chapter 4. Another measure of fluid viscosity, called the *kinematic viscosity* ν, is used in fluid mechanics analysis. It is defined as $\nu = \mu/\rho$, in which ρ is the mass density of the fluid.

5.2.2 Streamlines and Stream Tubes

A *streamline* is a trace of any point in a moving fluid. The direction of the line coincides with the direction of the motion of the moving point in the fluid. Streamlines have the following properties:

1. No flow can cross a streamline.
2. No two streamlines may intersect, except at a point where the velocity of the fluid is zero.
3. Any boundary line of the flow must also be a streamline.
4. In general, a streamline can change its position and its shape from time to time in a steady-state flow.

A *stream tube* is made up by a bundle of streamlines through all points of a closed curve.

5.2.3 Control Volumes and Control Surfaces

As fluids do not have fixed shapes of their own, it is not possible to define their volumes as we do with solids. Consequently the term *control volume* is adopted for computational purposes. The control volume of a bulk of fluid (v) is selected in such a way that the flow properties, e.g., the velocities, can be evaluated at locations where the mass of the fluid crosses the surface of this volume. The surface of a control volume is called a *control surface* (s).

5.2.4 Flow Patterns and Reynolds Number

The patterns of fluid flow can significantly influence the engineering results. There are generally two patterns of fluid flow: (1) laminar flow and (2) turbulent flow. Laminar flow is a gentle flow of the fluid that follows the streamlines. Turbulent flow, on the other hand, is violent. It does not follow any traceable pattern.

The Reynolds number, defined as follows, usually determines the fluid flow patterns

$$\mathrm{Re} = \frac{\rho L V}{\mu} \tag{5.3}$$

where ρ, V, and μ represent the corresponding mass density, velocity, and dynamic viscosity of the fluid. The characteristic length L is the dimension that is a principal factor that relates to the flow. For example, L might be the diameter, d of a pipeline or the length L of a flat plate on which the fluid flows.

Laminar fluid flows occur at Re $<$ 10 to 100 for compressible fluids, and Re $<$ 1000 for incompressible fluids. Turbulent flows result in large drag forces when the fluid flows over a solid surface. Fortunately, in almost all MEMS and nanodevice applications, fluid flow is always in the laminar flow regime.

5.3 | BASIC EQUATIONS IN CONTINUUM FLUID DYNAMICS

5.3.1 The Continuity Equation

The *continuity equation* of fluid flow is used to evaluate the volumetric flow rates. With reference to the situation as illustrated in Figure 5.5, a control volume of the fluid flowing through a stream tube is selected. This volume is enclosed in the region

Figure 5.5 | Continuous fluid flow in a stream tube.

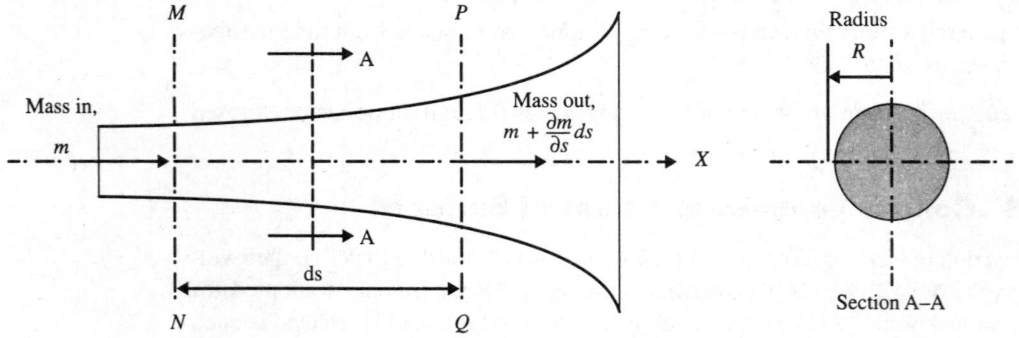

designated by *MNPQ* in the figure. The two control surfaces *MN* and *PQ* are separated by an infinitesimally small distance *ds*. The rates of the fluid flow at the control surfaces *MN* and *PQ* are *m* and $(m + \Delta m = m + (\partial m/\partial s)\, ds)$ respectively. We may readily account for the total volume of fluid entering and leaving the two control surfaces during a small time interval, Δt to be $m\, \Delta t$ at *MN* and $(m + (\partial m/\partial s)\, ds)\, \Delta t$ at *PQ*. The change of mass in the control volume is $(A\, ds)\, \Delta\rho$, in which *A* is the average cross-sectional area of the stream tube of the control volume. By the law of conservation of mass, we should have the following relation:

$$m\, \Delta t - \left(m + \frac{\partial m}{\partial s}\, ds\right) \Delta t = (A\, ds)\, \Delta\rho$$

For the case with $\Delta t \to 0$, we have $\Delta\rho/\Delta t \to \partial\rho/\partial t$. Consequently, we have the following equation for the equation of continuity for a one-dimensional flow:

$$A\frac{\partial \rho}{\partial t} + \frac{\partial m}{\partial s} = 0 \qquad [5.4]$$

For steady-state fluid flows, $\partial\rho/\partial t = 0$, which leads to $\partial m/\partial s = 0$ as in Equation (5.4). We will thus have the *rate of the mass flow* in the steady-state flow condition with $\dot{m}$ = constant, or

$$\dot{m} = \rho_1 V_1 A_1 = \rho_2 V_2 A_2 \qquad [5.5]$$

where A_1 and A_2 are cross-sectional areas through which the fluid flows at the respective velocities V_1 and V_2.

If the units of kg/m^3, m/s, and m^2 are used for the corresponding mass density ρ, the velocity *V* and the cross-sectional area *A*, then the rate of mass flow $\dot{m}$ in Equation (5.5) will have units of kg/s.

For incompressible fluids, the above relationship can be used to compute the volumetric flow rate:

$$Q = V_1 A_1 = V_2 A_2 \qquad [5.6]$$

in which *Q* has a unit of m^3/s.

EXAMPLE 5.1

A noncompressible fluid is used in a microfluidic system. It flows through a tube with a diameter of 1 mm at a rate of 1 microliter (μl) per minute. A reducer is used to connect this tube to the microconduits in the fluidic system. The reducer has an outlet diameter of 20 μm. Determine the velocity of the fluid at the inlet and outlet of the reducer. The system is illustrated in Figure 5.6.

Figure 5.6 | Fluid supply to a microfluidic system.

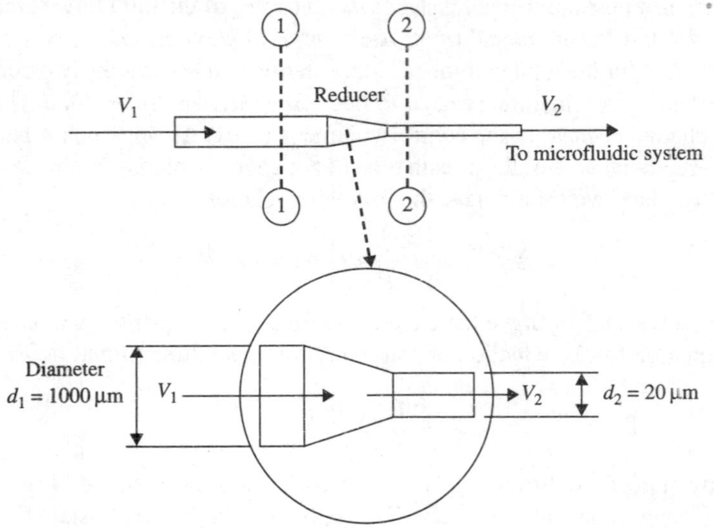

■ Solution

It is safe to assume a laminar flow situation because of the low flow rate. We further assume that a steady-state flow is maintained. Thus, from the continuity equation in Equation (5.6), the rate of fluid flow through the control surface 1–1 in Figure 5.6 should be the same as that through surface 2–2.

From Equation (5.6), the volumetric mass flow rate is:

$$Q = A_2 V_2 = A_1 V_1$$

in which $Q = 1 \times 10^{-6}$ cm^3/min $= 1.67 \times 10^{-14}$ m^3/s and the entrance and the exit cross-sectional areas of the reducer are:

$$A_1 = \pi(1000 \times 10^{-6})^2/4 = 0.785 \times 10^{-6}\ \text{m}^2$$

$$A_2 = \pi(20 \times 10^{-6})^2/4 = 314 \times 10^{-12}\ \text{m}^2$$

We can thus compute the velocities of the fluid at both entrance and exit of the reducer by using the continuity relation as follows:

$$V_1 = \frac{Q}{A_1} = \frac{1.67 \times 10^{-14}}{0.785 \times 10^{-6}} = 0.0213\ \mu\text{m/s}$$

$$V_2 = \frac{Q}{A_2} = \frac{1.67 \times 10^{-14}}{314 \times 10^{-12}} = 53.2\ \mu\text{m/s}$$

5.3.2 The Momentum Equation

The momentum equation is used to compute the fluid-induced forces on the solid on which the fluid flows. This equation is derived on the bases of conservation of momentum and Newton's law of dynamic equilibrium.

Consider a two-dimensional steady-state flow situation as illustrated in Figure 5.7. The selected control volume of $ABCD$ makes a small movement to a new position designated as $A'B'C'D'$.

Figure 5.7 | A moving fluid in a stream tube.

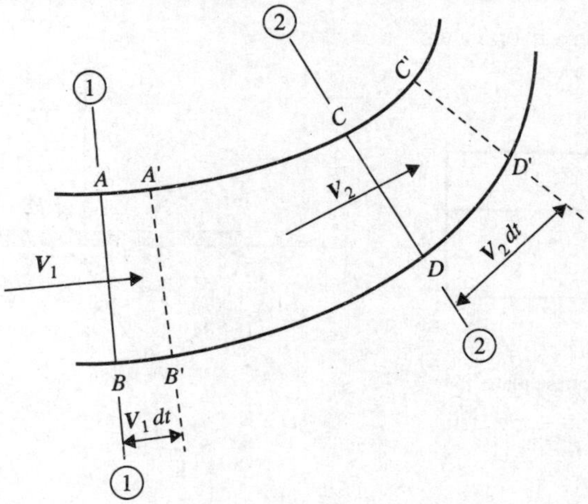

The change of momentum in the control volume during this short period dt is:

$$\boxed{\begin{array}{c}\text{Change of momentum}\\ \text{in the control volume}\end{array}} = \boxed{\begin{array}{c}\text{Momentum}\\ \text{of } A'B'C'D'\end{array}} - \boxed{\begin{array}{c}\text{Momentum}\\ \text{of } ABCD\end{array}}$$

$$= \boxed{\begin{array}{c}\text{Momentum}\\ \text{of } CDC'D'\end{array}} - \boxed{\begin{array}{c}\text{Momentum}\\ \text{of } ABA'B'\end{array}}$$

$$= (\dot{m}_2 dt)V_2 - (\dot{m}_1 dt)V_1$$

in which $\dot{m}_1$ and $\dot{m}_2$ are the rate of mass flow across the respective control surfaces 1–1 and 2–2 in Figure 5.7.

In the case of steady-state flow condition, the rate of mass flow remain constant, i.e. $\dot{m}_1 = \dot{m}_2 = \dot{m}$. We can obtain the induced forces by using the relationships between the impulse forces and change of momentum as follows:

$$\Sigma F = \dot{m}(\mathbf{V}_2 - \mathbf{V}_1) \tag{5.7}$$

in which $\mathbf{V}_1$ and $\mathbf{V}_2$ are the respective velocity vectors at control surface 1–1 and 2–2.

EXAMPLE 5.2

A micromachined silicon valve utilizing electrostatic actuation is constructed. The valve unit has a similar configuration as that reported in Ohnstein et al. [1990] as illustrated in Figure 5.8a. The thin closure plate is used as the valve with a dimension of 300 μm wide × 400 μm long × 4 μm thick. The plate is bent to open or close by electrostatic actuation to regulate the hydrogen gas flow. The maximum opening of the closure plate is 15° tilt from the horizontal closed position. Determine the force induced by the flow of the gas at a velocity of 60 cm/min and a volumetric rate of 30,000 cm³/min. Also, calculate the split of mass flow over the lower surface of the plate.

Figure 5.8 | Fluid induced force on a microvalve.

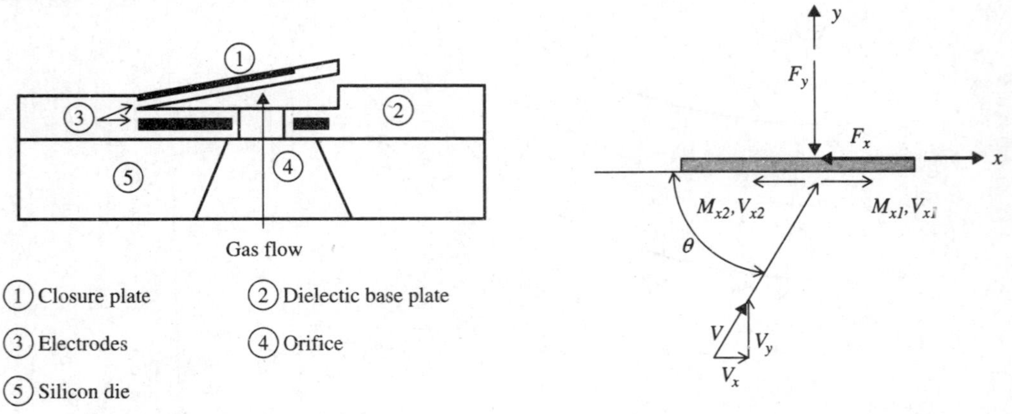

(**a**) Microvalve unit (**b**) Fluid-induced forces on closure plate

■ Solution

We assume that the gas stream impinges on the plate at an angle $\theta = 75°$, at the maximum opened position. The gas stream splits into two components, i.e. M_{x1} induced by velocity V_{x1} and M_{x2} by velocity V_{x2} as shown in Figure 5.8b. We designate M_{x1} and M_{x2} to be the respective components of the rate of mass flow of the gas, $\dot{m}$.

The volumetric flow of the gas Q is 30,000 cm³/min or 500×10^{-6} m³/s. The rate of mass flow of the hydrogen gas involves the mass density of the gas, $\rho = 0.0826$ kg/m³ [Janna 1993] with $\dot{m} = \rho Q = 0.0826 \times (500 \times 10^{-6}) = 41.3 \times 10^{-6}$ kg/s.

By referring to the force diagram shown in Figure 5.8b and using Equation (5.7), we have the induced force components expressed in the following forms:

$$\Sigma F_y = \dot{m}(V_{y2} - V_y) \quad \textbf{[a]}$$

$$\Sigma F_x = (M_{x1}V_{x1} - M_{x2}V_{x2}) + \dot{m}V_x \quad \textbf{[b]}$$

where

V_{y2} = fluid velocity along the y axis at the plate surface (= 0 in the present case)

$V_x = V\cos\theta$ = the velocity component along the x axis in the gas stream

$V_y = V\sin\theta$ = the velocity component along the y axis in the gas stream

Thus, by substituting the values of $\theta = 75°$ and $V = 60$ cm/min or 10^{-2} m/s into Equation (a), we obtain the force $F_y = 40 \times 10^{-8}$ kg-m/s², or 40×10^{-8} N.

The horizontal force component F_x on the plate exists only if the coefficient of friction between the gas and the contacting plate surface is known. However, we may reasonably assume a frictionless gas flow at that surface, which will then lead, according to Equation (b), to the following relationship:

$$(M_{x1}V_{x1} - M_{x2}V_{x2}) + \dot{m}V\cos\theta = 0$$

It is also reasonable to assume that $V_{x1} = V_{x2} = V$ in a frictionless flow. Consequently, the split of mass flow at the lower surface of the plate can be obtained by solving the following simultaneous equations:

$$M_{x1} - M_{x2} = -\dot{m}\cos\theta \qquad \text{[c]}$$

$$M_{x1} + M_{x2} = \dot{m} \qquad \text{[d]}$$

from which we obtain the following split mass flow rates:

$$M_{x1} = \frac{\dot{m}}{2}(1 - \cos\theta) = \frac{41.3 \times 10^{-6}}{2}(1 - \cos 75°) = 15.3 \times 10^{-6} \text{ kg/s}$$

$$M_{x2} = \frac{\dot{m}}{2}(1 + \cos\theta) = \frac{41.3 \times 10^{-6}}{2}(1 + \cos 75°) = 26 \times 10^{-6} \text{ kg/s}$$

5.3.3 The Equation of Motion

The equation of motion in fluid dynamics is used to evaluate the relationship between the motion of the fluid and the required driving force, namely the pressure. It is a very important part of fluid dynamics in assessing the pumping power required for moving the fluid. The following derivation of the equation is based on an assumption that the friction between the fluid and the contacting surface of the conduit is negligible.

Figure 5.9 illustrates a fluid element moving along a streamline with a unit thickness in a two-dimensional domain. The small fluid element is situated in a cartesian coordinate system (x, y) with the two sides, ds and dn, along the respective tangential (s) and normal (n) directions that define the streamline. The movement of the fluid element is maintained by the application of pressure p at the backside, i.e., the left face of the element. The weight of the fluid, dw, is always pointed vertically downward. By summing up all forces acting on both the tangential and the normal faces of the element, we may establish the relationship of the applied pressure and the velocities of the element in motion by using Newton's law expressed as $\Sigma F_s = ma_s$ and $\Sigma F_n = ma_n$, in which F_s and F_n are respective inertia forces in the s and n directions in Figure 5.9, m is the mass of the fluid element, and a_s and a_n are the respective accelerations in both the s and n directions. Both these acceleration components can be related to the tangential velocity of the fluid element, $V(s, t)$ by the following expressions:

Figure 5.9 | A fluid element travelling along a streamline.

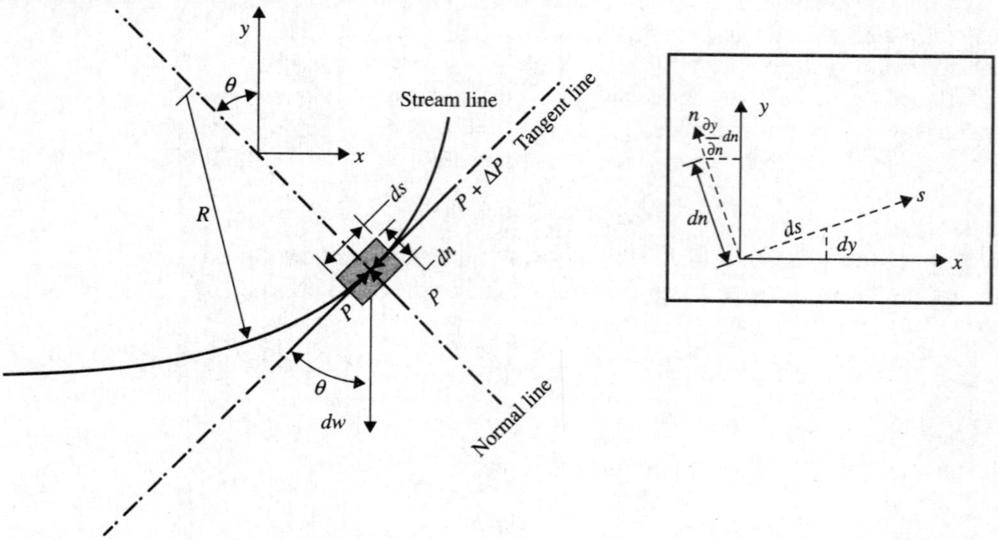

$$a_s = \frac{\partial V(s, t)}{\partial t} + V(s, t)\frac{\partial V(s, t)}{\partial s}$$

$$a_n = \frac{\partial V_n(s, t)}{\partial t} + V(s, t)^2/R$$

in which $V_n(s, t)$ is the component of the tangential velocity $V(s, t)$ in the normal direction.

Thus, by using the above relationships in Newton's law, we may derive the equations of motion in the following forms:

$$\rho\left(\frac{\partial V}{\partial t} + V\frac{\partial V}{\partial s}\right) = -\frac{\partial P}{\partial s} - \rho g \sin \alpha \qquad \textbf{[5.8a]}$$

$$\rho\left(\frac{\partial V_n}{\partial t} + \frac{V^2}{R}\right) = -\frac{\partial P}{\partial n} - \rho g \cos \alpha \qquad \textbf{[5.8b]}$$

Equation (5.8) is called *Euler's equation.*

The equations of motion in a three-dimensional cartesian coordinate system, where the x coordinate coincides with the streamline, can be expressed as:

$$\rho\left(\frac{\partial u}{\partial t} + u\frac{\partial u}{\partial x} + v\frac{\partial u}{\partial y} + w\frac{\partial u}{\partial z}\right) = -\frac{\partial P}{\partial x} + X \qquad \textbf{[5.9a]}$$

$$\rho\left(\frac{\partial v}{\partial t} + u\frac{\partial v}{\partial x} + v\frac{\partial v}{\partial x} + w\frac{\partial v}{\partial z}\right) = -\frac{\partial P}{\partial y} + Y \qquad \textbf{[5.9b]}$$

$$\rho\left(\frac{\partial w}{\partial t} + u\frac{\partial w}{\partial x} + v\frac{\partial w}{\partial y} + w\frac{\partial w}{\partial z}\right) = -\frac{\partial P}{\partial z} + Z \qquad \textbf{[5.9c]}$$

where

$u = u(x, y, z)$ = the fluid velocity along the x direction
$v = v(x, y, z)$ = the fluid velocity along the y direction
$w = w(x, y, z)$ = the fluid velocity along the z direction
X, Y, and Z = components of body force of the fluid (e.g., the weight) along the respective x, y, and z directions.

The above equations of motion can be deduced into the well-known *Bernoulli equation:*

$$\frac{V_1^2}{2g} + \frac{P_1}{\rho g} + y_1 = \frac{V_2^2}{2g} + \frac{P_2}{\rho g} + y_2 \qquad [5.10]$$

which depicts the situation as illustrated in Figure 5.10.

Figure 5.10 | Fluid flow with varying state properties.

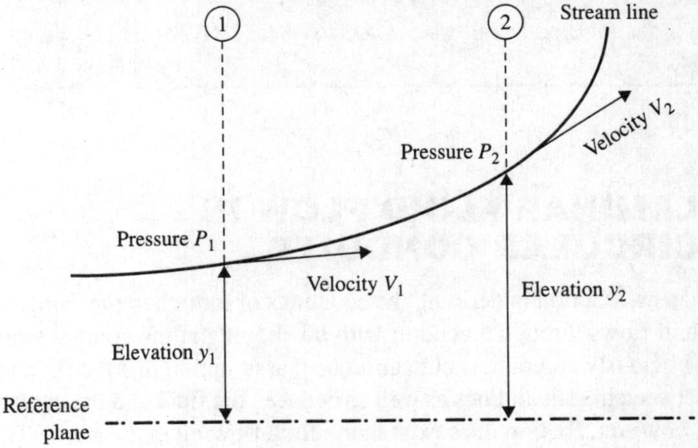

EXAMPLE 5.3

Determine the pressure drop in a minute stream of alcohol flowing through a section of a tapered tube 10 cm in length. The inlet velocity is 600 μm/s. The mass density of alcohol is 789.6 kg/m³ [*Marks' Standard Handbook for Mechanical Engineers* 1996]. The tube is inclined 30° from the horizontal plane as illustrated in Figure 5.11.

Figure 5.11 | Flow in a tapered tube.

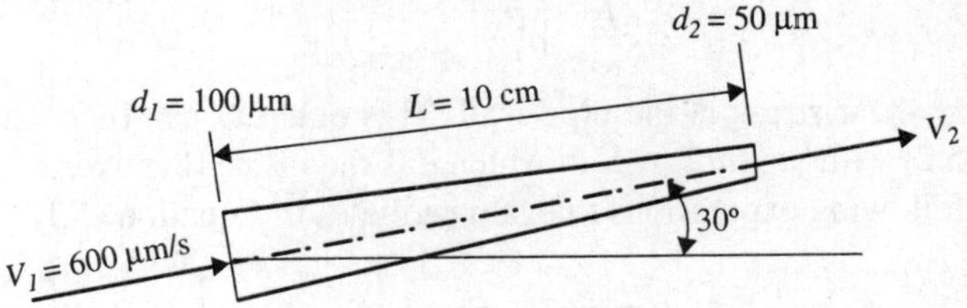

Solution

The cross-sectional areas of the tube at the entrance, A_1, and the exit, A_2, are first calculated to be 0.784×10^{-8} m^2 and 0.1964×10^{-8} m^2 respectively. We are given the entrance velocity as 600 μm/s, or 600×10^{-6} m/s. The exit velocity of the flow can be determined from Equation (5.6) as:

$$V_2 = \frac{V_1 A_1}{A_2} = 2.4 \times 10^{-3} \text{ m/s}$$

By rearranging the Bernoulli equation in Equation (5.10), we have the following expression for the pressure drop ΔP:

$$\Delta P = (P_1 - P_2) = \rho g \left[\frac{V_2^2 - V_1^2}{2g} + (y_2 - y_1) \right] \quad \textbf{[5.11]}$$

The difference of elevation of the tube between the exit and entrance, $(y_2 - y_1)$ in Equation (5.11), is equal to $L \sin 30° = 5 \times 10^{-2}$ m.

The pressure drop is thus calculated to be 387.3 N/m^2, or 387 Pa, from Equation (5.11).

5.4 | LAMINAR FLUID FLOW IN CIRCULAR CONDUITS

We made an assumption in deriving the equations of motion in the foregoing section that the fluid flows through a conduit with no slipping at the conduit wall. This assumption obviously contradicts our subsequent assumption that the fluid is also frictionless between the streamlines as well as between the fluid and the contacting wall. In reality, however, friction does exist in the fluid flowing in a conduit. This friction factor inevitably results in additional pressure drop in these cases. The equivalent "head" loss due to the friction in the fluid and between the fluid and the contacting wall in a circular pipe with diameter d and length L can be accounted for by the Darcy–Weisbach equation as expressed below [White 1994]:

$$h_f = f \frac{L}{d} \frac{V^2}{2g} \quad \textbf{[5.12]}$$

where f is the *Darcy friction factor*:

$$f = \frac{8\tau_w}{\rho V^2} \quad \textbf{[5.13]}$$

in which τ_w is the shear stress at the pipe wall. This quantity can be evaluated by using Equation (5.2) with $y = d/2 = a$, in which d is the inside diameter of the pipe. We may use the following expressions to evaluate the τ_w in Equation (5.13):

$$\tau_w = \frac{a}{2} \left| \frac{d}{dx}(P + \rho g y) \right| \quad \textbf{[5.14]}$$

For laminar flow, which is almost always the case in microsystems, the Darcy friction factor is deduced to a simple expression as:

$$f_l = \frac{64}{\text{Re}} \quad [5.15]$$

in which Re is the Reynolds number as expressed in Equation (5.3), with the characteristic length L replaced by the diameter of the pipe d.

The head loss due to friction, h_f, in Equation (5.12) is included in the right-hand side of the Bernoulli equation in Equation (5.10) in evaluating the pressure drop in a pipeline flow.

EXAMPLE 5.4

Estimate the equivalent head loss due to friction in Example 5.3.

■ Solution

We will approach the problem with average values: average diameter of the tube $d = 75\ \mu\text{m}$ and the average velocity between the entrance and exit of the tube, $V = 1.5 \times 10^{-3}$ m/s. We further obtain the dynamic viscosity of alcohol, $\mu = 1199.87 \times 10^{-6}$ N-s/m² from Table 4.3.

Thus, we obtain the Reynolds number Re from Equation (5.3) to be:

$$\text{Re} = \frac{\rho V d}{\mu} = \frac{789.6 \times (1.5 \times 10^{-3}) \times (75 \times 10^{-6})}{1199.87 \times 10^{-6}} = 0.074$$

The friction factor in Equation (5.15) leads to the following:

$$f = \frac{64}{\text{Re}} = \frac{64}{0.074} = 864.86$$

The equivalent head loss h_f can thus be determined by using Equation (5.12) as:

$$h_f = (864.86)\frac{10^{-1}}{75 \times 10^{-6}}\frac{(1.5 \times 10^{-3})^2}{2 \times 9.81} = 0.1322\ \text{m}$$

Figure 5.12 illustrates a laminar flow of a fluid in a circular tube of radius a. The Hagen-Poiseuille equations for the volumetric flow Q and the pressure drop ΔP can

Figure 5.12 | Fluid flow in a circular tube.

be derived from the equations of motion in Equations (5.8) and (5.9) by using a cylindrical polar coordinate system as [White 1994]:

$$Q = \frac{\pi a^4}{8\mu}\left[-\frac{d}{dx}(P + \rho g y)\right] \qquad [5.16]$$

where y is the elevation of the tube from a reference plane.

The pressure drop in the fluid over the tube length L is:

$$\Delta P = \frac{8\mu L Q}{\pi a^4} \qquad [5.17]$$

The equivalent head loss in relation to the volumetric flow Q is:

$$h_{f,l} = \frac{128\mu L Q}{\pi \rho g d^4} \qquad [5.18]$$

The parameter d in the above expression stands for the diameter of a circular pipe filled with the fluid. It is replaced by the *hydraulic diameter* d_h for conduits of circular cross section with partial fluid flow, or channels with geometry other than circular cross sections. A hydraulic diameter is defined as:

$$d_h = \frac{4A}{p} \qquad [5.19]$$

In Equation (5.19) A is the cross-sectional area of fluid flow and p is the "wet perimeter," meaning the perimeter that is in contact with the fluid. We may readily find the hydraulic diameter d_h for a conduit of rectangular cross section with width w and height h to be:

$$d_h = \frac{4A}{p} = \frac{4(wh)}{2(w+h)} = \frac{2wh}{w+h}$$

One striking phenomenon that one may observe from Equations (5.17) and (5.18) is that both the pressure drop and the friction head loss for laminar flows are inversely proportional to the fourth power of the conduit diameter. This means that a 16 times higher pumping power would be required to pump the same amount of volumetric flow of fluid with the conduit diameter reduced by half.

5.5 | COMPUTATIONAL FLUID DYNAMICS

The governing equation for fluid dynamics is the well-known *Navier–Stokes equations* on which modern computational fluid dynamics (CFD) codes are built. Derivation of Navier–Stokes equations is beyond the scope of this book. They can be found in many fluid mechanics books such as [White 1994]. Here we will present only the equations for the solutions for velocity vectors u, v, and w along the respective x, y, and z coordinates in a moving fluid in a three-dimensional space:

$$\rho g - \frac{\partial P}{\partial x} + \mu\left(\frac{\partial^2}{\partial x^2} + \frac{\partial^2}{\partial y^2} + \frac{\partial^2}{\partial z^2}\right)u(x, y, z) = \rho\frac{\partial u(x, y, z)}{\partial t}$$

$$\rho g - \frac{\partial P}{\partial x} + \mu\left(\frac{\partial^2}{\partial x^2} + \frac{\partial^2}{\partial y^2} + \frac{\partial^2}{\partial z^2}\right)v(x, y, z) = \rho\frac{\partial v(x, y, z)}{\partial t} \quad \textbf{[5.20a, b, c]}$$

$$\rho g - \frac{\partial P}{\partial x} + \mu\left(\frac{\partial^2}{\partial x^2} + \frac{\partial^2}{\partial y^2} + \frac{\partial^2}{\partial z^2}\right)w(x, y, z) = \rho\frac{\partial w(x, y, z)}{\partial t}$$

where ρ = mass density of the fluid, g = gravitational acceleration, P = the driving pressure, μ = dynamic viscosity of the fluid, and t = time.

The stress components in fluid in a small cubical element of the *control volume* of the fluid (Fig. 5.13) can be obtained by differentiating various velocity components from the Navier–Stokes equations as follows:

$$\tau_{xx} = 2\mu\frac{\partial u(x, y, z)}{\partial x}$$

$$\tau_{yy} = 2\mu\frac{\partial v(x, y, z)}{\partial y}$$

$$\tau_{zz} = 2\mu\frac{\partial w(x, y, z)}{\partial z}$$

$$\tau_{xy} = \tau_{yx} = \mu\left(\frac{\partial u(x, y, z)}{\partial y} + \frac{\partial v(x, y, z)}{\partial x}\right) \quad \textbf{[5.21]}$$

$$\tau_{xz} = \tau_{zx} = \mu\left(\frac{\partial w(x, y, z)}{\partial x} + \frac{\partial u(x, y, z)}{\partial z}\right)$$

$$\tau_{yz} = \tau_{zy} = \mu\left(\frac{\partial v(x, y, z)}{\partial z} + \frac{\partial w(x, y, z)}{\partial y}\right)$$

Figure 5.13 | Notation for stresses in a fluid element.

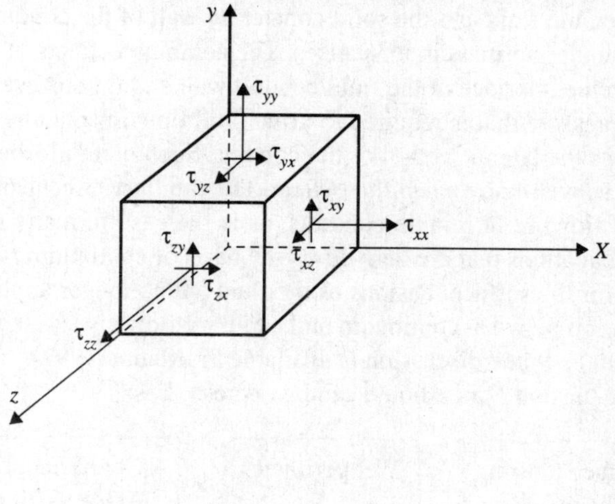

All stress components in a moving fluid can be expressed as follows:

$$\sigma_{ij} = \begin{bmatrix} -P + \tau_{xx} & \tau_{yx} & \tau_{zx} \\ \tau_{xy} & -P + \tau_{yy} & \tau_{zy} \\ \tau_{xz} & \tau_{yz} & -P + \tau_{zz} \end{bmatrix} \qquad [5.22]$$

Various CFD computer codes are available commercially for solving dynamic behavior of fluids in microsystems.

5.6 | INCOMPRESSIBLE FLUID FLOW IN MICROCONDUITS

A well-recognized phenomenon in fluid dynamics is that the velocity profile of a fluid flowing in a conduit has its velocity increasing toward the center of the conduit, partly due to the nonslip boundary at the conduit wall. Another common phenomenon that has been observed by many is the spherical surface of a droplet of viscous fluid on a flat surface. It has been further observed that such a phenomenon exists only for very small droplets. The formation of these small droplets is due to the fact that there exists a *tensile force* in the spherical surface of the droplet that exceeds the hydrostatic pressure in the contained fluid. This tensile force is called the *surface tension* of incompressible fluids.

5.6.1 Surface Tension

Surface tension in liquids relates to the cohesion forces of molecules. When a liquid is in contact with air or a solid, e.g., a pipe wall, intermolecular forces as described in Chapter 3 bond the molecules beneath the surface of the liquid. The molecules at the interface, on the other hand, are not bonded in the same way by its contacting neighbor medium. For liquid in contact with air, the surface is subjected to a tensile force that causes the contact surface to be concave upward, as observed in a capillary.

The nonslip boundary condition is principally due to the maximum *friction* existing between the fluid and the conduit wall. Surface tension also exists at the interface between the fluid and the solid contacting wall of the conduit, but is not a dominant force under normal circumstances. The combined effects of the maximum friction between the interface of the fluid/conduit wall and the surface tension in the fluid make the pressure that is required to drive fluid flow in conduits nonuniformly distributed across the lateral section of the conduit. Such nonuniformity is the principal cause for the well-known *capillary effect*. The capillary effect can affect the dynamics of fluid flowing in minute conduits, as is the case in many microsystems. Navier–Stokes equations that are derived on the basis of continuum fluid flow without accounting for the surface tensions of the fluid are no longer applicable in these cases. Demarcation between continuum and capillary flows is less than clear for incompressible fluids. Some discussion is available in [Madou 1997].

The surface tension F_s in a liquid can be expressed as:

Surface tension F_s	=	Wet perimeter S	×	Coefficient of surface tension γ

The *coefficient of surface tension* γ, with units of N/m, is a measure of the magnitude of the surface tension. As shown in White [1994], the γ value for water can be obtained by the following empirical formula:

$$\gamma(T) = 0.07615 - 1.692 \times 10^{-4}\,T \qquad \textbf{[5.23]}$$

where T is the temperature in °C and γ has units of N/m.

Surface tension has significant effect on liquid flow in microconduits, as additional pressure must be applied to overcome the surface tension of the fluid, which becomes more dominant in the flow at this scale. The following formulations [White 1994] may be used to assess such additional pressure drop due to surface tension.

Figure 5.14 illustrates the pressure change due to surface tension in a liquid cylinder and in the interior of a spherical droplet. The following relationships are derived on the principles of equilibrium of the pressure changes and the surface tensions in the liquid volumes as shown in Figure 5.14.

Figure 5.14 | Pressure change due to surface tension across liquid volumes.

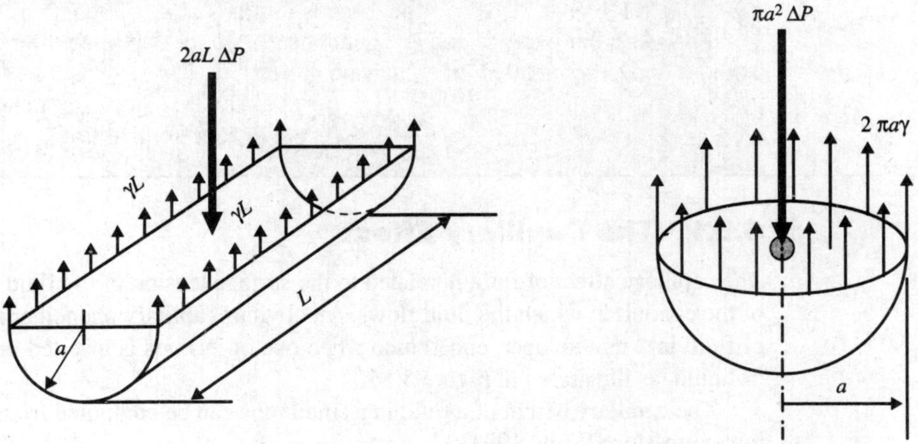

In the liquid cylinder in Figure 5.14a, the total force $(2aL)\,\Delta P$ is equal to the wet perimeter $(2L)$ times the coefficient of surface tension γ, from which we have:

$$\Delta P = \frac{\gamma}{a} \qquad \textbf{[5.24a]}$$

In the liquid sphere in Figure 5.14b, the total force $(\pi a^2)\,\Delta P$ is equal to the wet perimeter $(2\pi a)$ times the coefficient of surface tension γ, from which we have:

$$\Delta P = \frac{2\gamma}{a} \qquad \textbf{[5.24b]}$$

Consequently, we may estimate the total pressure change in a free-standing liquid inside a small tube with diameter $d \approx 2a$, as shown in Figure 5.15, to be the sum of the pressure changes in Equation (5.24a) and (5.24b) as:

$$\Delta P = \frac{3\gamma}{a}$$

Figure 5.15 | Fluid volume in a small tube.

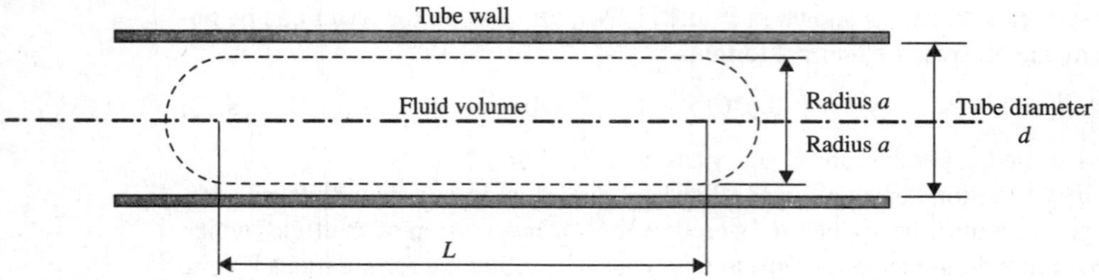

EXAMPLE 5.5

Determine the pressure required to overcome the surface tension of water in a small tube of 0.5 mm inside diameter. Assume that the water is at 20°C.

Solution

We first determine the surface tension coefficient of water at 20°C from Equation (5.23) to be $\gamma = 0.073$ N/m. The tube has a radius of $a = 250\ \mu m = 250 \times 10^{-6}$ m. Following the expressions in Equations (5.24a, b), we have the pressure required to overcome the surface tension as:

$$\Delta P = \frac{3\gamma}{a} = \frac{3 \times 0.073}{250 \times 10^{-6}} = 876\ \text{N/m}^2, \text{ or } 876\ \text{Pa}$$

5.6.2 The Capillary Effect

The capillary effect of fluid is related to the surface tension of the fluid and the size of the conduit in which the fluid flows. An obvious capillary phenomenon is the rise of fluid in a minute open-ended tube when one of its ends is inserted into a volume of liquid, as illustrated in Figure 5.16.

The capillary height of a fluid in a small tube can be computed from the following expression [White 1994]:

$$h = \frac{2\gamma \cos\theta}{wa} \qquad [5.25]$$

where $w = \rho g$ is the specific weight of the fluid, θ is the angle between the edge of the free fluid surface and the tube wall, and a is the radius of the capillary tube.

EXAMPLE 5.6

Find the height of water, h, rising in the small tube in Figure 5.16. The tube has a diameter of 1 mm.

Solution

We may assume that the angle θ of the water surface is very small, so that $\theta \approx 0°$. The radius of the tube is 0.5 mm, or 500×10^{-6} m. The surface tension coefficient of water at 20°C is $\gamma = 0.073$ N/m and the mass density of water is $\rho =$ 1000 kg/m^3, which leads to the specific gravity to be 9810 kg_f.

Figure 5.16 | Capillary effect on a fluid in a small conduit.

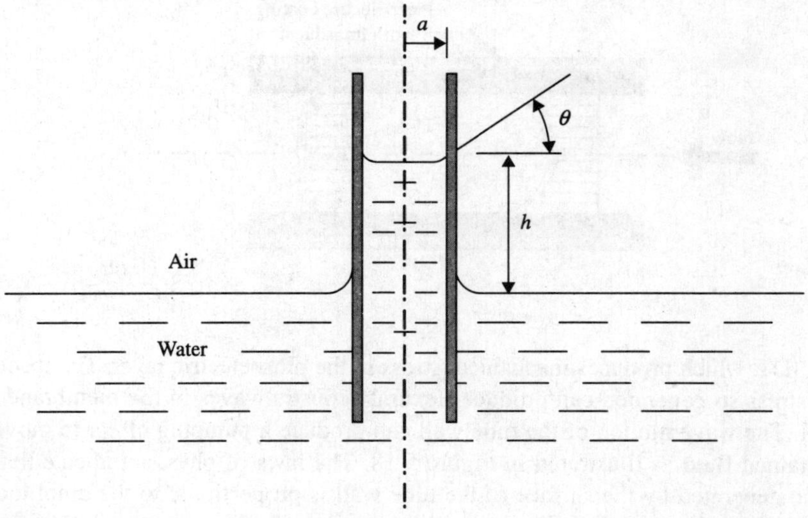

Thus by using the expression in Equation (5.25), we have the height of water column in the tube as:

$$h = \frac{2\gamma \cos \theta}{wa} = \frac{2(0.073) \cos (0)}{9810(500 \times 10^{-6})} = 0.02976 \text{ m, or } 2.9 \text{ cm}$$

5.6.3 Micropumping

A serious physical limitation on pumping liquid through microconduits is indicated in Equation (5.17), in which the pressure drop for a liquid flow in circular conduits (ΔP) is inversely proportional to the fourth power of the radius a and thus the diameter d of the conduit. As the diameter reduces to half, the required pumping power is increased 16 times. To make the situation worse, the surface tension effect also becomes more pronounced in liquid flow in minute conduits, as indicated in Equation (5.24). Conventional pumping methods for volumetric flow of liquids in micro conduits are thus not feasible in an engineering sense.

There are a number of ways that effective pumping of liquids in microscale conduits can be achieved [Madou 1997]. Electrohydrodynamic pumping with electro-osmosis and electrophoretic pumping methods such as presented in Chapter 3 are common practices in the industry. Following is another effective micropumping technique called *piezoelectric pumping.* It uses surface forces instead of the volumetric pressure to prompt the flow of liquids in minute conduits.

Figure 5.17 illustrates the working principle of piezoelectric pumping for microflow in a minute tube [Nyborg 1965, Madou 1997]. The tube usually has a thin wall in the order of a few micrometers. The thin membrane wall makes the tube highly flexible. The outside wall is coated with a piezoelectric film such as ZnO with aluminum interdigital transducers (IDTs). A radio-frequency (RF) voltage is applied to

Figure 5.17 | Capillary tube for microflow.

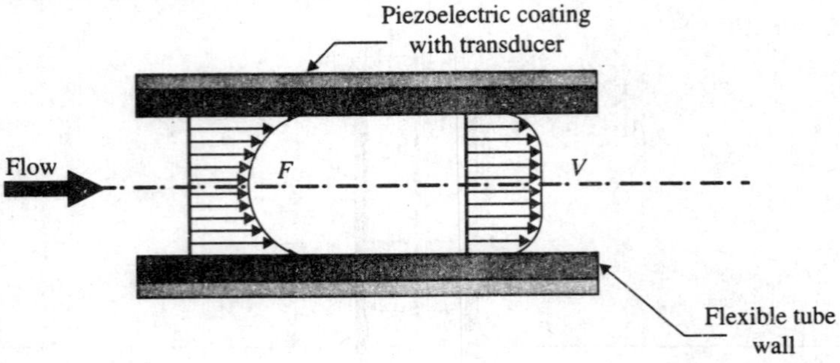

one IDT, which produces mechanical stress in the piezoelectric layer. The mechanical stress so generated can produce flexural acoustic waves in the membrane tube wall. The wave motion of the tube wall can produce a pumping effect to move the contained fluid as illustrated in Figure 5.18. The laws of physics indicate that the force generated by the surface of the tube wall is proportional to the amplitude of the acoustic wave generated in the wall by the piezoelectric effect, and it decays exponentially toward the center of the tube (see the variation of F in Fig. 5.17). This variation of the force results in a more uniform velocity of the fluid flow inside the tube (see the variation of V in Fig. 5.17).

Figure 5.18 | Microflow in wave stream in a capillary tube.

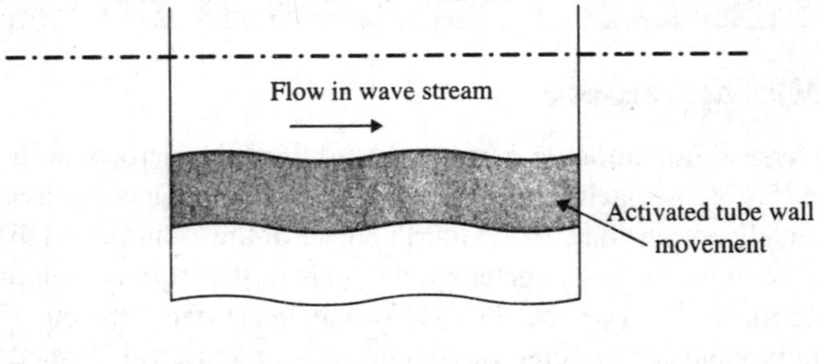

Acoustic streaming by using piezoelectricity as described above was demonstrated by Moroney et al. [Moroney 1990, 1991] to move air at a speed of 30 mm/s and water at 0.3 mm/s with 5 V supplied by a RF drive.

5.7 | FLUID FLOW IN SUBMICROMETER AND NANOSCALE

In Chapter 2, we predicted that bioMEMS that include miniature biomedical instruments and equipment and biosensors would be the next widely used MEMS products after pressure sensors. Indeed, the development of new bioMEMS products for biotechnology and the pharmaceutical industry has been reported frequently. Many of these applications involve micro fluidics for genome analyses or for separation or mixing microingredients in biomedical and pharmaceutical processes. Microfluidics in such applications typically involve fluid flow at low Reynolds numbers with low volumetric flow rates from a few microliters down to a few hundred nano liters per minute. The size of conduits for the flow varies from a few micrometers to a few hundred micrometers. Many of the fluids involved are gases, which behave quite differently from liquids because of their compressibility, and the more unique phenomenon of *rarefaction* in micro- and nanoscales. In this section, our attention will be focused on the behavior of gas flow in an extremely small space at low pressure and with low velocity.

5.7.1 The Rarefied Gas

The meaning of the word *rarefaction* as found in a dictionary [Merriam Webster 1995] is "a state or region of minimum pressure in a medium traversed by compression waves." The medium in the above statement is a gas that exists in a state of minimum pressure.

As we learned at the beginning of this chapter, fluids are aggregations of molecules. The molecules are much more widely spaced in gases than in liquids. Even in the natural state, gas molecules are randomly translating in the space and colliding with each other, or with the surface of an immersed body or the container. The phenomenon is much like particles in random motion [Patterson 1971]. For a given gas in a given state, there is the *mean free path* (MFP) length, λ, in which the gas molecules move without colliding with any obstacle. A value $\lambda = 65$ nm has been mentioned as a good approximation for most gases at atmospheric conditions [Beskok and Karniadakis 1999]. The MFP for liquids is about twice the size of the molecules.

5.7.2 The Knudsen and Mach Numbers

Apparently the validity of using the continuum fluid mechanics theories for fluid flow as stipulated in the foregoing sections depends on the magnitude of the *Knudsen number,* which is defined as:

$$K_n = \frac{\lambda}{L} \qquad [5.26]$$

in which L is the characteristic length scale, which is chosen to include the gradients of density, velocity, and temperature within the flow domain.

With MFP, $\lambda \approx 65$ nm, one may readily estimate the magnitude of the Knudsen number to be very small at 1.3×10^{-4} for the liquid flow in Example 5.5. On the other hand, $K_n = 1.3$ for the air entrapped in a space between a read/write head flying at a height of 50 nm and the surface of a fast-spinning disk in a computer data storage system.

The *Mach number* (Ma) of a moving gas is the measure of its velocity, as well as the compressibility of the gas. It is defined as:

$$\text{Ma} = V/\alpha \qquad [5.27]$$

where V is the velocity of the moving gas and α is the speed of sound in that gas. The corresponding speed of the sound of an ideal gas can be obtained by the expression:

$$\alpha^2 = \frac{c_p RT}{c_v} \qquad [5.28]$$

in which c_p and c_v are the respective specific heats of the gas at constant pressure and constant volume, R is the gas constant, and T is the absolute temperature of the gas.

For air at room temperature, the corresponding speed of sound is about 343 m/s. A gas is considered as incompressible if its Ma < 0.3.

We thus consider a rarefied gas flow at extremely low pressure to be incompressible because of its high Knudsen number ($K_n > 0.1$) and low Mach number (Ma < 0.3). Collision of gas molecules, which is characterized by the mean free path λ is expected to be dominant in the behavior of gas flow.

5.7.3 Modeling of Micro Gas Flow

A critical question that engineers involved in the design of microfluidics would ask is how far can they "stretch" the use of the classical governing Navier–Stokes equations, Equation (5.20), in the micro or nano regime. The spectrum of gas flow in Figure 5.19 may provide some useful clue for the answer. From this spectrum, we envisage that the continuum theories and thus the Navier–Stokes equations can be applied to model gas flows with Knudsen number $K_n < 0.01$. This application can be further extended to the range $0.01 < K_n < 0.1^+$ with proper modifications to accommodate the slip boundaries. Beyond the range, i.e., $K_n > 0.1^+$, other equations such as Burnett and modified Boltzmann equations will be required [Patterson 1971, Beskok and Karniadakis 1999]. These equations are beyond the scope of this book. Consequently, they will not be included in this chapter. The reader is referred to special books on gas dynamics such as Patterson [1971] for detailed description of these equations.

Figure 5.19 | Spectrum of gas flow regimes.

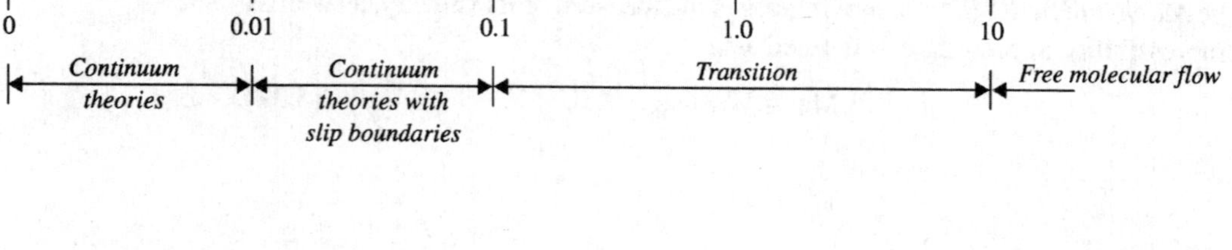

As indicated in Figure 5.19, gas flow is in the form of free molecule movement for $K_n > 10$. The mass flow rate for free gas molecules in a pipe can be obtained by using the following formula [Beskok and Karniadakis 1999]:

$$\dot{m}_{fm} = \frac{4d^3\,\Delta P}{3L}\sqrt{\frac{2\pi}{RT}} \qquad [5.29]$$

where d and L are the respective diameter and the length of the pipe, with L in the range of $d \leq L \leq \lambda$, the mean free path of gas molecules. Both d and L have units of meters (m). The unit for the pressure drop, ΔP is Pascal (Pa) or kg_f/m^2. The units for the specific gas constant R and the temperature T are J/kg-K and kelvins (K), respectively. The units for $\sqrt{RT}$ in Equation (5.29), however, are m/s. The units for the mass flow rate in the above equation are thus kg/s.

EXAMPLE 5.7

Estimate the flow rate of a nitrogen gas in a section of minute tube 30 nm in diameter × 50 nm long. A pressure difference of 0.5 Pa is applied to drive the flow. The flow is conducted at room temperature, 20°C.

■ Solution

We will need the following parameters for using Equation (5.29) for the mass flow rate of the gas: The assumption of the specific gas constant R = 286 J/kg-K for air is close enough for the nitrogen gas in question. By using the average mean free path length λ = 65 nm for gas in the present case, we have the Knudsen number $K_n = \lambda/d = 65/30 = 2.17$. Other parameters for the case are: L = 50 nm, d = 30 nm, ΔP = 0.5 Pa, T = 293 K.

Thus, by substituting the values of the above parameters into Equation (5.29), we have the rate of mass flow of the gas as:

$$\dot{m}_m = \frac{4(30 \times 10^{-9})^3(0.5)}{3(50 \times 10^{-9})}\sqrt{\frac{2\pi}{286 \times 293}} = 3.116 \times 10^{-18}\ \text{kg/s}$$

This extremely low mass flow rate practically means a stagnant gas in the nanoscale.

The same authors of the above reference [Beskok and Karniadakis 1999] proposed a unified model for two-dimensional gas flow that covers the entire regime of $0 \leq K_n \leq \infty$ for both pipes and channels.

For pipe flow, the normalized rate of gas flow can be obtained from the following expression:

$$\frac{\dot{m}}{\dot{m}_{fm}} = \frac{3\pi}{64\,K_n}(1 + \alpha\bar{K}_n)\left(1 + \frac{4\bar{K}_n}{1 - b\bar{K}_n}\right) \qquad [5.30]$$

where $\bar{K}_n$ is the Knudsen number evaluated at the average pressure, i.e., at $\bar{P} = (P_{in} + P_{out})/2$ in which P_{in} and P_{out} are the respective gas pressures at the inlet and outlet of the pipe. The parameter α is related to the *rarefaction coefficient* $C_r(K_n) = 1 + \alpha K_n$ and is determined by experiments.

For pipe flow, the approximate value of α can be obtained from the following expression:

$$\alpha \approx \left[\frac{64}{3\pi(1 - 4/b)}\right] \qquad [5.31]$$

The coefficient b is related to the slip boundary condition. The following values are used for b: $b = 0$ for nonslip boundaries, and $b < 0$ for slip boundaries. A value of $b = -1$ is used for gas flow in pipes, which leads to $\alpha = 1.358$ from Equation (5.31). Consequently, Equation (5.30) takes the following form:

$$\frac{\dot{m}}{\dot{m}_{fm}} = \frac{3\pi}{64\,K_n}(1 + 1.358\bar{K}_n)\left(1 + \frac{4\bar{K}_n}{1 + \bar{K}_n}\right) \qquad [5.32]$$

We will find that the above equation will lead to the ratio $\dot{m}/\dot{m}_{fm} = 1.001$ with an assumption of $K_n \approx \bar{K}_n = 2.17$ in Example 5.7.

EXAMPLE 5.8

Estimate the rate of airflow in the section of a small tube 10 μm in diameter and 1 cm in length. Assume that a pressure difference of 5 Pa is maintained between the inlet and outlet of the tube section. The airflow takes place at room temperature.

■ Solution

We will first evaluate the Knudsen number as:

$$K_n = \frac{\lambda}{d} = \frac{65 \times 10^{-9}}{10 \times 10^{-6}} = 0.0065$$

The equivalent free molecular mass flow rate in this case can be obtained from Equation (5.29) as:

$$\dot{m}_{fm} = \frac{4 \times (10 \times 10^{-6})^3 \times 5}{3 \times (10{,}000 \times 10^{-6})}\sqrt{\frac{2 \times 3.14}{286 \times 293}} = 5.77 \times 10^{-15}\ \text{kg/s}$$

Assume that $\bar{K}_n \approx K_n = 0.0065$. From Equation (5.32), we then have:

$$\frac{\dot{m}}{\dot{m}_{fm}} = \frac{3 \times 3.14}{64 \times 0.0065}(1 + 1.385 \times 0.0065)\left(1 + \frac{4 \times 0.0065}{1 + 0.0065}\right) = 23.44$$

from which we have the mass flow rate of the air as:

$$\dot{m} = 23.44 \times 5.77 \times 10^{-15} = 0.13524 \times 10^{-12}\ \text{kg/s, or } 0.487\ \mu\text{g/h}$$

5.8 | OVERVIEW OF HEAT CONDUCTION IN SOLIDS

We have learned from Chapter 2 that many MEMS devices are actuated by thermal means. Key issues involved in the design of these actuators are

1. The amount of heat that is required to invoke the desired action
2. The time required for initiating and terminating the action
3. The associated thermal stresses or distortions induced in the device
4. The possible damage to the delicate components of the device due to the heating

All these issues are less a problem for devices in macroscale, as there are established theories and formulations available to engineers to address these issues. However, the situation in submicrometer and nanoscale can be significantly different, and most theories and formulations derived for macroscale require significant modifications, as the mechanisms for heat transmission in solids in these scales are radically different. As in the section on fluid mechanics, we will begin the subject of heat transfer with a review of the formulations for macroscale. Necessary modifications of these formulations for submicrometer scale will follow thereafter.

5.8.1 General Principle of Heat Conduction

Let us first take a look at heat conduction in a solid slab as illustrated in Figure 5.20. In the figure is a solid slab with the temperature at the left side wall maintained at T_a and the right side wall maintained at T_b, with $T_a > T_b$. The temperature difference between the two walls causes the heat to flow from the left side to the right side of the slab. One would envisage that the total amount of heat flow through the slab, Q, is proportional to the cross-sectional area A, the temperature difference between the two faces, and the time t allowed for the heat to flow. However, the amount of heat flow is inversely proportional to the distance that the heat has to travel, i.e., the thickness of the slab, d. Mathematically, one may express the above qualitative correlation in the form:

Figure 5.20 | Conduction of heat in a solid slab.

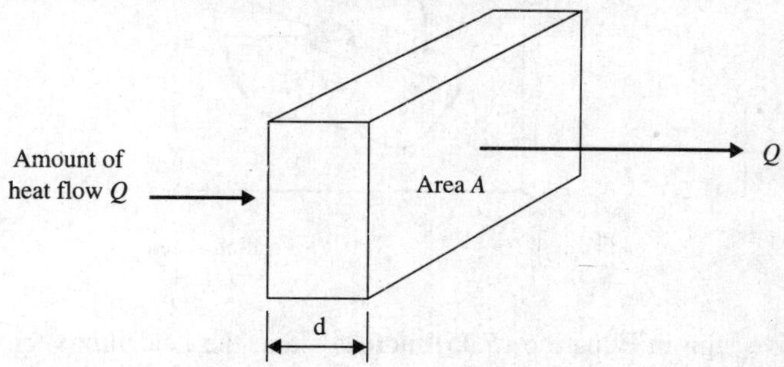

$$Q \propto \frac{A(T_a - T_b)t}{d}$$

Expressing the above relationship in the form of an equation, we have:

$$Q = k\frac{A(T_a - T_b)t}{d} \qquad [5.33]$$

where the proportionality constant k is the *thermal conductivity* of the solid, which has units of Btu/in-s-°F in the imperial system, or W/m-°C in the SI system.

Thermal conductivity is a material property. The value of k for solids generally increases with the temperature. However, for most engineering materials at realistic operating temperature ranges, k can be regarded as constant. The value of k is a measure of how conductive the material is to heat. Thus gold, silver, copper, and aluminum have heat conducting capability in that order. These materials have higher k values than many other materials, and they are much better heat conducting materials than ceramics such as SiC, SiO_2, or Al_2O_3.

5.8.2 Fourier Law of Heat Conduction

Equation (5.33) provides us with a way to account for the total heat flow in a planar slab. A more realistic quantity to account for in a heat conduction analysis is *heat flux q*, defined as the heat flow per unit area and time. Thus, from Equation (5.33), we may express the heat flux in the slab as:

$$q = \frac{Q}{At} = k\frac{(T_a - T_b)}{d} \qquad [5.34]$$

Heat flux is a vector quantity, as illustrated in Figure 5.21. It can be related to the associated temperature gradient by the *Fourier law of heat conduction* as follows.

With reference to Figure 5.21, the heat flux in a solid situated in a space defined by **r**: (x, y, z) coordinate system can be expressed as:

$$q(\mathbf{r}, t) = -k\,\nabla\,T(\mathbf{r}, t) \qquad [5.35]$$

Figure 5.21 | Heat flux in a solid.

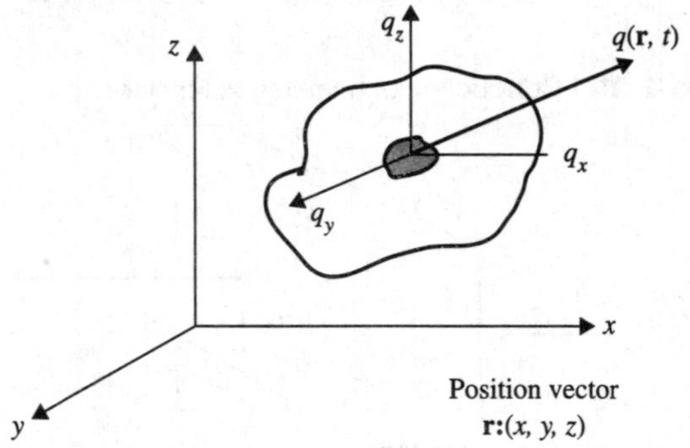

The negative sign in Equation (5.35) indicates that the heat flux vector, $q(\mathbf{r}, t)$ is along the outward normal to the surface of the solid. From Equations (5.33) and (5.34), we may assign the unit for heat flux to be W/m² or J/m²-s or N/m-s.

Equation (5.35) is the mathematical expression of the Fourier law of heat conduction. Expanding the equation in cartesian coordinates will lead to the following expression for heat flux in a solid:

$$q(x, y, z, t) = \sqrt{q_x^2 + q_y^2 + q_z^2} \qquad [5.36]$$

in which

$$q_x = -k_x \frac{\partial T(x, y, z, t)}{\partial x} \quad [5.37a]$$

$$q_y = -k_y \frac{\partial T(x, y, z, t)}{\partial y} \quad [5.37b]$$

$$q_z = -k_z \frac{\partial T(x, y, z, t)}{\partial z} \quad [5.37c]$$

where q_x, q_y, and q_z are the respective heat flux components in the x, y, and z directions as shown in Figure 5.21. The terms k_x, k_y, and k_z are the respective thermal conductivities of the solid in the x, y, and z directions. For isotropic materials, $k_x = k_y = k_z$.

5.8.3 The Heat Conduction Equation

When a solid is subjected to a heat input from a heat source, or dissipates heats to the surrounding medium which acts as a heat sink, the temperature field, i.e., the temperature distribution in the solid, will follow a state of nature after reaching a state of equlibrium. The temperature field, $T(\mathbf{r}, t)$, with $\mathbf{r}$: (x, y, z) in cartesian coordinates, or (r, θ, z) in cylindrical polar coordinates, can be obtained by solving the following heat conduction equation:

$$\nabla^2 T(\mathbf{r}, t) + \frac{Q(\mathbf{r}, t)}{k} = \frac{1}{\alpha} \frac{\partial T(\mathbf{r}, t)}{\partial t} \quad [5.38]$$

where the Laplacian is defined as follows:

$$\nabla^2 = \frac{\partial^2}{\partial x^2} + \frac{\partial^2}{\partial y^2} + \frac{\partial^2}{\partial z^2}$$

in cartesian coordinates, or

$$\nabla^2 = \frac{\partial^2}{\partial r^2} + \frac{1}{r}\frac{\partial}{\partial r} + \frac{1}{r^2}\frac{\partial^2}{\partial \theta^2} + \frac{\partial^2}{\partial z^2}$$

in cylindrical polar coordinates.

The term $Q(\mathbf{r}, t)$ in Equation (5.38) is the heat generated by the material per unit volume and time. A common heat source in microsystems is electric resistance heating in which the electric current is allowed to pass through an ohmic conductor such as a copper or aluminum film. The heat generated by electric resistance heating can be evaluated by the following relationship:

$$\boxed{\text{Power } P \text{ in watts (W)}} = \boxed{\text{Current, } i \text{ in amperes } (A)}^2 \times \boxed{\text{Resistance } R \text{ in ohms } (\Omega)}$$

Power in the above expression has units of watts, which is equivalent to 1 J/s. It is also equivalent to 1 N-m/s in SI units.

The constant α in Equation (5.38) is called *thermal diffusivity* of the material, with units of m^2/s. It has an important physical meaning as a measure of how fast heat can conduct in solids. Mathematically, it is equal to:

$$\alpha = \frac{k}{\rho c} \tag{5.39}$$

in which ρ and c are respectively the mass density and specific heat of the solid. The units for ρ are g/cm^3, and the units for c are J/g-°C.

It is apparent that a solid with higher value of α can conduct heat faster than the one with a lower value of α. Thus, a thermally-actuated microdevice made of materials with higher α values will respond to actuation more rapidly than those made of materials with low α values.

5.8.4 Newton's Cooling Law

We have learned that heat transmission in solids is in the mode of conduction, and that Fourier's law of heat conduction in Equation (5.35) governs such transmission. Heat transfer in fluids is quite different; convection is the mode for heat flow in fluids. *Newton's cooling law* is used as the basis for convective heat transfer analysis in fluids.

With reference to the fluid in Figure 5.22, the heat flux between any two points with respective temperatures T_a and T_b is proportional to the temperature difference between these two points, i.e., $q \propto (T_a - T_b)$, from which we can express Newton's cooling law by the following equation:

$$q = h(T_a - T_b) \tag{5.40}$$

Figure 5.22 | Heat flux in a fluid.

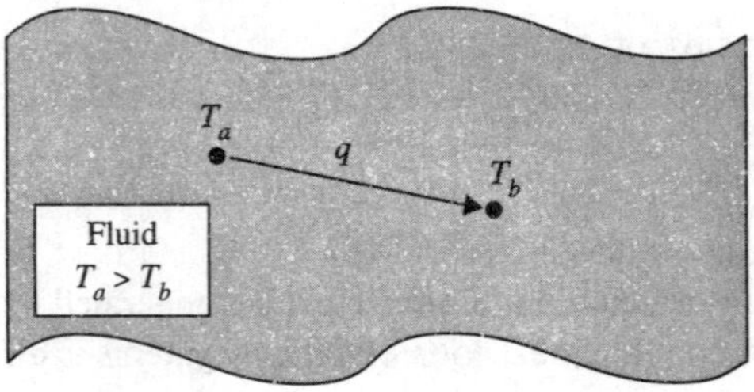

The constant h in Equation (5.40) has several different names. We will call it *heat transfer coefficient* in our analyses.

The value of h is evaluated by dimensional analysis, with the analytical parameters determined by experiments. It is usually embedded in the *Nusselt number*, $Nu = hL/k$, with L the characteristic length. The following relations are normally used for determining the numerical values of the Nusselt number:

$$Nu = \alpha(Re)^{\beta}(Pr)^{\gamma}$$

for forced convection, and

$$\text{Nu} = \alpha(\text{Re})^{\beta}(\text{Pr})^{\gamma}(\text{Gr})^{\delta}$$

for free convection with low velocity.

The parameters α, β, γ, and δ in the above expressions are determined by experiments. Other numbers involved in the above expressions are Re = *Reynolds number,* Pr = *Prandtl number,* and Gr = *Grashoff number.* Of these numbers in the above expressions, the Reynolds number has the dominant effect on the values of h. The Reynolds number was defined in Equation (5.3) as:

$$\text{Re} = \frac{\rho L V}{\mu} \qquad \textbf{[5.3]}$$

where ρ and μ are the respective mass density and dynamic viscosity of the fluid, and V is the velocity of fluid flow.

The Prandtl number and the Grashoff number are defined as follows:

$$\text{PR} = \frac{c_p\,\mu}{k} \qquad \textbf{[5.41a]}$$

$$\text{Gr} = \frac{L^3\,\rho^2\,g}{\mu^2\,(\beta\,\Delta t)} \qquad \textbf{[5.41b]}$$

in which c_p is the specific of heat of fluids under constant pressure, β is the volumetric coefficient of thermal expansion, Δt is the duration, and g is the gravitational acceleration.

The Grashoff number usually appears in analyses that involve natural convection.

For most convective heat transfers, the h value is related to the velocity of the fluid, or $h \propto V^{\phi}$, where ϕ is a constant determined by experiments.

5.8.5 Solid–Fluid Interaction

As described in Chapter 2, many thermally-actuated MEMS devices involve transferring heat from solid members to the surrounding fluids in contact, or vice versa. We will also learn from several microfabrication processes in Chapter 8, e.g., chemical vapor deposition, that gaseous fluids are made to flow over the surfaces of substrate solids and thereby transfer heat between the two media. Since heat transmission in solids differs from that in fluids, it is thus necessary to learn how these two modes of heat transfer interact at the solid/fluid interfaces.

Let us take a look at the situation illustrated in Figure 5.23. Heat is being dissipated from the solid with a temperature field $T(\mathbf{r}, t)$ into the surrounding fluid at a temperature T_f. The reverse situation, of course, is also possible.

Let us first take a closer look at the interface of the solid and the fluid. We realize that a *boundary layer* is built in the fluid immediately adjacent to the solid surface. The thickness of this layer depends on the velocity of the fluid over the solid surface and the property of the fluid. This layer creates a barrier for freer heat transfer between the solid and the fluid. A resistance to heat flow is generated in this layer.

Figure 5.23 | Heat flow between solid and fluid.

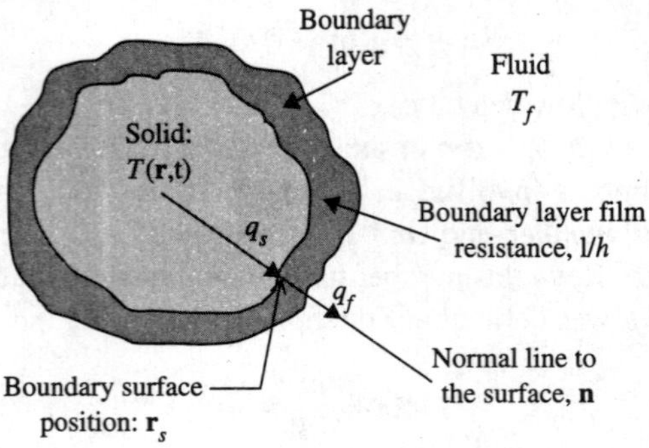

Thus, because of this resistance, the surface temperature of the solid is not equal to the bulk fluid temperature, T_f. Such thermal resistance may be numerically evaluated to be $1/h$, in which h is the heat transfer coefficient of the fluid.

By equating the heat flux vector q_s entering the interface from the solid side and the heat flux q_f leaving the interface into the bulk fluid, as illustrated in Figure 5.23, we obtain the following relation:

$$-k\frac{\partial T(\mathbf{r}, t)}{\partial \mathbf{n}}\bigg|_{\mathbf{r}_s} = h[T(\mathbf{r}_s, t) - T_f] \qquad [5.42]$$

at the interface. The vector **n** is the normal vector at the point of interest on the specific boundary between the solid and the fluid.

5.8.6 The Boundary Conditions

Equation (5.38) is used to determine the temperature distribution $T(\mathbf{r}, t)$ in a MEMS device component. The temperature distribution is then used to identify the location and the magnitude of the maximum temperature in the component, as well as to determine the associate thermal stresses as demonstrated in Section 4.4.3. Both the maximum temperature and the temperature-induced thermal stresses are primary design considerations of these components. They are thus an important part of the procedure in the design of microsystem involving thermal effects.

The solution of Equation (5.38) requires proper formulation of boundary conditions. There are three types of boundary conditions that can be used in thermal analyses. These are as presented below:

1. Prescribed Surface Temperature This type of boundary condition is applied to a solid with the temperature at specific locations on the surface specified. With reference to Figure 5.24a, the boundary conditions at the location $\mathbf{r}_s$ can be expressed as:

$$T(\mathbf{r}, t)|_{\mathbf{r} = \mathbf{r}_s} = f(\mathbf{r}_s, t) = F(t) \qquad [5.43]$$

Of course, the function $F(t)$ in Equation (5.43) can be a constant in special cases.

Figure 5.24 | Prescribed surface temperature or heat flux on solid surfaces.

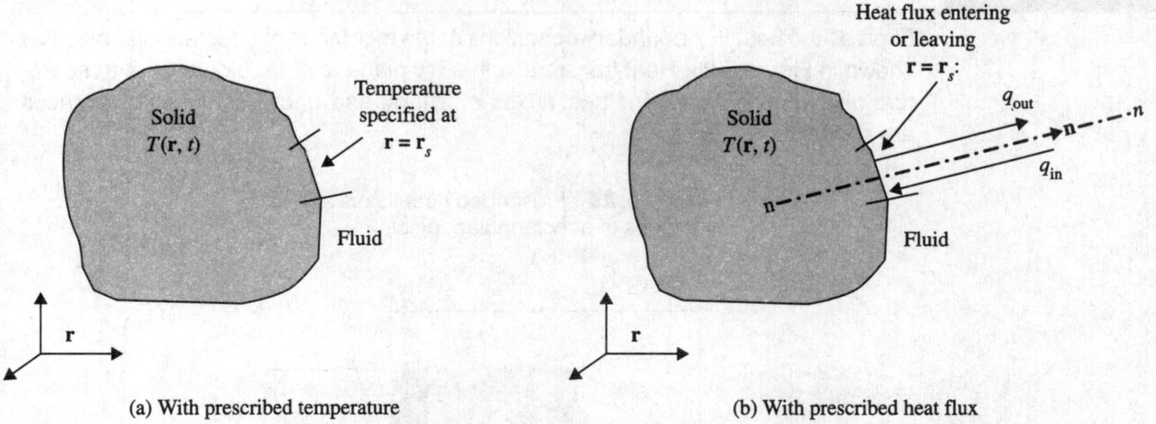

2. Prescribed Heat Flux at the Boundary Figure 5.24b illustrates the case in which the heat flux entering or leaving the part of the boundary at $\mathbf{r} = \mathbf{r}_s$ is specified. As we have learned from Equations (5.35) and (5.37), the heat flux in a solid can be expressed by the Fourier law of heat conduction. Thus, by assuming the line *n–n* to be the *outward normal* line to the surface at which the heat flux enters or leaves, we may express the corresponding boundary conditions with q_{in} and q_{out} specified as follows:

$$q_{\text{in}} = +k \left. \frac{\partial T(\mathbf{r}, t)}{\partial \mathbf{n}} \right|_{\mathbf{r}=\mathbf{r}_s} \quad \text{[5.44a]}$$

or

$$q_{\text{out}} = -k \left. \frac{\partial T(\mathbf{r}, t)}{\partial \mathbf{n}} \right|_{\mathbf{r}=\mathbf{r}_s} \quad \text{[5.44b]}$$

The reader will notice the plus and minus signs attached to the heat flux in Equation (5.44). These signs are attached to designate the direction of the heat flux along the outward normal line to the surface. Table 5.1 below will help the reader to assign a proper sign to the temperature gradients in the above equations.

Table 5.1 | Signs for specified heat flux boundary conditions

Sign of outward normal *n*	Is *q* along the same direction as *n*?	Sign of *q* in Equation (5.44)
+	Yes	–
+	No	+
–	Yes	+
–	No	–

EXAMPLE 5.9

Express the heat flux boundary conditions at the four faces of a rectangular block as shown in Figure 5.25. Heat transfers in the x–y plane and the block has a temperature distribution $T(x, y)$. The heat fluxes q_1, q_2, q_3, and q_4 crossing these four faces are specified.

Figure 5.25 | Prescribed heat fluxes across four faces of a rectangular object.

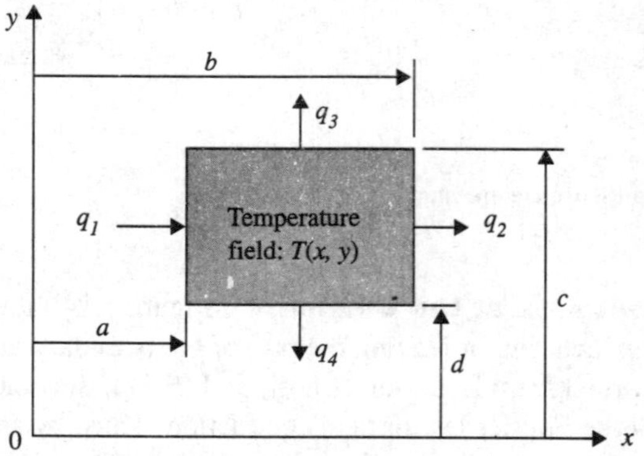

■ Solution

By using Equation (5.44) and Table 5.1, we may specify the following boundary conditions:

At the left face:

$$\left.\frac{\partial T(x, y)}{\partial x}\right|_{x=a} = -\frac{q_1}{k} \quad \text{[a]}$$

At the right face:

$$\left.\frac{\partial T(x, y)}{\partial x}\right|_{x=b} = -\frac{q_2}{k} \quad \text{[b]}$$

At the top face:

$$\left.\frac{\partial T(x, y)}{\partial y}\right|_{y=c} = -\frac{q_3}{k} \quad \text{[c]}$$

At the bottom face:

$$\left.\frac{\partial T(x, y)}{\partial y}\right|_{y=d} = +\frac{q_4}{k} \quad \text{[d]}$$

in which k is the thermal conductivity of the solid.

3. Convective Boundary Conditions This type of boundary condition applies to the situation when the solid boundary is in contact with fluid at bulk fluid temperature T_f as shown in Figure 5.26. The boundary conditions for this situation are expressed by rearranging Equation (5.42) to give:

$$\left.\frac{\partial T(\mathbf{r}, t)}{\partial \mathbf{n}}\right|_{\mathbf{r}=\mathbf{r}_s} + \frac{h}{k}\, T(\mathbf{r}, t)\big|_{\mathbf{r}=\mathbf{r}_s} = \frac{h}{k}\, T_f \qquad [5.45]$$

Figure 5.26 | A solid surrounded by fluid.

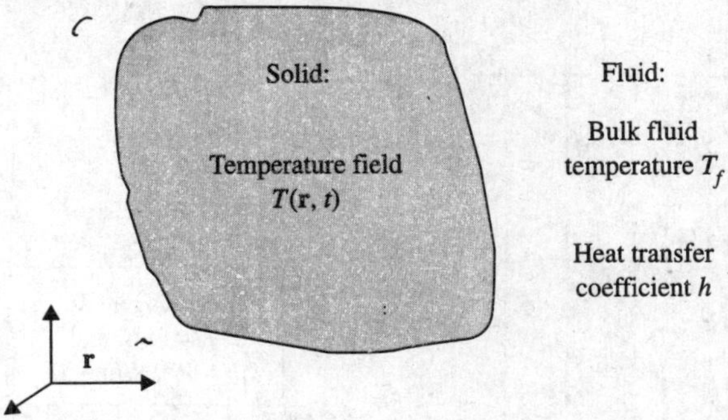

One will realize from the above formula that the case $h \to \infty$ is equivalent to the prescribed surface boundary conditions in Equation (5.43), whereas the case of $h = 0$ leads to the insulated boundary condition in which

$$\left.\frac{\partial T(\mathbf{r}, t)}{\partial \mathbf{r}}\right|_{\mathbf{r}=\mathbf{r}_s} = 0$$

in Equation (5.44).

EXAMPLE 5.10

Show the differential equation and the appropriate initial and boundary conditions for a thermally actuated microbeam as illustrated in Figure 5.27. A thin copper film is attached to the top surface of the silicon beam and used as a resistant heater. The actuator is initially at 20°C. Consider two cases for the contacting air at the bottom surface of the beam: (1) still air, (2) the air has a bulk temperature of 20°C and a heat transfer coefficient of 10^{-4} W/m²-°C.

Figure 5.27 | A thermally actuated beam.

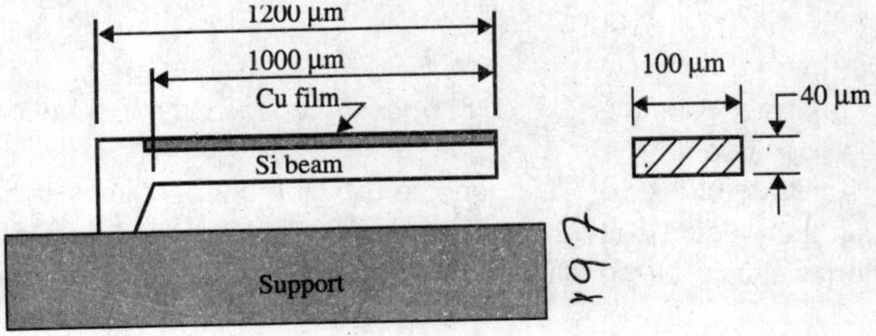

■ **Solution**

We may consider that the induced temperature field in the beam predominantly varies in the thickness of the beam. It is thus reasonable to assume a temperature function $T(x, t)$ in the beam, with x being the coordinate in the thickness direction as illustrated in Figure 5.28.

Figure 5.28 | Heat conduction in a silicon beam.

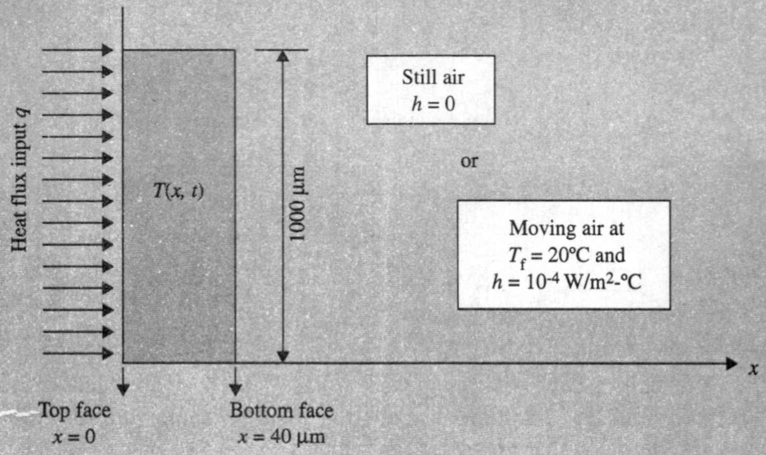

The governing differential equation, from the general form in Equation (5.38), for the present case is:

$$\frac{\partial^2 T(x, t)}{\partial x^2} = \frac{1}{\alpha}\frac{\partial T(x, t)}{\partial t} \qquad \textbf{[5.46]}$$

The thermal diffusivity α in Equation (5.46) for silicon beam can be obtained by the thermophysical properties listed in Table 7.3 in Chapter 7, with the thermal conductivity k = 1.57 W/cm-°C, the mass density ρ = 2.3 g/cm^3, and the specific heat c = 0.7 J/g-°C, which gives α = 0.9752 cm^2/s.

The initial condition is

$$T(x, t)|_{t=0} = 20°\text{C} \qquad \textbf{[a]}$$

For the boundary condition at the top surface, x = 0, we have the heat flux input condition:

$$\left.\frac{\partial T(x, t)}{\partial x}\right|_{x=0} = -\frac{q}{k}$$

where q = heat flux generated by the copper film and k is the thermal conductivity of the silicon beam.

The magnitude of the heat flux generated by the copper film is $q = i^2R/A$, in which i is the electric current passing through the film (amperes) and R is the electric resistance of the film (Ω). The top surface area of the beam A is

1000 μm × 100 μm, or 10^{-3} cm². The boundary condition in the above expression for the present case is thus equal to:

$$\left.\frac{\partial T(x,t)}{\partial x}\right|_{x=0} = -\frac{i^2R}{1.57 \times 10^{-3}}\ °\text{C/cm} \qquad \text{[b]}$$

We may apply a factor of 10^{-2} to the right-hand side of Equation (b) to convert to °C/m for consistency with the units for other parameters.

Two possible boundary conditions can be applied at the bottom surface of the beam; (1) the insulated boundary condition with still air and (2) the convective boundary condition:

1. The *insulated boundary condition* (i.e., $h = 0$) at the bottom surface of the beam ($x = 40$ μm). This boundary condition makes the solution of Equation (5.46) relatively easy. However, it is valid on the assumption that the temperature of the entrapped air beneath the bottom surface of the beam has not changed. Thus, the temperature solution $T(x, t)$ derived from using this boundary condition is applicable for the very early stage of the heating process before any significant temperature rise takes place in the entrapped air. Mathematically, we can express this condition in the following form:

$$\left.\frac{\partial T(x,t)}{\partial x}\right|_{x=40\times10^{-6}\,\text{m}} = 0 \qquad \text{[c]}$$

2. The *convective boundary condition* at the bottom surface of the beam ($x = 40$ μm). The mathematical expression for the boundary condition is represented by Equation (5.45) as:

$$\left.\frac{\partial T(x,t)}{\partial x}\right|_{x=40\times10^{-6}\,\text{m}} + \frac{10^{-4}}{157}\,T(x,t)\Big|_{x=40\times10^{-6}\,\text{m}} = \frac{10^{-4}}{157}\times 20 \qquad \text{[d]}$$

The reader is reminded that the value $k = 157$ W/m-°C is used in Equation (d) for consistency of the units used in the problem.

The temperature field in the beam, $T(x, t)$, can be obtained by solving Equation (5.46) with the conditions in Equations (a), (b), and (c), or the conditions in Equations (a), (b), and (d). Various ways are available for solving Equation (5.46), e.g., the separation of variables technique or the integral transformation method [Ozisik 1968]. There are also various numerical techniques available for the same solution.

5.9 | HEAT CONDUCTION IN MULTILAYERED THIN FILMS

Many MEMS structures are made of multilayered thin films of different materials. These structures are often used in devices fabricated by surface micromachining, as will be described in Chapter 9. Thin films are typically in the sandwich form and are subjected to thermal loads as in the case of actuators and microvalves. A heat conduction analysis is required not only to assess the sensitivity of the structure to the thermal loading, but is required also as input to thermal stress analysis as presented in Chapter 4.

Complete heat conduction analysis of multilayered media requires complex mathematical derivations, and is beyond the scope of this book. What we will present below is a simple one-dimensional formulation for such analyses. Layers in the structure are assumed to be in perfect thermal contact at the interfaces. Detailed derivation of the formulation is available in a few books on heat conduction, such as [Ozisik 1968].

Figure 5.29 | Temperatures in a multilayered structure.

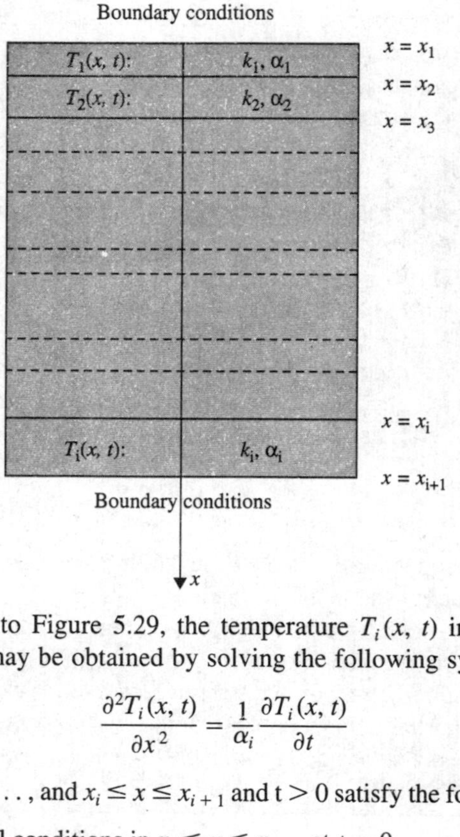

With reference to Figure 5.29, the temperature $T_i(x, t)$ in each layer of the i-layered structure may be obtained by solving the following system of equations:

$$\frac{\partial^2 T_i(x, t)}{\partial x^2} = \frac{1}{\alpha_i}\frac{\partial T_i(x, t)}{\partial t} \quad \text{[5.47]}$$

in which $i = 1, 2, 3, \ldots$, and $x_i \leq x \leq x_{i+1}$ and $t > 0$ satisfy the following conditions:

Prescribed initial conditions in $x_i \leq x \leq x_{i+1}$ at $t = 0$

Prescribed boundary conditions at $x = 0$ and $x = x_{i+1}$ for $t > 0$

In Figure 5.29 and Equation (5.47), k_i and d_i with $i = 1, 2, 3, \ldots$ denote the respective thermal conductivity and thermal diffusivity of the material in layer i.

Other conditions applicable to the problem involve the continuities of temperatures and heat fluxes at the interfaces, i.e.:

$$T_i(x_{i+1}, t) = T_{i+1}(x_{i+1}, t) \qquad \text{for } i = 1, 2, 3, \ldots$$

$$k_i \frac{\partial T_i(x_{i+1}, t)}{\partial x} = k_{i+1}\frac{\partial T_{i+1}(x_{i+1}, t)}{\partial x} \qquad \text{for } i = 1, 2, 3, \ldots$$

EXAMPLE 5.11

The structure of a thermal actuator is made of a compound beam involving silicon and SiO_2 as illustrated in Figure 5.30. A thin copper film is deposited on the top of the SiO_2 layer as the resistant heater. This heater will provide a maximum temperature of 50°C at the top surface of the SiO_2 layer. Determine the time required for the silicon beam to reach the input surface temperature of 50°C. Since heat will predominantly flow through the thickness of the compound beam because of the short distance of the passage, a one-dimensional heat conduction analysis along the thickness direction is justified.

Figure 5.30 | A thermal actuator made of a compound beam.

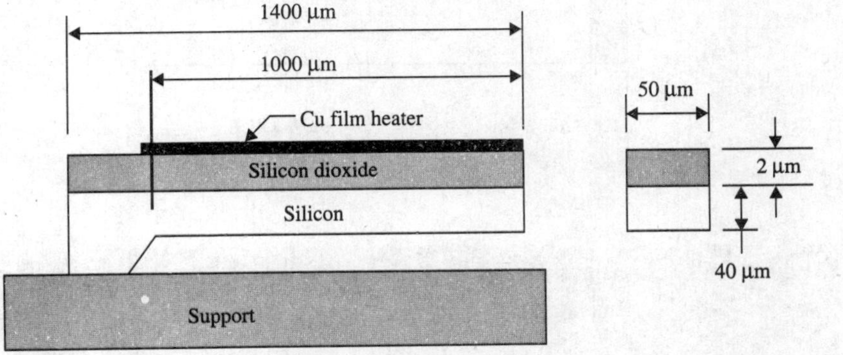

Given material properties are:

Thermal conductivities: $k_1 = 1.4$ W/m-°C for SiO_2 and $k_2 = 157$ W/m-°C for silicon

Thermal diffusivities: $\alpha_1 = 0.62 \times 10^{-6}$ m²/s for SiO_2 and $\alpha_2 = 97.52 \times 10^{-6}$ m²/s for silicon.

■ Solution

We assume that the temperatures $T_1(x, t)$ and $T_2(x, t)$ in the respective SiO_2 and silicon layers can be computed from Equation (5.47) and the similar arrangement shown in Figure 5.29. We may use the model illustrated in Figure 5.31 for this specific problem.

The following system of differential equations is used to solve the problem:

$$\frac{\partial^2 T_1(x,t)}{\partial x^2} = \frac{1}{\alpha_1}\frac{\partial T_1(x,t)}{\partial t} \quad \text{for the SiO}_2 \text{ layer with } 0 \le x \le a \qquad [5.48a]$$

$$\frac{\partial^2 T_2(x,t)}{\partial x^2} = \frac{1}{\alpha_2}\frac{\partial T_2(x,t)}{\partial t} \quad \text{for the silicon beam with } a \le x \le b \qquad [5.48b]$$

The initial conditions are:

$$T_1(x,t)|_{t=0} = F_1(x) = 20°\text{C} \qquad [5.49a]$$

$$T_2(x,t)|_{t=0} = F_2(x) = 20°\text{C} \qquad [5.49b]$$

Figure 5.31 | Heat conduction through a two-layer compound beam.

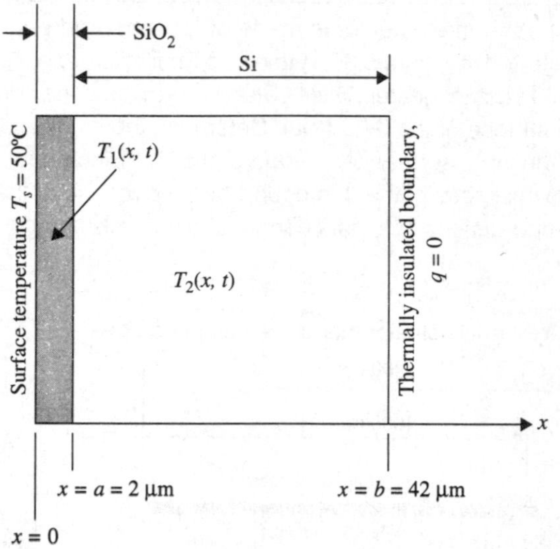

The boundary conditions are:

$$T_1(x, t)|_{x=0} = 50°\text{C} \quad \textbf{[5.49c]}$$

$$\left.\frac{\partial T_2(x, t)}{\partial x}\right|_{x=b=42\mu m} = 0 \quad \textbf{[5.49d]}$$

The compatibility conditions at the interface are:

$$T_1(x, t)|_{x=a=2\mu m} = T_2(x, t)|_{x=a=2\mu m} \quad \textbf{[5.49e]}$$

$$k_1 \left.\frac{\partial T_1(x, t)}{x}\right|_{x=a=2\mu m} = k_2 \left.\frac{\partial T_2(x, t)}{x}\right|_{x=a=2\mu m} \quad \textbf{[5.49f]}$$

There are several ways to solve the temperature distribution functions $T_1(x, t)$ and $T_2(x, t)$ from Equation (5.48) with the conditions specified in Equation (5.49). The following solution is obtained by using the integral transform method presented in Ozisik [1968]:

$$T_i(x, t) = \sum_{n=1}^{\infty} \frac{e^{-\beta_{nt}^2}}{N_n} \Psi_{in}(x)\left[\frac{k_1}{\alpha_1} \int_0^a \Psi_{1n}(x)F_1(x)\,dx + \frac{k_2}{\alpha_2} \int_a^b \Psi_{2n}(x)F_2(x)\,dx\right] \text{ with } i = 1, 2 \quad \textbf{[5.50]}$$

where the functions $F_1(x)$ and $F_2(x)$ are the respective initial conditions in SiO_2 and silicon. The norm N_n is defined as:

$$N_n = \frac{k_1}{\alpha_1} \int_0^a \Psi_{1n}^2(x)\,dx + \frac{k_2}{\alpha_2} \int_a^b \Psi_{2n}^2(x)\,dx$$

The functions $\Psi_{in}(x)$, with $i = 1, 2$, have the following forms:

$$\Psi_{1n}(x) = \sin\left(\frac{\beta_n}{\sqrt{\alpha_1}}x\right)$$

and

$$\Psi_{2n}(x) = A_{2n}\sin\left(\frac{\beta_n}{\sqrt{\alpha_1}}x\right) + B_{2n}\cos\left(\frac{\beta_n}{\sqrt{\alpha_2}}x\right)$$

The coefficients A_{2n} and B_{2n} have the forms:

$$A_{2n} = \sin\omega\sin\left(\frac{a\lambda}{b}\right) + \mu\cos\omega\cos\left(\frac{a\lambda}{b}\right)$$

and

$$B_{2n} = -\mu\cos\omega\sin\left(\frac{a\lambda}{b}\right) + \sin\omega\cos\left(\frac{a\lambda}{b}\right)$$

with

$$\omega = \frac{\beta_n a}{\sqrt{\alpha_1}} \qquad \lambda = \frac{\beta_n b}{\sqrt{\alpha_2}} \qquad \mu = \frac{k_1}{k_2}\frac{\sqrt{\alpha_2}}{\sqrt{\alpha_1}}$$

The eigenvalue β_n in the above formulations can be obtained by solving the following transcendental equation:

$$\begin{vmatrix} \sin\omega & -\sin\left(\frac{a\lambda}{b}\right) & -\cos\left(\frac{a\lambda}{b}\right) \\ \mu\cos\omega & -\cos\left(\frac{a\lambda}{b}\right) & \sin\left(\frac{a\lambda}{b}\right) \\ 0 & \cos\lambda & -\sin\lambda \end{vmatrix} = 0$$

Numerical solutions for the temperature distributions in Equation (5.50) for both SiO_2 and silicon layers are obtained by using a commercial software package, MathCad. The temperature variations in both layers at selected instants are plotted as shown in Figure 5.32, from which we can determine the time required for the silicon layer to reach the input temperature of 50°C as 600 μs. This information will enable the design engineer to assess the sensitivity of the thermally-actuated device. The expression of temperature distributions in Equation (5.50) is also used as input to thermal stress and the interfacial fracture analyses such as illustrated in Sections 4.4 and 4.5.

Figure 5.32 | Temperature variation in a bilayer actuator.

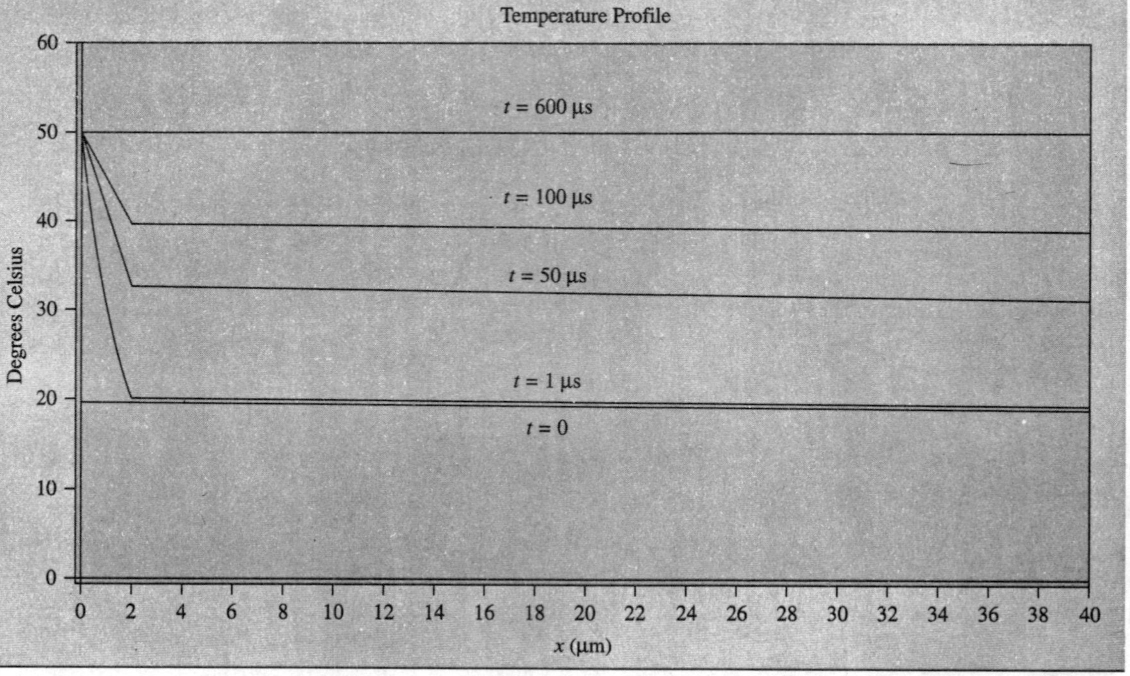

5.10 | HEAT CONDUCTION IN SOLIDS IN SUBMICROMETER SCALE

A well-known fact is that heat is a form of energy. As such, the flow of heat requires a carrier that is similar to other energy carriers. The carriers for heat vary from substances to substance. For example, the carriers for heat in metals are electrons and phonons, whereas the carrier in dielectric and semiconducting materials is predominantly phonons. We have learned from Chapter 3 that phonons may be viewed as a group of "virtual" mass particles that characterize the state of energy of the lattice in molecules. The vibration of the lattices that bond the atoms supplies the energy to phonons, and thus the heat in the substance. Consequently, the transmission of heat in a substance relies on the movement of phonons in semiconductors, or the movement of electrons in the case of metals.

The study of movements of electrons and phonons associated with heat transmission in a substance is an extremely complicated subject, and is beyond the scope of this book. What we hope to learn from this section is the basic science of molecular heat conduction in solids only. The reader is referred to special books [Tzou 1997, Tien et al. 1998] and publications [Flik et al. 1992, Tien and Chen 1994] for detailed descriptions of this subject.

One can well imagine that there are millions of phonon "particles" in a solid. The movement of these phonon particles carrying heat within the solid will inevitably result in numerous collisions among themselves. Figure 5.33 illustrates how

Figure 5.33 | Collision of a traveling phonon.

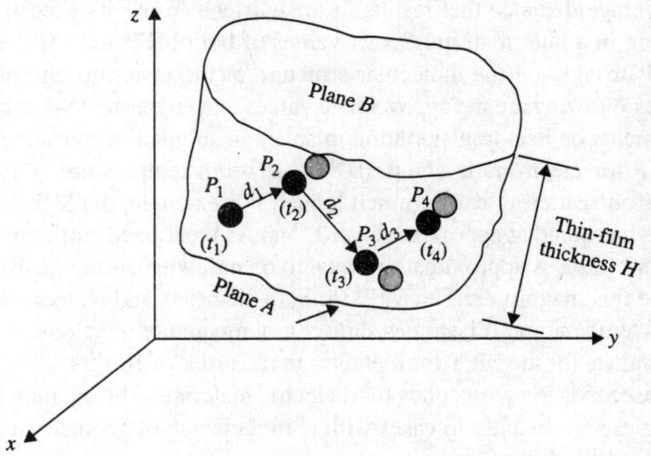

heat is transmitted from plane A to plane B. As heat transportation in a solid is predominantly done by the flow of phonon particles, we will hypothesize the traveling of phonons for this process. Because there are a great many phonons present, collisions of free phonon particles traveling inside the solid are unavoidable.

With reference to Figure 5.33, a phonon (in black) begins its journey of transmitting heat at position P_1 at plane A at time t_1. It collides with another phonon (in gray), which happens to be on its course of travel, after the energy carrier phonon has traveled a distance d_1 to position P_2 at time t_2. The collision with a neighboring phonon results in altering the direction of its further course of travel (we call this *phonon scattering*), as can be seen in the figure. Similar forms of collisions and scattering of the phonon with other nearby phonon particles take place in its remaining journey to the top plane B. The following phenomena for individual phonons are worth noting:

1. The traveling phonon changes its course after each collision—the scattering effect.
2. The distance of free travel, called the *free path,* varies between collisions.
3. Because of the change of free path, the time that the phonon travels between collisions also changes.

We will thus define the following key terminologies used in molecular heat transfer as illustrated in Figure 5.33:

1. The average *mean free path* (MFP):

$$\lambda = \frac{d_1 + d_2 + d_3}{3}$$

2. The average *mean free time* (MFT):

$$\tau = \frac{(t_2 - t_1) + (t_3 - t_2) + (t_4 - t_3)}{3} = \frac{t_4 - t_1}{3}$$

The average MFP in the above expression for a heat carrier in solids physically means the average distance that results in the carrier's losing its excessive energy while traveling in a bulk material. Exact values of both MFP and MFT are scarce. They depend strongly on the molecular structure of the materials and the thermal environment. Following are the approximate values that engineers may use for qualitative assessments on heat transportation in solids in submicrometer and nanoscale.

The MFP for electrons is about 10^{-8} m at room temperature. The MFP for phonon collision/scattering can be much longer. For example, the MFP in thin diamond films is measured at approximately 10^{-7} m. As mentioned earlier in this chapter, the MFP for gases is approximately equal to 65 nm, whereas the MFP for liquids is about twice this magnitude. The MFP for both electrons and phonons is strongly dependent on temperature. It becomes shorter as temperature increases.

Typical values for the MFT for metals is in the order of 10^{-12} s. They are in the range of nanoseconds to picoseconds for dielectric materials. The magnitude of MFT obviously varies significantly in cases with abrupt change of geometry, e.g., cracks and interfaces of thin films.

In macroscale heat transmission, for example, in the film of thickness H in Figure 5.33, the space is large enough to allow hundreds and thousands phonon collisions to take place. These collisions imply sufficiently long time for the transportation. The MFP and MFT are not critical in the process, as these effects of collisions are averaged out over the size of the solid, as well as the time frame that is involved in macro heat transfer. However, the phonon collision/scattering effect becomes critical when the size of the solid shrinks to submicrometer and nanoscales. Consequently, heat transfer in submicrometer- and nano-sized solids needs to be considered in both *length* and *time* scales.

The summary in Table 5.2 may be useful in understanding the principal heat carriers in substances [Tien and Chen 1994]:

Table 5.2 | General features of heat carriers

	Free electrons	Phonons	Photons
Substance of dominance	Heat conduction in metals	Heat conduction in dielectric and semiconductive materials	Thermal radiation
Sources of generation	Valence or excited electrons	Lattice vibration	Atomic or molecular transition
Propagation media	In vacuum or media	In media	In vacuum or media
Approximate velocity, m/s	10^6	10^3	10^8

There are two significant changes in heat conduction in the submicrometer and nanoscale domains. These are (1) the change in thermal conductivity and (2) the change in the heat conduction equation in Equation (5.38). The approximate demarcation between submicrometer and micro/macro heat conduction is at $H < 7\lambda$ [Flik et al. 1992, Tien and Chen 1994], in which H is the thickness of thin films and λ is the mean free path of phonons.

5.10.1 Thermal Conductivity of Thin Films

There are several models proposed for estimating the thermal conductivity of materials in submicrometer scales. The following simple model for thermal conductivity k for thin films was derived from the kinetic theory of molecular heat transfer [Rohsenow and Choi 1961]:

$$k = \frac{1}{3} cV\lambda \qquad [5.51]$$

The definitions of c, V, and λ in Equation (5.51) for metals and dielectrics and semiconductors are presented in Table 5.3. Also indicated in the table are the approximate values of these parameters for qualitative assessment purposes.

Table 5.3 | Parameters for thermal conductivity of thin films

	Metals	Dielectrics and semiconductors
Specific heat c	Specific heat of electrons c_e	Specific heat of phonons c_s
Molecular velocity V	Electron Fermi velocity, $V_e \approx 1.4 \times 10^6$ m/s	Velocity of phonons (sound velocity), $V_s \approx 10^3$ m/s
Average mean free path λ	Electron mean free path $\lambda_e \approx 10^{-8}$ m	Phonon mean free path, $\lambda_s \approx$ from 10^{-7} m and up

Source: Flik et al. [1992] and Tien and Chen [1994].

A more straightforward model for estimating thermal conductivity of thin films was proposed as follows [Flik and Tien 1990]:

$$\frac{k_{\text{eff}}}{k} = 1 - \frac{\lambda}{3H} \qquad [5.52a]$$

for thermal conductivity normal to the thin films, and

$$\frac{k_{\text{eff}}}{k} = 1 - \frac{2\lambda}{3\pi H} \qquad [5.52b]$$

for thermal conductivity along the surface of the thin film. Here k is the thermal conductivity of the same bulk material in the macro scale.

Equation (5.52) offers a more straightforward way for engineers to evaluate thermal conductivity in heat conduction analysis of solids in submicrometer and nanoscales. As expected, there are some discrepancies between the predicted values with the above formula and those measured by experiments. A 5 percent error is observed for thin films with $H < 7\lambda$ for thermal conductivity normal to the film, and the same error for thin films with $H < 4.5\lambda$ with thermal conductivity along the films.

EXAMPLE 5.12

Estimate the thermal conductivity of silicon films of 0.2 μm thick.

Solution

From Table 5.3, we have the average mean free path length of phonons, $\lambda_e = 10^{-7}$ m. Thus, by using Equation (5.52a), we obtain the following thermal conductivity for the thin silicon film with 0.2 μm thickness:

$$k_{\text{eff}} = \left(1 - \frac{\lambda_e}{3H}\right)k = \left(1 - \frac{10^{-7}}{3 \times (0.2 \times 10^{-6})}\right)k = 0.833k$$

With $k = 1.57$ W/cm-°C from Table 7.3. The thermal conductivity of the thin silicon film normal to the film surface is $k_{\text{eff}} = 0.833 \times 1.57$ W/cm-°C $= 1.308$ W/cm-°C.

One may use Equation (5.52b) to estimate the thermal conductivity of the same silicon film along the film surface to be $k_{\text{eff}} = 0.894k = 1.404$ W/cm-°C.

5.10.2 Heat Conduction Equation for Thin Films

A common perception of heat transportation by most engineers is that heat flux $q(\mathbf{r}, t)$ crossing the boundary of a solid results in a temperature difference or temperature gradient $\nabla T(\mathbf{r}, t)$ in the same solid. Likewise, a temperature gradient maintained in a solid can result in heat flow in the solid. A critical question is how long does it take to produce the resulting temperature gradient or heat flow by the respective causes mentioned above. We assume intuitively that these results take immediate effect with the input causes, so there is no time lag between the causes and the resulting effects. This perception serves as the basis for our derivation of the classical Fourier law of heat conduction in Equation (5.35). However, this hypothesis of the simultaneous occurrence of the causes and the corresponding results between the heat flux and the temperature gradient is a clear contradiction to what we have postulated on the heat transmission in solids as illustrated in Figure 5.33. As we have envisaged from the graphical illustration in that figure, finite time is required for phonons to carry heat from one location in the solid to another. The average mean free time (MFT) is used to establish the required time for heat to be transmitted from one location to another in a solid. The MFT, in essence, results in the delay of heat transportation in the solid and thus the delay of the resulting temperature gradient in applying heat flux across the boundaries of the solid.

The delay or the *lag time,* between the heat flux and the corresponding temperature gradient is not a significant factor in heat transportation in solids in macroscale, as a relatively large time scale is used in those analyses. However, for solids in submicrometer and nanoscale, such delay can introduce significant error in the analysis if it is not properly compensated. Consequently, the heat conduction equation in Equation (5.38) needs to be modified with an additional term to account for the lagging time. The modified heat conduction equation is expressed as:

$$\nabla^2 T(\mathbf{r}, t) + \frac{Q}{k} = \frac{1}{\alpha}\frac{\partial T(\mathbf{r}, t)}{\partial t} + \frac{\tau}{\alpha}\frac{\partial^2 T(\mathbf{r}, t)}{\partial t^2} \qquad [5.53]$$

The last term on the right-hand side of Equation (5.53) represents the velocity of heat transmission in solids as illustrated in Figure 5.33. It is in the form of wave propagation of the temperature $T(\mathbf{r}, t)$, and it is called *thermal wave propagation* in the solid. This term becomes insignificant when $H \gg \lambda$. The variable τ in the above equation is referred to as the *relaxation time,* which is the average time that a carrier, e.g., a phonon, travels between collisions. The following relationship is used for evaluating τ:

$$\tau = \frac{\lambda}{V} \tag{5.54}$$

where λ is the average mean free path and V is the average velocity of the heat carrier.

One may readily envisage that τ is in the order of 10^{-10} s from Table 5.3 for semiconductors.

According to Flik et al. [1992], the time to use Equation (5.53) is when the observation time t is shorter than H^2/α, in which H is the thickness of the thin film and α is the thermal diffusivity of the material.

PROBLEMS

Part 1. Multiple Choice

1. Viscosity of a fluid is a measure of a fluid's resistance to motion created by (1) pressure, (2) driving forces, (3) shear stress.

2. Newtonian fluids are defined by their relationship between the shear stress and shear strain rates, which exhibits (1) linear, (2) nonlinear, (3) combined linear and nonlinear characteristics.

3. Reynolds number is related to the characteristics of (1) a stationary, (2) a moving, (3) any state of the fluid.

4. Reynolds number is proportional to (1) the traveling distance, (2) the velocity, (3) the pressure of a fluid.

5. Control volume in a fluid dynamic analysis means (1) a conveniently selected volume of the fluid for the analysis, (2) the volume of the fluid in which the Reynolds number is constant, (3) the volume of the fluid in which the fluid properties are constant.

6. A laminar fluid flow means a (1) low velocity, (2) high velocity, (3) quasi-stagnant fluid flow.

7. Laminar flow of compressible fluids normally takes place with Reynolds number in the range of (1) 0 to 10, (2) 10 to 100, (3) 100 to 1000.

8. In general, fluid flows in microsystems are (1) laminar, (2) turbulent, (3) neither laminar nor turbulent.

9. The continuity equation is used to evaluate (1) volumetric flow rate, (2) relationship between the motion and the driving forces, (3) the induced forces in a moving fluid.

10. The momentum equation is used to evaluate (1) volumetric flow rate, (2) relationship between the motion and the driving forces, (3) the induced forces in a moving fluid.
11. We will use (1) continuity equation, (2) momentum equation, (3) equation of motion to assess the fluid induced forces on microsystem components.
12. Hydraulic diameter is used to evaluate the cross-sectional area of fluid flowing in (1) circular, (2) rectangular, (3) any shape of conduits.
13. CFD stands for (1) critical fluid dynamics, (2) computational fluid dynamics, (3) computerized fluid dynamics.
14. Navier–Stokes equations relate (1) pressure–velocity, (2) pressure–density change, (3) pressure–viscosity in a moving fluid.
15. Surface tension in a fluid is a form of (1) applied tension at the surface of the fluid, (2) an existing tension at the fluid surface, (3) tension that makes the surface of the fluid.
16. Surface tension is the principal cause for (1) capillary, (2) newtonian, (3) laminar flow of a fluid.
17. The coefficient of surface tension of a fluid is a measure of the magnitude of the (1) inherent strength, (2) surface tension, (3) topology of a fluid surface.
18. The capillary height of a fluid in a small tube is (1) equal to, (2) directly proportional to, (3) inversely proportional to the diameter of the tube.
19. Pressure drop at two points in a fluid drives the flow of the fluid in a circular conduit. It is inversely proportional to the (1) second, (2) third, (3) fourth power of the diameter of the conduit.
20. The driving force in the piezoelectric pumping of fluids in minute conduits is of the (1) linear, (2) surface, (3) volumetric nature.
21. A rarefied gas means the gas is at (1) extremely low pressure, (2) intermediate pressure, (3) vacuum.
22. A good estimate of the MFP for gases is (1) 65 nm, (2) 75 nm, (3) 130 nm.
23. A good estimate of the MFP for liquids is (1) 65 nm, (2) 75 nm, (3) 130 nm.
24. The Knudsen number is defined as (1) the density of the gas over the physical size of the confinement, (2) the mean free path over the physical size of the confinement, (3) the velocity of the gas over the speed of sound.
25. Conventional fluid dynamics theories are valid for (1) very small, (2) large, (3) very large Knudsen numbers.
26. The Mach number is defined as (1) the density of the gas over the physical size of the confinement, (2) the mean free path over the physical size of the confinement, (3) the velocity of the gas over the speed of sound.
27. A gas is considered to be compressible if the Mach number is (1) less than, (2) equal to, (3) greater than 0.3.
28. The larger the Knudsen number, the confinement of the gas becomes (1) smaller, (2) larger, (3) remains the same.
29. A rule of thumb to classify a rarefied gas is (1) $Kn < 0.1$, $Ma < 0.3$; (2) $Kn > 0.1$, $Ma < 0.3$; (3) $Kn > 0.1$, $Ma > 0.3$.

30. The Navier–Stokes equations can be reasonably used for gas flow with a Knudsen number that is less than (1) 0.01, (2) 0.1, (3) 1.0.

31. One could use the Navier–Stokes equation for gas flow between the values of 0.01 and 0.1 by using (1) any, (2) nonslip, (3) slip boundary condition.

32. Thermal conductivity of a material is a measure of its (1) conductance to heat, (2) resistance to conducting heat and electricity, (3) speed of heat conduction.

33. Metals are (1) better than, (2) worse than, (3) about the same as semiconductors and ceramics in conducting heat.

34. Thermal diffusivity of a material is a measure of its (1) conductance to heat, (2) resistance to conducting heat and electricity, (3) speed of heat conduction.

35. Heat flux is a measure of heat conduction in a solid per unit (1) length, (2) area, (3) volume for a given period of time.

36. Heat flux is a (1) scalar, (2) vector, (3) tensor quantity.

37. Heat generation in a solid by electric resistance is related to (1) current and resistance, (2) voltage and resistance, (3) inductance and resistance.

38. For micro thermal actuators, one would choose the actuating materials with (1) high thermal conductivity, (2) high thermal diffusivity, (3) neither of the above.

39. Newton's cooling law is used for microsystem components in contact with (1) fluids, (2) another solid component, (3) any substance.

40. The heat transfer coefficient of a fluid in contact with a solid is (1) equal, (2) directly proportional to, (3) inversely proportional to the velocity of the fluid flow.

41. The natural convective heat transfer is prompted by (1) the driving forces of the fluid, (2) the pressure drop in the fluid, (3) the change of density of the fluid due to heating of the fluid.

42. The surface of the solid becomes virtually impermeable to heat if the surrounding fluid is moving at (1) high, (2) low, (3) stagnant velocity.

43. The larger the Nusselt number, the (1) larger, (2) smaller, (3) same the heat transfer coefficient in the fluid.

44. In thermally-actuated micropumps, one needs to pay attention to the (1) conductive, (2) convective, (3) radiative heat transfer between the actuating elements and the contacting fluid.

45. When a thermally-actuated element is in contact with a working fluid, the interface temperature is normally (1) lower than, (2) about the same as, (3) higher than the bulk fluid temperature.

46. There are generally (1) 3, (2) 4, (3) 5 types of boundary conditions involved in heat conduction analysis of solids.

47. When designing a thermally-actuated beam element, one will be primarily concerned with (1) the weight of the beam, (2) the mechanical strength of the beam, (3) the thermal response of the beam.

48. Heat conduction analysis is necessary for the subsequent (1) thermal stress, (2) heat dissipation, (3) safety analysis in the design of a thermally-actuated micro device.

49. Heat transportation in solids in macroscale is by (1) vibration of the lattices of molecules, (2) the movement of phonons and photons in molecules, (3) the movement of electrons in molecules.
50. Heat transportation in solids in submicrometer and nanoscales is dominated by (1) the vibration of the lattices of molecules, (2) the movement of phonons and photons in molecules, (3) the movement of electrons in molecules.
51. MFP stands for (1) molecular free path, (2) minimum free path, (3) mean free path for energy carriers in a substance.
52. MFT stands for (1) molecular free time, (2) minimum free time, (3) mean free time for energy carriers in a substance.
53. One needs to be concerned with the validity of using continuum heat transfer theories when the size of the solids is (1) greater than, (2) about equal to, (3) less than 1 μm.
54. The thermal conductivity of solids of submicrometer and nanoscale is (1) less than, (2) about equal to, (3) greater than the value of the same material at macro scale.
55. The additional term in the heat conduction equation for solids of submicrometer and nanoscales is related to (1) the extra time, (2) the extra velocity, (3) the extra heat in heat transmission.

Part 2. Computational Problems

1. List the dynamic viscosity, in units of N-s/m^2, for the following fluids at standard conditions, i.e. 1 atm and 20°C: hydrogen, argon, helium, oxygen, nitrogen, water, hydrogen peroxide, and silicon oil.
2. Repeat the problem in Example 5.1 with $d_1 = 500$ μm and $d_2 = 50$ μm.
3. Estimate the opening of the check valve in Example 5.2, as illustrated in Figure 5.8, and the velocity of gas flow in that microdevice. Will this valve allow a flow rate of 30,000 cm^3/min?
4. Estimate the required electrostatic forces and the corresponding power supply to provide the opening of the closure plate described in Example 5.2. Assume that the dielectric material between the electrodes is air. The dimension of the top electrode is shown in Figure P5.1. Hint: You must determine the appropriate value of gap *d* for the gas passage first.
5. Estimate the pressure drop in the situation described in Example 5.3 using the Hagen-Poiseuille equation in Equation (5.17) instead, and compare the results.
6. Water is to flow through a capillary tube 20 μm in diameter over a length of 1000 μm. Use the Hagen-Poiseuille equation to determine the pressure drop in the flow, and also estimate the additional pressure drop due to the surface tension in the water.
7. When one end of the tube in the last problem is inserted vertically into a pot of water, what will be the rise of water level in the tube due to the capillary effect?
8. Compare the mass flow rate of the gas in Example 5.7 with what you will get by using the Hagen-Poiseuille equation. Make an observation on the results.

Figure P5.1 | Dimensions of electrodes for electrostatic forces on a microvalve.

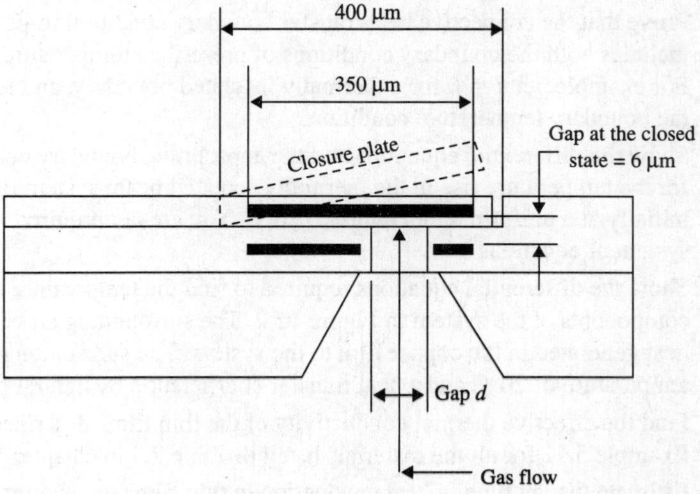

9. Compare the mass flow rate of the gas in Example 5.8 with what you will get by using the Hagen–Poiseuille equation. Make an observation on the results.
10. Rank all the materials common to MEMS as listed in Table 7.3 in Chapter 7 for their suitability to be thermally-actuated materials. This sensitivity is measured by how fast they can be heated or cooled under the influence of thermal forces.
11. Specify appropriate boundary conditions for a thermal analysis in a thermally-actuated beam as illustrated in Figure P5.2. Electric current (i) and voltage (V) are applied to the copper film to generate the heat for actuation. Use material

Figure P5.2 | A thermally actuated beam.

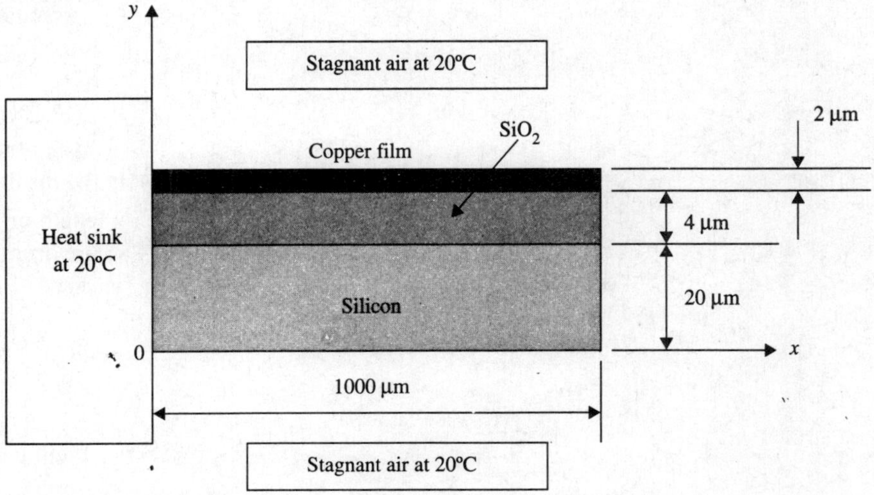

properties in Table 7.3 in Chapter 7 if necessary. Heat flow in the normal direction to the x–y plane is negligible.

12. Prove that the convective heat transfer boundary condition in Equation (5.45) includes both the boundary conditions of prescribed temperature and heat flux. For example, let $h = 0$ for a thermally insulated boundary, and let $h \to \infty$ for the boundary temperature condition.
13. Show the differential equations and the appropriate boundary conditions for the temperature rise in the thermally actuated beam system in Figure P5.2, initially at a uniform temperature of 20°C. You are not required to solve the system of equations.
14. Show the differential equations required to find the temperature fields in the components of the system in Figure P5.2. The surrounding air carries away heat generated in the copper film to the system. The surrounding air has a bulk temperature of 20°C and a heat transfer coefficient h by natural convection.
15. Find the effective thermal conductivity of the thin films described in Example 5.12 for all the materials listed in Table 7.3 in Chapter 7.
16. Estimate the lag time in heat conduction in thin films made of gold, silver and copper.

Chapter 6

Scaling Laws in Miniaturization

CHAPTER OUTLINE

6.1 | INTRODUCTION TO SCALING

The needs for and the advantages of miniaturization of machines and devices have been described in Chapter 1. Typically, a successful industrial product requires meeting consumer expectations to be intelligent and multifunctional, among many others. Consequently, a great many sensors, actuators, and microprocessors have to be systematically integrated and packaged in many of these products. The constraints on the size and geometry of the product require substantial miniaturization of these sensors, actuators, and microprocessors. It is thus essential that engineers make increasing efforts in miniaturizing new industrial products for reasons of physical appearance, volume, weight, and economy. The objective of this chapter is to provide engineers with a few selected scaling laws that will make them aware of the physical consequences of downscaling machines and devices. It should be kept in mind that some miniaturization either may be physically unfeasible, or may not make economic sense.

There are generally two types of scaling laws that are applicable to the design of microsystems. The first type of scaling law is strictly dependent on the size of physical objects, such as the scaling of geometry. Also in this category is the behavior of objects as governed by the law of physics. Examples of this type of scaling law include the scaling of rigid-body dynamics and electrostatic and electromagnetic forces. The second type of scaling law involves the scaling of phenomenological behavior of microsystems. We will realize that both the size and material properties of the system are involved in these latter scaling laws, which deal mostly with thermofluids in microsystems. The scaling of material properties in submicrometer and nanoscale is described in Section 5.10 in Chapter 5.

6.2 | SCALING IN GEOMETRY

Volume and surface are two physical quantities that are frequently involved in microdevice design. Volume relates to the mass and weight of device components, which are, for example, related to both mechanical and thermal inertia. Thermal inertia is related to the heat capacity of the solid, which is a measure of how fast we can heat or cool a solid. Such a characteristic is important in the design of a thermally actuated device, as described in Chapter 5. Surface properties, on the other hand, are related to pressure and the buoyant forces in fluid mechanics, as well as heat absorption or dissipation by a solid in convective heat transfer. We have learned from the last chapter that surface pumping is a more practical method in microfluidics than the conventional volumetric pumping in macrosystems. When the physical quantity is to be miniaturized, the design engineer must weigh the magnitudes of the possible consequences of the reduction on both the volume and surface of the particular device. Equal reduction of volume and surface of an object is not normally achievable in a scale-down process, as will be shown below.

Let us look at the example of a solid of rectangular geometry illustrated in Figure 6.1. The rectangular solid has three sides, $a > b > c$. It is readily seen that the volume $V = abc$ and the surface area is $S = 2 \times (ac + bc + ab)$. If we let l represent the linear dimension of a solid, then the volume $V \propto l^3$ and the surface $S \propto l^2$, which leads to the following relationship:

$$S/V = l^{-1} \tag{6.1}$$

Figure 6.1 | A solid rectangle.

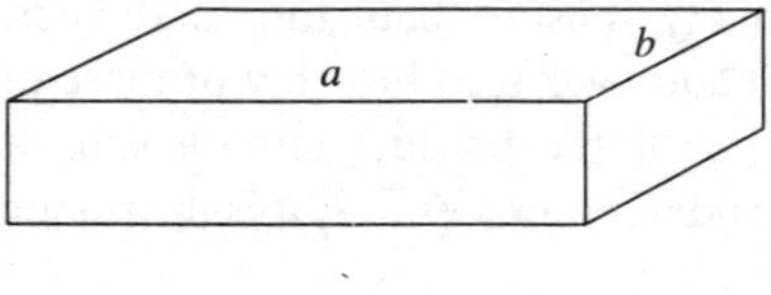

Figure 6.2 shows an interesting contrast of two living objects, an elephant and a dragonfly, with respective approximate S/V ratios of 10^{-4}/mm and 10^{-1}/mm. These distinct S/V ratios explain why a dragonfly requires little energy and power, thus low consumption of food and water, to fly, whereas an elephant has a huge appetite for food in order to generate sufficient energy (thus power) to make even slow movements.

Figure 6.2 | The distinct surface-to-volume ratios of two objects.

(**a**) An elephant ($S/V \approx 10^{-4}$/mm)

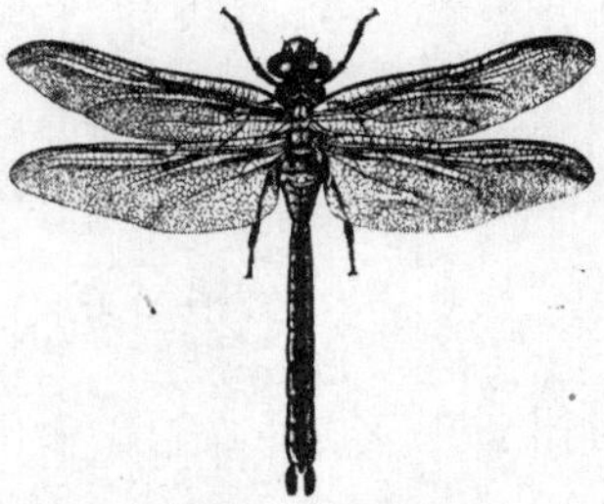

(**b**) A dragonfly ($S/V \approx 10^{-1}$/mm)

One may thus conclude from the scaling formula in Equation (6.1) that a reduction of size of 10 times (i.e., $l = 0.1$) will mean a $10^3 = 1000$ times reduction in volume, but only $10^2 = 100$ times reduction in surface area. A reduction of volume by 1000, of course, means a 1000 times reduction in weight. One may readily prove that the same scaling relations apply to other geometries of solids.

EXAMPLE 6.1

As mentioned in Chapter 1, micromirrors are essential parts of microswitches used in fiber optic networks in telecommunication. These mirrors are expected to rotate to a tightly controlled range at high rates. The angular momentum is a dominating factor in both the rotation control and the rate of rotation. In this example, we will estimate the reduction of torque required in turning a micro mirror with a reduction of 50 percent in the dimensions. The arrangement and the dimensions of the mirror are illustrated in Figure 6.3.

■ Solution

The torque that is required to turn the mirror about the y–y axis is related to the mass moment of inertia of the mirror, I_{yy}, which can be expressed as [Beer and Johnston 1988]:

$$I_{yy} = \frac{1}{12} Mc^2$$

where M = the mass of the mirror and c = the width of the mirror.

Figure 6.3 | A micro mirror.

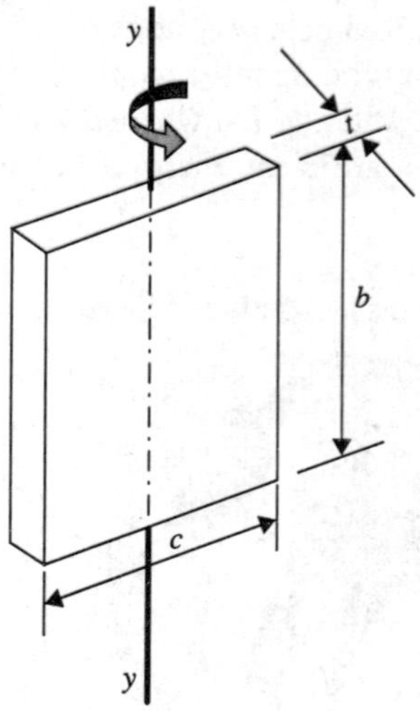

Since the mass of the mirror is $M = \rho V = \rho(bct)$, with ρ = mass density of the mirror material, the mass moment of inertia of the mirror is thus:

$$I_{yy} = \frac{1}{12}\rho bc^3 t \tag{6.2}$$

The mass moment of inertia of the mirror with a 50 percent reduction in the size becomes:

$$I'_{yy} = \frac{1}{12}\rho\left[\left(\frac{1}{2}b\right)\left(\frac{1}{2}c\right)^3\left(\frac{1}{2}t\right)\right] = \frac{1}{32}\left[\frac{1}{12}\rho bc^3 t\right] = \frac{1}{32}I_{yy}$$

It is clear from the above simple calculation that a reduction of a factor of 32 is achieved in the mass moment of inertia, and thus the required torque for rotating the mirror with a 50 percent reduction of the dimension.

6.3 | SCALING IN RIGID-BODY DYNAMICS

We demonstrated in Chapter 4 the important role that engineering mechanics plays in the design of MEM devices and microsystems. Forces are required to make parts move, whereas power is the source for the generation of forces. The amount of force required to move a part and how fast the desired movements can be achieved, as well as how readily a moving part can be stopped, depend on the inertia of the part. The inertia of a solid is related to its mass and the acceleration that is required to initiate or stop the motion of a solid device component. In miniaturizing these components, one needs to understand the effect of reduction in the size on the power P, force F, and pressure p, and the time t required to deliver the motion.

6.3.1 Scaling in Dynamic Forces

For a rigid solid traveling from one position to another, the distance that the solid travels, s, can be shown to be $s \propto l$, where l stands for the linear scale. The velocity $v = s/t$, and hence $v \propto (l)t^{-1}$, in which t is the time required for the travel.

Since, from particle kinematics, we have:

$$s = v_0 t + \frac{1}{2} a t^2 \tag{6.3}$$

where v_0 = initial velocity and a = acceleration.

By letting $v_0 = 0$, we may express the acceleration by using Equation (6.3):

$$a = \frac{2s}{t^2} \tag{6.4}$$

The dynamic force F, from the Newton's second law, can be expressed as:

$$F = Ma = \frac{2sM}{t^2} \propto (l)(l^3)\, t^{-2} \tag{6.5}$$

The reader will notice that the particle mass M is proportional to l^3, which is the linear scale of volume as in Example 6.1.

6.3.2 The Trimmer Force Scaling Vector

Trimmer [1989] proposed a unique matrix to represent force scaling with related parameters of acceleration a, time t and power density P/V_0 that is required for the scaling of systems in motion. This matrix has the generic name of *force scaling vector, F.*

The *force scaling vector* is defined as:

$$F = [l^F] = \begin{bmatrix} l^1 \\ l^2 \\ l^3 \\ l^4 \end{bmatrix} \tag{6.6}$$

We will now derive other quantities based on the above vector.

Acceleration *a* Let us use the first part of Equation (6.5), i.e., $F = Ma$, from which $a = F/M$. The following scaling is obtained:

$$a = [l^F][l^3]^{-1} = [l^F][l^{-3}] = \begin{bmatrix} l^1 \\ l^2 \\ l^3 \\ l^4 \end{bmatrix} [l^{-3}] = \begin{bmatrix} l^{-2} \\ l^{-1} \\ l^0 \\ l^1 \end{bmatrix} \tag{6.7}$$

Time *t* Now if we take a look at the second part of Equation (6.5) and express the transient time as the subject of the formula, the following relation is obtained:

$$t = \sqrt{\frac{2sM}{F}} \propto ([l^1][l^3][l^{-F}])^{1/2} = [l^2][l^F]^{-1/2}$$

$$= \begin{bmatrix} l^1 \\ l^2 \\ l^3 \\ l^4 \end{bmatrix}^{-1/2} [l^2] = \begin{bmatrix} l^{-1/2} \\ l^{-1} \\ l^{-1.5} \\ l^{-2} \end{bmatrix} [l^2] = \begin{bmatrix} l^{1.5} \\ l^1 \\ l^{0.5} \\ l^0 \end{bmatrix} \quad \textbf{(6.8)}$$

Power Density P/V_o It is apparent that no substance, whether it is a solid or fluid, can move without a power supply. Power is a very important parameter in the design of microsystems. Insufficient power supply to a microsystem can result in inactivity of the system. On the other hand, the system may suffer structural damage such as overheating with excessive power supply. Excessive power requirements by microsystems will increase the operational cost, as well as reduce the operational lives of biomedical devices that are implanted in human bodies. Here, we will deal with *power density* rather than power supply per se, which is defined as the power supply P per unit volume V_0.

We will begin our derivation of power supply with the work that is required to move a solid with mass M for a distance s. Mathematically, the work done is equal to force times distance traveled; i.e., $W = F \times s$. Thus, the power defined as the work done per unit time is $P = W/t$, and the power density can be expressed as:

$$\frac{P}{V_0} = \frac{Fs}{tV_0} \quad \textbf{(6.9)}$$

We can thus relate the power density to the force-scaling vector as follows:

$$\frac{P}{V_0} = \frac{[l^F][l^1]}{\{[l^1][l^3][l^{-F}]\}^{1/2}\,[l^3]} = [l^{1.5F}][l^{-4}] = [l^F]^{1.5}[l^{-4}]$$

$$= \begin{bmatrix} l^1 \\ l^2 \\ l^3 \\ l^4 \end{bmatrix}^{1.5} [l^{-4}] = \begin{bmatrix} l^{-2.5} \\ l^{-1} \\ l^{0.5} \\ l^2 \end{bmatrix} \quad \textbf{(6.10)}$$

By combining Equations (6.6), (6.7), (6.8) and (6.10), we can establish a set of scaling laws for rigid-body dynamics as presented in Table 6.1.

Table 6.1 | Scaling laws for rigid-body dynamics

Order	Force scale F	Acceleration a	Time t	Power density P/V_o
1	l^1	l^{-2}	$l^{1.5}$	$l^{-2.5}$
2	l^2	l^{-1}	l^1	l^{-1}
3	l^3	l^0	$l^{0.5}$	$l^{0.5}$
4	l^4	l^1	l^0	l^2

Table 6.1 is very useful in scaling down devices in a design process. The order number in the table means the order of reduction in relation to the *force-scaling vector* shown in the second column from the left.

EXAMPLE 6.2

Estimate the associated changes in the acceleration *a* and the time *t* and the power supply to actuate a MEMS component if its weight is reduced by a factor of 10.

■ Solution

Since the weight of a solid is equal to the mass times gravitational acceleration, and the mass is proportional to the cubic power of the linear scale, we will have the weight $W \propto l^3$, which means an order of 3 in Table 6.1. The same table provides the following information:

1. There will be no reduction in the acceleration (l^0).
2. There will be an ($l^{0.5}$) = $(10)^{0.5}$ = 3.16 reduction in the time to complete the motion.
3. There will be an ($l^{0.5}$) = 3.16 times reduction in power density (P/V_0). The reduction of power consumption is $P = 3.16\ V_0$. Since the volume of the component is reduced by a factor of 10, the power consumption after scaling down reduces by $P = 3.16/10 = 0.3$ times.

6.4 | SCALING IN ELECTROSTATIC FORCES

The mathematical expressions for electrostatic forces have been presented in Section 2.3.4. Here, let us revisit the configuration of a parallel plate capacitor as illustrated in Figure 6.4. The electric potential energy induced in the parallel plates is:

$$U = -\frac{1}{2}CV_2 = -\frac{\varepsilon_0 \varepsilon_r WL}{2d} V^2 \qquad \textbf{(2.7)}$$

where ε_o and ε_r are respectively the permittivity and the relative permittivity of the dielectric medium between the two electrodes and *V* is the applied voltage. This applied voltage is often called the *electric breakdown voltage* of a capacitor.

Figure 6.4 | Electrically charged parallel plates.

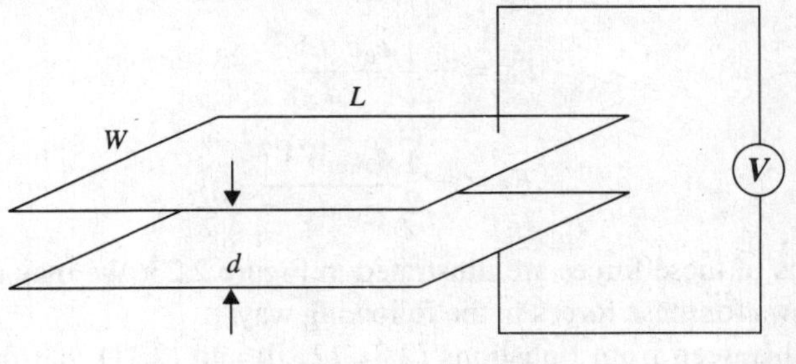

The breakdown voltage *V* in Equation (2.7) varies with the gap between the two plates, following the Paschen effect [Madou 1997]. This effect is illustrated in Figure 6.5. We can see from Figure 6.5 that the breakdown voltage *V* drops drastically with the increase of the gap for $d < 5\ \mu m$. The trend, however, is slowed down

Figure 6.5 | Paschen's effect.

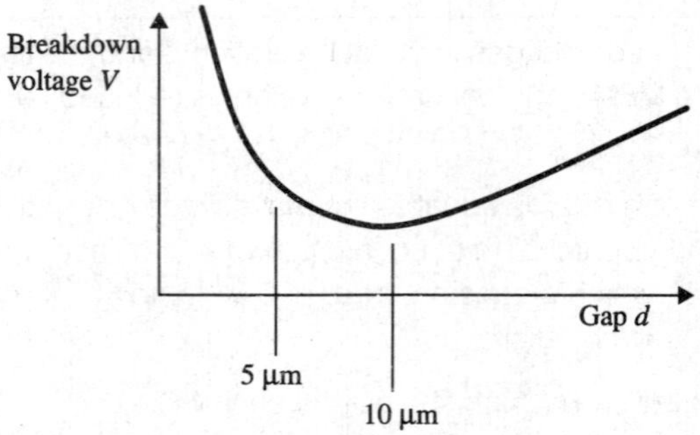

significantly after the gap widens from $d > 5\ \mu m$. Variation of the voltage reverses at $d \approx 10\ \mu m$. The breakdown voltage continues to increase linearly with further increase of the gap.

We can thus assert that applied voltage $V \propto d$, or in scaling, $V \propto l^1$ for the working range of $d > 10\ \mu m$. The scaling of ε_0 and ε_r in Equation (2.7) is neutral, i.e., ε_0, $\varepsilon_r \propto l^0$. We can thus express the scaling of the electrostatic potential energy in Equation (2.7) in the following form:

$$U \propto \frac{(l^0)(l^0)(l^1)(l^1)(l^1)^2}{l^1} = (l^3) \tag{6.11}$$

The scaling in Equation (6.11) means that a factor of 10 decrease in linear dimensions (i.e., W, L, and d simultaneously) will decrease the potential energy by a factor of $(10)^3 = 1000$.

Now, let us take a look at the scaling laws for electrostatic forces. We learned in Chapter 2 that electrostatic forces can be produced in three directions in parallel plate arrangements. Expressions for these forces are shown in Equations (2.8), (2.10), and (2.11). We will repeat these expressions below:

$$F_d = -\frac{1}{2}\frac{\varepsilon_0\varepsilon_r WLV^2}{d^2} \tag{2.8}$$

$$F_W = -\frac{1}{2}\frac{\varepsilon_0\varepsilon_r LV^2}{d} \tag{2.10}$$

$$F_L = -\frac{1}{2}\frac{\varepsilon_0\varepsilon_r WV^2}{d} \tag{2.11}$$

The directions of these forces are illustrated in Figure 2.23. We may readily derive the scaling laws for these forces in the following way.

It is readily seen from Equations (2.8), (2.10) and (2.11) that the three force components F_d, F_W, and $F_L \propto (l^2)$, which means that electrostatic forces are of order 2 in the force scaling in Table 6.1. Physically, it indicates, for example, that a 10 times reduction the size of the parallel plates will mean a 100 times decrease in the induced electrostatic forces.

Figure 6.6 | Electrostatic forces in charged parallel plates.

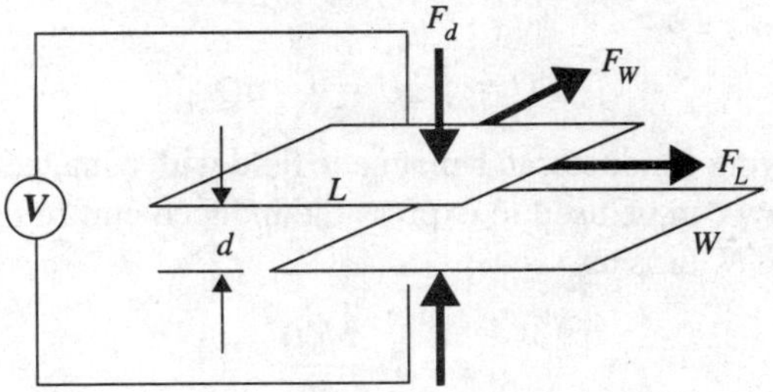

EXAMPLE 6.31

Find the reduction of electrostatic forces generated by a pair of parallel-plate electrodes as illustrated in Figure 6.6 if both the length L and the width W of these plates are reduced by a factor of 10.

Solution

We realize that the gap d between the plate electrodes remain unchanged. Consequently, the following relationships are obtained from Equations (2.8), (2.10), and (2.11) for the respective electrostatic force components:

$F_d \propto l^2$ for the normal force component

$F_W \propto l^2$ for the force component along the width

$F_L \propto l^2$ for the force component along the length

One will thus find a reduction of $(10)^2 = 100$ in the electrostatic forces in all three directions.

6.5 | SCALING IN ELECTROMAGNETIC FORCES

Although electromagnetic forces are the principal means of actuation in most machines in macroscale, working principles of electromagnetic force–actuated devices were not included in Chapter 2 for the reason that electromagnetic actuation is not scaled down nearly as favorably as electrostatic forces, as will be demonstrated in this section. However, readers should not rule out the use of electromagnetic actuation completely. It is used in a few machine tools built in mesoscale. One such application is in driving miniaturized cars as shown in Figure 1.19 [Teshigahara et al. 1995].

The classical electromagnetic theory has shown that electromagnetic forces F can be induced in a conductor or a conducting loop in a magnetic field **B** by passing current i in the conductor. In the following mathematical formulation, we will let **B** stand for the magnitude of the magnetic field with a unit of webers per square meter (Wb/m^2) and i is the passing current in amp. The electric current is the flow of electrons in a conductor per unit area and time, and it can be mathematically expressed as $i = Q/t$, in which Q represents the charges per unit area of the conductor and t is the time. The electromotive force (emf) is the "force" that drives the electrons through the conductor. The energy that is required to drive these charges can be expressed as:

$$U = \int dU = \int e\, dQ \tag{6.12}$$

For a current-carrying conductor in a magnetic field with a magnetic flux ϕ, the well-known Faraday law can be used to express the induced emf in the conductor in the form of a coil with N turns as:

$$e = N\frac{d\phi(t)}{dt}$$

By substituting the above relationship into Equation (6.12), along with $Q = it$, we have:

$$U = \int \frac{d\phi(t)}{dt} i\, dt = \int i\, d\phi(t) = \frac{1}{L}\int \phi(t)\, d\phi(t) \tag{6.13}$$

The inductance $L = \phi/i$ can be derived with $N = 1$ in the above equation.

By integrating Equation (6.13), we can obtain the following relationships:

$$U = \frac{1}{2}\frac{\phi^2}{L} \tag{6.14a}$$

or

$$U = \frac{1}{2}Li^2 \tag{6.14b}$$

The induced electromagnetic force will change with the relative position of the conductor in the magnetic field. We can thus derive the expression of these forces as follows:

$$F = \left.\frac{\partial U}{\partial x}\right|_{\phi\, =\, \text{constant}} \tag{6.15a}$$

or

$$F = \left.\frac{\partial U}{\partial x}\right|_{i\, =\, \text{constant}} \tag{6.15b}$$

If we consider the case with constant current flow, i.e., Equation (6.15b), the induced electromagnetic force will be expressed as:

$$F = \frac{1}{2}i^2\frac{\partial L}{\partial x} \tag{6.16}$$

A close examination of the expression in Equation (6.16) indicates that the current i, i.e., the amount of electrons flowing through a conductor, depends on the cross-sectional area of the conductor; i.e., $i \propto l^2$. The term $\partial L/\partial x$ is dimensionless. We can thus conclude that the scaling of electromagnetic force F is:

$$F \propto (l^2)(l^2) = l^4 \tag{6.17}$$

We can see from the above scaling that a 10 times reduction of size (l) will lead to a $10^4 = 10{,}000$ reduction of the electromagnetic force. This is a sharp contrast to the scaling of electrostatic force with l^2, which means a 100 times reduction in magnitude with the same reduction of linear dimensions. One will thus conclude that electromagnetic forces are 100 times less favorable in scale-down than electrostatic forces. Contrasting the scaling law in Equation (6.17) with that in Section 6.4 for electrostatic forces explains why almost all micromotors and actuators have been built based on electrostatic actuation, rather than electromagnetic actuation commonly used in macrosized motors and actuators. Another obvious reason is that there is just not sufficient space in microdevices to accommodate coil inductance for generating sufficient magnetic field for actuation power.

Despite what has been said above, engineers should not overlook the disadvantage of using electrostatic actuation forces. One obvious weakness of electrostatic actuation is its inherent low magnitude. However, not much force is normally required to actuate most microdevices.

6.6 | SCALING IN ELECTRICITY

Electricity is a dominant source of power to MEMS and microsystems. Principal applications of electricity include the actuation of many microsystems by electrostatic, piezoelectric, and thermal resistance heating as presented in Chapter 2. Electrodynamic pumping is currently a widely used pumping method for microsystems. Electromechanical transduction is another common application of electricity in micro systems. Scaling of electricity thus is a very important design issue.

Some of the scaling laws related to electricity may be derived from simple laws of physics as shown below:

Electric resistance $$R = \frac{\rho L}{A} \propto (l)^{-1} \quad \textbf{(6.18)}$$

in which ρ, L, and A are respectively electric resistivity of the material, the length, and the cross-sectional area of the conductor.

Resistive power loss $$P = \frac{V^2}{R} \propto (l)^1 \quad \textbf{(6.19)}$$

where V is the applied voltage $\propto (l)^0$.

Electric field energy $$U = \frac{1}{2}\,\varepsilon E^2 \propto (l)^{-2} \quad \textbf{(6.20)}$$

where ε is the permittivity of dielectric $\propto (l)^0$ and E is the electric field strength $\propto (l)^{-1}$.

The above scaling laws prove useful for miniaturizing devices. The central issue, however, is the scaling of electric power supply for miniaturization. For example, the power loss due to resistivity of the material in Equation (6.19) follows the first-order law; i.e., $P \propto l^1$. For a system that carries its own power supply such as an electrostatic actuation circuit, the available power is directly related to the system's volume, i.e., $E_{av} \propto (l)^3$. The ratio of power loss to available energy or power for performing the designed functions can be expressed as:

$$\frac{P}{E_{av}} = \frac{(l)^1}{(l)^3} = (l)^{-2} \tag{6.21}$$

The relationship in Equation (6.21) indicates a significant disadvantage of scaling down of power supply systems. It means, for example, a 10 times reduction in the size (l) of the power supply system (e.g., the linear dimension of the material for passing electric current for the power supply) would lead to a 100 times greater power loss due to the increase of resistivity.

6.7 | SCALING IN FLUID MECHANICS

We have learned from Chapter 5 that fluids flow under the influence of shear forces (or shear stresses), and that the continuum fluid mechanics based on Navier-Stokes equations breaks down in the flows at submicrometer and nanoscales. The capillary effect is the major reason for this breakdown. Here, we will learn why the capillary flow does not scale down favorably, and what are good alternatives for microflow.

Let us take a look at a volume of fluid as illustrated in Figure 6.7. The volume of the fluid V can be expressed as the area (A) × height (h). Assume that the fluid is originally in a rectangular shape between two imaginary parallel plates. Moving the top plate in the x direction to the right induces the motion of the fluid. The original rectangular-shaped fluid volume is now turned into a parallelogram as shown in the figure. The shear force that is applied to the fluid volume for this motion is represented by the shear stress τ. The action of the shear force on the fluid results in an instantaneous velocity profile that varies from the maximum value along the top wall to zero value at the bottom wall. For newtonian flow, the variation of the velocity between the top and bottom walls follows a linear relation as illustrated in Figure 6.7. The viscosity of the fluid, which was defined in Equations (5.1) and (5.2) in Chapter 5,

Figure 6.7 | Velocity profile of a volume of moving fluid.

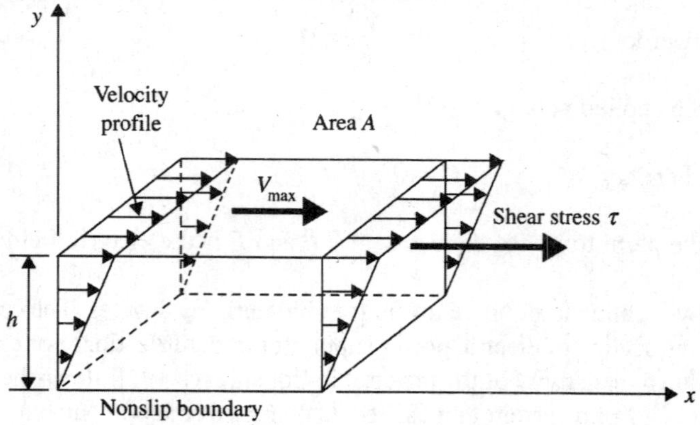

causes the variation of the velocities in the moving fluid. Let us express the viscosity in a different way as follows:

$$\mu = \frac{\tau}{R_s} \tag{6.22}$$

where μ = dynamic viscosity of the fluid with units Pa-s and $R_s = V_{max}/h$ = shear rate. The shear stress $\tau = F_s/A$, where F_s is the shear force.

The average velocity and the cross-sectional area of the passage are used to obtain the rate of volumetric fluid flow:

$$Q = A_s V_{ave} \tag{6.23}$$

in which A_s = cross-sectional area for the flow and V_{ave} = average velocity of the fluid.

We have learned from Chapter 5 that almost all fluid flows in microscale are in the laminar flow regime. Consequently, formulations in Section 5.4 for laminar fluid flow in circular conduits could be used to derive the scaling law for liquid flow in microscale.

Figure 6.8 illustrates a fluid flowing through a small circular conduit of length L and radius a. The pressure drop ΔP over the length L of the conduit can be computed by using the Hagen-Poiseuille law expressed in Equation (5.17). The rate of volumetric flow of the fluid, Q can be expressed by the same equation, but in a different form as:

$$Q = \frac{\pi a^4 \Delta P}{8\mu L} \tag{6.24}$$

where a = the radius of the tube and ΔP = pressure drop over the length of the tube, L.

Figure 6.8 | Fluid flow in a small circular conduit.

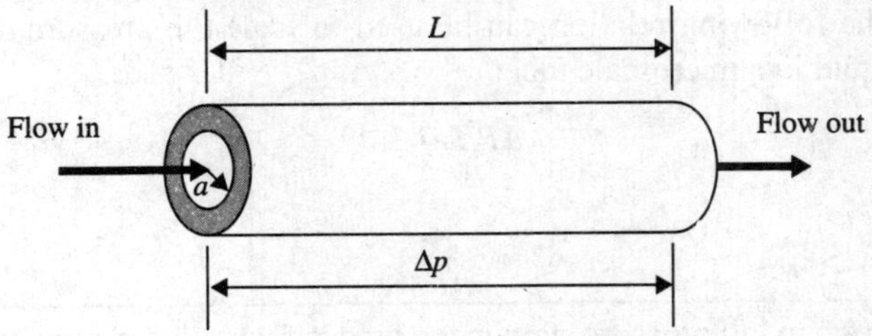

From Equation (6.23), the average velocity V_{ave} can be shown to equal:

$$V_{ave} = \frac{Q}{\pi a^2}$$

and the pressure gradient is

$$\frac{\Delta P}{\Delta x} = \frac{8\mu V_{ave}}{a^2}$$

We may compute the pressure drop for a section of a capillary tube of length L to be:

$$\Delta P = \frac{8\mu V_{ave} L}{a^2} \tag{6.25}$$

The scaling laws for fluid flows in capillary tubes are thus derived as $Q \propto a^4$ for volumetric flow from Equation (6.24), and $\Delta P/L \propto a^{-2}$ for pressure drop per unit length from Equation (6.25), in which a is the radius of the tube.

EXAMPLE 6.4

Use the scaling laws to estimate the variations of the volumetric flow and pressure drop in a circular tube if the radius of the tube is reduced by a factor of 10. Make an observation on this scaling practice.

■ Solution

From Equation (6.24), we have the volumetric flow $Q \propto a^4$, in which a is the radius of the tube. From this scaling law, we have the volumetric flow reduced by 10^4 = 10,000 times. Likewise, the scaling law in Equation (6.25) indicates a pressure drop per unit length increase by 10^2 = 100 times.

This is obviously a highly undesirable situation for a scaled-down device. Alternative mechanisms would thus be desirable for micro fluid flow.

In regard to scaling of fluid flow in capillary tubes, for cases when the radius of the conduit is very small in the microscale, a more complete scaling for micro fluid flow would include the capillary effect, which occurs to liquid flow in minute conduits as described in Section 5.6. This effect is attributed to the dominance of surface tension of the fluids in small scale.

The pressure that is required to overcome the surface tension is expressed in Equation (5.24a, b), from which we have $\Delta P \propto a^{-1}$, in which a is the radius of the tube. Thus, the following relation can be used to scale the pressure drop per unit length of a liquid in a microscale tube:

$$\Delta P/L \propto l^{-3} \tag{6.26}$$

EXAMPLE 6.5

What will happen to the pressure drop in the fluid in Example 6.4 if the tube radius is microscale?

■ Solution

The pressure drop per unit length of the tube, from the scaling law in Equation (6.26) increases 1000 times with a 10 times reduction of the tube radius. The situation is thus one order of magnitude more severe than the case in meso- or macroscale in Example 6.4.

The significantly adverse effect in scaling down of fluid flows in micrometer and submicrometer scales has prompted engineers to search for alternative fluid propulsion mechanisms to that used in conventional volumetric pumping. These mechanisms include piezoelectric, electro-osmotic, electrowetting, and electrohydrodynamic pumping [Madou 1997]. Many of these special pumping techniques are based on surface pumping forces, which scale more favorably, in microdomains than conventional volumetric pumping forces. A typical pumping technique based on surface forces is the piezoelectric pump.

The principle of piezoelectric pumping is to use the forces generated on tube wall to drive the fluid flow instead of the conventional pressure difference. This way of moving fluid is similar to the squeezing of toothpaste from a plastic tubular container. The surface force F, which is proportional to the surface area of the inner wall of the tube, scales much more favorably than the backpressure, such as given in Equation (6.25), required to drive the equivalent volume of the fluid. The reason surface force is favored over the volumetric pumping pressure is that the surface area of the inner wall of the section of the small tube in Figure 6.8 is $S = 2\pi aL$ and the equivalent volume of the fluid is $V = \pi a^2 L$. These relations result in the surface area to volume ratio $S/V = 2/a$. Since the surface force that is required to pump a volume of the fluid is proportional to the surface area, we may readily see from the S/V ratio that $F \propto l^{-1}$. Consequently, we can assert that scaling down the tube radius will result in the increase of the surface force available for pumping the unit volume of fluid. The feasibility of the piezoelectric pumping technique for microflow was described in Section 5.6.3 in Chapter 5.

The working principles of two principal electrohydrodynamic pumping methods, namely electro-osmosis and electrophoresis, were presented in Section 3.8.2. These pumping techniques are widely used in microfluidics systems designed and constructed by biotechnology and pharmaceutical industries.

6.8 | SCALING IN HEAT TRANSFER

Heat transfer is an essential part of design for many microsystems. For most cases, heat transmission in microsystems is in the modes of conduction and convection. In some special cases, heat is also transferred by radiation, such as in laser treatments involved in some manufacturing techniques for microsystems. Formulation of heat conduction and convection in solids and fluids was presented in the second part of Chapter 5. Here, we will take an overview of the scaling of heat transmission in these two modes. We will offer scaling laws in two regimes, one being in the meso- and microscale and the cther in the submicrometer scale, meaning the size of microsystems is 1 μm or less. In the latter regime, some thermophysical properties vary with the size of the MEMS or microsystems components. Therefore, these properties should also be included in the scaling law in addition to the size parameter.

6.8.1 Scaling in Heat Conduction

Scaling of Heat Flux Heat conduction in solids is governed by the Fourier law, which was expressed in a general form in Equation (5.35). For one-dimensional heat conduction along the x coordinate, we have:

$$q_x = -k\frac{\partial T(x, y, z, t)}{\partial x}$$

where q_x is the heat flux along the x coordinate, k is the thermal conductivity of the solid, and $T(x, y, z, t)$ is the temperature field in the solid in a cartesian coordinate system at time t. A more generic form for the rate of heat conduction in a solid is:

$$Q = qA = -kA\frac{\Delta T}{\Delta x} \tag{6.27}$$

It is readily seen from Equation (6.27) that the scaling law for heat conduction for solids in meso- and microscales is $Q \propto (l^2)(l^{-1}) = (l^1)$. It is easy to see from this simple scaling law that reduction in size leads to the decrease of total heat flow in a solid.

Scaling in Thermal Conductivity in Submicrometer Regime In Section 5.10 we learned about the variation of thermal conductivity k in solids in the submicrometer regime. Values of thermal conductivity can be estimated by the expression in Equation (5.51), which leads to the following scaling law for the thermal conductivity in the submicrometer regime:

$$k = \frac{1}{3}cV\lambda \propto (l^{-3})(l^1)(l^3) = (l^1) \tag{6.28}$$

We have used the relationship $\lambda \propto 1/\rho$ in the above derivation, in which ρ is the mass density of the solid with an order that is similar to the volume, i.e., l^3. The reader will notice that the mass density of the solid, ρ, is not treated as a material constant in the submicrometer scale; rather, its value is related to the size of the solid, namely the volume.

The scaling of heat flow in a solid in the submicrometer regime can thus be obtained by combining Equations (6.27) and (6.28):

$$Q \propto (l^1)(l^1) = (l^2) \tag{6.29}$$

Equation (6.29) represents a significant factor in that a reduction in size of 10 would lead to a reduction of total heat flow by 100.

Scaling in Effect of Heat Conduction in Solids of Meso- and Microscales A dimensionless number, called the Fourier number, F_0 is frequently used to determine the time increments in a transient heat conduction analysis. Mathematically, it is defined as:

$$F_0 = \frac{\alpha t}{L^2} \tag{6.30}$$

where α is the thermal diffusivity of the material as defined in Equation (5.39) and t is the time for heat to flow across the characteristic length L.

Physically, the Fourier number is the ratio of the rate of heat transfer by conduction to the rate of energy storage in the system [Kreith and Bohn 1997]. By rearranging Equation (6.30), we will obtain the scaling in the time for heat conduction in a solid to be:

$$t = \frac{F_0}{\alpha} L^2 \propto (l^2) \tag{6.31}$$

We assign both F_0 and α in Equation (6.31) to be constants.

EXAMPLE 6.6

Estimate the variation of the total heat flow and the time required to transmit heat in a solid with a reduction of size by a factor of 10.

Solution

From the scaling laws in Equations (6.29) and (6.31), the total heat flow and the time required for heat transmission are both reduced by $(10)^2 = 100$ times with a reduction of size by a factor of 10.

EXAMPLE 6.7

What will happen to the total heat flow and the time for heat transmission if the solid in Example 6.6 is of submicrometer size?

Solution

We will review the effect of the modification of the scaling law for total heat flow in Equation (6.29); i.e., $Q \propto (l)^2$. The time required for heat flow, however, remains the same as shown in Equation (6.31).

Thus, we will have the total heat flow Q reduced by $(10)^2 = 100$ times and the time for heat flow reduced by $(10)^2 = 100$ times with a reduction of linear dimensions of the solid by a factor of 10 in submicrometer scale.

6.8.2 Scaling in Heat Convection

We have seen from Figure 5.23 that boundary layers are present at the interfaces of solids and fluids. Heat transfer in fluid is in the mode of convection, which is governed by Newton's cooling law as expressed in Equation (5.40), or in a more generic form as:

$$Q = qA = hA\,\Delta T \tag{6.32}$$

where Q is the total heat flow between two points in the fluid, q is the corresponding heat flux, A is the cross-sectional area for the heat flow, h is the heat transfer coefficient, and ΔT is the temperature difference between these two points.

As described in Section 5.8.4, the heat transfer coefficient h depends primarily on the velocity of the fluid, which does not play a significant role in the scaling of the heat flow. Consequently, the total heat flow from Equation (6.32) primarily depends on the cross-sectional area A, which is of the order (l^2). We thus have the scaling of heat transfer in convection to be $Q \propto (l^2)$ for fluids in meso- and microregimes.

The scaling of heat convection for fluids in submicrometer regime is not quite as simple as that described above. Here we will consider only convective heat transfer of gases in that regime.

For the cases in which gases pass in narrow channels at submicrometer scale, the classical heat transfer theories based on continuum fluids, as presented in Chapter 5, break down. The seemingly convective heat transfer has in fact become conduction of heat among the gas molecules as the effect of the boundary layer becomes a dominant factor. Figure 6.9 illustrates such a situation. The passage is represented by a gap $H < 7\lambda$, in which λ is the mean free path of the gas. As mentioned in Chapter 5, we may use an approximate value of $\lambda = 65$ nm, or 0.65 μm for gases, and 1.3 μm for liquids. The transfer of heat between the two plates is predominantly by conduction through the gas between the plates rather than convection.

Figure 6.9 | Gas flow in a micro channel.

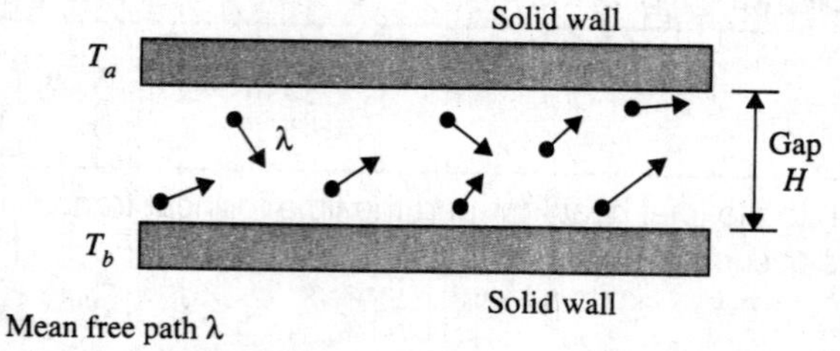

As described in Chapter 5, the thermal conductivity of the gas depends on the mean free path of the gas molecules, λ. This mean free path is proportional to the reciprocal of the mass density of the gas [Madou 1997]:

$$\lambda \propto 1/\rho$$

and the thermal conductivity of the gas is presented in Equation (5.51):

$$k = \frac{1}{3} cV\lambda \tag{5.51}$$

where c is the specific heat of the gas at constant volume and the mean velocity V is obtained from the following expression:

$$V = \sqrt{\frac{8kT}{\pi m}} \tag{6.33}$$

in which T is the mean temperature of the gas and m is the molecular weight of the gas.

The effective heat flux can be calculated from the following equation:

$$q_{\text{eff}} = \frac{k\,\Delta T}{H + 2\varepsilon} \tag{6.34}$$

where ΔT is the temperature difference between the two plates.

Equation (6.34) differs from the traditional Fourier law of heat conduction in Equation (6.27), by a factor of 2ε in its denominator on the right-hand side of the equation. The value ε depends on the gases entrapped between the two plates. For example: $2.4\lambda < \varepsilon < 2.9\lambda$ for air, O_2, N_2, CO_2, methane, and He, and $\varepsilon = 11.7\lambda$ with $H > 7\lambda$ for H_2, where λ is the mean free path of the gas.

PROBLEMS

Multiple Choice

1. The application of the scaling laws in miniaturization is to assess the consequences on (1) the physical effect, (2) the economic effect, (3) the market effect on miniaturized products.
2. Scaling laws are derived from (1) design engineers' experience, (2) the laws of physics, (3) the market demands.
3. Scaling in geometry is critical in miniaturizing (1) moving components, (2) sensing components, (3) overall dimensions of MEMS products.
4. The Trimmer's force scaling vector is used to assess miniaturization relating to (1) heat flow in solids, (2) fluid flow, (3) rigid-body dynamics in the design of MEMS.
5. For order 1 scaling such as surface-to-volume scaling, the acceleration varies (1) linearly, (2) to the square power, (3) to the cubic power.
6. The reason why electrostatic forces are favored over the electromagnetic forces in microactuation is that electrostatic forces scale (1) better, (2) worse, (3) about the same as electromagnetic forces.
7. Electromagnetic forces scale (1) 2, (2) 3, (3) 4 orders of magnitude worse than electrostatic forces.
8. The power loss due to electric resistivity is (1) much more severe, (2) less severe, (3) about the same in small-sized systems.
9. Pressure drop in a fluid flowing through a smaller circular conduit is (1) much greater, (2) about equal to, (3) much less than that in a larger conduit.
10. The volumetric flow of fluid in a smaller circular conduit is (1) much greater, (2) about equal to, (3) much smaller than that in a larger conduit.
11. The effect of surface tension on fluid flowing in a capillary tube makes the pressure drop (1) much greater, (2) about equal to, (3) much less than the same flow in mesosize tubes.
12. The effect of surface tension on fluid flowing in a capillary tube makes the volumetric flow (1) much greater than, (2) about equal to, (3) much less than the same flow in mesosize tubes.
13. Heat flows (1) faster, (2) slower, (3) about the same in a smaller solid than in a larger solid.
14. The mode of heat transmission in gas in extremely narrow passages is (1) conduction, (2) convection, (3) radiation.
15. Heat transmission in gases is drastically different in a narrow passage of size less than (1) 5λ, (2) 7λ, (3) 9λ where λ is the mean free path of gas molecules.

Chapter 7

Materials for MEMS and Microsystems

CHAPTER OUTLINE

7.1 | INTRODUCTION

In Chapter 1, we maintained that the current technologies used in producing MEMS and microsystems are inseparable from those of microelectronics. This close relationship between microelectronics and microsystems fabrication often misleads engineers to a common belief that the two are indeed interchangeable. It is true that many of the current microsystem fabrication techniques are closely related to those used in microelectronics. Design of microsystems and their packaging, however, is significantly different from that for microelectronics. Many microsystems use microelectronics materials such as silicon, and gallium arsenide (GaAs) for the sensing or actuating elements. These materials are chosen mainly because they are dimensionally stable and their microfabrication and packaging techniques are well established in microelectronics. However, there are other materials used for MEMS and microsystems products—such as quartz and Pyrex, polymers and plastics, and ceramics—that are not commonly used in microelectronics. Plastics and polymers are also used extensively in the case of microsystems produced by the LIGA processes, as will be described in Chapter 9.

7.2 | SUBSTRATES AND WAFERS

The frequently used term *substrate* in microelectronics means a flat macroscopic object on which microfabrication processes take place [Ruska 1987]. In microsystems, a substrate serves an additional purpose: it acts as signal transducer besides supporting other transducers that convert mechanical actions to electrical outputs or vice versa. For example, in Chapter 2, we saw pressure sensors that convert the applied pressure to the deflection of a thin diaphragm that is an integral part of a silicon die cut from a silicon substrate. The same applies to microactuators, in which the actuating components, such as the microbeams made of silicon in microaccelerators, are also called substrates.

In semiconductors, the substrate is a single crystal cut in slices from a larger piece called a *wafer*. Wafers can be of silicon or other single crystalline material such as quartz or gallium arsenide. Substrates in microsystems, however, are somewhat different. There are two types of substrate materials used in microsystems: (1) active substrate materials and (2) passive substrate materials, as will be described in detail in the subsequent sections.

Table 7.1 presents a group of materials that are classified as electric *insulators* (or dielectrics), *semiconductors,* and *conductors* [Sze 1985]. The same reference classifies the insulators to have electrical resistivity ρ in the range of $\rho > 10^8$ Ω-cm; semiconductors with 10^{-3} Ω-cm $< \rho < 10^8$ Ω-cm; and conductors with $\rho < 10^{-3}$ Ω-cm. We will find that common substrate materials used in MEMS such as silicon

(Si), germanium (Ge), and gallium arsenide (GaAs) all falı in the category of semiconductors. One major reason for using these materials as principal substrate materials in both microelectronics and microsystems is that these materials are at the borderline between conductors and insulators, so they can be made either a conductor or an insulator as needs arise. Indeed, the doping techniques that were described in Chapter 3 can be used to convert the most commonly used semiconducting material, silicon, to an electrically conducting material by doping it with a foreign material to form either p- or n-type silicon for conducting electricity. All semiconductors are amenable to such doping. Another reason for using semiconductors is that the fabrication processes, such as etching, and the equipment required for these processes have already been developed for these materials.

A checklist of factors that help the designer in selecting substrate materials for microsystems is available in Madou [1997].

Table 7.1 | Typical electrical resistivity of insulators, semiconductors, and conductors

Materials	Approximate electrical resistivity ρ, Ω-cm	Classification
Silver (Ag)	10^{-6}	Conductors
Copper (Cu)	$10^{-5.8}$	
Aluminum (Al)	$10^{-5.5}$	
Platinum (Pt)	10^{-5}	
Germanium (GE)	10^{-3}–$10^{1.5}$	Semiconductors
Silicon (Si)	10^{-3}–$10^{4.5}$	
Gallium arsenide (GaAs)	10^{-3}–10^{8}	
Gallium phosphide (GaP)	10^{-2}–$10^{6.5}$	
Oxide	10^{9}	Insulators
Glass	$10^{10.5}$	
Nickel (pure)	10^{13}	
Diamond	10^{14}	
Quartz (fused)	10^{18}	

7.3 | ACTIVE SUBSTRATE MATERIALS

Active substrate materials are primarily used for sensors and actuators in a microsystem (Fig. 1.5) or other MEMS components (Fig. 1.7). Typical active substrate materials for microsystems include silicon, gallium arsenide, germanium, and quartz. We realize that all these materials except quartz are classified as semiconductors in Table 7.1. These substrate materials have basically a cubic crystal lattice with a tetrahedral atomic bond. [Sze 1985]. These materials are selected as active substrates primarily for their dimensional stability, which is relatively insensitive to environmental conditions. Dimensional stability is a critical requirement for sensors and actuators with high precision.

As indicated in the periodic table (Fig. 3.3), each atom of these semiconductor materials carries four electrons in the outer orbit. Each atom also shares these four electrons with its four neighbors. The force of attraction for electrons by both nuclei holds each pair of shared atoms together. They can be doped with foreign materials to alter their electric conductivity as described in Section 3.5.

7.4 | SILICON AS A SUBSTRATE MATERIAL

7.4.1 The Ideal Substrate for MEMS

Silicon is the most abundant material on earth. However, it almost always exists in compounds with other elements. Single-crystal silicon is the most widely used substrate material for MEMS and microsystems. The popularity of silicon for such application is primarily for the following reasons:

1. It is mechanically stable and it can be integrated into electronics on the same substrate. Electronics for signal transduction, such as a p- or n-type piezoresistor, can be readily integrated with the Si substrate.
2. Silicon is almost an ideal structural material. It has about the same Young's modulus as steel (about 2×10^5 MPa), but is as light as aluminum, with a mass density of about 2.3 g/cm^3. Materials with a high Young's modulus can better maintain a linear relationship between applied load and the induced deformations.
3. It has a melting point at 1400°C, which is about twice as high as that of aluminum. This high melting point makes silicon dimensionally stable even at elevated temperature.
4. Its thermal expansion coefficient is about 8 times smaller than that of steel, and is more than 10 times smaller than that of aluminum.
5. Above all, silicon shows virtually no mechanical hysteresis. It is thus an ideal candidate material for sensors and actuators. Moreover, silicon wafers are extremely flat and accept coatings and additional thin-film layers for building microstructural geometry or conducting electricity.
6. There is a greater flexibility in design and manufacture with silicon than with other substrate materials. Treatments and fabrication processes for silicon substrates are well established and documented.

7.4.2 Single-Crystal Silicon and Wafers

To use silicon as a substrate material, it has to be pure silicon in a single-crystal form. The Czochralski (CZ) method appears to be the most popular method among several methods that have been developed for producing pure silicon crystal. The raw silicon, in the form of quartzite are melted in a quartz crucible, with carbon (coal, coke, wood chips, etc). The crucible is placed in a high-temperature furnace as shown in Figure 7.1 [Ruska 1987]. A "seed" crystal, which is attached at the tip of a puller, is brought into contact with the molten silicon to form a larger crystal. The puller is slowly pulled up along with a continuous deposition of silicon melt onto the seed crystal. As the puller is pulled up, the deposited silicon melt condenses and a large bologna-shaped boule of single-crystal silicon several feet long is formed. Figure 7.2 shows one of such boules produced by this method. The diameter of the boules ranges from 100 mm to 300 mm.

Figure 7.1 | The Czochralski method for growing single crystals. (Ruska [1987].)

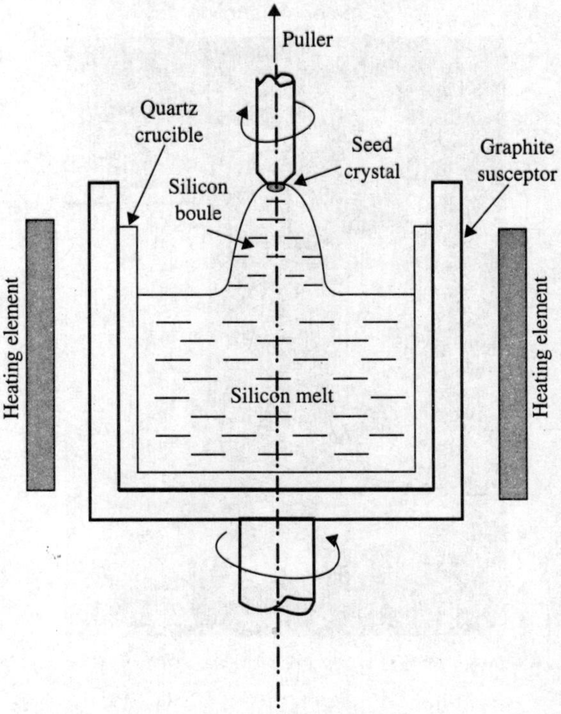

The silicon crystal boule produced by the CZ method is then ground to a perfect circle on its outside surface, then sliced to form thin disks of the desired thickness by fine diamond saws. These thin disks are then chemically-lap polished to form the finished wafers.

Principal materials in the silicon melt are silicon oxide and silicon carbide. These materials react at high temperature to produce pure silicon, along with other gaseous by-products as shown in the following chemical reaction:

$$SiC + SiO_2 \rightarrow Si + CO + SiO$$

The gases produced by the above reaction escape to the atmosphere and the liquid Si is left and solidifies to pure silicon. Circular pure-crystal silicon boules are produced by this technique in three standard sizes: 100 mm (4 in), 150 mm (6 in) and 200 mm (8 in) in diameters. A larger size of boule at 300 mm (12 in) in diameter is the latest addition to the standard wafer sizes. Current industry standard on wafer sizes and thicknesses are as follows:

100 mm (4 in) diameter $\times$ 500 μm thick
150 mm (6 in) diameter $\times$ 750 μm thick
200 mm (8 in) diameter $\times$ 1 mm thick
300 mm (12 in) diameter $\times$ 750 μm thick (tentative)

Figure 7.2 | A 300-mm single-crystal silicon boule cooling on a material-handling device.

(Courtesy of MEMC Electronic Materials Inc., St. Peters, Missouri.)

The size difference between a 200-mm and a 300-mm wafer is shown in Figure 7.3. The latter size wafer has 2.25 times more surface area than the 200 mm wafer and thus provides significant economic advantage for accommodating many more substrates on a single wafer.

Silicon substrates often are expected to carry electric charges, either in certain designated parts or in the entire area, as in the resonant frequency pressure sensors described in Section 4.3.6. Substrates thus often require p or n doping of the wafers. The doping of p- and n-type impurities, as described in Section 3.5, can be done either by ion implantation or by diffusion, as will be described in detail in Chapter 8. Common n-type dopants of silicon are phosphorus, arsenic, and antimony, whereas boron is the most common p-type dopant for silicon.

7.4.3 Crystal Structure

Silicon has an uneven lattice geometry for its atoms, but it has basically a face-centered cubic (FCC) unit cell as illustrated in Figure 7.4. A unit cell consists of atoms situated at fixed locations defined by imaginary lines called a *lattice*. The dimension b of the lattice is called the *lattice constant* in the figure. In a typical FCC crystal, atoms are situated at the eight corners of the cubic lattice structure, as well as at the center of each of the six faces. We have shown those "visible" atoms in black and the "invisible" or "hidden" ones in gray in Figure 7.4. For silicon crystals, the lattice constant $b = 0.543$ nm. In an FCC lattice, each atom is bonded to 12 nearest-neighbor atoms.

Figure 7.3 | Size difference between a 200-mm wafer and a 300-mm wafer.

(Courtesy of MEMC Electronic Materials Inc., St. Peters, Missouri.)

Figure 7.4 | A typical face-center-cubic unit cell.

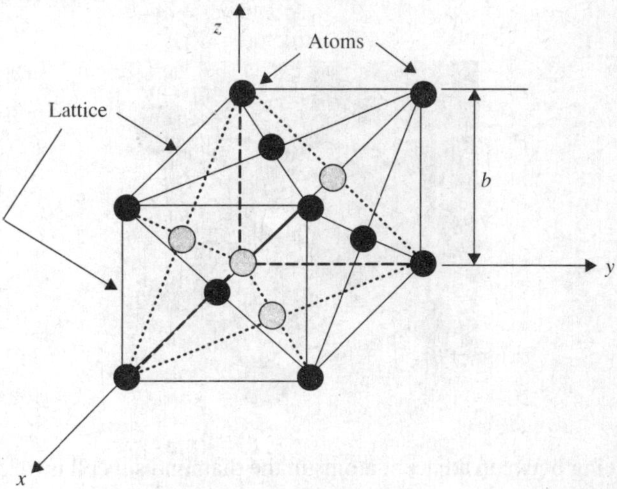

The crystal structure of silicon, however, is more complex than that of regular FCC structure illustrated in Figure 7.4. It can be considered the result of two interpenetrating face-centered cubic crystals, FCC A and FCC B, illustrated in Figure 7.5a [Angell et al. 1983]. Consequently, the silicon crystal contains an additional four atoms as shown in Figure 7.5 b.

Figure 7.5 | Structure of a silicon crystal.

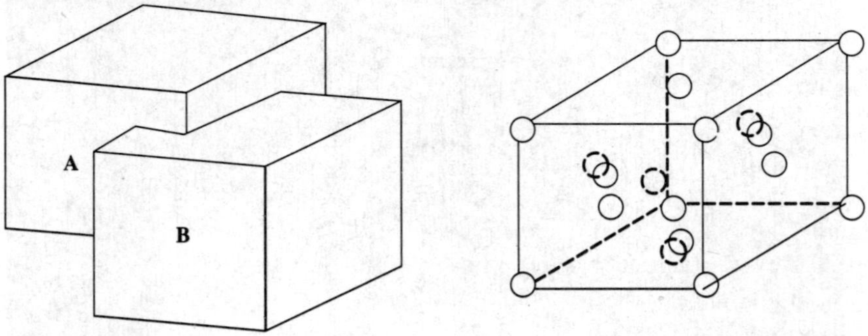

(a) Merger of two FCC crystals **(b)** Merged crystal structure

Figure 7.6 shows a three-dimensional model of the crystal structure of silicon. A closer look at this structure will reveal that these four additional atoms in the interior of the FCC (the white balls in Fig. 7.6) form a subcubic cell of the diamond lattice type as illustrated in Figure 7.7 [Ruska 1987]. A silicon unit cell thus has 18 atoms with 8 atoms at the corners plus 6 atoms on the faces and another 4 interior atoms. Many perceive the crystal structure of silicon to be a *diamond lattice* at a cubic lattice spacing of 0.543 nm [Kwok 1997].

Figure 7.6 | Photograph of a silicon crystal structure

The spacing between adjacent atoms in the diamond subcell is 0.235 nm [Brysek et al. 1991, Sze 1985, Ruska 1987]. Four equally spaced nearest-neighbor atoms that lie at the corners of a tetrahedron make the diamond lattice, as shown in the inset in

Figure 7.7 | A subcubic cell of the diamond lattice type in a silicon crystal.

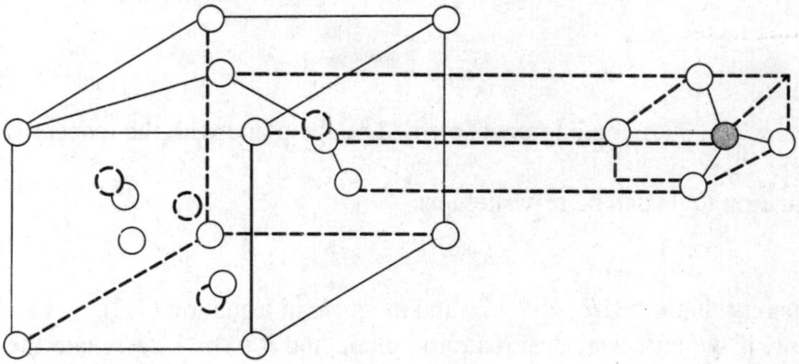

Figure 7.7. One may also perceive the silicon crystal as being stacked layers of repeating cubes. Each cube has an atom at each corner and at the center of each face (FCC structure). These cubes are interlocked with four neighboring cubes in bulk single-crystal silicon boules, and the wafers are sliced from the boules.

Because of the asymmetrical and nonuniform lattice distance between atoms, single-crystal silicon exhibits anisotropic thermophysical and mechanical characteristics that need to be understood for the benefits of handling and manufacturing. These orientation-dependent material characteristics can be better expressed by using the Miller indices [Ruska 1987, Sze 1985].

EXAMPLE 7.1

Estimate the number of atoms per cubic centimeter of pure silicon.

Solution

Since the lattice constant b = 0.543 nm = 0.543 × 10^{-9} m, and there are 18 atoms in each cubic cell, the number of atoms in a cubic centimeter, with 1 cm = 0.01m, is

$$N = \left(\frac{V}{v}\right)n = \left(\frac{0.01}{0.543 \times 10^{-9}}\right)^3 \times 18 = 1.12 \times 10^{23} \text{ atoms/cm}^3$$

In the above computation, *V* and *v* represent respectively the bulk volume of silicon in the question and the volume of a single crystal and *n* is the number of atoms in a single unit crystal of silicon.

7.4.4 The Miller Indices

Because of the skew distribution of atoms in a silicon crystal, material properties are by no means uniform in the crystal. It is important to be able to designate the principal orientations as well as planes in the crystal on which the properties are specified. A popular method of designating crystal planes and orientations is the *Miller indices*. These indices are effectively used to designate planes of materials in cubic crystal families. We will briefly outline the principle of these indices as follows.

Let us consider a point $P(x, y, z)$ in an arbitrary plane in a space defined by the cartesian coordinate system x-y-z. The equation that defines the point P is

$$\frac{x}{a} + \frac{y}{b} + \frac{z}{c} = 1 \tag{7.1}$$

where a, b, and c are the intercepts formed by the plane with the respective x, y, and z axes.

Equation (7.1) can be re-written as:

$$hx + ky + mz = 1 \tag{7.2}$$

It is apparent that $h = 1/a$, $k = 1/b$, and $m = 1/c$ in Equation (7.2).

Now, if we let (hkm) designate the plane and <hkm> designate the direction that is normal to the plane (hmk), then we may designate the three planes that apply to a cubic crystal. These three groups of planes are illustrated in Figure 7.8. We will assume that the cubic structure has unit length with the intercepts $a = b = c = 1.0$.

Figure 7.8 | Designation of the planes of a cubic crystal.

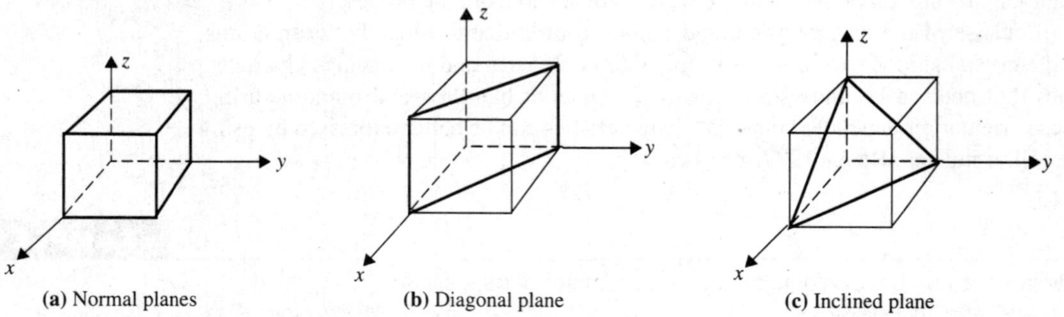

We can designate various planes in Figure 7.8 by using Equations (7.1) and (7.2) as follows:

Top face in Figure 7.8a:	(001)
Right face in Figure 7.8a:	(010)
Front face in Figure 7.8a:	(100)
Diagonal face in Figure 7.8b:	(110)
Inclined face in Figure 7.8c:	(111)

The orientations of these planes can be represented by <100>, which is perpendicular to plane (100), <111>, which is normal to the plane (111), etc. Figure 7.9 shows these three planes and orientations for a unit cell in single-crystal silicon.

The silicon atoms in the three principal planes, (100), (110) and (111), are illustrated in Figure 7.10. The atoms shown in open circles are those at the corners of the cubic, solid circles in gray are the ones at the center of the faces, and the gray circles in dotted lines are the atoms at the interior of the unit cell.

As shown in Figures 7.6 and 7.10, the lattice distances between adjacent atoms are shortest for those atoms on the (111) plane. These short lattice distances between

Figure 7.9 | Silicon crystal structure and planes and orientations.

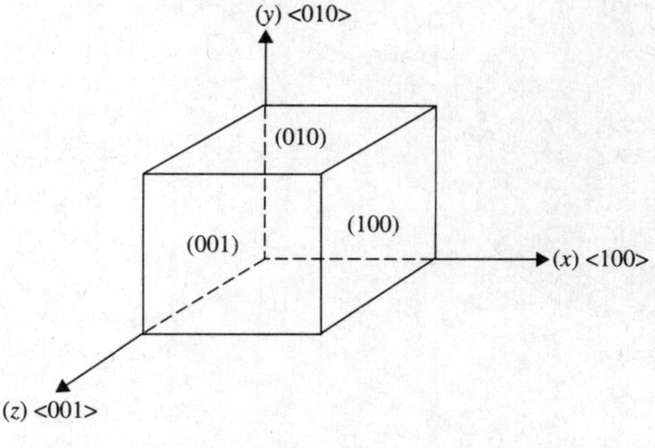

Figure 7.10 | Silicon atoms on three designated planes.

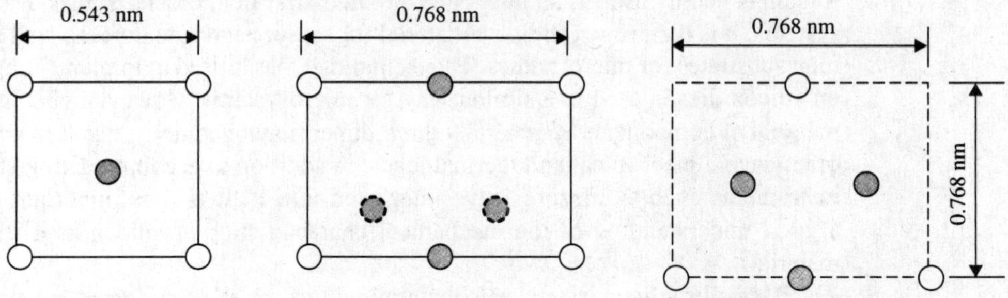

atoms make the attractive forces between atoms stronger on this plane than those on the other two planes. Also, this plane contains three of the four atoms that are situated at the center of the faces of the unit cell (in gray). Thus, the growth of crystal in this plane is the slowest and the fabrication processes, e.g., etching, as we will learn in Chapters 8 and 9, will proceed slowest.

Because of the importance of the orientation-dependent machinability of silicon substrates, wafers that are shipped by suppliers normally indicate in which directions, i.e., $<110>$, $<100>$, or $<111>$ the cuts have been made, by "flats" as illustrated in Figure 7.11. The edge of silicon crystal boules can be ground to produce single *primary flats,* and in some cases, with additional single *secondary flats.* The wafers that are sliced from these crystal boules may thus contain one or two flats as shown in Figure 7.11. The primary flats are used to indicate the crystal orientation of the wafer structure, whereas the secondary flats are used to indicate the dopant type of the wafer. For example, Figure 7.11a indicates a flat that is normal to the $<111>$ orientation with p-type of silicon crystal. The wafer with additional second flat at 45° from the primary flat in Figure 7.11b indicates an n-type doping in the wafer. Other arrangements of flats in wafers for various designations can be found in [van Zant 1997, Madou 1997].

Figure 7.11 | Primary and secondary flats in silicon wafers.

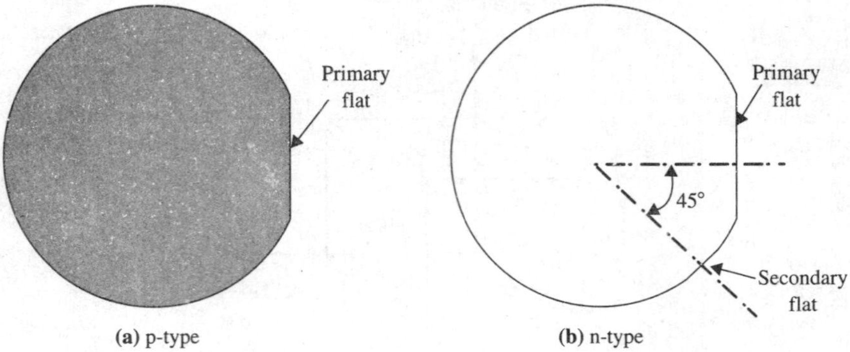

7.4.5 Mechanical Properties of Silicon

Silicon is mainly used as an integrated circuit carrier in microelectronics. For microsystems, it is the prime candidate material for sensors and actuators, as well as common substrates for microfluidics. The technology that is used in implementing the IC on silicon dies is used in a similar way for microsystems. However, silicon, as the material of components of generally three-dimensional geometry, needs to withstand often-severe mechanical and thermal loads, in addition to accommodating electrical instruments such as piezoresistive integrated into it. It is thus important to have a good understanding of the mechanical characteristics of silicon as a structural material.

Basically, silicon is an elastic material with no plasticity or creep below 800°C. It shows virtually no fatigue failure under all conceivable circumstances. These unique characteristics make it an ideal material for sensing and actuating in microsystems. However, it is a brittle material. Therefore, undesirable brittle fracture behavior with weak resistance to impact loads needs to be considered in the design of such microsystems. Another disadvantage of silicon substrates is that they are anisotropic. This makes accurate stress analysis of silicon structures tedious, since directional mechanical property must be included. For example, Table 7.2 indicates the different Young's modulus and shear modulus of elasticity of silicon crystals in different orientations [Madou 1997]. For most cases in microsystem design, the bulk material properties of silicon, silicon compounds, and other active substrate materials presented in Table 7.3 are used.

Table 7.2 | The diverse Young's moduli and shear moduli of elasticity of silicon crystals

Miller index for orientation	Young's modulus E, GPa	Shear modulus G, GPa
<100>	129.5	79.0
<110>	168.0	61.7
<111>	186.5	57.5

Table 7.3 I Mechanical and thermophysical properties of MEMS materials*

Material	σ_y, 10^9 N/m²	E, 10^{11} N/m²	ρ, g/cm³	c, J/g-°C	k, W/cm-°C	α, 10^{-6}/°C	T_M, °C
Si	7.00	1.90	2.30	0.70	1.57	2.33	1400
SiC	21.00	7.00	3.20	0.67	3.50	3.30	2300
Si_3N_4	14.00	3.85	3.10	0.69	0.19	0.80	1930
SiO_2	8.40	0.73	2.27	1.00	0.014	0.50	1700
Aluminum	0.17	0.70	2.70	0.942	2.36	25	660
Stainless steel	2.10	2.00	7.90	0.47	0.329	17.30	1500
Copper	0.07	0.11	8.9	0.386	3.93	16.56	1080
GaAs	2.70	0.75	5.30	0.35	0.50	6.86	1238
Ge		1.03	5.32	0.31	0.60	5.80	937
Quartz	0.5-0.7	0.76-0.97	2.66	0.82-1.20	0.067.0.12	7.10	1710

*Principal source for semiconductor material properties: *Fundamentals of Microfabrication*, Marc Madou, CRC Press, 1997

Legend: σ_y = yield strength, E = Young's modulus, ρ = mass density, c = specific heat, k = thermal conductivity, α = coefficient of thermal expansion, T_M = melting point.

EXAMPLE 7.2

As indicated in Chapter 5, the thermal diffusivity of a material is a measure of how fast heat can flow in the material. List the thermal diffusivities of silicon, silicon dioxide, aluminum, and copper, and make an observation on the results.

Solution

The thermal diffusivity α is a function of several properties of the material as shown in Equation (5.39):

$$\alpha = \frac{k}{\rho c}$$

where the properties k, ρ, and c for the four materials are given in Table 7.3. They are listed in slightly different units in Table 7.4.

Table 7.4 I Thermal diffusivity of selected materials for microsystems

Material	k, J/sce-m-°C	ρ, g/m³	c, J/g-°C	Thermal diffusivity, α, m²/s
Si	157	2.3×10^6	0.7	97.52×10^{-6}
SiO_2	1.4	2.27×10^6	1.0	0.62×10^{-6}
Aluminum	236	2.7×10^6	0.94	93×10^{-6}
Copper	393	8.9×10^6	0.386	114.4×10^{-6}

By substituting the material properties tabulated in the three left columns in Table 7.4 into Equation (5.39), we can compute the thermal diffusivities of the materials as indicated in the right column in the same table. It is not surprising to observe that copper has the highest thermal diffusivity, whereas silicon and aluminum have about the same value. Useful information from this exercise is that silicon oxide conducts heat more than 150 times slower than silicon and aluminum.

We may thus conclude that copper films are the best material for fast heat transmission in microsystems, whereas silicon dioxide can be used as an effective thermal barrier.

7.5 | SILICON COMPOUNDS

Three silicon compounds are often used in microsystems: silicon dioxide, SiO_2; silicon carbide, SiC; and silicon nitride, Si_3N_4. We will take a brief look at each of these compounds as to the roles they play in microsystems.

7.5.1 Silicon Dioxide

There are three principal uses of silicon oxide in microsystems: (1) as a thermal and electric insulator (see Table 7.1 for the low electric resistivity of oxides), (2) as a mask in the etching of silicon substrates, and (3) as a sacrificial layer in surface micromachining, as will be described in Chapter 9. Silicon oxide has much stronger resistance to most etchants than silicon. Important properties of silicon oxide are listed in Table 7.5.

Table 7.5 | Properties of silicon dioxide

Properties	Values
Density, g/cm^3	2.27
Resistivity, Ω-cm	$\geq 10^{16}$
Relative permittivity	3.9
Melting point, °C	~1700
Specific heat, J/g-°C	1.0
Thermal conductivity, W/cm-°C	0.014
Coefficient of thermal expansion, ppm/°C	0.5

Source: Ruska [1987].

Silicon dioxide can be produced by heating silicon in an oxidant such as oxygen with or without steam. Chemical reactions for such processes are:

$$Si + O_2 \rightarrow SiO_2 \tag{7.3}$$

for "dry" oxidation, and

$$Si + 2H_2O \rightarrow SiO_2 + 2H_2 \tag{7.4}$$

for "wet" oxidation in steam.

Oxidation is effectively a diffusion process, as described in Chapter 3. Therefore, the rate of oxidation can be controlled by similar techniques used for most other diffusion processes. Typical diffusivity of silicon dioxide at 900°C in dry oxidation is 4×10^{-19} cm^2/s for arsenic-doped (n-type) silicon and 3×10^{-19} cm^2/s for boron-doped (p-type) silicon [Sze 1985]. The process can be accelerated to much faster rates by the presence of steam; the highly activated H_2O molecules enhance the process. As in all diffusion processes, the diffusivity of the substance to be diffused into the base material is the key parameter for the effectiveness of the diffusion process.

7.5.2 Silicon Carbide

The principal application of silicon carbide (SiC) in microsystems is its dimensional and chemical stability at high temperatures. It has very strong resistance to oxidation even at very high temperatures. Thin films of silicon carbide are often deposited over MEMS components to protect them from extreme temperature. Another attraction of using SiC in MEMS is that dry etching (to be described in Chapters 8 and 9) with aluminum masks can easily pattern the thin SiC film. The patterned SiC film can further be used as a passivation layer (protective layer) in micromachining for the underlying silicon substrate, as SiC can resist common etchants such as KOH and HF.

Silicon carbide is a by-product in the process of producing single crystal silicon boules as described in Section 7.4.2. As silicon exists in the raw materials of carbon (coal, coke, wood chips, etc.), the intense heating of these materials in the electric arc furnace results in SiC sinking to the bottom of the crucible. Silicon carbide films can be produced by various deposition techniques. Pertinent thermophysical properties of SiC are given in Table 7.3.

7.5.3 Silicon Nitride

Silicon nitride (Si_3N_4) has many superior properties that are attractive for MEMS and microsystems. It provides an excellent barrier to diffusion of water and ions such as sodium. Its ultrastrong resistance to oxidation and many etchants makes it suitable for masks for deep etching. Applications of silicon nitride include optical waveguides, encapsulants to prevent diffusion of water and other toxic fluids into the substrate. It is also used as high-strength electric insulators and ion implantation masks.

Silicon nitride can be produced from silicon-containing gases and NH_3 in the following reaction:

$$3SiCl_2H_2 + 4\,NH_3 \rightarrow Si_3N_4 + 6HCl + 6H_2 \quad \mathbf{(7.5)}$$

Selected properties of silicon nitride are listed in Table 7.6. Both chemical vapor deposition processes [*low-pressure chemical vapor deposition* (LPCVD) and the *plasma-enhanced chemical vapor deposition* (PECVD)] in Table 7.6 will be described in detail in Chapter 8. Additional material properties are given in Table 7.3.

7.5.4 Polycrystalline Silicon

Silicon in polycrystalline form can be deposited onto silicon substrates by chemical vapor deposition (CVD) as illustrated in Figure 7.12. It has become a principal material in surface micromachining, as will be described in Chapter 9.

Table 7.6 | Selected properties of silicon nitride

Properties	LPCVD	PECVD
Deposition temperature, °C	700–800	250–350
Density, g/cm^3	2.9–3.2	2.4–2.8
Film quality	Excellent	Poor
Relative permittivity	6–7	6–9
Resistivity, Ω-cm	10^{16}	10^6–10^{15}
Refractive index	2.01	1.8–2.5
Atom % H	4–8	20–25
Etch rate in concentrated HF	200 Å/min	
Etch rate in boiling HF	5–10 Å/min	
Poisson's ratio	0.27	
Young's modulus, GPa	385	
Coefficient of thermal expansion, ppm/°C	1.6	

Source: Madou [1997].

Figure 7.12 | Polysilicon deposits on a silicon substrate.

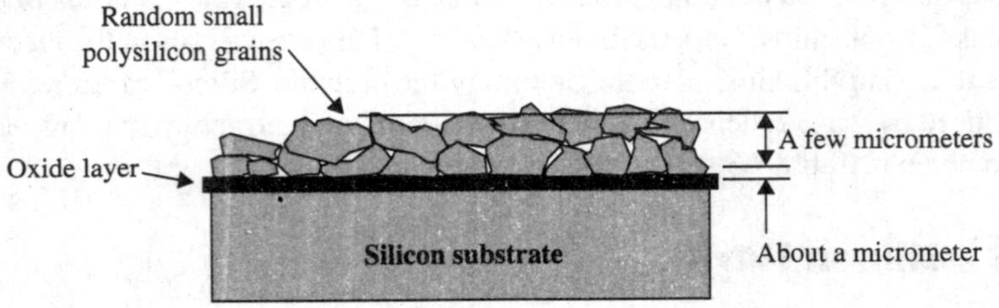

The low pressure chemical vapor deposition (LPCVD) process is frequently used for depositing polycrystalline silicon onto silicon substrates. The temperature involved in this process is about 600 to 650°C. Polysilicon (an abbreviation of polycrystalline silicon) is widely used in the IC industry for resistors, gates for transistors, thin-film transistors, etc. Highly doped polysilicon, (with arsenic and phosphorous for n type or boron for p type) can drastically reduce the resistivity of polysilicon to produce conductors and control switches. They are thus ideal materials for microresistors as well as easy ohmic contacts. A comparison of some key properties of polysilicon and other materials is presented in Table 7.7. Being a congregation of single silicon crystals in random sizes and orientations, polysilicon can be treated as isotropic material in thermal and structural analyses.

7.6 | SILICON PIEZORESISTORS

Piezoresistance is defined as a change in electrical resistance of solids when subjected to stress fields. Silicon piezoresistors that have such characteristics are widely used in microsensors and actuators.

We have learned from Chapter 3 and Section 7.4.2 that doping boron into the silicon lattice produces p-type silicon crystal while doping arsenic or phosphorus

Table 7.7 | Comparison of mechanical properties of polysilicon and other materials

Materials	Young's modulus, GPa	Poisson's ratio	Coefficient of thermal expansion, ppm/°C
As substrates:			
Silicon	190	0.23	2.6
Alumina	415		8.7
Silica	73	0.17	0.4
As thin films:			
Polysilicon	160	0.23	2.8
Thermal SiO_2	70	0.2	0.35
LPCVD SiO_2	270	0.27	1.6
PECVD SiO_2			2.3
Aluminum	70	0.35	25
Tungsten	410	0.28	4.3
Polymide	3.2	0.42	20–70

Source: Madou [1997].

results in n-type silicon. Both p- and n-type silicon exhibit excellent piezoresistive effect. Charles Smith in 1954 discovered the piezoresistance of p- and n-type silicon.

The fact that silicon crystal, whether it is p type or n type, is anisotropic has made the relationship between the change of resistance and the existent stress field more complex. This relationship is shown below:

$$\{\Delta R\} = [\pi]\{\sigma\} \tag{7.6}$$

where $\{\Delta R\} = \{\Delta R_{xx} \quad \Delta R_{yy} \quad \Delta R_{zz} \quad \Delta R_{xy} \quad \Delta R_{xz} \quad \Delta R_{yz}\}^T$ represents the change of resistance in an infinitesimally small cubic piezoresistive crystal element with corresponding stress components $\{\sigma\} = \{\sigma_{xx} \quad \sigma_{yy} \quad \sigma_{zz} \quad \sigma_{xy} \quad \sigma_{xz} \quad \sigma_{yz}\}^T$ as shown in Figure 7.13. Of the six independent stress components in the stress tensor $\{\sigma\}$, there are three normal stress components, σ_{xx}, σ_{yy}, and σ_{zz}, and three shearing stress components, σ_{xy}, σ_{xz}, and σ_{yz}. The vector $[\pi]$ in Equation (7.6) is referred to as *piezoresistive coefficient matrix.* It has the following form:

Figure 7.13 | A silicon piezoresistance subjected to a stress field.

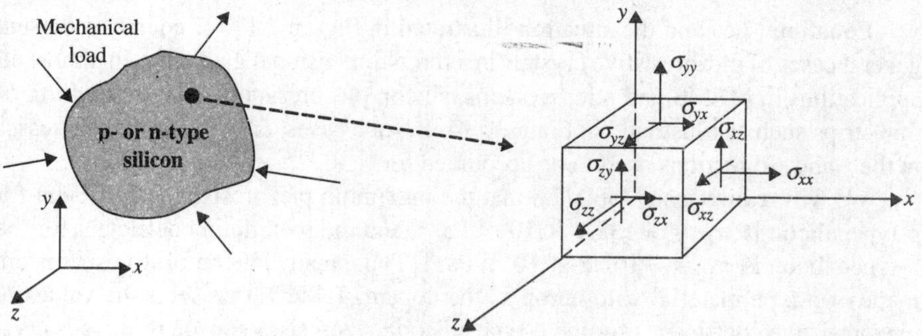

$$[\pi] = \begin{bmatrix} \pi_{11} & \pi_{12} & \pi_{12} & 0 & 0 & 0 \\ \pi_{12} & \pi_{11} & \pi_{12} & 0 & 0 & 0 \\ \pi_{12} & \pi_{12} & \pi_{11} & 0 & 0 & 0 \\ 0 & 0 & 0 & \pi_{44} & 0 & 0 \\ 0 & 0 & 0 & 0 & \pi_{44} & 0 \\ 0 & 0 & 0 & 0 & 0 & \pi_{44} \end{bmatrix} \tag{7.7}$$

We notice from Equation (7.7) that only three coefficients, π_{11}, π_{12}, and π_{44}, appear in the matrix. By expanding the matrix equation in Equation (7.6) with the appropriate piezoresistive coefficients in Equation (7.7), we will have the following relations:

$$\Delta R_{xx} = \pi_{11}\sigma_{xx} + \pi_{12}(\sigma_{yy} + \sigma_{zz})$$

$$\Delta R_{yy} = \pi_{11}\sigma_{yy} + \pi_{12}(\sigma_{xx} + \sigma_{zz})$$

$$\Delta R_{zz} = \pi_{11}\sigma_{zz} + \pi_{12}(\sigma_{xx} + \sigma_{yy})$$

$$\Delta R_{xy} = \pi_{44}\sigma_{xy}$$

$$\Delta R_{xz} = \pi_{44}\sigma_{xz}$$

$$\Delta R_{yz} = \pi_{44}\sigma_{yz}$$

It is thus apparent that the coefficients π_{11} and π_{12} are associated with the normal stress components, whereas the coefficient π_{44} is related to the shearing stress components.

The actual values of these three coefficients depend on the angles of the piezoresistor with respect to the silicon crystal lattice. The values of these coefficients at room temperature are given in Table 7.8.

Table 7.8 | Resistivity and piezoresistive coefficients of silicon at room temperature in <100> orientation

Materials	Resistivity, Ω-cm	π_{11}*	π_{12}*	π_{44}*
p silicon	7.8	+6.6	−1.1	+138.1
n silicon	11.7	−102.2	+53.4	−13.6

*in 10^{-12} cm²/dyne or in 10^{-11} m²/N (or Pa⁻¹)
Source: French and Evans [1988].

Equation (7.6) and the situation illustrated in Figure 7.13, of course, represent general cases of piezoresistive crystals in a three-dimensional geometry. In almost all applications in MEMS and microsystems, silicon piezoresistors exist in the form of thin strips such as illustrated in Figure 7.14. In such cases, only the in-plane stresses in the x and y directions need to be accounted for.

We will realize from Table 7.8 that the maximum piezoresistive coefficient for p-type silicon is $\pi_{44} = +138.1 \times 10^{-11}$ Pa^{-1}, and the maximum coefficient for the n-type silicon is $\pi_{11} = -102.2 \times 10^{-11}$ Pa^{-1}. Thus, many silicon piezoresistors are made of p-type material with boron as the dopant. Table 7.9 presents the values of piezoresistive coefficients for the p-type silicon piezoresistors made in various crystal planes [Brysek et al. 1991].

Figure 7.14 | Silicon strain gages.

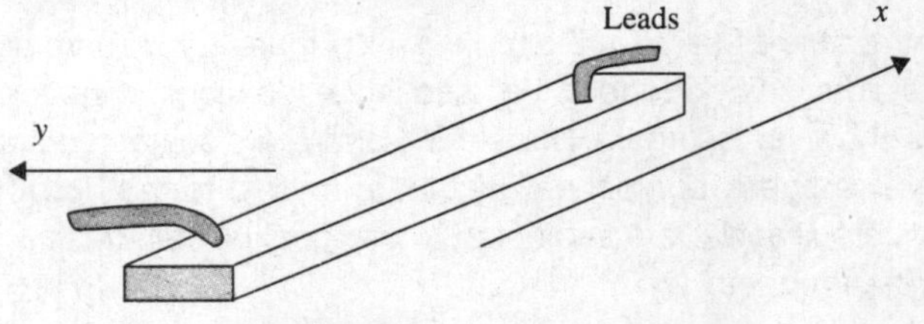

The values of π_L denote the piezoresistive coefficient along the longitudinal direction, i.e., along the <x> direction in Figure 7.14, whereas π_T represents the piezoresistive coefficient in the tangential direction, i.e., along the <y> direction in the same figure.

Table 7.9 | Piezoresistive coefficients of p-type silicon piezoresistors in various directions

Crystal planes	Orientation <x>	Orientation <y>	π_L	π_T
(100)	<111>	<211>	$+0.66\pi_{44}$	$-0.33\pi_{44}$
(100)	<110>	<100>	$+0.5\pi_{44}$	~0
(100)	<110>	<110>	$+0.5\pi_{44}$	$-0.5\pi_{44}$
(100)	<100>	<100>	$+0.02\pi_{44}$	$+0.02\pi_{44}$

Source: Brysek et al. [1991].

The change of electric resistance in a silicon piezoresistance gage can thus be expressed as:

$$\frac{\Delta R}{R} = \pi_L \sigma_L + \pi_T \sigma_T \tag{7.8}$$

in which ΔR and R are respectively the change of resistance and the original resistance of the silicon piezoresistive. The value of the original resistance R in Equation (7.8) can be obtained either by direct measurement, or by using the formula $R = \frac{\rho L}{A}$ in which ρ is the resistivity of the piezoresistor such as given in Figure 3.8, L and A are respective length and cross-sectional area of the piezoresistor. The piezoresistive coefficients in both longitudinal and tangential directions are given in Table 7.9 for p-type silicon piezoresistives. The stress components in the longitudinal and tangential directions, σ_L and σ_T, are the stresses that cause the change of the resistance in the piezoresistor.

EXAMPLE 7.3

Estimate the change of resistance in piezoresistors attached to the diaphragm of a pressure sensor described in Example 4.4 in Chapter 4.

Solution

Let us reillustrate the situation in Example 4.4 in Figure 7.15. There are four identical piezoresistors, *A*, *B*, *C*, and *D*, diffused in the locations at the top face of the silicon die as shown in the figure. Resistors *A* and *D* are subjected predominantly to the transverse stress component σ_T that is normal to the horizontal edges, whereas resistors *B* and *C* are subjected to the longitudinal stress σ_L that is normal to the vertical edges.

Figure 7.15 | Piezoresistors in a pressure sensor die.

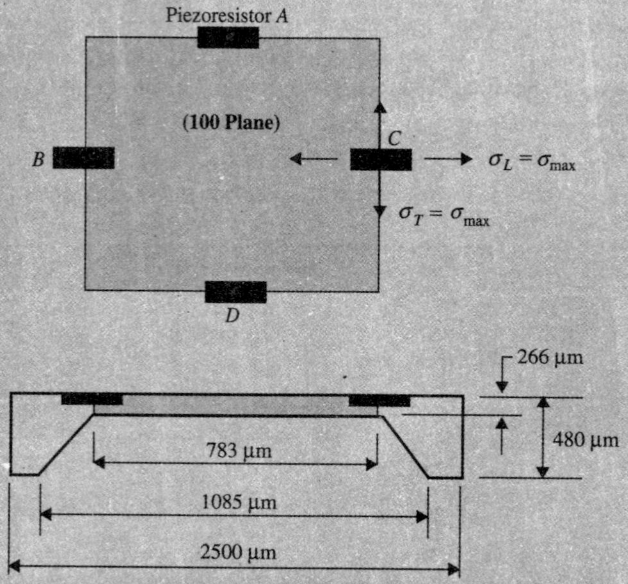

Because of the square plane geometry of the diaphragm that is subjected to uniform pressure loading at the top surface, the bending moments normal to all edges are equal in magnitudes. Thus, we may let $\sigma_L = \sigma_T = \sigma_{max} = 186.8$ MPa $= 186.8 \times 10^6$ Pa (from Example 4.4).

We assume that the diaphragm is on the (100) plane and both stresses are along the <100> directions (why?), we thus obtain the piezoresistive coefficient $\pi_L = \pi_T = 0.02\pi_{44}$ from Table 7.9 for p-type piezoresistors. The value of the piezoresistive coefficient π_{44} is available from Table 7.8; it is $\pi_{44} = 138.1 \times 10^{-11}$ Pa^{-1}.

We can thus estimate the change of electric resistance in the piezoresistors to be:

$$\frac{\Delta R}{R} = \pi_L\sigma_L + \pi_T\sigma_T = 2 \times 0.02\pi_{44}\,\sigma_{max} = 2 \times 0.02(138.1 \times 10^{-11})(186.8 \times 10^6)$$

$$= 0.01032\Omega/\Omega$$

Determining the net change of resistance in the resistors requires knowledge of the length of the resistor in the stress-free state and the resistivity of the resistor material as given in Figure 3.8.

One major drawback of silicon piezoresistors is the strong temperature dependence of their piezoresistivity. The sensitivity of piezoresistivity to the applied stress deteriorates rapidly with increase of temperature. Table 7.10 presents the variation of piezoresistive coefficients with reference to those at room temperature.

Table 7.10 | Temperature dependence of resistivity and piezoresistivity of silicon piezoresistors

Doping concentration, 10^{18}/cm^3	p-type TCR, % per °C	p-type TCP, % per °C	n-type TCR, % per °C	n-type TCP, % per °C
5	0.0	−0.27	0.01	−0.28
10	0.01	−0.27	0.05	−0.27
30	0.06	−0.18	0.09	−0.18
100	0.17	−0.16	0.19	−0.12

Source: [French and Evans 1988].
TCR = temperature coefficient of resistance; TCP = temperature coefficient of piezoresistivity.

Take for example, a p-type silicon piezoresistor with a doping concentration of 10^{19} per cm^3; the loss of piezoresistivity is 0.27% per °C. The same piezoresistor operating at 120°C would have lost (120 – 20) × 0.27% = 27% of the value of the piezoresistivity coefficient. Appropriate compensation for this loss must be considered in the design of signal conditioning systems.

The doping concentration for piezoresistives normally should be kept below 10^{19}/cm^3 because the piezoresistive coefficients drop considerably above this dose, and reverse breakdown becomes an issue.

7.7 | GALLIUM ARSENIDE

Gallium arsenide (GaAs) is a compound semiconductor. It is made of equal numbers of gallium and arsenic atoms. Because it is a compound, it is more complicated in lattice structure, with atoms of both constituents, and hence is more difficult to process than silicon. However, GaAs is an excellent material for monolithic integration of electronic and photonic devices on a single substrate. The main reason that GaAs is a prime candidate material for photonic devices is its high mobility of electrons in comparison to other semiconducting materials, as shown in Table 7.11.

As we see from the table, GaAs has about 7 times higher electron mobility than silicon. The high electron mobility in this material means it is easier for electric current to flow in the material. The photoelectronic effect, as illustrated in Section 2.2.4 in Chapter 2, describes the electric current flow in a photoelectric material when it is energized by incoming photons. GaAs, being a material with high mobility of

Table 7.11 | Electron mobility of selected materials at 300 K

Materials	Electron mobility, $m^2/V\text{-}s$
Aluminum	0.00435
Copper	0.00136
Silicon	0.145
Gallium arsenide	0.850
Silicon oxide	~0
Silicon nitride	~0

Source: Kwok [1997].

electrons, thus can better facilitate the electric current flow when it is energized by photon sources.

Gallium arsenide is also a superior thermal insulator, with excellent dimensional stability at high temperature. The negative aspect of this material is its low yield strength as indicated in Table 7.3. Its yield strength, at 2700 MPa, is only one-third of that of silicon. This makes GaAs less attractive for use as substrates in microsystems. Because of its relatively low use in the microelectronics industry, GaAs is much more expensive than silicon.

In addition to the differences in thermophysical properties as indicated in Table 7.3, Table 7.12 gives a good comparison of these two substrate materials used in microsystems.

Table 7.12 | A comparison of GaAs and silicon in micromachining

Properties	GaAs	Silicon
Optoelectronics	Very good	Not good
Piezoelectric effect	Yes	No
Piezoelectric coefficient, pN/°C	2.6	Nil
Thermal conductivity	Relatively low	Relatively high
Cost	High	Low
Bonding to other substrates	Difficult	Relatively easy
Fracture	Brittle, fragile	Brittle, strong
Operating temperature	High	Low
Optimum operating temp., °C	460	300
Physical stability	Fair	Very good
Hardness, GPa	7	10
Fracture strength, GPa	2.7	6

Source: Madou [1997].

7.8 | QUARTZ

Quartz is a compound of SiO_2. The single-unit cell for quartz is in the shape of tetrahedron with three oxygen atoms at the apexes at the base and one silicon atom at the other apex of the tetrahedron. The axis that is normal to the base plane is called the *Z* axis. The quartz crystal structure is made up of rings with six silicon atoms.

Quartz is close to being an ideal material for sensors because of its near absolute thermal dimensional stability. It is used in many piezoelectric devices in the market, as will be described in Section 7.9. Commercial applications of quartz crystals include wristwatches, electronic filters, and resonators. Quartz is a desirable material in microfluidics applications in biomedical analyses. It is inexpensive and it works well in electrophoretic fluid transportation as described in Chapters 3 and 5 because of its excellent electric insulation properties. It is transparent to ultraviolet light, which often is used to detect the various species in the fluid.

Quartz is a material that is hard to machine. Diamond cutting is a common method, although ultrasonic cutting has been used for more precise geometric trimming. It can be etched chemically by HF/NH_4F into the desired shape. Quartz wafers up to 75 mm in diameter by 100 μm thick are available commercially.

Quartz is even more dimensionally stable than silicon, especially at high temperatures. It offers more flexibility in geometry than silicon despite the difficulty in machining. Some key properties are presented in Table 7.13.

Table 7.13 | Some properties of quartz

Properties	**Value ∥ Z**	**Value ⊥ Z**	**Temperature dependency**
Thermal conductivity, cal/cm-s°/C	29×10^{-3}	16×10^{-3}	↓ with T
Relative permittivity	4.6	4.5	↓ with T
Density, kg/m³	2.66×10^3	2.66×10^3	
Coefficient of thermal expansion, ppm/°C	7.1	13.2	↑ with T
Electrical resistivity, Ω/cm	0.1×10^{15}	20×10^{15}	↓ with T
Fracture strength, GPa	1.7	1.7	↓ with T
Hardness, GPa	12	12	

Source: Madou [1997].

7.9 | PIEZOELECTRIC CRYSTALS

One of the most commonly used nonsemiconducting materials in MEMS and microsystems is piezoelectric crystals. Piezoelectric crystals are the solids of ceramic compounds that can produce a voltage when a mechanical force is applied between their faces. The reverse situation, that is the application of voltage to the crystal, can also change its shape. This conversion of mechanical energy to electronic signals (i.e., voltage) and vice versa is illustrated in Figure 7.16. This unique material behavior is called the *piezoelectric effect*. Jacques and Pierre Curie discovered the piezoelectric effect in 1880. This effect exists in a number of natural crystals such as quartz, tourmaline, and sodium potassium tartrate, and quartz has been used in electromechanical transducers for many years. There are many other synthesized crystals such as Rochelle salt ($N_aKC_4H_4O_6$-$4H_2O$), barium titanate ($BaTiO_3$), and lead zirconate titanate (PZT).

For a crystal to exhibit the piezoelectric effect, its structure should have no center of symmetry. A stress applied to such a crystal will alter the separation between the positive and negative charge sites in each elementary cell, leading to a net polarization at the crystal surface [Waanders 1991]. An electric field with voltage potential is thus created in the crystal because of such polarization.

The most common use of the piezoelectric effect is for high voltage generation through the application of high compressive stress. The generated high-voltage field can be used as an impact detonation device. It can also be used to send signals for depth detection in a sonar system. The principal applications of the piezoelectric effect in MEMS and microsystems, however, are in actuators, as described in Chapter 2, and dynamic signal transducers for pressure sensors and accelerometers. The piezoelectric effect is also used in pumping mechanisms, as described in Chapters 5 and 6, for microfluidic flows, as well as for inkjet printer heads.

Figure 7.16 | Conversion of mechanical and electrical energies by piezoelectric crystals.

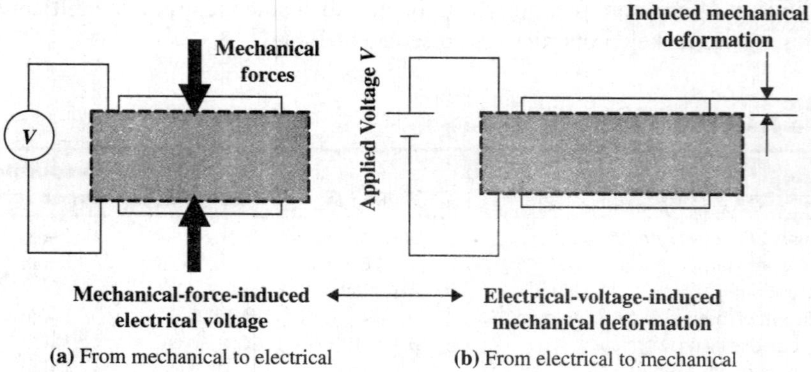

(From Kasap [1997].)

The effectiveness of the conversion of mechanical to electrical energy and vice versa can be assessed by the electromechanical conversion factor K, defined as follows [Kasop 1997]:

$$K^2 = \frac{\text{output of mechanical energy}}{\text{input of electrical energy}} \quad \textbf{(7.9a)}$$

or

$$K^2 = \frac{\text{output of electrical energy}}{\text{input of mechanical energy}} \quad \textbf{(7.9b)}$$

The following simple mathematical relationships between the electromechanical effects can be used in the design of piezoelectric transducers in unidirectional loading situations [Askeland 1994]:

1. The electric field produced by stress:

$$V = f\sigma \quad \textbf{(7.10)}$$

where V is the generated electric field in V/m and σ is the applied stress in pascals (Pa). The coefficient f is a constant.

2. The mechanical strain produced by the electric field:

$$\varepsilon = dV \tag{7.11}$$

in which ε is the induced strain and V is the applied electric field in V/m. The piezoelectric coefficient d for common piezoelectric crystals is given in Table 7.14.

The coefficients f and d in Equations (7.10) and (7.11) have the following relationship:

$$\frac{1}{fd} = E \tag{7.12}$$

where E is the Young's modulus of the piezoelectric crystal.

Table 7.14 | Piezoelectric coefficients of selected materials

Piezoelectric crystals	Coefficient d, 10^{-12} m/V	Electromechanical conversion factor K
Quartz (crystal SiO_2)	2.3	0.1
Barium titanate ($BaTiO_3$)	100–190	0.49
Lead zirconate titanate, PZT ($PbTi_{1-x}Zr_xO_3$)	480	0.72
$PbZrTiO_6$	250	
$PbNb_2O_6$	80	
Rochelle salt ($NaKC_4H_4O_6$-$4H_2O$)	350	0.78
Polyvinylidene fluoride, PVDF	18	

Source: Kasap [1997], Askeland [1994].

EXAMPLE 7.4

A thin piezoelectric crystal film of PZT is used to transduce the signal in a micro-accelerometer with a cantilever beam made of silicon as described in Example 4.7. The accelerometer is designed for maximum acceleration/deceleration of 10*g*. The PZT transducer is located at the support base of the cantilever beam where the maximum strain exists during the bending of the beam, as illustrated in Figure 7.17. Determine the electrical voltage output from the PZT film at the maximum acceleration/deceleration of 10*g*.

Figure 7.17 | Piezoelectric transducer in a beam-accelerometer.

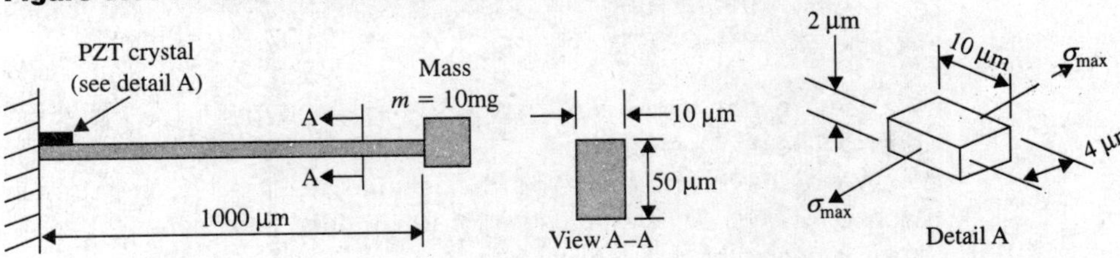

Solution

The solution can be obtained by determining the maximum bending stress, and thus the maximum bending strain in the beam due to the dynamic load of the attached mass accelerated to 10*g*. The maximum bending strain in the beam is assumed to give the same magnitude of strain to the attached thin PZT film. The voltage generated in the PZT can be computed from Equation (7.11).

We will first determine the equivalent bending load P_{eq} that is equivalent to that of the 10-mg mass accelerated or decelerated to 10*g*:

$$P_{eq} = ma = (10 \times 10^{-6}) \times (10 \times 9.81) = 981 \times 10^{-6}\ \text{N}$$

The beam-accelerometer is equivalent to a statically loaded cantilever beam subjected to the equivalent force acting at its free end as illustrated in Figure 7.18.

Figure 7.18 | Equivalent static bending of a cantilever beam in an accelerometer.

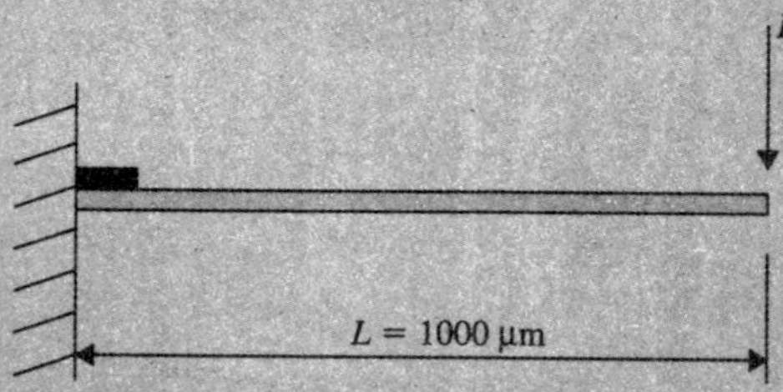

The maximum bending moment $M_{max} = P_{eq}L = (981 \times 10^{-6})(1000 \times 10^{-6}) = 0.981 \times 10^{-6}$ N-m.

We will need the moment of inertia I of the beam cross section to calculate the equivalent maximum bending stress σ_{max}. The value of I was computed in Example 4.7 to be 0.1042×10^{-18} m^4. We thus have the maximum bending stress at the support as:

$$\sigma_{max} = \frac{M_{max}\,C}{I} = \frac{(0.981 \times 10^{-6})(25 \times 10^{-6})}{(0.1042 \times 10^{-18})} = 235.36 \times 10^6\ \text{Pa}$$

where C is the half-depth of the beam cross section.

The associated maximum bending strain ε_{max} in the beam is:

$$\varepsilon_{max} = \frac{\sigma_{max}}{E} = \frac{235.36 \times 10^6}{1.9 \times 10^{11}} = 123.87 \times 10^{-5}\ \text{m/m}$$

We used Young's modulus $E = 1.9 \times 10^{11}$ Pa as given in Table 7.3 for the silicon beam in the above calculation.

We assumed that the strain in the beam will result in the same strain in the attached PZT film. Thus with a strain of 123.87×10^{-5} m/m in the PZT, the induced voltage per meter in the crystal is:

$$V = \frac{\varepsilon}{d} = \frac{\varepsilon_{max}}{d} = \frac{123.87 \times 10^{-5}}{480 \times 10^{-12}} = 0.258 \times 10^7 \text{ V/m}$$

We used the piezoelectric coefficient $d = 480 \times 10^{-12}$ m/V obtained from Table 7.14 in the above computation.

Since the actual length of the PZT crystal attached to the beam is $l = 4$ μm, we will expect the total voltage generated by the transducer at 10g load to be:

$$v = Vl = (0.258 \times 10^7)(4 \times 10^{-6}) = 10.32 \text{ V}$$

EXAMPLE 7.5

Determine the electric voltage required to eject a droplet of ink from an inkjet printer head with a PZT piezoelectric crystal as a pumping mechanism. The ejected ink will have a resolution of 300 dots per inch (dpi). The ink droplet is assumed to produce a dot with a film thickness of 1 μm on the paper. The geometry and dimension of the printer head is illustrated in Figure 7.19. Assume that the ink droplet takes the shape of a sphere and the inkwell is always refilled after ejection.

Figure 7.19 | Schematic diagram of an inkjet printer head.

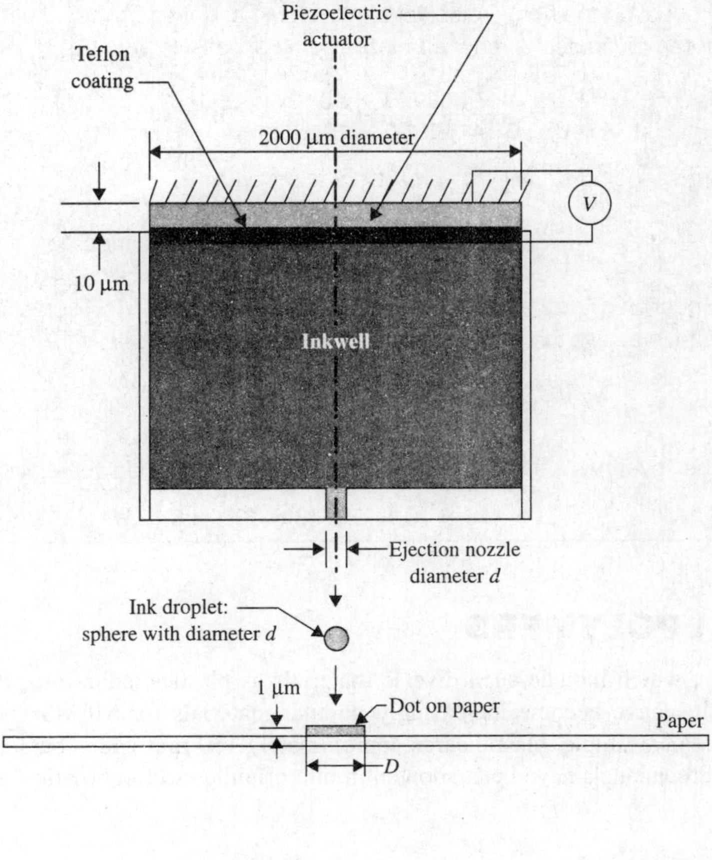

Solution

We will first have to determine the diameter of the ejection nozzle, d, corresponding to ink dots with 300-dpi resolution on the paper. We will first determine the diameter of the dot, D, on the paper. With 300 such dots in a linear space of 1 in, we can readily find the corresponding diameter to be D = 1 in/300 = 25.4 mm/300 = 0.084666 mm, or 84.67 μm.

Since the dots are produced from spherical droplets of diameter d, by letting the volume of the ink droplet from the nozzle be equal to that of the ink dot on the paper, the following relation can be used:

$$\frac{4}{3}\pi r^3 = \left(\frac{\pi}{4}D^2\right)(t)$$

in which r = radius of the spherical ink droplet and t is the thickness of the ink dot on the paper. We thus find the radius of the spherical ink droplet, $r = 11.04 \times 10^{-6}$ m, with D = 84.67 μm and t = 1 μm.

Next, we assume that the volume of an ink droplet leaving the inkwell is equivalent to the volume created by vertical expansion of the PZT piezoelectric cover at the back of the inkwell. Let the vertical expansion of the piezoelectric cover be W and the corresponding volume of the displaced ink be $(\pi/4)\,\Delta^2 W$, in which Δ is the diameter of the piezoelectric cover at 2000 μm.

By equating the above displaced ink volume and the volume of the ink dot, V_{dot}, we can compute the required piezoelectric expansion W to be:

$$W = \frac{4V_{dot}}{\pi\Delta^2} = \frac{4 \times 5629.21 \times 10^{-18}}{3.1416(2000 \times 10^{-6})^2} = 1791.83 \times 10^{-12}\text{ m}$$

The corresponding strain in the piezoelectric cover is:

$$\varepsilon = \frac{W}{L} = \frac{1791.83 \times 10^{-12}}{10 \times 10^{-6}} = 179.183 \times 10^{-6}\text{ m/m}$$

From Equation (7.11) and the piezoelectric coefficient of PZT crystals from Table 7.14, we have the required applied voltage per meter as:

$$V = \frac{\varepsilon}{d} = \frac{179.183 \times 10^{-6}}{480 \times 10^{-12}} = 0.3733 \times 10^{6}\text{ V/m}$$

Since the cover has a thickness of 10 μm, the required applied voltage is:

$$v = LV = (10 \times 10^{-6})(0.3733 \times 10^{6}) = 3.733\text{ V}$$

7.10 | POLYMERS

Polymers, which include such diverse materials as plastics, adhesives, Plexiglas, and Lucite, have become increasingly popular materials for MEMS and microsystems. For example, plastic cards approximately 150 mm wide containing over 1000 microchannels have been adopted in microfluidic electrophoretic systems by the biomedical industry [Lipman 1999] as described in Chapter 2. Epoxy resins and adhesives such as silicone rubber are customarily used in MEMS and microsystem packaging.

This type of material is made up of long chains of organic (mainly hydrocarbon) molecules. The combined molecules, i.e., polymer molecules, can be a few hundred nanometers long. Low mechanical strength, low melting point, and poor electrical conductivity characterize polymers. Thermoplastics and thermosets are two groups of polymers that are commonly used for industrial products. Thermoplastics can be easily formed to the desired shape for the specific product, whereas thermosets have better mechanical strength and temperature resistance up to 350°C. Because of the rapid increase of applications in industrial products, polymers and polymerization, which is the process of producing various kinds of polymers, constitute a distinct engineering subject. It is not realistic to offer a complete list of available polymers and plastics, as well as the many polymerization processes in this chapter. What will be presented in this section is information on the applications of polymers that are relevant to the design and packaging of MEMS and microsystems. Much of the materials presented here are available in greater detail in a special reference on polymers for electronics and optoelectronics [Chilton and Goosey 1995].

7.10.1 Polymers as Industrial Materials

Traditionally, polymers have been used as insulators, sheathing, capacitor films in electric devices, and die pads in integrated circuits. A special form of polymer, the plastics, has been widely used for machine and device components. Following is a summary of the many advantages of polymers as industrial materials:

Light weight
Ease in processing
Low cost of raw materials and processes for producing polymers
High corrosion resistance
High electrical resistance
High flexibility in structures
High dimensional stability

Perhaps the most intriguing fact about polymers is their variety of molecular structures. This unique feature has offered scientists and engineers great flexibility in developing "organic alloys" by mixing various ingredients to produce polymers that satisfy specific applications. Consequently, there are a great variety of polymers available for industrial applications in today's marketplace.

7.10.2 Polymers for MEMS and Microsystems

Polymers have become increasingly important materials for MEMS and microsystems. Some of these applications are:

1. Photoresist polymers are used to produce masks for creating desired patterns on substrates by photolithography, as will be described in Chapter 8.

2. The same photoresist polymers are used to produce the prime mold with the desired geometry of MEMS components in the LIGA process for manufacturing microdevice components, as will be described in Chapter 9. These prime molds are plated with metals such as nickel for subsequent injection molding for mass production of microcomponents.
3. As will be described later in this subsection, conductive polymers are used as organic substrates for MEMS and microsystems.
4. The ferroelectric polymers, which behave like piezoelectric crystals, can be used as a source of actuation in microdevices such as those for micropumping, as described in Section 5.6.3 in Chapter 5.
5. The thin Langmuir–Blodgett (LB) films can be used to produce multilayer microstructures, similar to the micromachining technique presented in Chapter 9.
6. Polymers with unique characteristics are used as a coating substances for capillary tubes to facilitate electro-osmotic flow in microfluidics as described in Section 3.8.2.
7. Thin polymer films are used as electric insulators in microdevices and as a dielectric substances in microcapacitors.
8. Polymers are widely used for electromagnetic interference (EMI) and radio-frequency interference (RFI) shielding in microsystems.
9. Polymers are ideal materials for the encapsulation of microsensors and packaging of other microsystems.

7.10.3 Conductive Polymers

For polymers to be used in certain applications in microelectronics, MEMS, and microsystems, they have to be made electrically conductive with superior dimensional stability. Polymers have been used extensively in the packaging of MEMS, but they have also been used as substrates for some MEMS components in recent years with the successful development of techniques for controlling the electric conductivity of these materials.

By nature, polymers are poor electric conductors. Table 7.15 shows the electric conductivity of various materials. One will readily see that polymers, represented by polyethylene, have the lowest electric conductance of all the listed materials.

Polymers can be made electrically conductive by the following three methods:

Pyrolysis A pyropolymer based on phthalonitrile resin can be made electrically conductive by adding an amine heated above 600°C. The conductivity of the polymer produced by this process can be as high as 2.7×10^4 S/m, which is slightly better than that of carbon.

Doping Doping with the introduction of an inherently conductive polymer structure, such as by incorporating a transition metal atom into the polymer backbone, can result in electrically conductive polymers. Doping of polymers depends on the dopants and the individual polymer. Following are examples of dopants used in producing electrically conductive polymers:

Table 7.15 | Electric conductivity of selected materials

Materials	Electric conductivity, S/m*
Conductors:	
Copper	10^6–10^8
Carbon	10^4
Semiconductors:	
Germanium	10^0
Silicon	10^{-4}–10^{-2}
Insulators:	
Glass	10^{-10}–10^{-8}
Nylon	10^{-14}–10^{-12}
SiO_2	10^{-16}–10^{-14}
Polyethylene	10^{-16}–10^{-14}

*S/m = siemens per meter. Siemens = Ω^{-1}= A^2-s^3/kg-m^2

For polyacetylenes (PA): Dopants such as Br_2, I_2, AsF_5, $HClO_4$, and H_2SO_4 are used to produce the p-type polymers, and sodium naphthalide in tetrahydrofuran (THF) is used for the n-type polymers.

For polyparaphenylenes (PPP): AsF_5 is used for the p-type and alkali metals are used for the n-type polymer.

For polyphenylene sulfide (PPS): The dopant used in this case is AsF_5.

Insertion of Conductive Fibers Incorporating conductive fillers into both thermosetting and thermoplastic polymer structures can result in electrically conductive polymers. Fillers include such materials as carbon, aluminum flakes, and stainless steel, gold, and silver fibers. Other inserts include semiconducting fibers, e.g., silicon and germanium. Fibers are in the order of nanometers in length.

7.10.4 The Langmuir–Blodgett (LB) Film

A special process developed by Langmuir as early as 1917, and refined by Blodgett, can be used to produce thin polymer films at molecular scale. This process is generally known as the *LB process*. It involves spreading volatile solvent over surface-active materials. The LB process can produce more than a single monolayer by depositing films of various compositions onto a substrate to create a multilayer structure. The process closely resembles that of the micromachining manufacturing technique presented in Chapter 9. It is thus regarded as an alternative micromanufacturing technique.

LB films are good candidate materials for exhibiting ferro (magnetic), pyro (heat-generating), and piezoelectric properties. LB films may also be produced with controlled optical properties such as refractive index and antireflectivity. They are thus ideal materials for microsensors and optoelectronic devices. Following are a few examples of LB film applications in microsystems:

1. *Ferroelectric polymer thin films.* Useful in particular is polyvinylidene fluoride (PVDF). Applications of this type of film include sound transducers in air and water, tactile sensors, biomedical applications such

as tissue-compatible implants, cardiopulmonary sensors, and implantable transducers and sensors for prosthetics and rehabilitation devices. The piezoelectric coefficient of PVDF is given in Table 7.14.

2. *Coating materials with controllable optical properties.* These are widely used in broadband optical fibers, which can transmit laser light at different wavelengths.
3. *Microsensors.* The sensitivity of many electrically conducting polymeric materials to the gases and other environmental conditions makes this material suitable for microsensors. Its ability to detect specific substances relies on the reversible and specific absorption of species of interest on the surface of the polymer layer and the subsequent measurable change of conductivity of the polymer. Figure 7.20 shows a schematic structure of such a sensor. These sensors work on the principle that the electric conductivity of the polymer sensing element will change when it is exposed to a specific gas. This principle is similar to that of chemical sensors as described in Chapter 2.

Figure 7.20 | Microsensor using polymers.

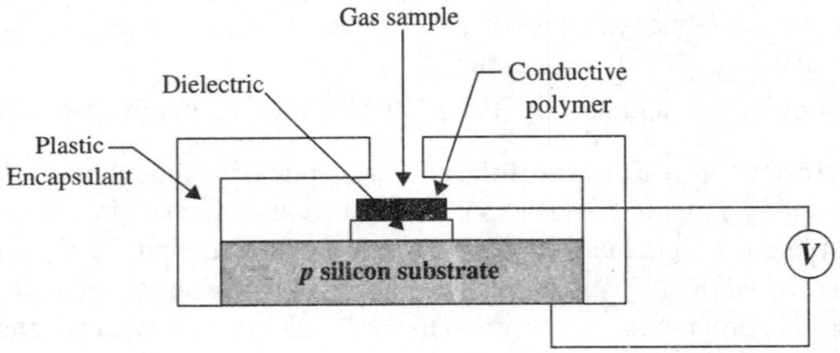

Conductive polymers are also used in electrochemical biosensors, such as microamperometric glucose and galactose biosensors.

7.11 | PACKAGING MATERIALS

We learned about the evolution of micromachining technology from the integrated circuits industry in Chapter 1. Consequently, many techniques that were developed for the IC packaging are now used for microsystems packaging. However, a major distinction between the two types of packaging is that in microelectronics packaging, it is usually sufficient to protect the IC die and the interconnects from the often hostile operating environment. In microsystem packaging, however, not only are the sensing or actuating elements to be protected, but they are also required to be in contact with the media that are the sources of actions. Many of these media are hostile to these elements. Special technologies that have been developed to package delicate aerospace device components are also used frequently in the packaging of microsystems.

Materials used in microsystem packaging include those used in IC packaging—wires made of noble metals at the silicon die level, metal layers for lead wires, solders for die/constraint base attachments, etc. Microsystems packaging also include metals and plastics. Let us take a look at a typical pressure sensor, schematically shown in Figure 7.21. At the die level, we use aluminum or gold metal films as the ohmic contacts to the piezoresistors that are diffused in the silicon diaphragm. Similar materials are used for the lead wires to the interconnects outside the casing. The casing can be made of plastic or stainless steel, depending on the severity of the environment. Glass such as Pyrex or ceramics such as alumina are often used as constraint bases. The adhesives that enable the silicon die to be attached to the constraint base can be tin–lead solder alloys or epoxy resins or adhesives such as the popular Room-temperature vulcanizing (RTV) silicone rubber. In the case of soldered attachments, thin metal layers need to be sputtered at the joints to facilitate the soldering. Copper and aluminum are good candidate materials for this purpose. We will learn from Chapter 11 that these adhesive materials for die/base attachments are very important in isolating the die from undesired interference from other components in the package. When part of a microsystem package, such as silicon diaphragm, needs to be in contact with a hostile medium, silicone gel or silicone oil is used to shield the part.

Figure 7.21 | A typical packaged micropressure sensor.

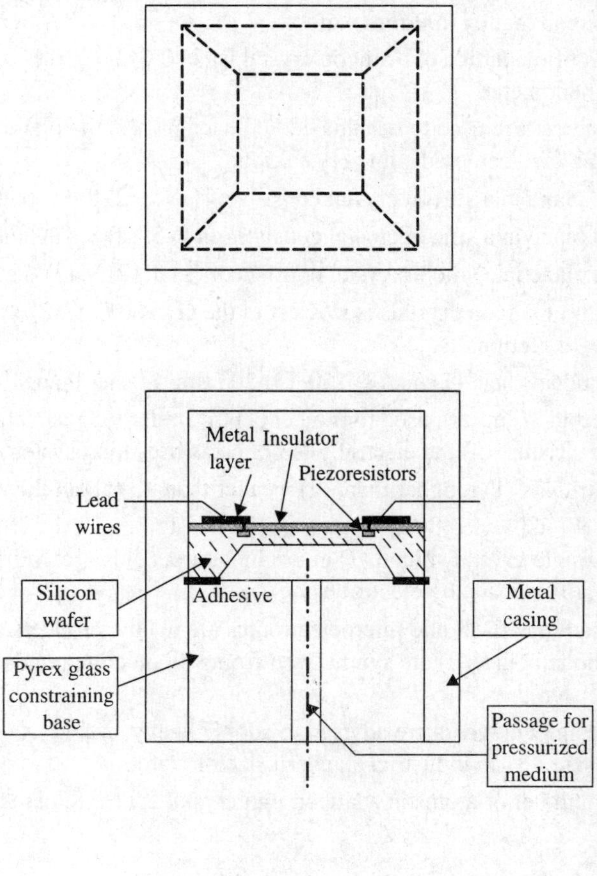

PROBLEMS

Part 1. Multiple Choice

1. A substrate is (1) a sublayer in MEMS, (2) a flat microscopic object, (3) a flat macroscopic object in microelectronics.
2. A semiconducting material can be made to become an electrically conducting material by (1) applying high electric voltage, (2) applying high current, (3) introducing the right kind of foreign atoms into the semiconducting material.
3. Silicon has a Young's modulus similar to that of (1) aluminum, (2) stainless steel, (3) copper.
4. Silicon has a mass density similar to that of (1) aluminum, (2) stainless steel, (3) copper.
5. The principal reason why silicon is an ideal material for MEMS is (1) its dimensional stability over a wide range of temperatures, (2) it is light and strong, (3) it is readily available.
6. Silicon has a coefficient of thermal expansion (1) higher than, (2) lower than, (3) about the same as that of silicon dioxide.
7. The 300-mm wafers offer (1) 2, (2) 2.25, (3) 2.5 times more area for substrates than that by 200-mm wafers.
8. The length of the lattice of a silicon crystal is (1) 0.543, (2) 0.643, (3) 0.743 nanometer.
9. Miller's indices are used to designate (1) the length, (2) the plane, (3) the volume of a face-centered cubic crystal.
10. The (100) plane in a silicon crystal consists of (1) 5, (2) 8, (3) 6 atoms.
11. The (110) plane in a silicon crystal consists of (1) 5, (2) 8, (3) 6 atoms.
12. The (111) plane in a silicon crystal consists of (1) 5, (2) 8, (3) 6 atoms.
13. The growth of silicon crystals is slowest in the (1) <100>, (2) <110>, (3) <111> direction.
14. Silicon conducts heat (1) 50, (2) 150, (3) 200 times faster than silicon oxide.
15. Silicon carbide films are used to protect (1) the underlying substrates, (2) the integrated circuits, (3) the electric interconnects in a microsystem.
16. Silicon nitride is (1) tougher than, (2) weaker than, (3) about the same as silicon in strength.
17. Pure and single-crystal silicon (1) exists in nature, (2) is grown from special processes, (3) is made by electrolysis.
18. Wafers used in MEMS and microelectronics are (1) the products of a single-crystal silicon boule, (2) are synthesized from silicon compounds, (3) exist in nature.
19. MEMS design engineers are advised to adopt (1) any size, (2) a custom-specified size, (3) an industrial standard size of wafer.
20. The total number of atoms in a silicon unit crystal is (1) 18, (2) 16, (3) 14.

21. The toughest plane for processing in a single silicon crystal is (1) the (100) plane, (2) the (110) plane, (3) the (111) plane.
22. The 54.74° slope in the cavity of a silicon die for a pressure sensor is (1) determined by choice, (2) a result of the crystal's resistance to etching in the (111) plane, (3) a result of the crystal's resistance to etching in the (110) plane.
23. Polysilicon is popular because it can easily be made as a (1) semiconductor, (2) insulator, (3) electrical conductor.
24. Polysilicon films are used in microsystems as (1) dielectric material, (2) substrate material, (3) electrically conducting material.
25. The electrical resistance of silicon piezoresistors varies in (1) all directions, (2) only in the preferred directions, (3) neither of the above applies.
26. It is customary to relate silicon piezoresistance change to (1) deformations, (2) strains, (3) stresses induced in the piezoresistors in MEMS and microsystems.
27. There are (1) three, (2) four, (3) six piezoresistive coefficients in silicon piezoresistors.
28. The single most serious disadvantage of using silicon piezoresistor is (1) the high cost of producing such resistors, (2) its strong sensitivity to signal transduction, (3) its strong sensitivity to temperature.
29. Gallium arsenide has (1) 6, (2) 7, (3) 8 times higher electron mobility than silicon.
30. Gallium arsenide is chosen over silicon for the use in micro-optical devices because of its (1) optical reflectivity, (2) dimensional stability, (3) high electron mobility.
31. Gallium arsenide is not as popular as silicon in MEMS application because of (1) its higher cost in production, (2) difficulty of mechanical work, (3) low mechanical strength.
32. Quartz crystals have the shape of (1) a cube, (2) a tetrahedron, (3) a body-centered cube.
33. It is customary to relate the voltage produced by a piezoelectric crystal to the (1) deformations, (2) temperature, (3) stresses induced in the crystal.
34. Application of mechanical deformation to a piezoelectric crystal can result in the production of (1) electric resistance change, (2) electric current change, (3) electric voltage change in the crystal.
35. Most piezoelectric crystals (1) exist in nature, (2) are made by synthetic processes, (3) are made by doping the substrate.
36. A polymer is a material that is made up of many (1) small-size, (2) large-size, (3) long-chain molecules.
37. In general, polymers are (1) electrically conductive, (2) semi electrically conductive, (3) insulators.
38. Polymers (1) can, (2) cannot, (3) may never be made electrically conductive.

39. The LB process is used to produce (1) thin films, (2) dies, (3) piezoelectric polymers in MEMS and microsystems.
40. MEMS and microsystem packaging materials are (1) restricted to microelectronics packaging materials, (2) just about all engineering materials, (3) semiconducting materials.

Part 2. Computational Problems

1. Estimate the maximum number of silicon dies of size of 2 mm wide × 4 mm long that can be accommodated in the four standard-sized wafers given in Section 7.4.2. The dies are laid out in parallel on the wafer with 0.25-μm gaps between dies. Make your observation on the results.
2. Prove that the 300-mm wafers do indeed have 2.25 times larger area for substrates than that of 200-mm wafers.
3. Estimate the number of atoms per cubic millimeter and cubic micrometer of pure silicon.
4. Find the change of electric resistance if a piezoresistor made of p-type silicon replaces the piezoelectric crystal in Example 7.4. The piezoresistor has a length of 4 μm.
5. What would be the voltage output of the piezoelectric transducer in Example 7.4 if PVDF polymer films were used instead?
6. Determine the length of lattices that bond the atoms in the three principal planes of the silicon crystal shown in Figure 7.8.
7. Determine the spacing between atoms in the three planes in a single silicon crystal as illustrated in Figure 7.10.
8. Determine the angle between the orientation <100> to the (111) plane in a single silicon crystal cell.
9. For homogeneous and isotropic solids, the relation between the three elastic constants is $E = 2(1 + \nu)G$ exists. In this relationship, E is the Young's modulus, ν is Poisson's ratio, and G is the shear modulus of elasticity. Determine the Young's modulus of silicon in three orientations in Table 7.2 using the tabulated values of ν and G, and compare the computed results with those tabulated.
10. Determine the electric voltage required to pump a droplet of ink from the well in Example 7.5 with a resolution of 600 dpi.

Chapter 8

Microsystem Fabrication Processes

CHAPTER OUTLINE

8.1 | INTRODUCTION

We have mentioned in several places in this book that MEMS technology evolves from microelectronics fabrication technology. Consequently, many of the fabrication techniques used in producing integrated circuits have been adopted to create the complex three-dimensional shapes of many MEMS and microsystems. As we will learn from the various microfabrication techniques presented in this chapter, they are radically different from those used in producing traditional machines and devices. Furthermore, almost all microfabrication techniques or processes involve physical and chemical treatment of materials whose effects are relatively unknown to many engineers with traditional engineering background. It is thus necessary for engineers who are involved in the design of these products to acquire a good knowledge of solid-state physics and the associated microfabrication techniques. Such knowledge and experience are not only required for the analyses in the design process, but also are necessary to ensure the manufacturability of whatever products are designed.

This chapter will provide the reader with an overview of various fabrication techniques involved in MEMS and microsystems manufacturing; each of these techniques will be treated separately. They will be integrated in micromanufacturing processes as described Chapter 9.

8.2 | PHOTOLITHOGRAPHY

Among the many distinct characteristics and features of microelectronics and microsystems as presented in Table 1.1, a fundamental difference between these two advanced technologies is that microsystems almost always involve complex three-dimensional structural geometry in microscale. Patterning of geometry with extremely high precision at this scale thus becomes a major challenge to engineers. *Photolithography* or *microlithography* appears to be the only viable way for producing high-precision patterning on substrates in microscale at the present time.

8.2.1 Overview

The word *lithography* is a derivation from two Greek words: litho (stone) and graphein (to write) [Madou 1997]. According to Ruska [1987], the photolithography process involves the use of an optical image and a photosensitive film to produce a pattern on a substrate. Photolithography is one of the most important steps in microfabrication. It appears to be the only technique that is available at present to create patterns on substrates with submicrometer resolution. In microelectronics, these patterns are necessary for the p-n junctions, diodes, capacitors, etc. for integrated circuits. In microsystems, however, photolithography is used to set patterns for masks for cavity etching in bulk micromanufacturing, or for thin film deposition and etching of sacrificial layers in surface micromachining, as well as for the primary circuitry of electrical signal transduction in sensors and actuators.

The general procedure of photolithography is outlined in Figure 8.1. Let us begin with the substrate at the top left of the figure. This substrate can be a silicon wafer as in microelectronics or other substrate material such as silicon dioxide or silicon nitride. A photoresist is first coated onto the flat surface of the substrate. The substrate with photoresist is then exposed to a set of lights through a transparent mask with the desired patterns. Masks used for this purpose are often made of quartz. Patterns on the mask are photographically reduced from macro- or mesosizes to the desired microscales. Photoresist materials change their solubility when they are exposed to light. Photoresists that become more soluble under light are classified as

Figure 8.1 | General procedure of photolithography.

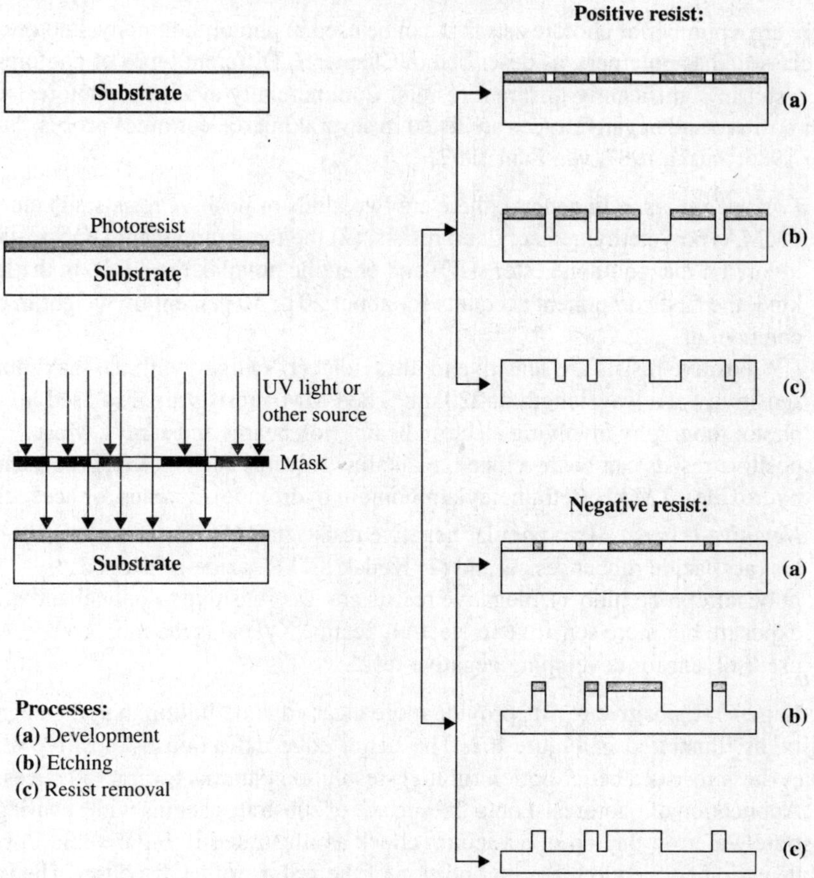

positive photoresists, whereas the *negative photoresists* become more soluble under the shadow. The exposed substrate after development with solvents will have opposite effects in these two types of photoresists, as illustrated in the right column of Figure 8.1. The retained photoresist materials create the imprinted patterns after the development [see step (a) in the figure]. The portion of the substrate under the shadow of the photoresists is protected from the subsequent etching [step (b) in the figure]. A permanent pattern is thus created in the substrate after the removal of the photoresist [step (c) in the figure].

Photolithography for MEMS and microsystems needs to be performed in a class-10 clean room or better. The class number of a clean room is a designation of the air quality in it. A class-10 clean room means that the number of dust particles 0.5 μm or larger in a cubic foot of air in the room is less than 10. Most other microfabrication processes presented in this chapter can tolerate a clean room of class-100. These requirements for clean room air quality are in sharp contrast to those of the air quality of class 5 million in a typical urban environment!

8.2.2 Photoresists and Application

There are a number of photoresists that can be used in photolithography. Photoresists are classified as polymers, as described in Chapter 7. Different types of photoresists can result in significantly different results. Commercially available photoresists of both positive and negative types are listed in several microelectronics process books [Sze 1985, Ruska 1987, van Zant 1997].

- *Positive resists.* In general, there are two kinds of positive resists: (1) the PMMA (polymethymethacrylate) resists, (2) the two-component DQN resist involving diazoquinone ester (DQ) and phenolic novolak resin (N). In the latter kind, the first component accounts for about 20 to 50 percent by weight in the compound.

  Positive resists are sensitive to ultraviolet (UV) light, with the maximum sensitivity at a wavelength of 220 nm. The PMMA resists are also used in photolithography involving electron beams, ion beams and x-rays. Most positive resists can be developed in alkaline solvents such as KOH (potassium hydroxide), TMAH (tetramethylammonium hydroxide), ketones, or acetates.
- *Negative resists.* Two popular negative resists are: (1) two-component bis (aryl)azide rubber resists and (2) Kodak KTFR (azide-sensitized polyisotroprene rubber). Negative resists are less sensitive to optical and x-ray exposure but more sensitive to electron beams. Xylene is the most commonly used solvent for developing negative resists.

In general, positive resists provide more clear edge definition than the negative resists, as illustrated in Figure 8.2. The better edge definition by positive resists makes these resists a better option for high-resolution patterns for microdevices.

Application of photoresist onto the surface of substrates begins with securing the substrate wafer on the top of a vacuum chuck as illustrated in Figure 8.3a. A resist puddle is first applied to the center portion of the wafer from a dispenser. The wafer

Figure 8.2 | Profiles of lines from photolithography.

is then subjected to high-speed spinning at a rotational speed that varies from 1500 to 8000 rpm for 10 to 60 seconds. The speed depends on the type of the resist, the desired thickness, and the uniformity of the resist coating. The centrifugal forces applied to the resist puddle spread the fluid over the entire surface of the wafer. Typically the thickness is between 0.5 and 2 μm with $\pm$5-nm variation. For some microsystem applications, the thickness is increased to as large as 1 cm. However, that is hardly a normal practice.

Figure 8.3 | Schematic arrangement of photoresist application.

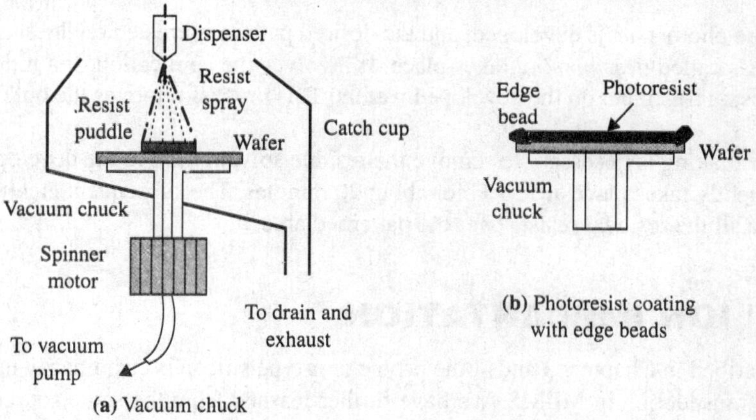

The spinner is stopped once the desired thickness is reached. A common problem is the bead of resist that occurs at the edge of the wafer as shown in Figure 8.3b. These edge beads can be several times of the thickness of the resist. The uniformity of the coating can be increased and the thickness of the edge beads can be reduced by controlling the spin speed, often by spinning slowly at first, followed by high-speed spinning.

8.2.3 Light Sources

Photoresists are sensitive to light with wavelengths ranging from 300 to 500 nm. The most popular light source for photolithography is the mercury vapor lamp. This light source provides a wavelength spectrum from 310 to 440 nm. Deep UV light has wavelengths of 150 to 300 nm, while the normal UV light has wavelengths between 350 and 500 nm. These are all suitable light sources for photolithography purposes. In special applications for extremely high resolution, x-rays are used. The wavelengths of x-rays are in the range from 4 to 50 Å (1 angstrom Å = 0.1 nm or 10^{-4} μm).

8.2.4 Photoresist Development

The same chamber used for the resist application in Figure 8.3 may be used for the development of negative resists. In such applications, the exposed wafer substrate is secured on the vacuum chuck and is spun at a high speed with spray of solvent, e.g., xylene, from the overhead nozzle. A rinsing process with distilled water follows the development.

The development of positive resists is a little more complicated than that of the negative resists. Wafers with positive resist are usually developed in batches in a tank. This development process requires a more controlled chemical reaction than just washing. A developer agent such as KOH or TMAH is projected onto the wafer surface in streams of mist at low velocity. After the development, the wafer is rinsed with a spray of distilled water, as with negative resists.

8.2.5 Photoresist Removal and Postbaking

After the photoresist is developed, and the desired pattern is created on the substrate, a process called *descumming* takes place. It involves the application of a mild oxygen plasma treatment on the developed wafers. This process removes the bulk of the resist.

Postbaking is necessary to remove the residue solvent used in the development. It frequently takes place at 120°C for about 20 minutes. The subsequent etching will remove all the residual resists over the patterned areas.

8.3 | ION IMPLANTATION

As described in Chapters 3 and 7, the p-type or n-type silicon is customarily used for signal transduction in MEMS. We have further learned from the same sources that doping boron atoms into silicon crystals can produce p-type silicon piezoresistors, whereas doping with arsenic or phosphorus atoms can produce the n-type silicon. There are generally two ways of doping a semiconductor such as silicon with foreign substances, and *ion implantation* is one of these two methods.

Ion implantation involves "forcing" free atoms, such as boron or phosphorus, with charged particles (i.e., ions) into a substrate, thereby achieving imbalance between the number of protons and electrons in the resulting atomic structure. The ions, whether they are boron ions or phosphorus ions, must carry sufficient kinetic energy to be implanted (that is, to penetrate) into the silicon substrate. For this reason, the ion implantation procedure always involves the acceleration of the ions in order to gain sufficient kinetic energy for the implantation.

As we will see from the schematic arrangement of ion implantation in Figure 8.4, ions to be implanted are created in a ion source, in which an ion beam is formed. Descriptions on two ion-generating sources are presented in Chapter 3. Figure 3.4 illustrates how ions can be produced by electron beams. Ions are extracted from the substance in the gaseous state, i.e., the plasma. The ion beam is then led into a beam controller in which the size and direction of the beam can be adjusted. The

Figure 8.4 | Ion implantation on a substrate.

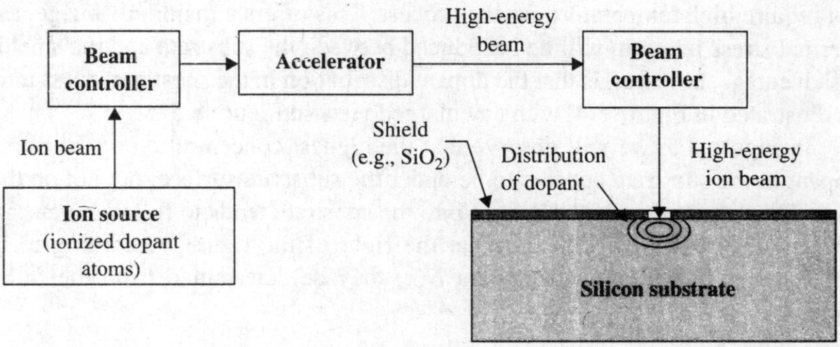

ions in the beam are then energized in the acceleration tube, or *accelerator,* in which the final energy with which the ions will impact the substrate surface is attained. The ions in the beam are focused onto the substrate, which is protected by a shield, or mask, usually made of SiO_2. The highly energized ions enter the substrate and collide with the electrons and nuclei of the substrate. The ions will transfer all their energy to the substrate upon collision, and finally come to a stop at a certain depth inside the substrate. The distribution of the implanted ions in silicon substrate is illustrated in Figure 8.5. The energy required for p- or n-type doping is presented in Table 8.1.

Figure 8.5 | Distribution of dopant through a shield. [Ruska 1987].

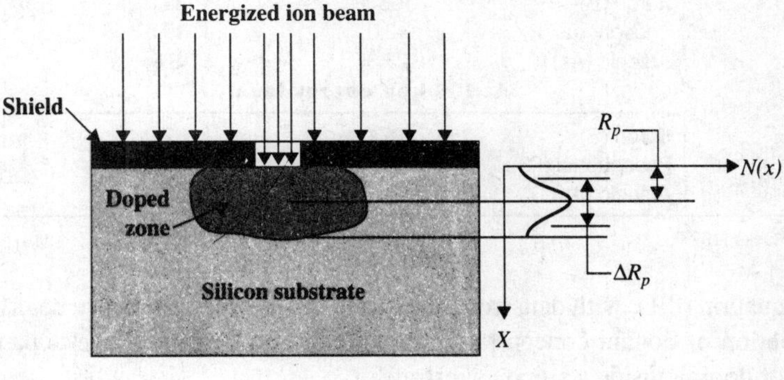

Table 8.1 | Ionization energy for dopants in silicon.

Dopants	Type	Ionization energy, eV
Phosphorus (P)	n	0.044
Arsenic (As)	n	0.049
Antimony (Sb)	n	0.039
Boron (B)	p	0.045
Aluminum (Al)	p	0.057
Gallium (Ga)	p	0.065
Indium (In)	p	0.160

Source: Ruska [1987].

Unlike the diffusion technique presented in Section 8.4, ion implantation does not require high temperature for the process. This offers a major advantage, as little thermal stress or strain will be introduced between the substrate and the shield. The disadvantage, however, is that the dopant distribution in the substrate is less uniform, as illustrated in Figure 8.4, with an enlarged view in Figure 8.5.

In Figure 8.5, we will observe that the highest concentration of the implanted dopant in the substrate appears to be under the substrate surface, but not on the surface. The distribution of the dopant into the substrate tends to follow a gaussian (or normal) distribution as illustrated at the right of the figure. According to Ruska [1987], the concentration of dopant $N(x)$ may be determined from the following equation:

$$N(x) = \frac{Q}{\sqrt{2\pi}\,\Delta R_p} \exp\left[\frac{-(x - R_p)^2}{2\,\Delta R_p^2}\right] \tag{8.1}$$

where R_p = projected range in μm, ΔR_p = scatter or "straggle" in μm, and Q = dose of the ion beam (atoms/cm^2).

The doping ranges for selected dopants in silicon substrate are given in Table 8.2.

Table 8.2 | Ion implantation of common dopants in silicon

Ion	Range R_p, nm	Straggle ΔR_p, nm
At 30 keV energy level		
Boron (B)	106.5	39.0
Phosphorus (P)	42.0	19.5
Arsenic (As)	23.3	9.0
At 100 keV energy level		
Boron (B)	307.0	69.0
Phosphorus (P)	135.0	53.5
Arsenic (As)	67.8	26.1

Source: Ruska [1987].

Equation (8.1), with data from Table 8.2, can assist engineers in determining the distribution of dopant concentration, the maximum density of the dopant, and the depth of doping inside a silicon substrate.

EXAMPLE 8.1

A silicon substrate is doped with boron ions at 100 keV. Assume the maximum concentration after the doping is 30×10^{18}/cm^3 (refer to Fig. 3.8). Find (1) the dose Q in Equation (8.1), (2) the dopant concentration at a depth 0.15 μm, and (3) the depth at which the dopant concentration is 0.1 percent of the maximum value.

Solution

From Table 8.2, we find the projected range for boron ion is 307 nm at 100 keV, or $R_p = 307 \times 10^{-7}$ cm at 100 keV, and the straggle, $\Delta R_p = 69 \times 10^{-7}$ cm.

1. We are given the maximum concentration $N_{max} = 30 \times 10^{18}/cm^3$ at $x = R_p$ in Equation (8.1), so from the same equation, we will have:

$$N_{max} = \frac{Q}{\sqrt{2\pi}\,\Delta R_p}$$

from which we have the dose $Q = (2\pi)^{0.5}(\Delta R_p)N_{max} = (6.28)^{0.5}(69 \times 10^{-7}\ cm)(30 \times 10^{18}\ cm^{-3}) = 5.2 \times 10^{14}/cm^2$

2. To find the concentration at $x = 0.15\ \mu m$, we may use the following relationship derived from Equation (8.1):

$$N(0.15\ \mu m) = N_{max} \exp\left[-\frac{(0.150 - 0.307)^2}{2(0.069)^2}\right] = (30 \times 10^{18})\exp\left(-\frac{0.0246}{0.009522}\right)$$

$$= 2.27 \times 10^{18}/cm^3$$

3. To find $x = x_0$ at which the concentration $N(x_0) = (0.1\%)N_{max} = 3 \times 10^{16}/cm^3$, we may solve for x_0 in the following equation:

$$N(x_0) = \frac{5.2 \times 10^{14}}{\sqrt{2 \times 3.14} \times 69 \times 10^{-7}} \exp\left[-\frac{(x_0 - 307 \times 10^{-7})^2}{2(69 \times 10^{-7})^2}\right] = 3 \times 10^{16}$$

$$x_0 = 563.5 \times 10^{-7}\ cm \text{ or } 0.5635\ \mu m$$

8.4 | DIFFUSION

The general principle of the diffusion process was presented in Section 3.6. The diffusion process is often used in microelectronics for the introduction of a controlled amount of foreign materials (dopants) into the selected regions of another material (the substrate).

Unlike ion implantation, diffusion is a slow doping process. It is also used as thin-film buildings in microelectronics and microsystems. Diffusion takes place at elevated temperatures. As illustrated in Figure 8.6, the atoms of the dopant gas move (diffuse) into the crystal vacancies, or interstitials, of the silicon substrate. The profile of the doped region in the substrate as shown in Figure 8.6 is also different from that of ion implantation.

The physics of the diffusion process is similar to that of heat conduction in solids as described in Chapter 5. The applicable mathematical model for diffusion is Fick's law, represented in Equation (3.2), which is similar to the Fourier law for heat conduction in solids in Equation (5-35). Fick's law gives the dopant flux in the substrate in the x direction during a diffusion process. We express the Fick's law, but in a different notation, for the diffusion of foreign atoms into the substrates of microsystems as follows:

$$F = -D\frac{\partial N(x)}{\partial x} \tag{8.2}$$

Figure 8.6 | Doping of a silicon substrate by diffusion.

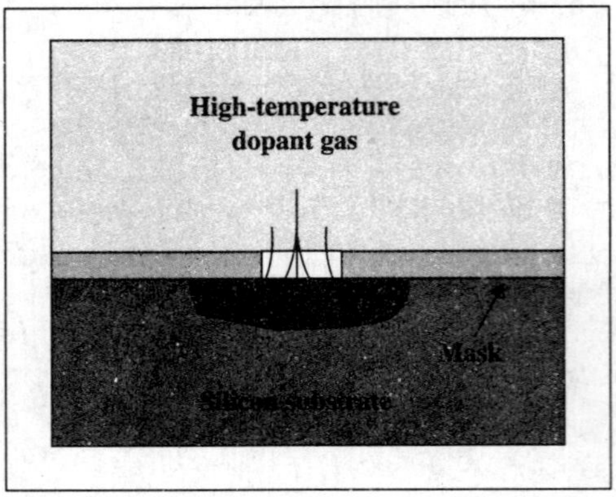

where

F = dopant flux, the number of dopant atoms passing through a unit area of the substrate in a unit time, atoms/cm^2-s.
D = diffusion coefficient, or diffusivity, of the dopant to the substrate, cm^2/s
N = dopant concentration in the substrate per unit volume, atoms/cm^3

The distribution of dopant in the substrate at any given time during the diffusion process can be obtained by the solution of the following Fick diffusion equation, which is similar to the heat conduction equation in Equation (5.38) in the direction along the x coordinate:

$$\frac{\partial N(x, t)}{\partial t} = D \frac{\partial^2 N(x, t)}{\partial x^2} \tag{8.3}$$

The solution of Equation (8.3) depends on the initial and boundary conditions involved in the process. One set of such conditions can be:

Initial condition: $N(x, 0) = 0$, i.e., there is no impurity in the substrate when the diffusion process begins.
Boundary conditions: (1) $N(0, t) = N_s$, which is the concentration at the surface exposed to the gaseous dopant. (2) $N(\infty, t) = 0$; i.e., the diffusion of foreign substance is localized, and the concentration far away from the exposed surface is negligible.

The solution of Equation (8.3) with the above initial and boundary conditions has the form:

$$N(x, t) = N_s \operatorname{erfc}\left[\frac{x}{2\sqrt{Dt}}\right] \tag{8.4}$$

where erfc (x) is the complementary error function, which is equal to:

$$\operatorname{erfc}(x) = 1 - \operatorname{erf}(x) = 1 - \frac{2}{\sqrt{\pi}} \int_0^x e^{-y^2}\, dy \tag{8.5}$$

The solutions expressed in Equations (8.4) and (8.5) are similar to those in Equations (3.4) and (3.5) for the general cases of diffusion.

Numerical values of complementary error function can be obtained either from the above integral, or from the values of error functions erf (x) that are available in mathematical tables with the corresponding values of the argument, x. Values of erfc (x) are available in Table 3.4. The term $\sqrt{Dt}$ in the solution in Equation (8.4) is the diffusion length.

The diffusivity D for a silicon substrate with common dopants such as boron, arsenic, and phosphorus can be found in references [Ruska 1987], or as given in Figure 3.12. We may use the expression in Equation (8.6), derived from the graphs in the above-cited reference, for estimating the values of D for the three common dopants boron, arsenic, and phosphorus:

$$\ln(\sqrt{D}) = aT' + b \tag{8.6}$$

where D is the diffusivity, $\mu m^2/h$.

The temperature $T' = 1000/T$ with T = diffusion temperature in K. The values of constants a and b in Equation (8.6) are given in Table 8.3.

Table 8.3 | Constants for Equation (8.6)

Dopants	Constant a	Constant b
Boron	−19.9820	13.1109
Arsenic	−26.8404	17.2250
Phosphorus ($N_s = 10^{21}/cm^3$)	−15.8456	11.1168
Phosphorus ($N_s = 10^{19}/cm^3$)	−20.4278	13.6430

EXAMPLE 8.2

A silicon substrate is subjected to diffusion of boron dopant at 1000°C with a dose $10^{11}/cm^2$. Find (1) the expression for estimating the concentration of the dopant in the substrate, (2) the concentration at 0.1 μm beneath the surface after one hour into the diffusion process. The substrate is initially free of impurity.

Solution

Let us first determine the diffusivity of boron in silicon. This can be done by using Equation (8.6) with the constants a and b obtained from Table 8.3. The temperature T' in Equation (8.6) is $T' = 1000/(1000 + 273) = 0.7855$, and $a = -19.982$ and $b = 13.1109$ from Table 8.3. We will thus find the square root of diffusivity of boron in silicon, $\sqrt{D} = 0.07534$, or $D = 0.005676\ \mu m^2/h = 1.5766 \times 10^{-6}\ \mu m^2/s$. This value of D is in the "ballpark" range of $D = 2 \times 10^{-6}\ \mu m^2/s$ from Figure 3.12.

1. Since the initial condition $N(x, 0) = 0$ and the boundary conditions are $N(0, t) = N_s = 10^{11}/cm^2$ and $N(\infty, t) = 0$, the solution presented in Equation (8.5) can be used to express the concentration of dopant in the silicon substrate:

$$N(x, t) = N_s \operatorname{erfc}\left[\frac{x}{2\sqrt{Dt}}\right] = 10^{11} \operatorname{erfc}\left[\frac{x}{2\sqrt{1.5766 \times 10^{-6}\, t}}\right] = 10^{11} \operatorname{erfc}\left(\frac{398.21x}{\sqrt{t}}\right)$$

$$N(x, t) = 10^{11}\left[1 - \operatorname{erf}\left(\frac{398.21x}{\sqrt{t}}\right)\right]$$

The unit for x is μm, and t is in seconds in the above expression.

2. For $x = 0.1\ \mu$m and $t = 1$ h $= 3600$ s:

$$N(0.1\ \mu\text{m}, 1\text{ h}) = 10^{11}\left[1 - \operatorname{erf}\left(\frac{398.21 \times 0.1}{\sqrt{3600}}\right)\right] = 10^{11}\,[1 - \operatorname{erf}(0.6637)]$$

$$= 10^{11}\,(1 - 0.6518) = 3.482 \times 10^{10}/\text{cm}^3$$

The value of the function erf (0.6637) is obtained from Table 3.4.

8.5 | OXIDATION

8.5.1 Thermal Oxidation

Oxidation is a very important process in both microelectronic and microsystem fabrication. According to Sze [1985], there are four types of thin films that are frequently used in microelectronics: (1) thermal oxidation for electrical or thermal insulation media, (2) dielectric layers for electrical insulation, (3) polycrystalline silicon for local electrical conduction, and (4) metal films for electrical (ohmic) contact and junctions. All these types of thin films are also widely used in MEMS and microsystems for similar purposes. We have learned that dielectric layers are used to separate electrodes in capacitance transducers, or electrically insulate the metal lead wires from embedded transducers. Materials for dielectric films involve ceramics such as alumina and quartz or those grown over the substrate's surface such as silicon dioxide and silicon nitride. The polycrystalline silicon films are necessary in providing localized electric conduction as described in Chapter 7. The metal films are used as leads for piezoresistive transducers, as well as the pads for soldering a silicon die to the constraint base, e.g., at the footprint in a pressure sensor. We will focus our attention on the thermal oxidation process commonly used for the production of silicon dioxide films in microsystems.

8.5.2 Silicon Dioxide

Silicon dioxide is used as an electric insulator as well as for etching masks for silicon and sacrificial layers in surface micromachining as will be described in detail in Chapter 9. There are several ways silicon dioxide can be produced on substrate surfaces. The least expensive way to produce SiO_2 film on the silicon substrate is by thermal oxidation. Chemical reactions used in this process are as follows:

$$Si_{\text{(solid)}} + O_{2\,\text{(gas)}} \rightarrow SiO_{2\,\text{(solid)}} \tag{8.7}$$

or

$$Si_{(solid)} + 2H_2O_{(steam)} \rightarrow SiO_{2\,(solid)} + 2H_{2\,(gas)} \quad \textbf{(8.8)}$$

Silicon oxide is produced by thermal oxidation in an electric resistance furnace. A typical furnace consists of a large fused quartz tube as illustrated in Figure 8.7. Resistance heating coils surround the tube to provide the necessary high temperature in the tube. The furnace tube used in industry is in the order of 30 cm in diameter and 3 m in length.

Figure 8.7 | Facility for thermal oxidation of silicon dioxide.

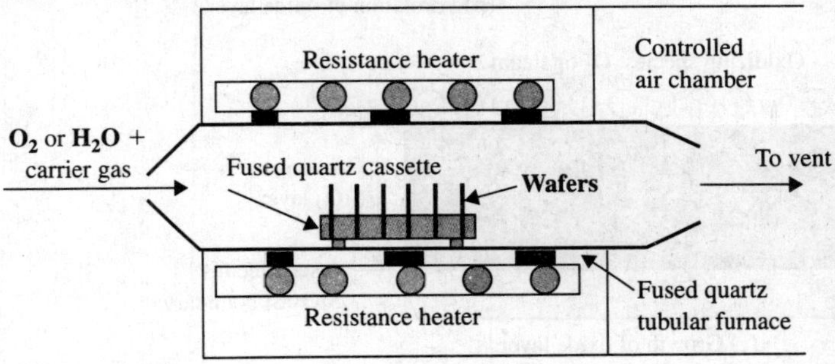

In the thermal oxidation process, wafers are placed in fused quartz cassettes that are pushed into the preheated furnace tube at a temperature in the range of 900 to 1200°C. Oxygen is blown into the tubular furnace for the oxidation of wafer surfaces. Often, steam is used instead of oxygen for accelerated oxidation. The timing, temperature, and gas flow are strictly controlled in order to achieve the desired quality and thickness of the SiO_2 film.

8.5.3 Thermal Oxidation Rates

The growth of an oxide layer on a silicon substrate is primarily a thermal diffusion process as presented in Section 8.4. However, because of the coupling of chemical reactions that take place simultaneously with the diffusion [see Eqs. (8.7) and (8.8)], the theory of diffusion as presented in Section 8.4 cannot be applied without significant modifications. The analysis is further complicated by the fact that both heat and mass transfer with moving silicon/SiO_2 boundaries need to be considered in the analysis. The so-called kinetics of thermal oxidation principle has to be used in assessing the growth of the oxidized SiO_2 layer in the substrate [Ruska 1987]. The growth of an SiO_2 layer on a silicon substrate is simulated in Figure 8.8.

The silicon substrate is first exposed to the oxidizing species such as oxygen in dry air or wet steam in a hot furnace as illustrated in Figure 8.7. The oxygen molecules in the oxidizing species begin to diffuse into the surface of the "fresh" silicon wafer as shown in Figure 8.8a and a SiO_2 layer (shown as a darker gray area in Fig. 8.8b) is formed, with SiO_2 molecules in solid circles in the figure. While the drive from the oxidizing species continues, the SiO_2 molecules that are already in the oxide layer diffuse and chemically react with the silicon molecules, crossing the layer's

Figure 8.8 | The kinetics of SiO_2 in silicon substrates.

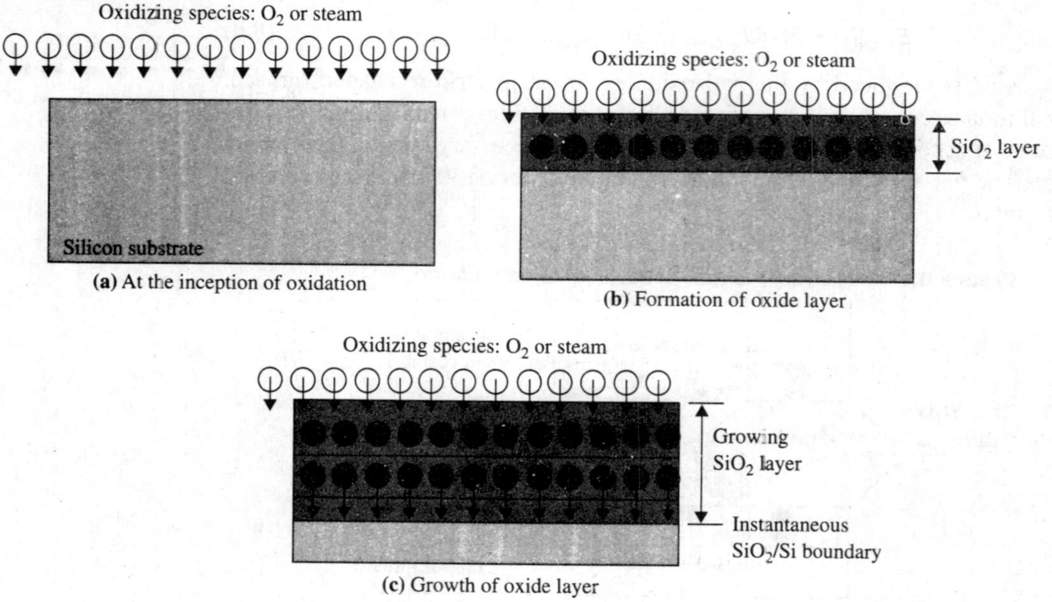

boundary as illustrated in Figure 8.8c. Thus the process can be viewed as one that begins with simple diffusion of oxidizing species into a silicon substrate. It is soon coupled with the diffusion of the same species into SiO_2. Meanwhile, there is a simultaneous diffusion of SiO_2 molecules and chemical reaction between the SiO_2 molecules and the silicon substrate, which creates new SiO_2/silicon boundaries.

Because of the highly complex oxidation process described in the foregoing paragraph, the prediction of the growth of an oxide layer over the silicon substrate is by no means easy. However, simple one-dimensional coupled diffusion/heat and mass transfer models were derived and they can be found in many publications [Sze 1985, Ruska 1987]. We will use the following simple expressions to estimate the growth of oxide layers over the silicon substrates in units of micrometers:

$$x = \frac{B}{A}(t + \tau) \tag{8.9}$$

for small time t, and

$$x = \sqrt{B(t + \tau)} \tag{8.10}$$

for large time t, where x is the thickness of the oxide layer in the silicon substrate in micrometers at time t, in hours. A and B are constants, and the parameter τ can be obtained by:

$$\tau = \frac{\left(\dfrac{d_0^2 + 2Dd_0}{k_s}\right) N_1}{2DN_0} \tag{8.11}$$

where

D = diffusivity of oxide in silicon, e.g., $D = 4.4 \times 10^{-16}$ cm^2/s at 900°C
d_0 = initial oxide layer (~ 200 Å in dry oxidation, = 0 for wet oxidation)
k_s = surface reaction rate constant
N_0 = concentration of oxygen molecules in the carrier gas
= 5.2×10^{16} molecules/cm^3 in dry O_2 at 1000°C and 1 atm
= 3000×10^{16} molecules/cm^3 in water vapor at the same temperature and pressure
N_1 = number of oxidizing species in the oxide
= 2.2×10^{22} SiO_2 molecules/cm^3 in dry O_2
= 4.4×10^{22} SiO_2 molecules/cm^3 in water vapor

It is readily seen that τ in Equation (8.11) represents the time coordinate shifts to account for the initial oxide layer, d_0, that may be in existence before the oxidation process begins. For wet oxidation, or in the case $d_0 = 0$, $\tau = 0$.

Equations (8.9) and (8.10) can be used to estimate the thickness of the oxide layer at time *t*.

The constant ratio *B*/*A* in Equation (8.9) is called the *linear rate constant*, whereas the constant *B* in Equation (8.10) is called the *parabolic rate constant*. Numerical values of these constants depend on the activation energies used in the oxidation as well as the orientations of the silicon crystal surfaces on which the oxidation takes place. We will offer the following equations to fit the graphical representations available in Sze [1985]:

For linear rate constants:

$$\log\left(\frac{B}{A}\right) = aT' + b \tag{8.12}$$

For parabolic rate constants:

$$\ln (B) = aT' + b \tag{8.13}$$

where $T' = 1000/T$, with the temperature T in K. The coefficients a and b in Equations (8.12) and (8.13) can be obtained from Table 8.4.

Table 8.4 | Coefficients for determining the rates of oxidation in silicon

Constants	Coefficient *a*	Coefficient *b*	Conditions
Linear rate	−10.4422	6.96426	Dry O_2, E_a = 2 eV, (100) silicon
constant,	−10.1257	6.93576	Dry O_2, E_a = 2 eV, (111) silicon
Eq. (8.12)	−9.905525	7.82039	H_2O vapor, E_a = 2.05 eV, (110) silicon
	−9.92655	7.948585	H_2O vapor, E_a = 2.05 eV, (111) silicon
Parabolic rate	−14.40273	6.74356	Dry O_2, E_a = 1.24 eV, 760 torr vacuum
constant,	−10.615	7.1040	H_2O vapor, E_a = 0.71 eV, 760 torr vacuum
Eq. (8.13)			

A question is which of the two equations, Equation (8.12) or Equation (8.13), is to be used for estimating the thickness of the oxide layer in the time scale. The linear nature of Equation (8.12) reflects the growth of the oxide layer in the early stage of oxidation. However, when the layer thickness increases, the coupling effect of diffusion and chemical reaction becomes dominant and the growth of the layer becomes nonlinear. Consequently, Equation (8.13) is a more realistic way for estimating the thickness of the oxide layer. No clear demarcation between the validity of either equation is available, as the definitions of "smaller" and "larger" times for these equations depends on the conditions involved in the oxidation processes.

EXAMPLE 8.3

Estimate the respective thickness of the SiO_2 layer over the (111) plane of a clean silicon wafer, resulting from both dry and wet oxidation at 950°C for one and half hours.

Solution

Since the wafer surface is free from initial oxidation, there is no need to consider the time-coordinate shift; i.e., $\tau = 0$ in Equations (8.9) and (8.10). Consequently, the oxide thickness from both dry and wet oxidation cases can be estimated by using the following expressions:

$$x = \frac{B}{A} t \tag{a}$$

for small time t, and

$$x = \sqrt{Bt} \tag{b}$$

for larger time t. The constants A and B in the above expressions can be obtained from Equations (8.12) and (8.13), with the coefficients a and b given in Table 8.4. Our selection of these coefficients is tabulated below.

	a	b	Conditions
Eq. (8.12)	−10.1257	6.9357	Dry O_2
Eq. (8.12)	−9.9266	7.9486	Wet steam
Eq. (8.13)	−14.4027	6.7436	Dry O_2
Eq. (8.13)	−10.6150	7.1040	Wet steam

We will find the temperature $T' = 1000/(950 + 273) = 0.8177$. On substituting T' and the coefficients into Equations (8.12) and (8.13), we have the constant B/A and B as follows:

	Dry oxidation	Wet oxidation
B/A, μm/h	0.04532	0.6786
B, μm/h	0.006516	0.2068

By using the above Equations (a) and (b), we may estimate the thickness of the oxide layers in the respective dry and wet processes as:

	Dry oxidation thickness, μm	Wet oxidation thickness, μm
Eq. (a) for small time	0.068	1.018
Eq. (b) for larger time	0.0989	0.5572

We may observe from the above summary of results that wet oxidation is indeed much more effective in the depth of penetration into the substrate than the dry oxidation processes.

For the cases that involve initial oxide layers, the prediction of the growth of the oxide layer becomes much more complex, as it will require the determination of the time parameter τ in Equation (8.11) with the value of k_s, the *surface reaction rate constant* in the process. In this particular case, it is also referred to as the *silicon oxidation rate constant.* This constant is a function of temperature, oxidant, crystal orientation, and doping. We will deal with the evaluation of k_s in Section 8.6.

8.5.4 Oxide Thickness by Color

Both SiO_2 and Si_3N_4 layers have a color distinct from that of the silicon substrates on which these layers grow. In the case of SiO_2 layers, they are essentially transparent but with a different light refraction index from that of the silicon substrate. Consequently, when the surface is illuminated by white light, one can view the different colors on the surface corresponding to the layer's thickness. The color of the surface of an SiO_2 layer is the result of the interference of the reflected light rays. It is thus not a surprise that the same color may repeat with different layer thicknesses. Table 8.5 offers a partial list of the colors of the surface of SiO_2 layers of selected thickness. A more complete color chart can be found in several prominent references [Ruska 1987, Madou 1997, and van Zant 1997].

Table 8.5 | Color of silicon dioxide layers of selected thickness

SiO_2 layer thickness, μm	0.050	0.075	0.275 0.465	0.310 0.493	0.50	0.375	0.390
Color	Tan	Brown	Red–violet	Blue	Green to yellow–green	Green–yellow	Yellow

8.6 | CHEMICAL VAPOR DEPOSITION

Depositing thin films over the surface of substrates and other MEMS and microsystem components is a common and necessary practice in micromachining. Unlike the diffusion and thermal oxidation processes that we learned about in the foregoing sections, deposition adds thin films to, instead of consuming, the substrates. There are abundant circumstances in which the need to add thin films of a wide range of materials on the substrate surfaces arises. These thin film materials can be organic or inorganic. They include a variety of metals; common ones are Al, Ag, Au, Ti, W, Cu, Pt, and Sn. These materials also include compounds such as the common shape-memory alloy NiTi and the piezoelectric ZnO, as used for coating tubes for microfluidics as described in Chapter 5. The application of shape-memory alloys in microdevices was covered in Chapter 2.

There are generally two classes of deposition used in microelectronics and micromachining. These are (1) *physical vapor deposition* (PVD) and (2) *chemical vapor deposition* (CVD). PVD involves the direct impingement of particles on the hot substrate surfaces. CVD, on the other hand, involves convective heat and mass transfer as well as diffusion with chemical reactions at the substrate surfaces. It is a much more complex process than PVD, but is a great deal more effective in terms of the rate of growth and the quality of deposition. Most CVD processes involve low gas pressures. There are some others that are carried out in high vacuum. We will focus on CVD in the subsequent subsections.

8.6.1 Working Principle of CVD

The working principle of CVD, simply put, involves the flow of a gas with diffused reactants over a hot substrate surface. The gas that carries the reactants is called the *carrier gas*. While the gas flows over the hot solid surface, the energy supplied by the surface temperature provokes chemical reactions of the reactants that form films during and after the reactions. The by-products of the chemical reactions are then vented. Thin films of desired composition can thus be created over the surface of the substrate. Various types of CVD reactors are built to perform the CVD processes. Two of these reactors are illustrated in Figure 8.9.

In either type of reactor, the substrate surface is exposed to the flowing gas with diffused reactants. Resistance heaters either surround the chamber as in Figure 8.9a or lie directly under the susceptor that holds the substrates as in Figure 8.9b.

8.6.2 Chemical Reactions in CVD

We will show the chemical reactions that are used in depositing three common thin films over silicon substrates: (1) silicon dioxide, (2) silicon nitride, and (3) polycrystalline silicon.

Silicon Dioxide Thin silicon dioxide films can be produced at the surface of silicon by the diffusion process as described in Section 8.4. Here, we will learn how SiO_2 thin films can be deposited to the surface of silicon substrates by chemical reaction. There are a number of chemical reactants that could be used to deposit silicon dioxide films on the silicon substrates. These include $SiCl_4$, $SiBr_4$, and SiH_2Cl_2 [Ruska 1987]. The carrier gases that can be used in these processes are O_2, NO, NO_2, and CO_2 with H_2. The most common reactant used for CVD is silane (SiH_4) together with oxygen. The chemical reaction in this process can be expressed as:

$$SiH_4 + O_2 \rightarrow SiO_2 + 2H_2 \quad \textbf{(8.14)}$$

Figure 8.9 | Two typical CVD reactors.

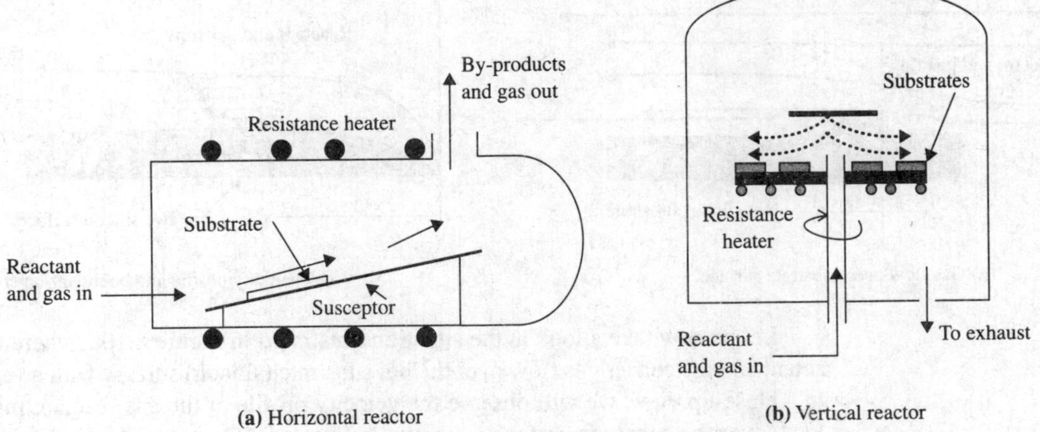

The chemical reaction takes place in a temperature range of 400 to 500°C with an activation energy around $E_a = 0.4$ eV.

Silicon Nitride (Si_3N_4) Ammonia is a common carrier gas for depositing silicon nitride on silicon substrates. There are three reactants that can produce the thin silicon nitride films [Ruska 1987]:

$$3SiH_4 + 4NH_3 \rightarrow Si_3N_4 + 12\,H_2 \quad \textbf{(8.15a)}$$

$$3SiCl_4 + 4NH_3 \rightarrow Si_3N_4 + 12\,HCl \quad \textbf{(8.15b)}$$

$$3SiH_2Cl_2 + 4NH_3 \rightarrow Si_3N_4 + 6HCl + 6H_2 \quad \textbf{(8.15c)}$$

The activation energy required for the above reactions is $E_a = 1.8$ eV. The temperature range for the reactions is 700 to 900°C for silane in Equation (8.15a); 850°C for silicon tetrachloride in Equation (8.15b); and 650 to 750°C for dichlorosilane in Equation (8.15c).

Polycrystalline Silicon As described in Section 7.5.4, polycrystalline silicon films consist of single silicon crystals of different sizes. Deposition of polycrystalline silicon is a *pyrolysis* process, which is a decomposition process using heat, as we will see from the following chemical reaction:

$$SiH_4 \rightarrow Si + 2H_2 \quad \textbf{(8.16)}$$

The process takes place in a temperature range of 600 to 650°C with activation energy of 1.7 eV.

8.6.3 Rate of Deposition

We learned from Sections 8.6.1 and 8.6.2 that CVD is a process by which films of a desired substance are produced by chemical reactions of the reactants and carrier gas at the hot substrate surfaces. The rate of buildup of these thin films obviously is a concern to process design engineers, as in many cases the production of three-dimensional MEMS structure relies on these build-ups.

Figure 8.10 | Thermal/fluid aspect of CVD.

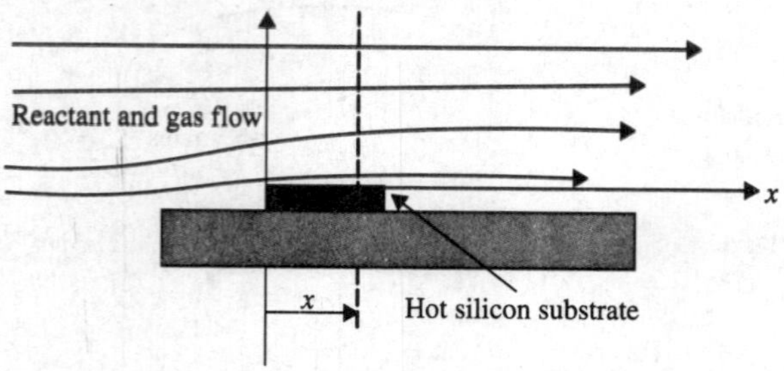

(a) Gas flow over substrate surface

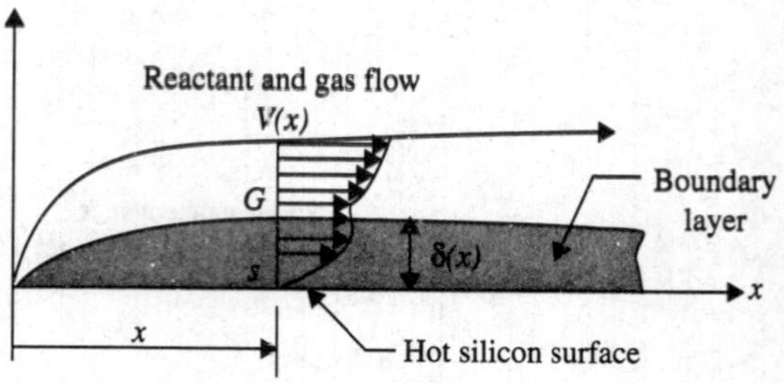

(b) Velocity profile and boundary layer

Let us now take a look at the situation illustrated in Figure 8.10a, where the reactant and the carrier gas flow over the hot substrate (silicon) surface with a velocity. In a close-up view, we will observe the velocity profile of the gas-reactant mixture, $V(x)$, over the substrate surface as shown in Figure 8.10b. As we learned from Section 5.8.5, a boundary layer with a thickness $\delta(x)$ exists at the solid/fluid interface whenever a fluid flows over a solid surface. This boundary layer is of great importance in heat transfer, as it acts as an additional thermal barrier. A similar effect occurs in a diffusion process, as additional time is required for the reactant to diffuse across this layer. Thus, the boundary layer is a retarding factor to the chemical reactions. Consequently, this layer affects the rate of deposition of the desired thin film on the substrate surfaces.

The approximate thickness of the boundary layer $\delta(x)$ at x distance away from the leading edge of the silicon substrate can be obtained from the following expression:

$$\delta(x) = \frac{x}{\sqrt{\text{Re}\,(x)}} \qquad \textbf{(8.17)}$$

where Re(x) is the Reynolds number of the gas mixture. By definition in Equation (5.3), the Reynolds number can be expressed as:

$$\text{Re}(x) = \frac{\rho L V(x)}{\mu} \qquad \textbf{(8.18)}$$

in which ρ is the mass density of the gas mixture, L is the characteristic length of the flow, and the dynamic viscosity of gases, μ, can be obtained from Table 8.6.

Table 8.6 | Dynamic viscosity of gases used in CVD processes

	Viscosity, μP			
Gas	**0°C**	**490°C**	**600°C**	**825°C**
Hydrogen, H_2	83	167	183	214
Nitrogen, N_2	153	337	–	419
Oxygen, O_2	189	400	–	501
Argon, Ar	210	448	–	563

1 poise (P) = 1 dyne-s/cm^2 = 0.1 N-s/m^2 = 0.1 kg/m-s

Source: Ruska [1987].

The diffusion flux of the reactant, $\vec{N}$ across the boundary layer thickness, with units of atoms or molecules/unit time/area (atoms or molecules/m^2-s) through the boundary layer can be expressed by the Fick's law described in Section 3.6. The expanded mathematical expression of the Fick's law for the present case is:

$$\vec{N} = \frac{D}{\delta}(N_G - N_S) \tag{8.19}$$

where D is the diffusivity of the reactant in the carrier gas with units of cm^2/s and N_G and N_S are the respective concentrations of the reactant at the top of the boundary layer, G and at the surface of the substrates, s. Both N_G and N_S have units of molecules/volume, or molecules/m^3.

The determination of the concentrations N_G and N_S can be carried out by the following procedure:

1. Use *Avogadro's theory*, which states that the volume occupied by all gases at standard conditions by a mole of gas is 22.4×10^{-3} m^3, which leads to a molar density of 44.643 moles/m^3. Also, *Avogadro's number* is defined as the number of molecules contained in 1 mole of any gas or substance, which is 6.022×10^{23}.
2. Find the molar mass of commonly used gases in Table 8.6.

Gases	Hydrogen, H_2	Nitrogen, N_2	Oxygen, O_2	Argon, Ar
Molar mass, g	2	28	32	40

3. Use the ideal gas law to determine the molar density of the gas at its temperature and pressure if they are different from the standard conditions.
4. The concentrations N_G and N_S can be determined by dividing the Avogadro's number, given in step 1, by the molar density obtained in step 3.

EXAMPLE 8.4

A CVD process involves a reactant being diluted to 2 percent in the carrier oxygen gas at 490°C. Find the number of molecules in a cubic meter volume of the carrier gas. Pressure variation in the process is negligible.

Solution

The ideal gas law follows the relationship

$$\frac{P_1 V_1}{P_2 V_2} = \frac{T_1}{T_2}$$

in which V_1 and V_2 are the respectively volumes of the gas at state 1 with pressure P_1 and temperature T_1 and state 2 with P_2 and T_2.

Since $P_1 = P_2$, and V_1, the molar volume of the gas at room temperature, 293 K, is 22.4×10^{-3} m^3/mol, we have the molar density of the gas at 490°C or 763 K as:

$$d_2 = \left(\frac{T_1}{T_2}\right) d_1 = \left(\frac{293}{763}\right)(44.643) = 17.1433 \text{ mol/m}^3$$

The reader will realize that $d_1 = 1/V_1 = 44.643$ mol/m^3 is used in the above computation.

The concentration, N_G = (the molecules of the gas per cubic meter) is thus:

$$N_G = (\text{Avogadro's number})\ d_2 = (6.022 \times 10^{23})(17.1433)$$
$$= 103.24 \times 10^{23} \text{ molecules/m}^3$$

Because most CVD processes take place at very low gas velocity, the corresponding Reynolds number Re is at a low value of about 100. The low velocity of the gas flow allows a significant amount of reactant to diffuse through the boundary layer and form the film by chemical reaction at the hot surface of the substrate. The following relationship exists:

$$\vec{N} = k_s N_S \tag{8.20}$$

where k_s is the surface reaction rate constant as shown in Equation (8.11). This rate can be expressed as:

$$k_s = k' \exp\left(-\frac{E_a}{kT}\right) \tag{8.21}$$

where k' = constant whose value depends on the reaction and the reactant concentration, E_a = activation energy, k = Boltzmann constant, and T = absolute temperature.

The flux of the carrier gas and the reactant, $\vec{N}$ in Equation (8.19) may be expressed in terms of the surface reaction rate k_s, with the substitution of Equation (8.20), as:

$$\vec{N} = \frac{D\, N_G\, k_s}{D + \delta\, k_s} \tag{8.22}$$

in which δ is the mean thickness of the boundary layer shown in Figure 8.10 as expressed in Equation (8.17).

The rate of the growth of the thin film over the substrate surface, r, in m/s, can be estimated by the following expressions [Ruska 1987]:

$$r = \frac{D\, N_G}{\gamma \delta} \tag{8.23a}$$

for $\delta k_s \geq D$, or

$$r = \frac{N_G\, k_s}{\gamma} \tag{8.23b}$$

for $\delta k_s \ll D$.

In both Equations (8.23a) and (8.23b), γ is the number of atoms or molecules per unit volume of the thin film. The value of γ may be estimated by a postulation that the thin film is "densely" packed by atoms or molecules in spherical shapes with the radius according to the selected materials as listed in Table 8.7.

The value of γ can thus be determined by the following expression:

$$\gamma = \frac{1}{v} = \frac{1}{\frac{4}{3}\pi a^3} \tag{8.24}$$

where a = the radius of atoms in meters based on Table 8.7. The units for γ are atoms or molecules/m^3.

Table 8.7 | Atomic radius of selected materials

Reactant materials	Atomic radius, nm	Ionic radius, nm
Hydrogen	0.046	0.154
Helium	0.046	0.154
Boron	0.097	0.02
Nitrogen	0.071	0.02
Oxygen	0.060	0.132
Aluminum	0.143	0.057
Silicon	0.117	0.198
Phosphorus	0.109	0.039
Argon	0.192	
Iron	0.124	0.067
Nickel	0.125	0.078
Copper	0.128	0.072
Gallium	0.135	0.062
Germanium	0.122	0.044
Arsenic	0.125	0.04

Source: Kwok [1997].

EXAMPLE 8.5

A CVD process is used to deposit SiO_2 film on a silicon substrate. Oxygen is used as the carrier gas in the chemical process shown in Equation (8.14). The CVD process is carried out in a horizontal reactor as illustrated in Figure 8.11. Other conditions are identical to those given in Example 8.4.

Figure 8.11 | CVD over silicon substrates.

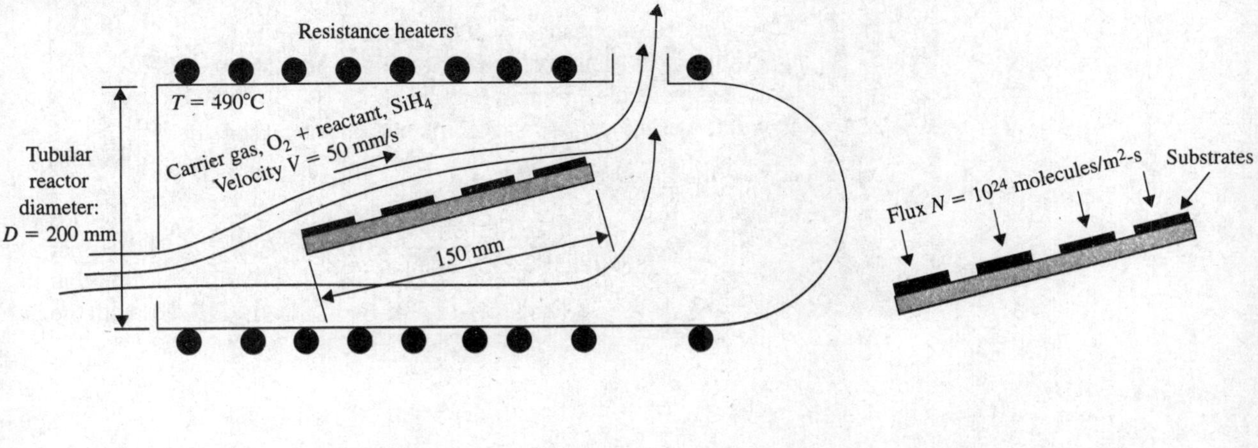

Determine the following:

1. The density of the carrier gas
2. The Reynolds number of the gas flow
3. The thickness of the boundary layer over the substrate surface
4. The diffusivity of the carrier gas and reactant in the silicon substrate
5. The surface reaction rate
6. The deposition rate

Solution

1. The density of the carrier gas, ρ, may be obtained by taking the product of the molar mass of the carrier gas and the molar density from Example 8.4. The molar mass of oxygen is 32 g. Hence the density of the oxygen carrier gas is:

$$\rho = (32 \text{ g/mol})(17.1433 \text{ mol/m}^3) = 548.586 \text{ g/m}^3$$

2. The Reynolds number Re is computed from Equation (8.18) with:

$$D = 200 \text{ mm} = 0.2 \text{ m}$$

$$V = 50 \text{ mm/s} = 0.05 \text{ m/s}$$

$$\mu = 400 \times 10^{-6} \text{ P (Table 8.6)} = 0.04 \text{ g/m-s}$$

$$\text{Re} = \frac{\rho D V}{\mu} = \frac{(548.586)(0.2)(0.05)}{0.04} = 137.147$$

3. The mean thickness of the boundary layer between the substrate surface and the carrier gas, δ, is, from Equation (8.17) with the length of the sled that carries the substrates, $L = 150$ mm $= 0.15$ m:

$$\delta = \frac{L}{\sqrt{\text{Re}}} = \frac{0.15}{\sqrt{137.147}} = 0.01281 \text{ m}$$

4. The diffusivity of the carrier gas and the reactant, D, is computed from Equation (8.19) with a modification to account for the dilution factor of the reactant, η = % of dilution (2% for the present case) is:

$$D = \frac{\delta \vec{N}}{\eta(N_G - N_s)}$$

with

$\vec{N}$ = given carrier gas flux = 10^{24} molecules/m^2-s
N_G = equivalent density of gas molecules = 103.24×10^{23} molecules/m^3 from Example 8.4.
N_S = 0 (an assumed value for complete diffusion within the film)

Thus, we have the diffusivity D to be:

$$D = \frac{(0.01281)(10^{24})}{(0.02)(103.24 \times 10^{23})} = 0.062 \text{ m}^2\text{/s}$$

5. The following expression is used to compute the surface reaction rate, k_s. It is derived from Equation (8.22):

$$k_s = \frac{D\vec{N}}{DN_G - \delta\vec{N}} = \frac{(0.062)(10^{24})}{(0.062)(103.24 \times 10^{23}) - (0.0128)(10^{24})} = 0.09884 \text{ m/sec}$$

6. The approximate rate of deposition, r, can be obtained by Equation (8.23a) or (8.23b), depending on the value of δk_s. This value is computed to be:

$$\delta k_s = (0.01281)(0.09884) = 0.0013 \ll D = 0.062$$

Hence Equation (8.23b) is used with a slight modification to account for the dilution factor of the reactant, i.e.,

$$r = \eta \frac{N_G k_s}{\gamma}$$

From Table 8.6, we get the radii of atoms of silicon, a_{Si}, and oxygen, a_{O2} to be:

$$a_{Si} = 0.117 \text{ nm} = 0.117 \times 10^{-9} \text{ m} \quad \text{and} \quad a_{O2} = 0.06 \text{ nm} = 0.06 \times 10^{-9} \text{ m.}$$

We will make a bold assumption that the radius of the SiO_2 molecules is the summation of the radii of the silicon and the oxygen atoms; i.e., $a_{SiO2} = 0.177 \times 10^{-9}$ m. The corresponding number of SiO_2 molecules per unit film volume can be computed from Equation (8.24) as:

$$\gamma = \frac{1}{\frac{4}{3}\pi(a_{SiO_2})^3} = \frac{1}{\frac{4}{3}(3.14)(0.177 \times 10^{-9})^3} = 4.3074 \times 10^{28}$$

This value leads to a deposition rate r of:

$$r = \eta \frac{N_G k_s}{\gamma} = (0.02)\frac{(103.24 \times 10^{23})(0.09884)}{4.3074 \times 10^{28}} = 0.4738 \times 10^{-6} \text{ m/s, or } 0.47 \ \mu\text{m/s}$$

It is useful for process engineers to be aware of the fact that the rate of chemical vapor deposition is affected by the following parameters:

- The temperature, $T^{3/2}$.
- The pressure of the carrier gas, P^{-1}.
- The velocity of gas flow, V^{-1}.
- The distance in the direction of gas flow, $x^{1/2}$, where x is as shown in Figure 8.10b.

8.6.4 Enhanced CVD

The CVD process we have described in the foregoing subsections involves elevated temperature, but at near atmospheric pressure. It is called *APCVD*, which stands for *atmospheric-pressure CVD*. There are several other CVD processes that are used for better results either for higher rate of growth or for better quality of the deposit films. We will mention two such popular CVD processes here.

1. *Low-pressure CVD (LPCVD).* As we can see from Equation (8.23), the rate of growth of the deposited film is inversely proportional to the thickness of the boundary layer (δ), but is directly proportional to the diffusivity of the reactant in the carrier gas (D). We have also learned from Equation (8.17) that the thickness δ of the boundary layer is inversely proportional to the square root of the Reynolds number Re. The diffusivity, on the other hand, varies inversely with pressure. The Reynolds number in Equation (8.18) depends on gas velocity V, viscosity μ, and density ρ.

   Consider now a 1000 times reduction in gas pressure in the process. We may expect an increase in D by 1000 times, but with the same 1000 times reduction in density. The velocity of the gas with a 1000-time reduction of pressure will reduce by 10 to 100 times in the process. These variations will result in a net reduction of Reynolds number by 10 to 100 times, which means a net increase of δ by $\sqrt{10} \approx 3$ to $\sqrt{100} = 10$ times. Despite the increase of δ by 3 to 10 times, the 1000 times increase of D will result in an increase in the diffusion flux $\vec{N}$ in Equation (8.19) by 10 to 30 times. It is therefore clear that reduction of gas pressure will definitely increase the rate of deposition. Consequently, the deposition rate is inversely proportional to the gas pressure as mentioned in the previous subsection.

   The LPCVD operates in vacuum at about 1 torr (1 mm of Hg). It uses a reactor chamber that is not much different from that used for APCVD. However, the chamber must be leakproof for the vacuum and structurally strong to withstand vacuum pressure in operation. The end product of the deposit film is typically more uniform. This technique allows the use of stacked wafers in the process, which is attractive from mass production point of view.
2. *Plasma-enhanced CVD (PECVD).* The CVD processes that we have learned thus far require the substrates and the carrier gases to be at elevated temperatures for sufficient activation energies to allow for chemical reactions to take place. This high temperature can damage substrates, especially the metallized ones. *Plasma-enhanced CVD* (PECVD) utilizes the *radio-frequency (RF) plasma* to transfer energy into the reactant gases, which allows the substrates to remain at lower temperature than that in APCVD or LPCVD. A radio-frequency source can be electromagnetic radiation in the frequency band between 3 KHz and 300 GHz, or alternating currents in the same frequency range. Precise temperature control of the substrate surfaces is necessary to ensure the quality of the deposit films. A typical PECVD reactor is illustrated in Figure 8.12.

A comprehensive summary and comparison of the three principal CVD processes is compiled in Table 8.8. It provides a useful guideline for process engineers in selecting a suitable CVD process for their microsystems.

Figure 8.12 | A PECVD reactor.

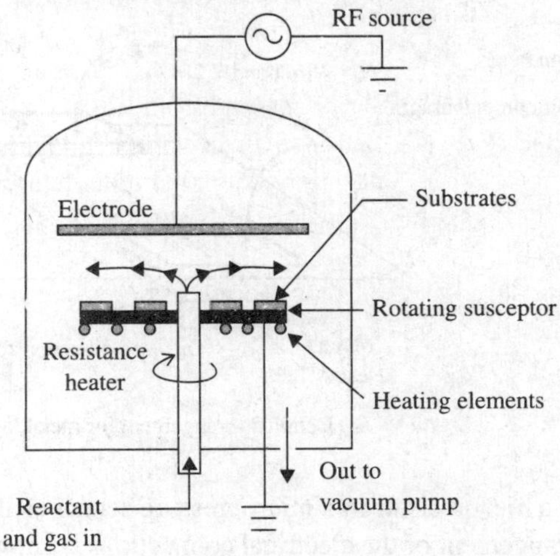

Table 8.8 | Summary and comparison of three principal CVD processes

CVD process	Pressure/ temperature	Normal deposition rates, 10^{-10} m/min	Advantages	Disadvantages	Applications
APCVD	100–10 kPa/ 350–400°C	700 for SiO_2	Simple, high rate, low temperature	Poor step coverage, particle contamination	Doped and undoped oxides
LPCVD	1–8 torr/ 550-900°C	50–180 for SiO_2 30–80 for Si_3N_4 100–200 for polysilicon	Excellent purity and uniformity, large wafer capacity	High temperature and high deposition rates	Doped and undoped oxides, silicon nitride, polysilicon, and tungsten
PECVD	0.2–5 torr/ 300–400°C	300–350 for Si_3N_4	Lower substrate temperature; fast, good adhesion	Vulnerable to chemical contamination	Low-temperature insulators over metals, and passivation

Source: Madou [1997].

8.7 | PHYSICAL VAPOR DEPOSITION—SPUTTERING

Sputtering is a process that is often used to deposit thin metallic films in the order of 100 Å thick (1 Å = 10^{-10} m) on substrate surfaces. Metallic films (or layers) are required to conduct electricity from signal generators in sensors, or for the supply of electricity to an actuator. For example, metallic layers are required to transmit the

Figure 8.13 | Metallic layers for signal transmission in a micropressure sensor.

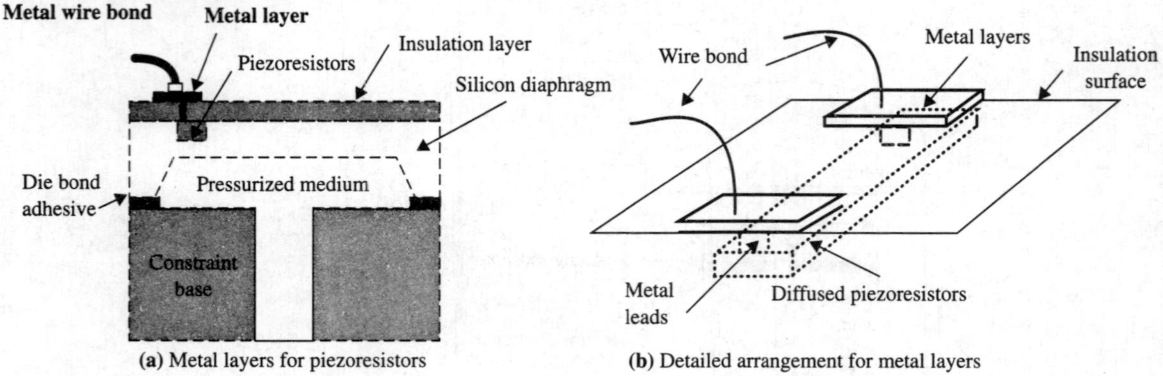

signals generated in a piezoresistor in a micropressure sensor as illustrated in Figure 8.13a. A detailed arrangement of the electrical connections is shown in Figure 8.13b.

The sputtering process is carried out with plasma under very low pressure (i.e., in high vacuum at around 5×10^{-7} torr). This process involves low temperature, which is contrary to CVD as presented in the foregoing section. At this temperature, little chemical reaction can take place. The process is thus regarded as physical deposition.

We have learned from Chapter 3 that plasma is made of positively charged gas ion, and plasma can be produced by either high-voltage dc sources or RF (radio-frequency) sources. Whatever the method, the positive ions of the metal in an inert argon gas carrier bombard the surface of the target at such a high velocity that the momentum transfer on impingement causes the metal ions to evaporate. The metal vapor is then led to the substrate surface and is deposited after condensation. This process is illustrated in Figure 8.14.

Figure 8.14 | Graphical illustration of sputtering process.

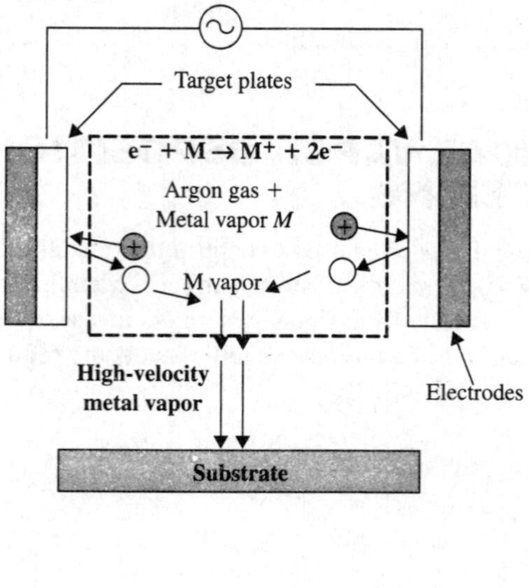

8.8 | DEPOSITION BY EPITAXY

Epitaxy is the extension of a single-crystal substrate by growing a film of the *same* single-crystal material. For example, one may use this process to deposit silicon films over the desired parts of the silicon substrate in order to build the thickness of the microstructure. This is one process that is frequently used in the microelectronics industry in the production of silicon diodes and transistors. For MEMS and microsystems, this technique is used to build the three-dimensional geometry of the devices.

Epitaxial deposition is very similar to the CVD processes in, for instance, the use of carrier gases with reactants involving the same substrate material. The main difference, however, is its ability to deposit not only the same substrate materials such as silicon, but also to deposit compounds such as GaAs over the surface of the same GaAs material. Because most MEMS and microsystems use silicon as substrate material, we will focus our attention only on the epitaxial deposition of silicon over silicon substrates.

There are several methods available for epitaxial deposition in microelectronics technology:

1. Vapor-phase epitaxy (VPE).
2. Molecular-beam epitaxy (MBE).
3. Metal-organic CVD (MOCVD).
4. Complementary metal oxidation of semiconductors (CMOS) epitaxy.

The vapor-phase epitaxy (VPE) technique appears to be the most popular one in the IC industry, although CMOS is also frequently used in fabricating MEMS components. The VPE technique involves the use of reactant vapors containing silicon, such as those listed in Table 8.9, diluted in hydrogen carrier gas.

Table 8.9 | Reactant vapors for epitaxial deposition

Reactant vapors	Normal process temperature, °C	Normal deposition rate, μm/min	Required energy supply, eV	Remarks
Silane (SiH_4)	1000	0.1–0.5	1.6–1.7	No pattern shift
Dichlorosilane (SiH_2Cl_2)	1100	0.1–0.8	0.3–0.6	Some pattern shift
Trichlorosilane ($SiHCl_3$)	1175	0.2–0.8	0.8–1.0	Large pattern shift
Silicon tetrachloride ($SiCl_4$)	1225	0.2–1.0	1.6–1.7	Very large pattern shift

Source: Ruska [1987].

The production of silicon film over silicon substrates using silane vapor in Table 8.9 is the simplest of all. Silicon can be produced by simple pyrolysis at about 1000°C as follows:

$$SiH_4 \rightarrow Si_{\,(solid)} + 2H_{2\,(gas)} \tag{8.25}$$

All the other three reactant vapors in Table 8.9 with the hydrogen carrier gas will react with the silicon substrate surface to produce more silicon on the surface but also with by-products such as $SiCl_2$ and HCl. The silicon dichloride $SiCl_2$ in the by-products in turn releases more silicon and $SiCl_4$. The hydrogen chloride (HCl) is highly erosive, and it has an etching effect on the newly produced silicon film. We thus can envision a picture in which the process continuously produces silicon crystals while some of the produced silicon is etched by the by-product HCl. Delicate control of the process is thus critically important in order to ensure that the rate of silicon production exceeds the rate of silicon etching.

Epitaxial deposition is carried out in reactors (or chambers) that have similar arrangement to some of the reactors used for chemical vapor deposition (see Fig. 8.9). A typical reactor is schematically illustrated in Figure 8.15.

Figure 8.15 | Horizontal epitaxial deposition reactor.

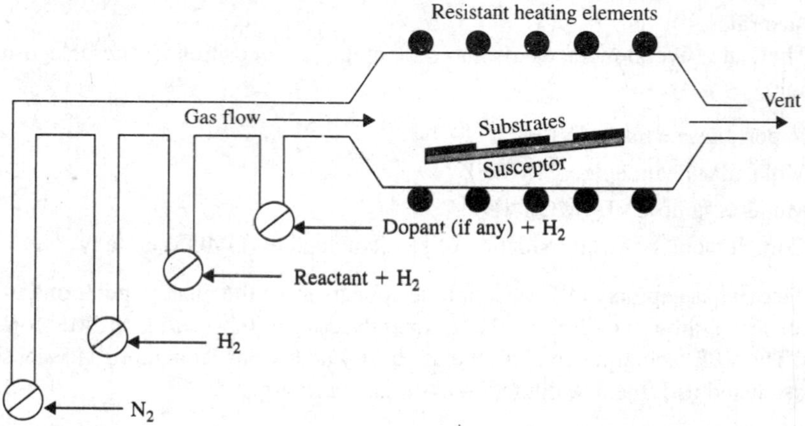

Epitaxial deposition has a high risk of explosion as a result of the use of hydrogen as the carrier gas at a high temperature. Therefore, the substrate surfaces must be thoroughly cleaned with solvents and remain clean during and after they are loaded and placed in the reactor. Once the substrates are in place in the reactor, purging of nitrogen gas takes place. The purging of this inert gas is necessary to drive out all the oxygen inside the chamber in order to avoid explosion from the mixing of oxygen and the hydrogen carrier gas at a high operating temperature. Hydrogen carrier gas is supplied to the chamber at an intermediate temperature of approximately 500°C. This step is followed by the supply of selected reactant from Table 8.9 and more hydrogen gas for the epitaxial deposition of silicon on the silicon substrates. As mentioned earlier, the by-product hydrogen chloride (HCl) can etch the silicon film product. Thus, careful control of the process is maintained to assure that the produced silicon outpaces the silicon that is etched away by the hydrogen chloride. The substrates with deposited silicon must be thoroughly cleaned before they are removed from the chamber.

Figure 8.16 illustrates a vertical reactor for epitaxial deposition. The multiple valves to control the mixing of the gases and reactant vapors shown in Figure 8.15 are used for this type of reactor.

Figure 8.16 | Vertical epitaxial deposition reactor.

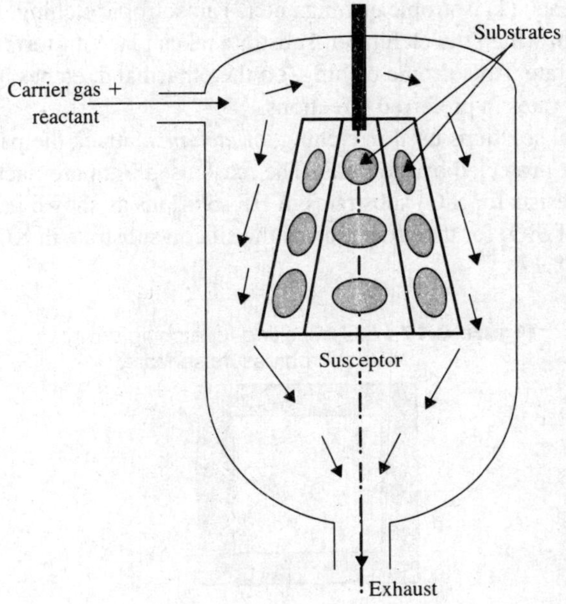

8.9 | ETCHING

Etching is one of the most important processes in microfabrication. It involves the removal of materials in desired areas by physical or chemical means. It is a way to establish permanent patterns developed at the substrate surface by photolithography as described in Section 8.2. In micromachining, etching is used to shape the geometry of microcomponents in MEMS and microsystems. For example, the cavity of the silicon die for a micropressure sensor such as illustrated in Figure 2.7 is produced by etching. A similar technique can be used to produce the silicon membranes and diaphragms of microvalves as illustrated in Figures 2.29 and 2.30.

Of the two common types of etching techniques mentioned above, the physical etching is usually referred to as *dry etching* or *plasma etching,* whereas the chemical etching is referred to as *wet* etching. We will present only the working principles of both these techniques in this chapter. Much of the detailed application of these techniques will be explained in the subsequent Chapter 9.

8.9.1 Chemical Etching

Chemical etching involves using solutions with diluted chemicals to dissolve substrates. For instance, diluted hydrofluoric (HF) solution is used to dissolve SiO_2,

Si_3N_4, and polycrystalline silicon, whereas potassium peroxide (KOH) is used to etch the silicon substrates. The rates of etching vary, depending on the substrate materials to be etched and the concentration of the chemical reactants in the solution, as well as the temperature of the solution.

There are generally two types of etching available for shaping the geometry of MEMS components: (1) isotropic etching and (2) anisotropic etching. Isotropic etching is a process in which the etching of substrate takes place uniformly in all direction at the same rate. Anisotropic etching, on the other hand, etches away substrate material at faster rates in preferred directions.

The chemical solutions used in etching, or *etchants,* attack the parts of the substrate that are not protected by the mask. The mask used in micromachining may be either the photoresists for SiO_2 substrates in HF solutions as shown in Figure 8.1, or the mask made of SiO_2 for the protection of the silicon substrate in KOH etchants as shown in Figure 8.17.

Figure 8.17 | SiO_2 masking for etching cavity in micropressure sensors.

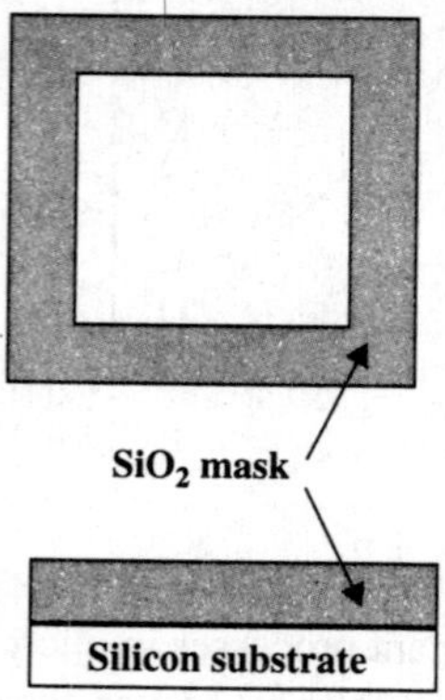

Wet etching is easy to apply, and it involves inexpensive equipment and facility for the process. It is also a faster etching process than the dry etching. The etching rates in wet etching range from a few micrometers to several tens of micrometers per minute for isotropic etchants and about 1 μm/min for anisotropic etchants, whereas only 0.1 μm/min is achievable in typical dry etching. Unfortunately, there are several disadvantages associated with wet etching; it often results in poor quality of etched surfaces due to bubbles and flow patterns of the solutions. No effective wet etching is available for some substrates such as silicon nitrides.

8.9.2 Plasma Etching

The plasma used for etching is a stream of positive-charge-carrying ions of a substance with a large number of electrons, diluted inert carrier gas such as argon. It can be generated by continuous applications of high-voltage electric charge, or by radio-frequency (RF) sources. Plasmas are usually generated in low-pressure environment, or vacuum.

As illustrated in Figure 8.18, the high-energy plasma containing gas molecules, free electrons, and gas ions bombards the surface of the target substrate and knock off the substrate material from its surface. This process in a way is like a reversed sputtering process at low temperature in the range of 50 to 100°C. It takes place in high vacuum.

Figure 8.18 | Plasma assisted etching.

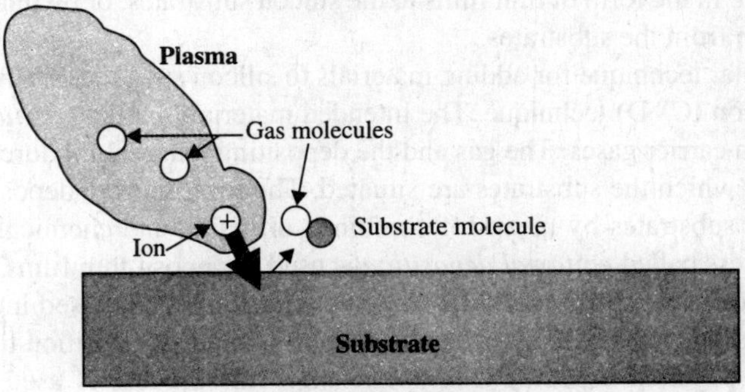

8.10 | SUMMARY OF MICROFABRICATION

This chapter has provided an overview of available microfabrication techniques for MEMS and microsystems. These techniques are used to produce the primarily three-dimensional geometry of most microdevices and microsystems.

Other than the special LIGA micromanufacturing technology, which will be described in Chapter 9, most silicon-based MEMS and microsystems are produced by adding material in the form of thin films to the silicon substrates, or by removing portions of silicon from the substrates.

One popular technique for adding materials to silicon substrates is the chemical vapor deposition (CVD) technique. The intended materials, called *reactants,* are diluted in certain carrier gases. The gas and the depositing material mixture is then led to a reactor in which the substrates are situated. The reactants are deposited on the surface of the substrates by means of combined diffusion and chemical reactions. A similar process called *epitaxial deposition* is used to deposit thin films of reactant over the surface of a substrate of the same material. It is widely used in depositing silicon crystals on the surface of silicon substrates. Another deposition technique is referred to as *physical deposition,* or *sputtering*. This process is used to deposit mostly thin metal films on silicon substrates by means of ionization of the depositing materials, energized by plasma with RF sources.

Removing portions of silicon from the substrate can also create three-dimensional geometry of MEMS components. The principal process used for this purpose is the *etching* process. There are generally two types of etching processes in microfabrication: dry etching and wet etching. The former process involves the use of plasma as a carving force for the material removal, whereas wet etching uses chemical reaction to dissolve a portion of the material from the substrate. Dry etching is slow in removing material, but usually produces better edge definition of the cavity. The design engineer will choose the etching process according to the needs for the intended MEMS product.

In all microfabrication processes, the geometry of the workpieces is defined and controlled by appropriate masks, which are produced by a photolithographic process. The layout of the masks is usually drawn in a macroscale. They are then reduced to the desired microscale by means of photoreduction procedures. The reduced microscale layout is then projected onto the base mask coated with photoresist materials. A permanent print of the layout is produced at the surface of the mask material after processes that are similar to those in photo development. An etching process is used to remove the unwanted portion of the mask material, and finished masks are thus produced.

Microfabrication processes also include the control of local material properties and characteristics by ion implantation, diffusion, and oxidation. Ion implantation and diffusion processes are frequently used in doping a silicon substrate to alter its electrical conductance, either in selected portions or for the entire volume. These processes produce most piezoresistors for signal transduction in microsensors.

PROBLEMS

Part 1. Multiple Choice

1. Microfabrication technologies are developed specifically to shape structures in (1) macro-, (2) meso-, (3) microscale.
2. Established microfabrication techniques primarily involve (1) electro-mechanical, (2) electrochemical, (3) physical-chemical means.
3. A class 1000 clean room is defined as one in which the number of dust particles smaller than 0.5 μm is less than 1000 per cubic (1) inch, (2) foot, (3) meter.
4. The dust particle size used in defining clean room air quality is (1) 0.05 μm, (2) 0.5 μm, (3) 5 μm.
5. The higher the class number of a clean room, the (1) cleaner, (2) dirtier, (3) neither will be the air in the room.
6. The air quality in a typical urban environment is equivalent to a clean room of class (1) 50,000, (2) 500,000, (3) 5,000,000.
7. Photolithography is used in microfabrication because (1) we need to take a photograph of the microdevice, (2) to create patterns in microscale on substrates, (3) to create pictures in microscale.
8. The photoresist that, after exposure to light, dissolves in development is (1) the positive type, (2) the negative type, (3) either positive or negative type.
9. Photolithography using positive-type photoresists results in (1) better, (2) poorer, (3) about the same effect than using negative photoresists.
10. Typical thickness of photoresists in a photolithographic process is (1) 0.1 to 1.0 μm, (2) 0.5 to 2.0 μm, (3) 1 to 2 μm.
11. Common light sources used in photolithographic process have wavelengths in the range of (1) 100 to 300 nm, (2) 300 to 500 nm, (3) 500 to 700 nm.
12. The development of positive photoresists is (1) more complex than, (2) less complex than, (3) about equally as complex as that of negative photoresists.
13. Ion implantation is one of (1) two, (2) three, (3) four techniques frequently used for doping semiconductors.
14. Ion implantation is implanting foreign substances by (1) melting, (2) insertion by force, (3) slow diffusion.
15. The ion implantation process takes place at (1) high temperature, (2) room temperature, (3) low temperature.
16. A common energy source used for ion implantation involves (1)an ion beam, (2) intense heating, (3) high-energy electromagnetic fields.
17. The implanted foreign substance beneath the substrate's surface exhibits (1) uniform distribution in density, (2) nonuniform distribution in density with the less near the surface, (3) a distribution that depends on the temperature in the process.
18. Diffusion is used for doping semiconductors. It is (1) slower than, (2) faster than, (3) about the same speed as the ion implantation process.

19. The diffusion process takes place at (1) high, (2) room, (3) low temperature.
20. The diffused foreign substance beneath the substrate's surface exhibits (1) uniform distribution in density, (2) nonuniform distribution in density with the highest near the surface, (3) a distribution that depends on the temperature in the process.
21. A mathematical model of the diffusion process is based on (1) Fourier's law, (2) Newton's law, (3) Fick's law.
22. Wet oxidation of silicon is often preferred because of (1) better quality of SiO_2, (2) faster oxidation, (3) lower cost.
23. Oxidation of silicon substrates is (1) desired for protection of the substrate surface, (2) needed for local electric and thermal insulation, (3) an unavoidable phenomenon.
24. The kinetics of thermal oxidation is used to assess the (1) growth, (2) erosion, (3) plating of silicon oxide layers in silicon substrates.
25. One torr is equal to (1) 1 in H_2O, (2) 1 cm Hg, (3) 1 mm Hg pressure.
26. The color of oxidized silicon observed under white light represents (1) only one, (2) two, (3) several specific thickness of the oxide layer.
27. The deposition process in microfabrication can deposit (1) only organic, (2) only inorganic, (3) any materials onto substrate surfaces.
28. There are generally (1) two, (2) three, (3) four types of deposition in microelectronics and micromachining.
29. CVD is effective in depositing foreign materials over silicon substrates because it is a process that (1) is thermally activated, (2) combines mechanical and chemical diffusion, (3) combines thermal diffusion and chemical reactions.
30. The necessary ingredients in CVD are (1) plasma and chemical reactants, (2) carrier gas and chemical reactants, (3) chemical reactants and charge-carrying ions.
31. CVD processes require the substrate's surface to be (1) cold, (2) moderately hot, (3) very hot.
32. Better results in CVD are achievable by (1) increasing the pressure, (2) decreasing the pressure, (3) maintaining high constant pressure in the process.
33. The boundary layer created between the flowing carrier gas and the substrate surface in a CVD process (1) retards, (2) enhances, (3) has no effect on the CVD process.
34. The thickness of the boundary layer in a CVD process (1) increases, (2) decreases, (3) remains constant with a decreased velocity of the carrier gas.
35. Avogadro's number of 6.022×10^{23} is defined as the number of (1) electrons, (2) atoms, (3) molecules contained in 1 mole of any gas.
36. The rate of CVD is (1) proportional to, (2) inversely proportional to, (3) independent of temperature.
37. The rate of CVD is (1) proportional to, (2) inversely proportional to, (3) independent of the carrier gas pressure.

38. The rate of CVD is (1) proportional to, (2) inversely proportional to, (3) independent of the carrier gas velocity.

39. The rate of CVD is (1) proportional to, (2) inversely proportional to, (3) independent of thickness of the boundary layer between the carrier gas and the substrate surface.

40. PECVD is popular because it offers (1) good adhesion, (2) a simple process, (3) relatively low operating temperature.

41. The process engineer would choose (1) APCVD, (2) LPCVD, (3) PECVD for lower process temperature.

42. The process engineer would choose (1) APCVD, (2) LPCVD, (3) PECVD for higher rate of deposition.

43. Sputtering is processed at (1) low, (2) elevated, (3) high temperature.

44. Sputtering is normally used for depositing (1) organic, (2) inorganic, (3) metal films over silicon substrates.

45. Epitaxy involves the growth of (1) single-crystal films, (2) organic films, (3) metallic films over a substrate made of the same material.

46. CMOS is a (1) CVD, (2) PVD, (3) epitaxial thin-film growth process.

47. A common carrier gas used in epitaxial deposition is (1) oxygen, (2) nitrogen, (3) hydrogen.

48. Wet etching involves the use of (1) distilled water, (2) mineral water, (3) chemical solutions to dissolve the materials intended for removal.

49. Dry etching involves the use of (1) dry air, (2) dry toxic gas, (3) plasma to remove the substrate material.

50. In general, wet etching is (1) 10, (2) 100, (3) 1000 times faster in removing materials from silicon substrate than dry etching.

Part 2. Computational Problems

1. Solve Example 8.1, but change the doping substance to phosphorus at 30 keV energy level.

2. Estimate the time required to dope a silicon substrate by ion implantation with boron ions at 100 keV. The required maximum concentration of the dopant is $20 \times 10^{20}/\text{cm}^3$ at a depth of 0.2 μm beneath the substrate surface. What will be the asymptotic depth of doping in this case?

3. Solve Example 8.2, but with the diffusion temperatures at 900°C and 800°C. What observation, if any, will you make from this exercise?

4. In Example 8.2, estimate the time required to reach a concentration of the dopant at 0.2 μm beneath the substrate surface.

5. In Example 8.3, estimate the required time to achieve a 1-μm-thick SiO_2 layer over the silicon substrate surface in both wet and dry processes.

6. A CVD process involves the reactant diluted at 1 percent in hydrogen gas at 800°C. The horizontal tubular reactor has a diameter of 20 cm, and the susceptor that holds the substrates is 20 cm long. Pressure variation in the reactor is negligible. Find the following:

(a) Number of molecules in a cubic meter volume of the gas mixture
(b) The molar density of the gas mixture
(c) The Reynolds number of the gas flow
(d) The thickness of the boundary layer over the substrates' surface
(e) The diffusivity of the carrier gas with the reactant to the silicon substrates
(f) The surface reaction rate
(g) The deposition rate

7. Determine the required time for depositing a 0.5-μm-thick film over the silicon substrates with conditions described in computational Problem 6.

8. What will be the deposition rate in computational Problem 6 if the process temperature is dropped to 490°C?

Chapter 9

Overview of Micromanufacturing

CHAPTER OUTLINE

9.1 | INTRODUCTION

For mechanical engineers, a major effort in manufacturing a product is the proper selection and application of fabrication techniques such as machining, drilling, milling, forging, welding, casting, molding, stamping, and peening. We will quickly realize that none of these traditional fabrication techniques can be used in manufacturing MEMS and microsystems because of the extremely small size of these products. Some of these traditional fabrication techniques, however, are used in the packaging of MEMS and microsystems products.

The microfabrication techniques that we have learned from the last chapter are process related. What we will learn from this chapter is how these processes can be used either individually or in an integrated nature in the manufacture of MEMS and microsystem products such as microsensors, accelerometers, and actuators as described in Chapter 2. The techniques used to produce these products are called *micromachining*, or *micromanufacturing*.

Generally speaking, there are three distinct micromachining techniques used by current industry. These are (1) *bulk micromanufacturing*, (2) *surface micromachining,* and (3) the *LIGA process.* The term LIGA is an acronym for the German term for lithography, electroforming, and plastic molding. A number of excellent articles have been published in recent years for the overview of micromachining and the MEMS products produced by these three micromanufacturing techniques [O'Connor 1992, Bryzek et al. 1994, and Pottenger et al. 1997].

There are other process-related micromachining techniques that have been developed in recent years. Laser drilling and "machining" appears to be gaining popularity. However, we will focus our attention on the aforementioned three principal micromanufacturing techniques in this chapter.

9.2 | BULK MICROMANUFACTURING

Bulk micromanufacturing is widely used in the production of microsensors and accelerometers. It was first used in microelectronics in the 1960s. Further improvement for producing three-dimensional microstructures took place in the 1970s.

Bulk micromanufacturing or micromachining involves the removal of materials from the bulk substrates, usually silicon wafers, to form the desired three-dimensional geometry of the microstructures. The technique is thus similar to that used by sculptors in shaping sculptures. Neomicrosculpts by handicraft was achievable and was well documented in ancient Chinese history. For instance, a vivid scene of a tea party involving a famous Chinese poet and his friend and two servants inside a small boat was carved out from an olive nut a little over 1 cm long by a microsculptor in 1737. The display of this minute sculpture can be seen at the Imperial Museum of Arts in Taipei, Taiwan. There are similar microsculptures displayed in other museums in China. However, all these sculptures are handicrafts, not industry products, and hardwood was exclusively used in them. Shaping of microsystem components of the size between 0.1 μm and 1 mm made of tough materials such as silicon is beyond any existing mechanical means. Physical or chemical techniques, either by dry or wet etching, are the only practical solutions. Substrates that can be treated this way involve silicon, SiC, GaAs and quartz as mentioned in Chapter 8. Etching, either the orientation-independent isotropic etching or the orientation-dependent anisotropic etching, is thus the key technology used in bulk micromanufacturing.

9.2.1 Overview of Etching

We will deal with wet (chemical) etching in this section. As we have learned from Section 8.9 in Chapter 8, etching involves the exposure of a substrate covered by an etchant protection mask to chemical etchants as illustrated in Figure 9.1a.

Figure 9.1 | Wet etching of substrates.

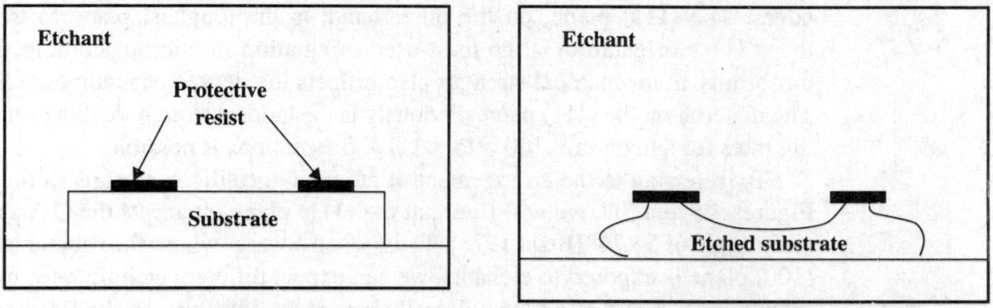

(a) Substrate in wet etching (b) Partially-etched substrate

The part of the substrate that is not covered by the protective mask is dissolved in the etchants and removed. However, as we can see from Figure 9.1b in an exaggerated way, the etching can undercut the part that is immediately under the protective mask after a lengthy period of time. Additionally, it may also damage the protective mask itself.

9.2.2 Isotropic and Anisotropic Etching

For substrates made of homogeneous and isotropic materials, the chemical etchants will attack the material uniformly in all directions, as illustrated in Figure 9.1b. This orientation-independent etching is referred to as *isotropic etching*.

Isotropic etching is hardly desirable in micromanufacturing because lack of control of the finished geometry of the workpiece. Fortunately, most substrate materials are not isotropic in their crystalline structures as described in Chapter 7. For example, silicon has a diamond cubic crystal structure. Therefore, some parts in the crystal are stronger, and thus more resistant to etching, than others. Again, as we have learned from Chapter 7, three planes of silicon crystals are of particular importance in micromachining. These are the (100), (110), and (111) planes as illustrated in Figure 9.2.

The three orientations <100>, <110>, and <111> are the respective normal lines to the (100), (110), and (111) planes. The two most common orientations used in the IC industry are the <100> and <111> orientations. However, in micromachining, the <110> orientation is the favored orientation. This is because, in this orientation, the wafer breaks or cleaves more cleanly than in the other orientations. The (110) plane is the only plane in which one can cleave the crystal in vertical

Figure 9.2 | The three principal planes in silicon crystal.

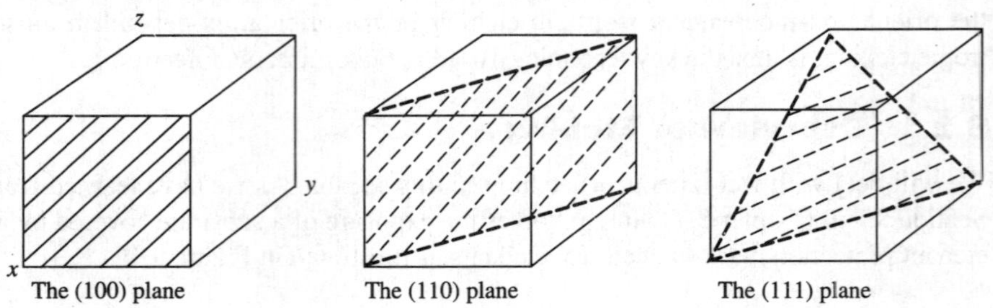

edges. The (111) plane, on the other hand, is the toughest plane to treat. Thus, the <111> orientation is the least-used orientation in micromachining. This nonuniformity in mechanical strength also reflects the degree of readiness for etching. The material on the (111) plane obviously is the hardest to etch. A 400:1 ratio in etching rates for silicon in <100> to <111> orientations is possible.

By referring to the arrangement of atoms in the silicon crystals as illustrated in Figures 7.5 and 7.7, we will find that the (111) plane intersects the (100) plane at a steep angle of 54.74° [Bean 1978]. Thus, when a wafer whose face coincides with the (100) plane is exposed to etchants, we can expect different etching rates in different orientations. A pyramid with sidewall slope at 54.74° exists in the finished product [Angell et al. 1983], as illustrated in Figure 9.3.

Figure 9.3 | Anisotropic etching of silicon substrate.

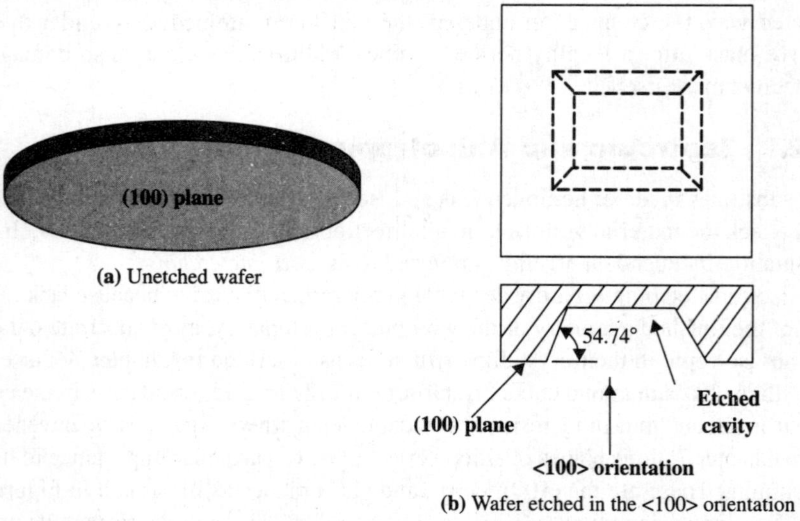

Despite the many advantages of anisotropic etching in controlling the shape of the etched substrates, there are several disadvantages: (1) It is slower than isotropic etching; the rate rarely exceeds 1 μm/min. (2) The etching rate is temperature sensitive. (3) It usually requires an elevated temperature around 100°C in the process, which precludes the use of many photoresistive masking materials.

9.2.3 Wet Etchants

There are a number of different types of etchants that can be used to etch different substrate materials. The common isotropic etchant for silicon is called HNA, which designates acidic agents such as $HF/HNO_3/CH_3COOH$. These etchants can be used effectively at room temperature. Alkaline chemicals with pH > 12, on the other hand, are used for anisotropic etching. Popular anisotropic etchants for silicon include potassium hydroxide (KOH), ethylene–diamine and pyrocatecol (EDP), tetramethyl ammonium hydroxide (TMAH), and hydrazine. Most etchants based on the above chemicals are diluted with water, normally 1:1 by weight. Typical ranges for etching rates for common substrate materials with these etchants are given in Table 9.1 [Wise 1991, Kovacs 1998].

Table 9.1 | Typical etching rates for silicon and silicon compounds

Material	Etchant	Etch Rate
Silicon in <100>	KOH	0.25–1.4 μm/min
Silicon in <100>	EDP	0.75 μm/min
Silicon dioxide	KOH	40–80 nm/h
Silicon dioxide	EDP	12 nm/h
Silicon nitride	KOH	5 nm/h
Silicon nitride	EDP	6 nm/h

We may observe from Table 9.1 that the etching rate for silicon dioxide with KOH is 1000 times slower than that for silicon, and silicon nitride is another order slower than that for silicon dioxide. Table 9.2 shows the selectivity ratio of etchants in different substrates. The *selectivity ratio* of a material is defined as the ratio of the etching rate of silicon to the etching rate of another material using the same etchant. For example, silicon dioxide has a selectivity ratio of 10^3 in Table 9.2, meaning this material has an etching rate in KOH that is 10^3 times slower than the etching rate for silicon. Thus, the higher the selectivity ratio of the material, the better the masking material it is.

Table 9.2 | Selectivity ratio of etchants to two silicon substrates

Substrate	Etchant	Selectivity ratio
Silicon dioxide	KOH	10^3
	TMAH	10^3–10^4
	EDP	10^3–10^4
Silicon nitride	KOH	10^4
	TMAH	10^3–10^4
	EDP	10^4

Source: Kovacs [1988].

Consequently, the high selectivity ratio of silicon dioxide and silicon nitride makes these materials suitable candidates for the masks for etching silicon substrates. However, the timing of etching and the agitated flow patterns of the etchants over the substrate surfaces need to be carefully controlled in order to avoid serious underetching and undercutting as illustrated in Figure 9.4.

Figure 9.4 | Definition of etched geometries.

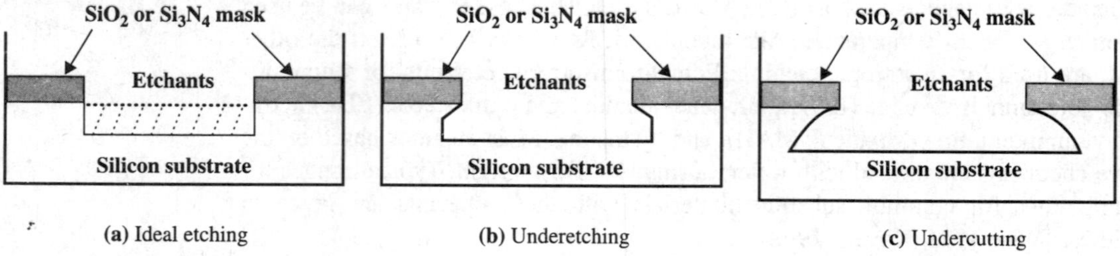

One also has to take special caution in selecting the masking materials. A common practice is to use an SiO_2 layer as mask for the silicon substrate in KOH etchants for trenches of modest depth. SiO_2 masking is relatively inexpensive in an etching process. However, even though etching is a slow process, the SiO_2 mask itself can be attacked by the etchants if the system is left in the etchant for a long period of time, as in the case of deep etching. In such cases, silicon nitride should be used as the mask instead.

An effective way to control the shape of the etched silicon substrate, as well as to achieve acceptably clean and accurate edge definition, is to apply etch stop, as will be presented below.

9.2.4 Etch Stop

There are two popular techniques used in etch stop. These are (1) dopant-controlled etch stop and (2) electrochemical etch stop.

Dopant-Controlled Etch Stop A peculiar phénomenon that can be used to control the etching of silicon is that doped silicon substrates, whether they are doped with boron for p-type silicon or phosphorus or arsenic for n-type silicon, show a different etching rate than pure silicon. In the case when the isotropic HNA etchants are used, the p- or n-doped areas are dissolved significantly faster than the undoped regions. However, excessive doping of boron in silicon for faster etching can introduce lattice distortion in the silicon crystal and thus produce undesirable internal (residual) stresses.

Electrochemical Etch Stop This technique is popular for controlling anisotropic etching. As illustrated in Figure 9.5, a lightly doped p–n junction is first produced in the silicon wafer by a diffusion process. The n-type is phosphorus doped at $10^{15}/cm^3$ and the p-type is boron doped at 30 Ω-cm (refer to Fig. 3.8 for the corresponding doping). The doped silicon substrate is then mounted on an inert substrate container made of a material such as sapphire. The n-type silicon layer is used as one of the electrodes in an electrolyte system with a constant voltage source as shown in Figure 9.5 [Madou 1997].

As we may observe from the arrangement in the figure, the unmasked part of the p-type substrate face is in contact with the etchant. Etching thus takes place as usual

Figure 9.5 | Illustrative arrangement for electrochemical etch stop.

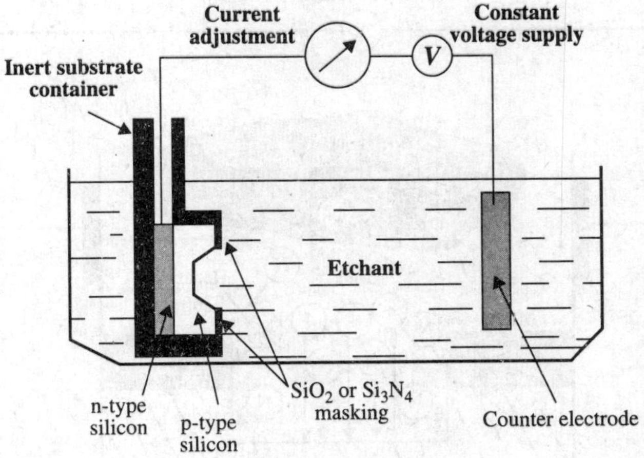

until it reaches the interface of the p–n junction, at which point etching stops because of the rate difference in p- and n-doped silicon. Consequently, one can effectively control the depth of etching simply by establishing the p–n silicon boundaries at the desired locations in a doped silicon substrate.

9.2.5 Dry Etching

Dry etching involves the removal of substrate materials by gaseous etchants without wet chemicals or rinsing. There are three dry etching techniques: *plasma, ion milling,* and *reactive ion etch* (RIE) [van Zant 1997]. We will focus our attention on plasma etching and a relatively new technique called *deep reactive ion etching* (DRIE) in this section.

Plasma Etching As we have learned from Chapter 3, plasma is a neutral ionized gas carrying a large number of free electrons and positively charged ions. A common source of energy for generating plasma is a radio-frequency (RF) source. The process involves adding a chemically reactive gas such as CCl_2F_2 to the plasma, one that contains ions and has its own carrier gas (inert gas such as argon gas). As illustrated in Figure 9.6, the reactive gas produces reactive neutrals when it is ionized in the plasma. The reactive neutrals bombard the target on both the sidewalls as well as the normal surface, whereas the charged ions bombard only the normal surface of the substrate. Etching of the substrate materials is accomplished by the high-energy ions in the plasma bombarding the substrate surface with simultaneous chemical reactions between the reactive neutral ions and the substrate material. This high-energy reaction causes local evaporation, and thus results in the removal of the substrate material. One may envisage that the etching front moves more rapidly in the depth direction than in the direction of the sidewalls. This is due to the larger number of high-energy particles involving both the neutral ions and the charged ions bombarding the normal surface, while the sidewalls are bombarded by neutral ions only.

Figure 9.6 | Plasma ion etching.

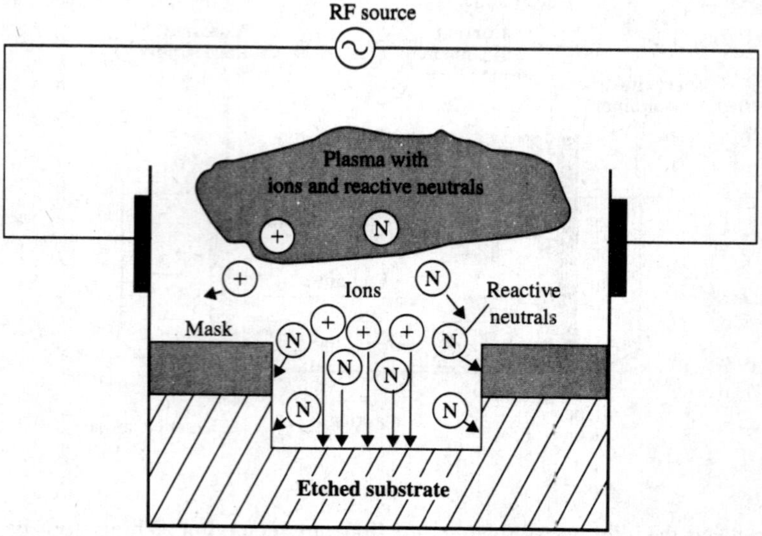

Conventional dry etching is a very slow process, at a rate of about 0.1 μm/min or 100 Å/min. Plasma etching can increase the etching rates in the order of 2000 Å/min. This increase in etching rate is primarily due to the increased mean free path of the reacting gas molecules in the depth to be etched. The working principle of the plasma etching process was presented in Section 8.9.2 in Chapter 8. In many ways, it can be viewed as a reversed sputtering process. Plasma etching is normally performed in high vacuum. Table 9.3 lists a number of etchants for selected substrate materials.

Table 9.3 | Plasma gas etchants for selected materials

Materials	Conventional chemicals	New chemicals
Silicon and silicon dioxide, SiO_2	CCl_2F_2 CF_4 C_2F_6 C_3F_8	CCl_2F_2 CHF_3/CF_4 CHF_3/O_2 CH_2CHF_2
Silicon nitride, Si_3N_4	CCl_2F_2 CHF_3	CF_4/O_2 CF_4/H_2 CHF_3 CH_3CHF_2
Polysilicon	Cl_2 or BCl_3/CCl_4 $/CF_4$ $/CHCl_3$ $/CHF_3$	$SiCl_4/Cl_2$ BCl_3/Cl_2 $HBr/Cl_2/O_2$ HBr/O_2 Br_2/SF_6 SF_6 CF_4
Gallium arsenide, GaAs	CCl_2F_2	$SiCl_4/SF_6$ $/HF_3$ $/CF_4$

Source: van Zant [1997].

Dry etching of silicon substrates, such as by plasma, typically is faster and cleaner than wet etching. A typical dry etching rate is 5 μm/min, which is about 5 times that of wet etching. Like wet etching, dry etching also suffers the shortcoming of being limited to producing shallow trenches. Consequently, both wet and dry etching processes are limited to producing MEMS with low *aspect ratios.* The aspect ratio (A/P) of a MEMS component is defined as the ratio of its dimension in the depth to those in the surface. For dry etching, the *A/P* is less than 15. Another problem with dry etching relates to the contamination of the substrate surface by residues.

Deep Reactive Ion Etching (DRIE) Despite the significant increase in the etching rate and the depth of the etched trench or cavity that can be achieved with the use of plasma, the etched walls in the trenches remain at a wide angle (θ) to its depth, as illustrated in Figure 9.7a. The cavity angle θ is critical in many MEMS structures, such as the comb electrodes in the microgrippers illustrated in Figures 10.21 and 10.25. These structures require the faces of the electrodes, or "fingers," to be parallel to each other. Etching processes produce most of these comb electrodes. It is highly desirable that the angle θ be kept at a minimum in deep-etched trenches that separate the plate electrodes. Obtaining deep trenches with vertical walls have been a major impediment of bulk manufacturing for a long while. Consequently, the bulk manufacturing technique has been generally regarded as suitable only for MEMS with low aspect ratios, and in many cases, with tapered cavity walls.

Figure 9.7 | The deep reactive ion etching process.

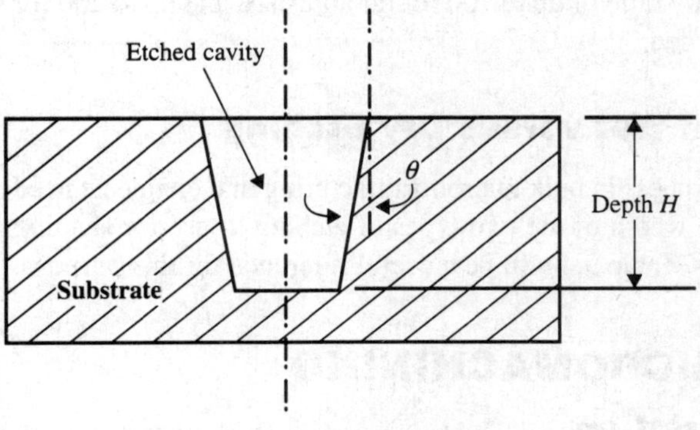

(a) A sidewall angle in an etched cavity

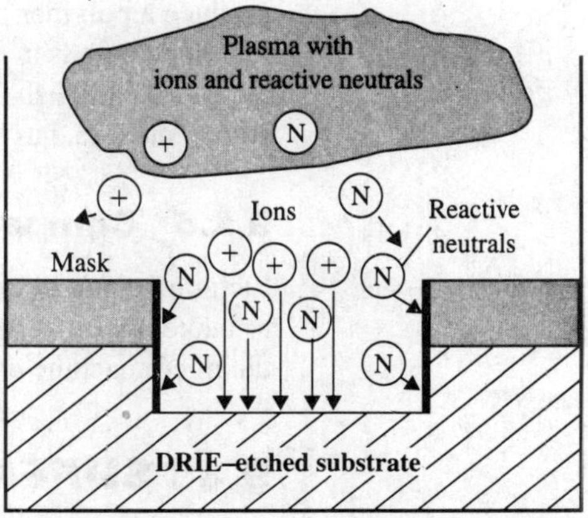

(b) The DRIE process

Deep reactive ion etching (DRIE) is a process that can overcome the problem described above. The DRIE process has since extended the use of the bulk manufacturing technique to the production of MEMS of high aspect ratio with virtually vertical walls; i.e., $\theta \approx 0$.

The DRIE process differs from dry plasma etching in that it produces thin protective films of a few micrometers on the sidewalls during the etching processes. It involves the use of a high-density plasma source, which allows alternating processes of plasma (ion) etching of the substrate material and the deposition of etching-protective material on the sidewalls as illustrated in Figure 9.7b. Suitable etching-protective materials (shown in black in the figure) are those materials of high selectivity ratio, such as silicon dioxide in Table 9.2. Polymers are also frequently used for this purpose. Polymeric materials such as photoresists are produced by polymerization during the plasma etching process.

The DRIE process with polymeric sidewall protection has been used to produce MEMS structures with $A/P = 30$ with virtually vertical walls of $\theta = \pm 2°$ for several years. Recent developments have substantially improved the performance of DRIE with better sidewall protecting materials. For example, silicon substrates with A/P over 100 was achieved as presented in the following summary with $\theta = \pm 2°$ at a depth of up to 300 μm. The etching rate, however, was reduced to 2 to 3 μm/min. A more recent report [Williams 1998] indicated that a trench depth of up to 380 μm was obtained by this technique.

Sidewall protection materials	Selectivity ratio	Aspect ratio *A/P*
Polymer		30:1
Photoresists	50:1	100:1
Silicon dioxide	120:1	200:1

There are a number of reactant gases that could be used in DRIE. One of these reactants is fluoropolymers (nCF_2) in the plasma of Argon gas ions. This reactant can produce a polymer protective layer on the sidewalls while etching takes place. The rate of etching is in the range of 2 to 3 μm/min, which is higher than what wet etching can accomplish. A selectivity ratio of up to 100 for photoresists, and up to 200 for silicon dioxide, has been recorded.

9.2.6 Comparison of Wet versus Dry Etching

Etching is such an important process in bulk micromanufacturing that engineers need to make intelligent choices on which of the two types of etching to use for shaping the micromachine components. Table 9.4 will be a useful reference for this purpose.

9.3 | SURFACE MICROMACHINING

9.3.1 General Description

In contrast to bulk micromanufacturing in which substrate material is removed by physical or chemical means, the surface micromachining technique builds microstructure by adding materials layer by layer on top of the substrate. Deposition techniques, in particular the low pressure chemical vapor deposition (LPCVD) technique, such as those described in Section 8.6.4 in Chapter 8 are used for such buildups, and polycrystalline silicon (polysilicon) is a common material for the layer material. *Sacrificial layers,* usually made of SiO_2, are used in constructing the MEMS components

Table 9.4 | Comprehensive comparison of wet versus dry etching

Parameters	Dry etching	Wet etching
Directionality	Good for most materials	Only with single-crystal materials (aspect ratio up to 100)
Production-automation	Good	Poor
Environmental impact	Low	High
Masking film adherence	Not as critical	Very critical
Selectivity	Poor	Very good
Materials to be etched	Only certain materials	All
Process scale-up	Difficult	Easy
Cleanliness	Conditionally clean	Good to very good
Critical dimensional control	Very good ($< 0.1\ \mu$m)	Poor
Equipment cost	Expensive	Less expensive
Typical etch rate	Slow (0.1 μm/min) to fast (6 μm/min)	Fast (1 μm/min and up)
Operational parameters	Many	Few
Control of etch rate	Good in case of slow etch	Difficult

Source: Madou [1997].

but are later removed to create necessary void space in the depth, i.e., in the thickness direction. Wet etching is the common method used for that purpose.

We will thus see that, although we will still deal with single-crystal silicon as the substrate in most cases, the added layers need not be single crystals or silicon compounds. The overall height of the structure therefore is no longer limited by commercially available wafer thickness. Layers that are being added in surface micromachining are typically 2 to 5 μm thick each. In special applications, this range can be extended to 5 to 20 μm. They are thus regarded as thin films in micromanufacturing. We will realize that there will indeed be problems associated with structures built with thin films.

Figure 9.8 illustrates the difference between bulk micromanufacturing and surface micromachining. In Figure 9.8a we see a microcantilever beam that can be used either as a microaccelerometer (Fig. 2.33), or as an actuator (Figs. 2.17 and 2.19). The cantilever beam is made of single-crystal silicon with a significant amount of material etched away as illustrated in Figure 9.9. The same cantilever beam structure can be produced by polysilicon with a surface micromachining technique as illustrated in

Figure 9.8 | Microcantilever beams produced by two micromachining techniques.

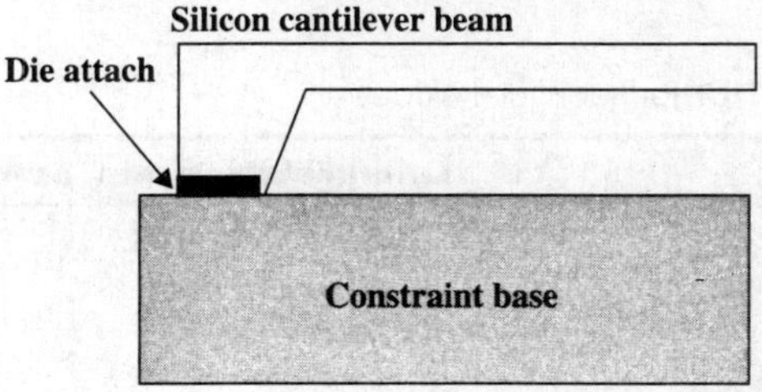

(a) By bulk micromachining

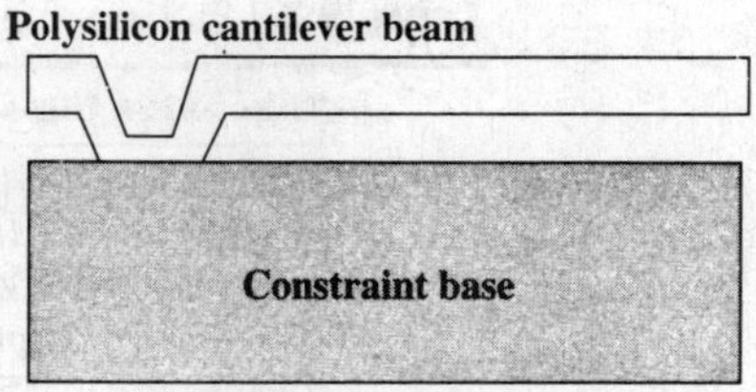

(b) By surface micromachining

Figure 9.9 | Waste of material in bulk micromanufacturing.

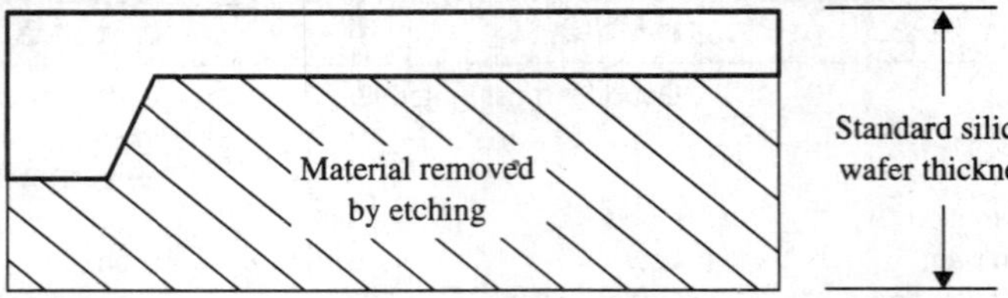

Figure 9.8b. We can see from this figure that surface micromachining not only saves material, but also eliminates the need for a die attach, as the polysilicon beam can be built on the top of the constraint base directly.

9.3.2 Process in General

Surface-micromachined devices are typically made up of three types of components: (1) a sacrificial component (also called a spacer layer), (2) a microstructural component, and (3) an insulator component.

The sacrificial components are usually made of phosphosilicate glass (PSG) or SiO_2 deposited on substrates by LPCVD techniques. PSG can be etched more rapidly than SiO_2 in HF etchants. These components in the form of films can be as long as 1 to 2000 μm and 0.1 to 5 μm thick. Both microstructural and insulator components can be deposited in thin films. Polysilicon is a popular material. The etching rates for the sacrificial components must be much higher than those for the two other components.

In Figure 9.10, we demonstrate how a microcantilever beam such as the one shown in Figure 9.8b is produced by the surface micromachining technique. We begin in step 1 with a silicon substrate base with a PSG deposited on its surface. A mask (mask 1) is made in step 2 to cover the surface of the PSG layer for the subsequent etching to allow for the attachment of the future cantilever beam as shown in step 3. Another mask (mask 2) is made for the deposition of polysilicon microstructural material in step 4. The PSG that remains in step 5 is subsequently etched away to produce the desired cantilever beam as shown in step 6. The most suitable etchant used in the last step for the sacrificial PSG layer is 1:1 HF, which is made of 1:1 $HF:H_2O$ + 1:1 $HCl:H_2O$. After etching, the structure is rinsed in deionized water thoroughly followed by drying under infrared lamps. The etching rates of using these etchants for various sacrificial materials are presented in Table 9.5.

Table 9.5 | Etching rate in HF/HCl for sacrificial oxides

Thin oxide films	Lateral etching rate, μm/min
CVD SiO_2 (densified at 1050°C for 30 min)	0.6170
Ion-implanted SiO_2 (at $8 \times 10^{15}/cm^2$, 50 keV)	0.8330
Phosphosilicate (PSG)	1.1330
5%-5% Boronphosphosilicate (BPSG)	4.1670

Source: Madou [1997].

Figure 9.10 | Surface micromachining process.

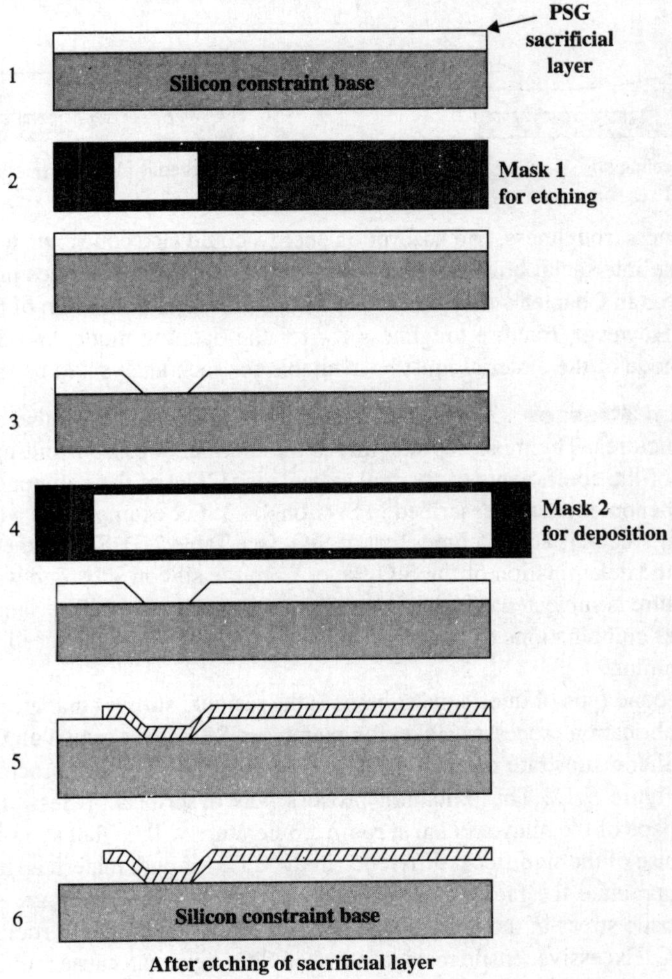

9.3.3 Mechanical Problems Associated with Surface Micromachining

There are three major problems of mechanical nature that result from surface micromachining. These are (1) adhesion of layers, (2) interfacial stresses, and (3) stiction.

Adhesion of Layers Whenever two layers of materials, whether similar or dissimilar, are bonded together, a possibility of delamination exists. A bilayer structure can delaminate at the interface either by peeling of one layer from the other or by shear that causes the severing of the interfaces locally along the interface. Figure 9.11 illustrates both these failures.

Of the many causes for interfacial failures, excessive thermal and mechanical stress is the main cause. However, other causes including the surface conditions, e.g.,

Figure 9.11 | Interfacial failure of bilayer materials.

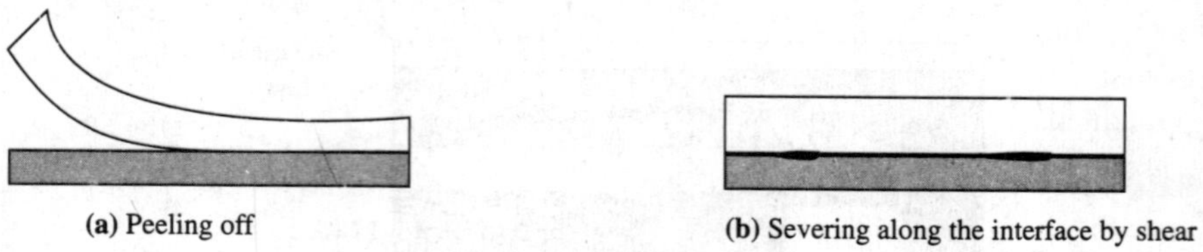

the cleanliness, roughness, and adsorption energy, could also contribute to the weakening of the interfacial bonding strength. Fracture mechanics theories presented in Section 4.5.3 in Chapter 4 may be used to assess the fracture strength of the bonded structure. However, fracture toughness K_{IC} for the opening mode, and K_{IIC} for the shearing mode of the material must be available for such analysis to be meaningful.

Interfacial Stresses There are typically three types of stresses that exist in the bilayer structures. The most obvious one is the thermal stresses resulting from the mismatch of the coefficients of thermal expansion (CTE) of the component materials. This phenomenon was described in Section 4.4.3. For example, we will find that the CTE for silicon is about 5 times that of SiO_2 (see Table 7.3). Severe thermal stress can cause the delamination of the SiO_2 layer from the silicon substrate when the bilayer structure is subjected to high enough operating temperature. The same can happen in other combinations of materials in multilayer structures produced by surface micromachining.

The second type of interfacial stresses is the residual stresses that are inherent in the microfabrication processes. Take, for example, a SiO_2 layer grown on the top surface of a silicon substrate beam at 1000°C by a thermal oxidation process as illustrated in Figure 9.12. The oxidation process was described in Section 8.5. The resultant shape of the bilayer beam at room temperature will be that shown in Figure 9.12b because of the significant difference in the CTE for both materials. It is not difficult to appreciate the fact that associated with the residual strain are significant residual tensile stress in the SiO_2 layer after it is cooled down to the room temperature of 20°C. Excessive tensile residual stress in SiO_2 layer can cause multiple cracks in the layer. Hsu and Sun [1998] reported the analysis of residual stresses in the oxide diaphragm of a pressure sensor.

Figure 9.12 | Residual stress and strain in a bilayer beam.

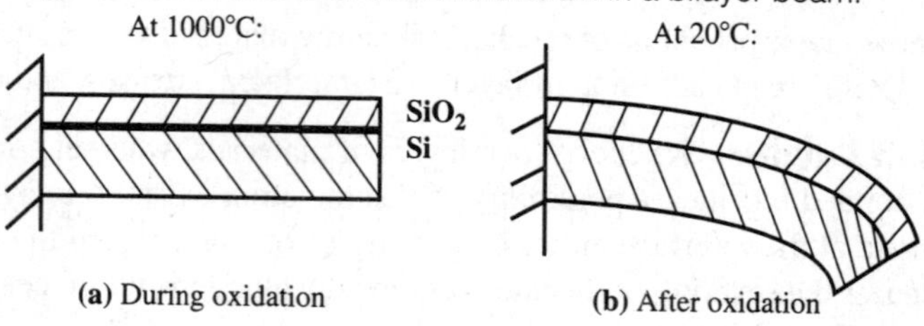

The third type of stress that could be introduced in thin-film structures is the intrinsic stress due to local change of atomic structure during microfabrication processes. Excessive doping, for instance, could introduce substantial residual stresses in the structure after surface micromachining. The exact causes and the quantitative assessment of the intrinsic stresses in thin films are far from being clear to engineers. A qualitative description of this stress was presented in Section 4.6 in Chapter 4.

Stiction Many have experienced the difficulty in separating two transparencies after the thin dividing paper is pulled out. A similar phenomenon occurs in surface micromachining. This phenomenon of two separated pieces sticking together is called *stiction*.

Stiction is the most serious problem for engineers to deal with in surface micromachining. It often occurs when the sacrificial layer is removed from the layers of the material that it once separated (e.g., step 6 in the case illustrated in Fig. 9.10). The thin structure that was once supported by the sacrificial layer may collapse on the other material. Take, for example, the production of a thin beam as illustrated in Figure 9.13; stiction could happen with the thin polysilicon beam dropping onto the top surface of the silicon substrate (the constraint base) after the removal of the sacrificial PSG layer (Fig. 9.13b). The two materials would then stick together after the joint. Considerable mechanical forces are required to separate the two stuck layers again and these excessive forces can break the delicate microstructure. Stiction is the main cause for the large amount of scraps in surface micromachining.

Figure 9.13 | The collapse of a thin cantilever beam due to stiction.

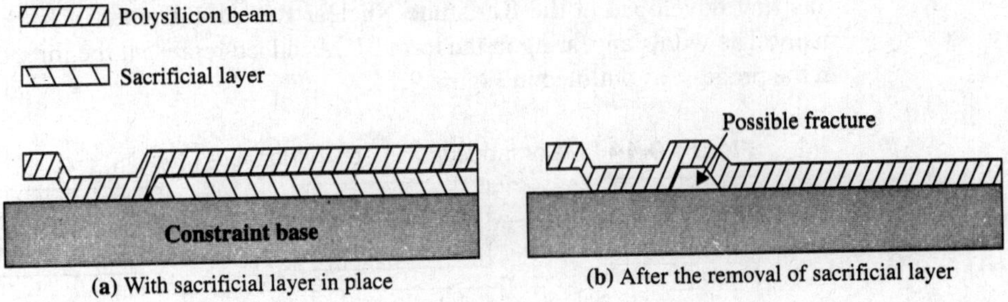

Stiction occurs presumably as a result of hydrogen bonding of surfaces during rinsing of the interface after the etching of the PSG sacrificial layer, or by forces such as the van der Waals forces described in Chapter 3. Various ways of avoiding stiction have been suggested by several researchers such as Madou [1997]. Among the proposed remedial actions are temporary spacers using polysilicon and sacrificial polymer columns that can be removed by etching with oxygen plasma afterward. Whatever the remedial action, the cost and the time required for production are major concerns to the industry.

9.4 | THE LIGA PROCESS

Both micromanufacturing techniques—bulk manufacturing and surface micromachining—involve microfabrication processes evolved from microelectronics technology. Consequently, much of the developed knowledge and experience as well as equipment used for the production of microelectronics and integrated circuits can be adapted for MEMS and microsystems manufacturing with little modifications. Unfortunately, these inherited advantages are overshadowed by two major drawbacks: (1) the low geometric aspect ratio, and (2) the use of silicon-based materials. Geometric aspect ratio of a microstructure is the ratio of the dimension in the depth to that of the surface. Most silicon-based MEMS and microsystems use wafers of standard sizes and thicknesses as substrates, on which etching or thin-film deposition takes place to form the desired three-dimensional geometry. Severe limitations of the depth dimension is thus unavoidable. The other limitation is on the materials. Silicon-based MEMS preclude the use of conventional materials such as polymers and plastics, as well as metals for the structures and thin films.

The LIGA process for manufacturing MEMS and microsystems is radically different from these two manufacturing techniques. This process does not have the two aforementioned major shortcomings in silicon-based micromanufacturing techniques. This process offers a great potential for manufacturing non-silicon-based microstructures. The single most important feature of this process is that it can produce "thick" microstructures that have extremely flat and parallel surfaces such as microgear trains (Fig. 1.9), motors and generators (Fig. 1.10), and microturbines (Fig. 1.11) made of metals and plastics. These unique advantages are the primary reasons for its increasing popularity in the MEMS industry.

The term LIGA is an acronym for the German terms *Lithography* (Lithographie), *electroforming* (Galvanoformung), and *molding* (Abformung). The technique was first developed at the Karlsruhe Nuclear Research Center in Karlsruhe, Germany. The words appearing in the term LIGA indeed represent the three major steps in the process, as outlined in Figure 9.14.

Figure 9.14 | Major fabrication steps in the LIGA process.

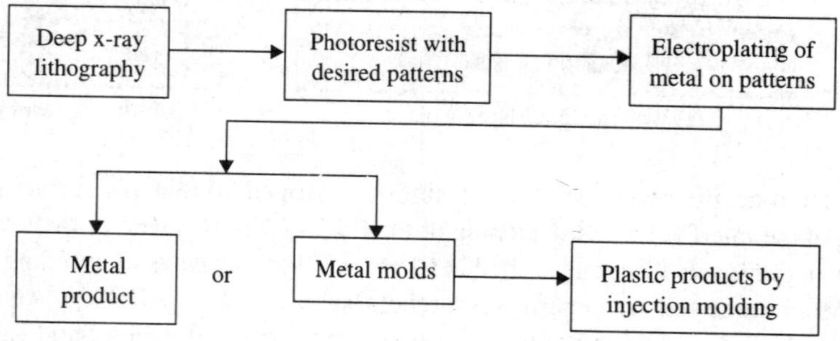

9.4.1 General Description of the LIGA Process

As shown in Figure 9.14, the LIGA process begins with deep x-ray lithography that sets the desired patterns on a thick film of photoresist. X-rays are used as the light source in photolithography because of their short wavelength, which provides higher penetration power into the photoresist materials. This high penetration power is necessary for high resolution in lithography, and for a high aspect ratio in the depth. The short wavelength of x-ray allows a line width of 0.2 μm and an aspect ratio of more than 100:1 to be achieved. The x-rays used in this process are provided by a synchrotron radiation source, which allows a high throughput because the high flux of collimated rays shortens the exposure time.

The LIGA process outlined in Figure 9.14 may be demonstrated by a specific example as illustrated in Figure 9.15. The desired product in this example is a microthin-wall metal tube of square cross-section. We may begin the process by depositing a thick film of photoresist material on the surface of a substrate as shown in Figure 9.15a. A popular photoresist material that is sensitive to x-ray is polymethylmethacrylate (PMMA). Masks are used in the x-ray lithography. Most masking materials are transparent to x-rays, so it is necessary to apply a thin film of gold to the area that will block x-ray transmission. The thin mask used for this purpose is silicon nitride with a thickness varying from 1 to 1.5 μm. The deep x-ray lithography will cause the exposed area to be dissolved in the subsequent development of the resist material (see Fig. 9.15b). The PMMA photoresist after the development will have the outline of the product, i.e. the outside profile of the tube. This is followed by electroplating of the PMMA photoresist with a desired metal, usually nickel, to produce the tubular product of the required wall thickness (see Fig. 9.15c). The desired tubular product is produced after the removal of the photoresist materials (i.e., PMMA in this case) by oxygen plasma or chemical solvents.

Figure 9.15 | Major steps in the LIGA process.

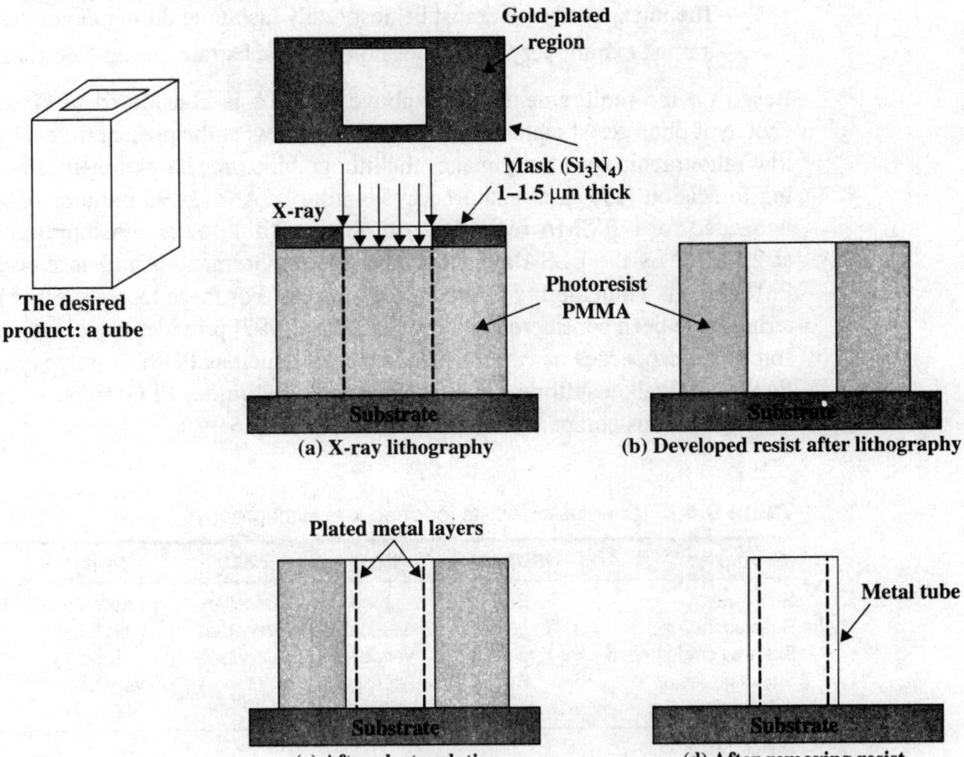

For most applications the desired product is metal molds for subsequent injection molding of microplastic products as shown in Figure 9.14.

9.4.2 Materials for Substrates and Photoresists

Substrate Materials for the LIGA Process The substrate used in the LIGA process is often called the *base plate.* It must be an electrical conductor, or an insulator coated with electrically conductive materials. Electrical conduction of the substrate is necessary in order to facilitate electroplating, which is a part of the LIGA process. Suitable materials for the substrates include: austenite steel; silicon wafers with a thin titanium or Ag/Cr top layer; and copper plated with gold, titanium, and nickel. Glass plates with thin metal plating could also be used as the substrate.

Photoresist Materials Basic requirements for photoresist materials for the LIGA process include the following:

—It must be sensitive to x-ray radiation.
—It must have high resolution as well as high resistance to dry and wet etching.
—It must have thermal stability up to 140°C.
—The unexposed resist must be absolutely insoluble during development.
—It must exhibit very good adhesion to the substrate during electroplating.

Based on the requirements listed above, PMMA is considered to be an optimal choice of photoresist material for the LIGA process at the present time. However, its low lithographic sensitivity makes the lithographic process extremely slow. According to Madou [1997], at a short wavelength of 5Å, over 90 minutes of irradiation is needed for a PMMA resist 500-μm thick with a power consumption of 2 MW at 2.3 GeV by the ELSA synchrotron in Bonn, Germany. Another shortcoming of PMMA is its vulnerability to crack due to stress. For these reasons, other resist materials have been considered and used. Madou [1997] provides a qualitative comparison of the properties of various resist materials such as POM = polyoxymethylene, PAS = polyalkensulfone, PMI = polymethacrylimide, PLG = poly (lactide-co-glycolide). This comparison is presented in Table 9.6.

Table 9.6 | Properties of resists for deep x-ray lithography.

Property	PMMA	POM	PAS	PMI	PLG
Sensitivity	Bad	Good	Excellent	Reasonable	Reasonable
Resolution	Excellent	Reasonable	Very bad	Good	Excellent
Sidewall smoothness	Excellent	Very bad	Very bad	Good	Excellent
Stress corrosion	Bad	Excellent	Good	Very bad	Excellent
Adhesion on substrate	Good	Good	Good	Bad	Good

Source: Madou [1997].

9.4.3 Electroplating

Electroplating is an important step in the LIGA process. Electroforming of metal films onto the surface of the cavities in the photoresist after x-ray lithography has been performed as illustrated in Figure 9.15c. Nickel is the common metal to be electroplated on the photoresist walls. Other metals and metallic compounds that could be used for electroplating include Cu, Au, NiFe and NiW. The conductive substrate and the carrying photoresist structure form the cathode in an electroplating process is illustrated in Figure 9.16.

Figure 9.16 I An electroplating process of nickel.

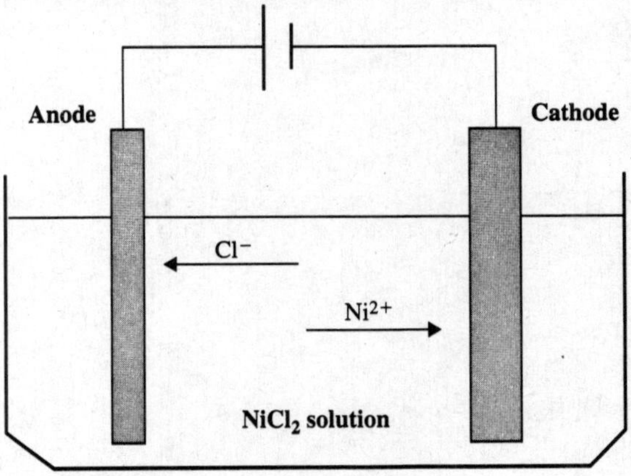

(Schulze [1998].)

Electroplating works on the principle that the nickel ions Ni^{2+} from the nickel chloride ($NiCl_2$) solution react with the electrons at the cathode to yield nickel as shown in the process below:

$$Ni^{2+} + 2e^- \rightarrow Ni$$

However, we should be aware of the presence of H_2 on the surface of the cathode that may cause non-uniform Ni plating. The H_2 gas is the product of the $H_2{}^+$ ions produced by the electrolysis of the solution. The chemical reaction of $H_2{}^+$ ions to H_2 gas at the cathode is presented below:

$$2H^+ + 2e^- \rightarrow H_2$$

To avoid damage of the plated surfaces by hydrogen bubbles, we need to control the pH of the solution, the temperature, and the current density in the electrolysis.

9.4.4 The SLIGA Process

We have seen from Figures 9.14 and 9.15, the finished product, whether it is a microstructure or a metal mold, is attached to the substrate, or base plate. The attachment to the electrically conductive substrate is necessary for the electroplating process.

However, this attachment is considered as a redundancy in the LIGA process. For instance, the hollow square tube produced in the LIGA process as described in Figure 9.15 would not be separated after electroplating of metal film on the inner walls. A modified process called sacrificial LIGA (SLIGA) has been developed to solve this problem. The principle of SLIGA is to introduce a sacrificial layer between the PMMA resist and the substrate thereby to allow the separation of the finished mold from the subtrate after the electroplating. The separation is achieved by the removal of the sacrificial layer by etching. Polyimide with a metal-film coating is used as a common sacrificial layer material for that purpose.

9.5 | SUMMARY OF MICROMANUFACTURING

We have learned in the foregoing sections the three principal micromanufacturing techniques that are currently available. Following is a summary of these three techniques, which will be useful for engineers in selecting the optimal manufacturing technique for their specific design projects.

9.5.1 Bulk Micromanufacturing

- Straightforward, involving well-documented fabrication processes.
- Less expensive in the process, but material loss is high.
- Suitable for simple geometry, e.g., micropressure sensors dies and some actuating elements.
- Limited to low-aspect ratio in geometry; i.e., the surface dimensions are much greater than that of the depth. This is because the overall height of the microstructure is limited by the thickness of commercially available silicon wafers.

9.5.2 Surface Micromachining

- Requires the building of layers of materials on the substrate.
- Complex masking design and productions.
- Etching of sacrificial layers is necessary.
- The process is tedious and more expensive.
- There are serious engineering problems such as interfacial stresses and stiction.
- Major advantages: (1) not constrained by the thickness of silicon wafers; (2) wide choices of thin film materials to be used; (3) suitable for complex geometries such as microvalves and actuators.

9.5.3 The LIGA Process

- The most expensive process of all.
- Requires a special synchrotron radiation facility for deep x-ray lithography.
- Requires the development of microinjection molding technology and a facility for mass production purposes.
- Major advantages are: (1) virtually unlimited aspect ratio of the microstructure geometry; (2) flexible microstructure configurations and geometry; (3) the only one of the three techniques that allows the production of metallic microstructures; (4) the best of the three manufacturing processes for mass production, with the provision for injection molding.

PROBLEMS

Part 1. Multiple Choice

1. *Micromanufacturing* is (1) synonymous to, (2) antonymous to, (3) unrelated to *microfabrication.*
2. In general, there are (1) two, (2) three, (3) four distinct micromanufacturing techniques.
3. Bulk manufacturing involves primarily (1) adding, (2) subtracting, (3) both adding and subtracting portions of material from the substrate.
4. The principal microfabrication process used in bulk manufacturing is (1) etching, (2) deposition, (3) diffusion.
5. Isotropic etching is hardly desirable in micromanufacturing because (1) the etching rate is too low, (2) the cost is too high, (3) it is hard to control the direction of etching.
6. The most favored orientation for micromachining is the (1) <100>, (2) <110>, (3) <111> orientation.
7. The least-used orientation for micromachining is the (1)<100>, (2) <110>, (3) <111> orientation.
8. A ratio of 400:1 etching rate was observed for silicon between the (1) <100> and <111>, (2) <110> and <111>, (3) <110> and <100> orientations.
9. The (111) plane intersects the (100) plane in a silicon crystal with an angle of (1) 50.74°, (2) 54.74°, (3) 57.47°.
10. Anisotropic etching is (1) faster, (2) slower, (3) about the same speed in comparison to isotropic etching.
11. Silicon dioxide with KOH etchant is (1) 100, (2) 1000, (3) 20,000 times slower than silicon.
12. Silicon dioxide with EDP etchant is (1) 100, (2) 1000, (3) 20,000 times slower than silicon.
13. Silicon nitride is (1) a stronger, (2) a weaker, (3) about the same, in etching resistance as in silicon oxide.
14. The higher the selectivity ratio of a material, (1) the better, (2) the worse, (3) neither better nor worse is the material as an etching mask.
15. Doped silicon has resistance to etching (1) stronger than, (2) weaker than, (3) about the same as that of undoped silicon.
16. Excessive doping of silicon introduces (1) residual stresses, (2) residual strains, (3) fracture of atomic bonds in a silicon substrate.
17. Wet etching can be stopped at the boundaries of (1) p–n doped silicon, (2) p silicon/silicon, (3) n silicon/silicon.
18. There are (1) one, (2) two, (3) three dry etching techniques available for microelectronics.
19. Isotropic etching is (1) 2, (2) 5, (3) 10 times faster than anisotropic etching.
20. When selecting materials for masks in deep etching process, one would select materials with (1) high, (2) low, (3) medium selectivity ratio.

21. DRIE stands for (1) dry etching, (2) dry reactive ion etching, (3) deep reactive ion etching.
22. DRIE is the best means of (1) dry etching, (2) fast etching, (3) deep etching.
23. PSG stands for (1) polysilicon glass, (2) phosphosilicate glass, (3) phosphorus silicon glass.
24. PSG is a common material for (1) active substrates, (2) passive substrates, (3) sacrificial layers.
25. Sacrificial layers in surface micromachining are used to (1) strengthen the microstructure, (2) create necessary geometric voids in the microstructure, (3) be part of the structure.
26. PSG is a more popular sacrificial layer material than silicon dioxide because it can be etched (1) more rapidly, (2) more slowly, (3) more cheaply in HF etchant.
27. The most popular structural material in surface micromachining is (1) PSG, (2) polysilicon, (3) silicon dioxide.
28. In surface micromachining, the etch rate for sacrificial layers must be (1) much slower, (2) about the same, (3) much faster than etch rates for other layers.
29. Stiction occurs in (1) bulk micromanufacturing, (2) surface micromachining, (3) laser microfabrication.
30. Stiction in finished microstructures made by surface micromachining is a result of (1) layers of dissimilar materials, (2) thin films, (3) atomic forces between layers.
31. The geometric aspect ratio in MEMS structures is defined as the ratio of dimensions in (1) depth to surface, (2) surface to depth, (3) width to length.
32. The LIGA process produces MEMS with materials (1) limited to silicon, (2) limited to ceramics, (3) virtually no limitation on materials.
33. Synchrotron x-rays are used in photolithography in the LIGA process because (1) they are more effective with the photoresist, (2) they are a cheaper source of light, (3) they can penetrate deep into the photoresist material.
34. The photoresist materials commonly used in the LIGA process are (1) any photoresist materials, (2) positive photoresist materials, (3) negative photoresist materials.
35. One of the principal advantages of the LIGA process is its ability to produce (1) microstructures with high aspect ratio, (2) microstructures with low cost, (3) microstructures with precise dimensions.
36. X-ray lithography in a LIGA process produces (1) the outline, (2) the actual geometry, (3) the duplicate of a MEMS component.
37. The x-ray lithography photoresists provide (1) the outline, (2) the actual geometry, (3) the duplicate of a MEMS component.
38. An electrically conductive base plate must be used in a LIGA process because of the need for (1) signal transduction, (2) electroplating of metals, (3) required electric heating of the mold.

39. Micromolds for injection molding of MEMS are produced from (1) the x-ray lithography photoresist, (2) metals that are electroplated to the photoresist outline, (3) the removal of the base plate in the process.
40. The optimal choice of photoresist in a LIGA process is (1) POM, (2) PMI, (3) PMMA.
41. The photoresist that is most sensitive to x-radiation is (1) POM, (2) PAS, (3) PMMA.
42. SLIGA is an improvement of the LIGA process with the provision of (1) slicing the base plate from the finished product, (2) a sacrificial layer for ready separation of the base plate from the product, (3) a clean product.
43. The least expensive micromanufacturing technique is (1) bulk manufacturing, (2) surface micromachining, (3) the LIGA process.
44. The most flexible micromanufacturing technique is (1) bulk micromanufacturing, (2) surface micromachining, (3) the LIGA process.
45. The most costly micromanufacturing technique is (1) bulk micromanufacturing, (2) surface micromachining, (3) the LIGA process.

Part 2. Use no more than 20 words to answer the following questions

1. What are the limitations of the height (depth) of microstructures that can be produced by bulk manufacturing technique?
2. What is your major criterion in selecting materials for the masks used in etching, e.g., in etching silicon substrate with moderate depth and also for deeper etching?
3. Describe the DRIE process. How can DRIE achieve virtually perfect vertical etching?
4. What is the principal advantage of DRIE? Why is it necessary to have near perfectly vertical walls in microstructures?
5. What is the principal difference between bulk manufacturing and surface micromachining?
6. Give one advantage and one disadvantage of using surface micromachining.
7. Describe the role of sacrificial layers in surface micromachining.
8. Describe the phenomenon of stiction, and possible ways to avoid it.
9. List the principal advantages and disadvantages of the LIGA process.
10. Why is electroplating necessary in a LIGA process?

Part 3. Design of Micromanufacturing Processes

1. Following the illustrative examples, such as Figure 9.15 offered in this chapter, design micromanufacturing processes for the production of a thick gear made of silicon or plastics with a root diameter of 500 μm and pitch diameter of 750 μm. The gear is 1000 μm thick. Use (a) bulk micromanufacturing, (b) the LIGA process. Draw a conclusion on what you have learned from this exercise.
2. Following the illustrative example in Figure 9.10, design a surface micromachining process for the production of a comb-drive actuator as outlined in Figure 2.25.

Chapter 10

Microsystem Design

CHAPTER OUTLINE

10.1 | INTRODUCTION

The working principles of many MEMS devices were presented in Chapter 2. Examples of how various MEMS components can be designed were given in Chapters 4, 5, 6, and 7. The many ways that these components can be manufactured were described in Chapters 8 and 9. However, it requires reliable mechanical engineering design to configure the optimum geometry and dimensions of the components that are to be assembled into microsystems. Structural integrity and reliability of these products are primary requirements, but manufacturability of the systems is also a critical consideration in the design process.

A major difference between mechanical engineering design of microsystems and that of other products is that the design for microsystems requires the integration of the related manufacturing and fabrication processes. Mechanical engineering design of traditional products and systems rarely requires the consideration of the consequences of the manufacturing process. For example, components such as gears, bearings, and fasteners in a mechanical system can be purchased from suppliers without the knowledge of how these components are produced. In microsystems, which involve MEMS components, however, the situation is quite different. Components for MEMS are fabricated by various physical-chemical means as described in Chapters 8 and 9. These fabrication and manufacturing processes often involve high temperature and harsh physical and chemical treatments of delicate materials used for the components. These processes can have serious repercussions in the performance of microsystems and hence must be taken into design considerations. Tolerance of the finished components and the intrinsic effects such as residual stresses and strains inherent from microfabrication processes are just two obvious examples of such repercussions.

In general, microsystem design involves three major tasks that are mutually coupled: (1) *process flow design,* (2) *electromechanical and structural design,* and (3) *design verifications* that include packaging and testing. Material selections in microsystem design are also much more complex than those involved in traditional products. Selections of materials for microsystems not only involve the materials for the basic structure of the systems, but also the materials in the process flow, such as proper etchants and thin films for depositions.

The intricacy of microsystem design as described above is the main reason for the long design-cycle time that has been a major stumbling block for wider acceptance of MEMS and microsystems in the marketplace. Few engineers have knowledge and experience in this multidisciplinary practice. We will introduce computer-aided design (CAD) specifically developed for MEMS and microsystems later in the chapter. CAD appears to be a viable way to alleviate such problems.

10.2 | DESIGN CONSIDERATIONS

Before we begin with specific design considerations, let us take an overview of the necessary ingredients that we will be involved with in microsystem design as outlined in Figure 10.1. The diagram indicates that once the product is properly defined by a specification, a number of particular items need to be considered. These items include

1. Design constraints
2. Selection of materials

Figure 10.1 | An overview of mechanical design of microsystems.

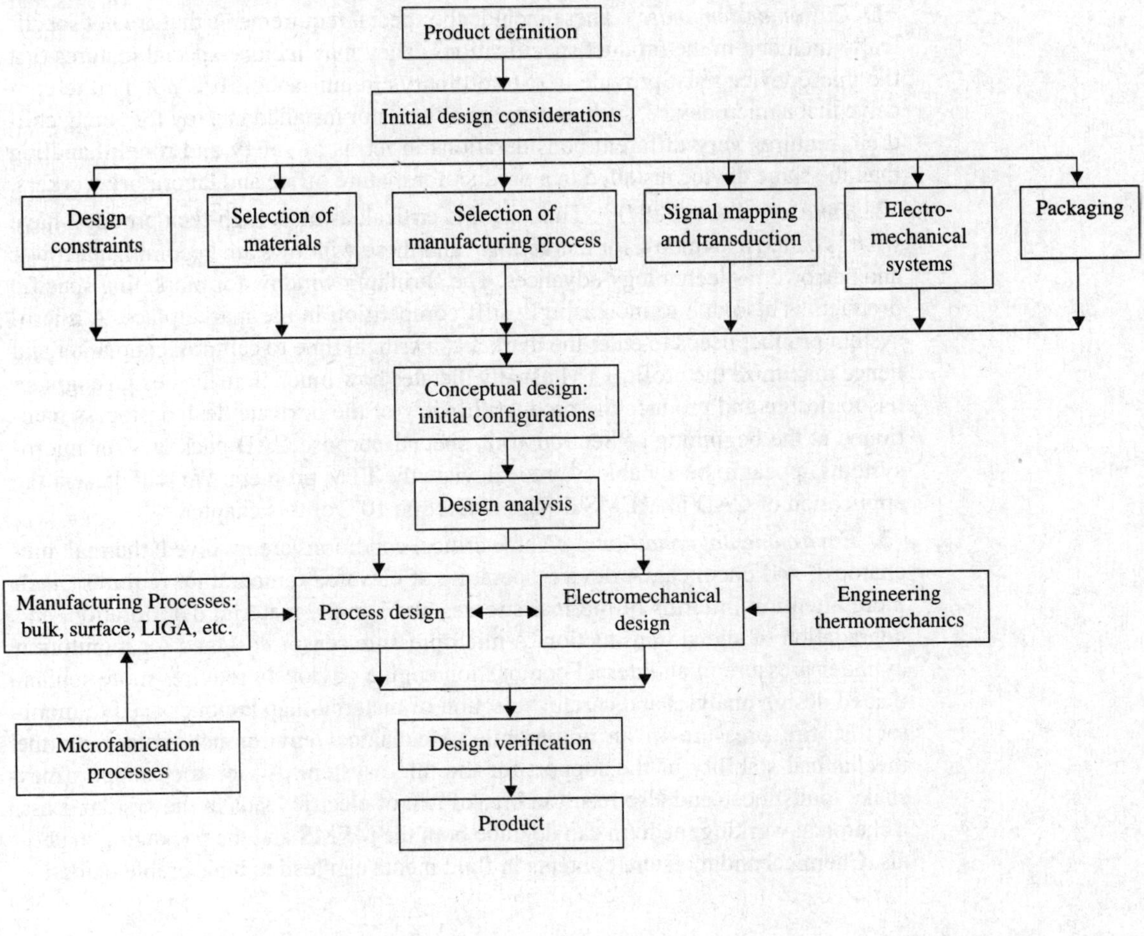

3. Selection of manufacturing process
4. Signal mapping and transduction
5. Electromechanical systems
6. Packaging of the product

After these considerations, engineers will be in a position to develop the initial configuration of the product in terms of geometry, dimensions, materials, and the fabrication and packaging methods. A more detailed and in-depth design of both the manufacturing process and the electromechanical systems follows, to confirm the feasibility of the initial configuration. Design verification on the prototype usually can be carried out by computer simulation to ensure the structure's integrity, and more importantly, the desired functions of the microsystem.

10.2.1 Design Constraints

Design constraints can be many or they can be only a few. They vary from case to case. Many of these constraints are nontechnical, and they may be related to the marketing of the product. Following are a few typical constraints for microsystems.

1. *Customer demands.* These include the special requirements that are not specifically included in the product specifications. They may include special features that the microdevice must provide in extraordinary circumstances. It is not hard to conceive that a microdevice, such as a sensor or actuator installed in a toy for young children, requires very different considerations in terms of safety and rough-handling than the same device installed in a product for mature office and laboratory workers.
2. *Time to market (TTM).* This factor is critical, as most high-tech products have what is called a "window for marketing," and these windows are becoming narrower and narrower as technology advances. The shrinking window for marketing specific products is also due to increasingly stiff competition in the marketplace. A microsystem product needs to enter the market at a critical time to capture that market and hence maximize the profit. TTM usually dictates how much time the design engineer has to design and produce the product. In view of the intricate design process mentioned at the beginning of Section 10.2, special-purpose CAD packages for microsystems appear to be a viable way to alleviate the TTM problem. We will present the application of CAD in MEMS design in Section 10.9 of this chapter.
3. *Environmental conditions.* Three critical conditions are involved: thermal, mechanical, and chemical. A device operating at elevated temperature requires much more attention in terms of thermal stresses and strains, material deterioration, and degradation of signal transduction. A micropressure sensor designed for monitoring cylinder pressure in an internal combustion engine obviously requires more sophisticated design analysis and careful selection of material than the ones used to monitor the tire pressure in an automobile. Mechanical environment relates to the mechanical stability of the support for the microsystem. A vibratory support may shake joints loose and also result in breakdown of electric leads in the system. Last, a chemical working medium can degrade both the MEMS and the packaging materials. Chemical and moisture contents in fluid media can lead to undesirable oxidation

and corrosion of the components in contact. Moisture is also the main reason for the stiction of microswitches in optoelectronic network systems. Clogging of microchannels in microvalves and pumps in microfluidics is a possibility if the system is not properly designed and manufactured.

4. *Physical size and weight limitations.* These constraints are normally covered in the product specification. They can affect the overall configurations of the product with imposition of limitations on some key design parameters.

5. *Applications.* It is important to know whether the microsystem is intended for once-only application, or for repeated usage. If the latter is the case, then one needs to design for the life expectancy of the product, as well as the possibility of creep and fatigue failure of the components.

6. *Fabrication facility.* This relates to the selection of manufacturing methods for the product. The availability of a fabrication facility for the intended product is a critical factor in meeting TTM as well as the cost for the production of the product.

7. *Costs.* This factor can dictate the overall direction of the design. In today's competitive marketplace, cost is a critical factor for the marketability of the product. In this early design stage of the product, engineers should be seriously involved in the cost analysis of the product, which will be translated into constraints on many design parameters such as selection of materials and fabrication methods.

10.2.2 Selection of Materials

We learned in Chapter 7 about the many materials that can be used in microsystems. Following is a summary of the unique characteristics of these materials. Design engineers may use it in selecting appropriate materials for various parts of the microsystem. As *process flow* is an integrated part of the design process, materials such as etchants and thin films for depositions need to be carefully evaluated and selected for the systems design.

Principal Substrate Materials There are two types of substrate materials: (1) *passive substrate* materials for support only. These include polymers, plastics, ceramics, etc., and (2) *active substrate* materials such as silicon, GaAs, and quartz for the sensing or actuating components in a microsystem.

Silicon:

- Mechanically stable, inexpensive and ready machinability.
- An excellent candidate material for microsensors and accelerometers.

GaAs:

- Has fast response to an externally applied influence, such as photons in light rays even at elevated temperature.
- Can be used as thermal insulation.
- Its high piezoelectricity makes this material suitable for precision microactuation.
- Suitable for surface micromachining.

- A good candidate material for optical shutters, choppers, and actuators.
- A desirable material for both microdevices and microcircuits.
- Unfortunately, it is more expensive than other substrate materials such as silicon.

Quartz:

- More mechanically stable than silicon or silicon family compounds even at high temperature.
- Virtually immune to thermal expansion. It is thus the ideal material for high-temperature applications.
- Excellent resonance capability for precision microactuation.
- Unfortunately, it is hard to shape into desirable configurations.

Polymers:

- Used primarily as passive substrate material.
- Low cost in both materials and the production processes.
- Easily formed into the desired shapes.
- Has flexibility in "alloying" for specific purposes.
- Sensitive to environmental conditions such as temperature and moisture.
- Vulnerable to chemical attacks.
- Most polymers age; i.e., they deteriorate with time.

Other Substrates in the Silicon Family

Silicon dioxide (SiO_2):

- Can be easily grown on a silicon substrate surface or by deposition as described in Chapter 8.
- Excellent for both thermal and electrical insulation.
- Can be used as good masking material for wet etching of silicon substrates.

Silicon carbide (SiC):

- Dimensionally and chemically stable even at high temperature.
- Dry etching with aluminum masks can easily pattern it.
- An excellent passivation material for deep etching.

Silicon nitride (Si_3N_4):

- An excellent barrier for water and sodium ions in diffusion processes.
- A good masking material for deep etching and ion implantation.
- An excellent material for optical wave guidance.
- A good protective material for high strength electric insulation at high temperature.

Polycrystalline silicon:

- Widely used as resistors, gates for transistors, and for thin-film transistors.
- A good material for controlling the electrical characteristics of substrates.

Packaging Materials

- Ceramics (alumina, silicon carbide)
- Glasses (Pyrex, quartz)
- Adhesives (solder alloys, epoxy resins, silicone rubbers)
- Wire bonds (gold, silver, aluminum, copper, and tungsten).
- Headers and casings (plastics, aluminum, stainless steel)
- Die protectors (silicone gel, silicone oil)

10.2.3 Selection of Manufacturing Processes

In Chapter 9, we presented three principal manufacturing processes that are available for producing microdevices and systems. Following is a summary of these manufacturing processes and their advantages and disadvantages.

Bulk Micromanufacturing

- Relatively straightforward in operation. It involves well-documented fabrication processes, mainly the etching processes.
- The least expensive of the three manufacturing techniques.
- Suitable for simple geometry; e.g., the dies for micropressure sensors.
- A major drawback is the low aspect ratio. (The ratio of the dimension in depth to that of the plane defines the *aspect ratio* in MEMS industry. The height of microstructures in bulk micromachining is limited by standard silicon wafer thickness.)
- The process involves removal of material from bulk substrates—resulting in high material consumption.

Surface Micromachining

- Requires the building layers of materials over the substrate.
- Requires the design and fabrication of complex masks for deposition and etching in the processes.
- Etching of sacrificial layers is necessary after layer buildings—a wasteful practice.
- More expensive than the bulk manufacturing technique because of its complex fabrication procedures.
- Major advantages are: (1) it is less constrained by the thickness of silicon wafers than that in bulk manufacturing, (2) it provides wide choices of materials to be used in layer buildings, and (3) it is suitable for complex geometries such as microvalves and comb-driven actuators.

LIGA and SLIGA and Other High-Aspect-Ratio Processes

- The most expensive micromanufacturing techniques of all.
- Both the LIGA and SLIGA processes require a special synchrotron radiation facility for deep x-ray lithography. This facility is not readily accessible to most of the MEMS industry.
- These processes also require the development of microinjection molding technology and facilities.

- Major advantages are: (1) they offer great flexibility in aspect ratio of structure geometry. A high aspect ratio of 200 is achievable by the LIGA process. (2) They offer the most flexibility in microstructure configurations and geometry. (3) There is virtually no restriction on the materials for the microstructure including metals by the LIGA process. (4) These are the best of the three manufacturing processes for mass production.

10.2.4 Selection of Signal Transduction

Signal transduction is necessary in both microsensors and actuators. In either case, there is a need to convert chemical, optical, thermal, or mechanical energy such as motion or other physical behavior of MEMS components into an electrical signal, or vice versa. Figure 10.2 illustrates such conversions of signals. We will see the various means for signal transduction available for either type of microdevice from the diagram.

Figure 10.2 | Options for signal transduction in microsystems.

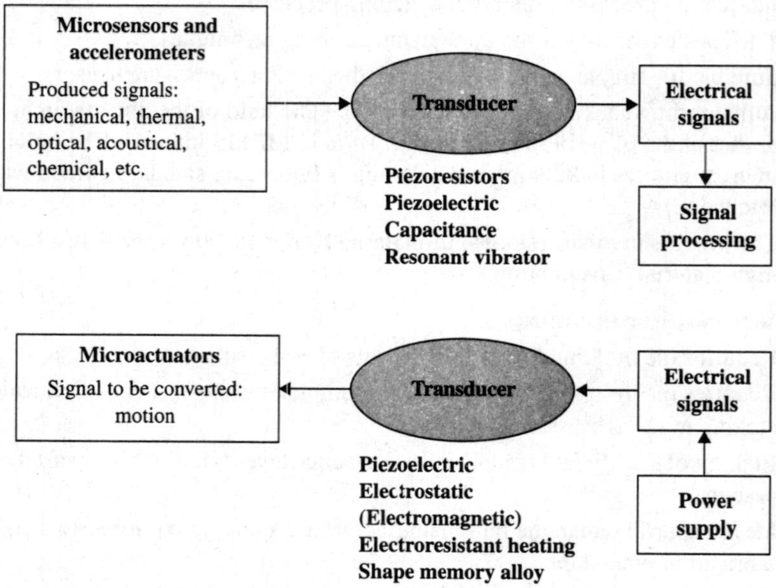

A brief description of each of the signal transduction techniques is presented below. Engineers will select specific techniques that will best serve the purpose for the intended product.

Another critical consideration in the design of the transduction system is signal mapping, which involves a strategy that selects the optimal locations for the transducers and the circuits that transmit the signals. One example is the choice of proper locations for signal transduction for pressure sensors, as illustrated in Figures 2.8 and 7.15. The four piezoresistors for signal transduction appear to be in optimal positions. However, should the pressurized medium be present on the top face of the pressure sensor die, the engineer would be faced with two options:

1. Leave the transduction arrangement as it is in Figure 2.8, but protect the piezoresistors, the wire bond, and the metal circuitry and film pads from direct contact with the pressurized medium.
2. Place the transduction system at the bottom face of the die. In this case, a much more complex process for implanting piezoresistors and the wiring needs to be arranged.

Neither of these two options is attractive from the manufacturing point of view.

Following is a summary of materials and signal transduction techniques that are available for engineers in the design of microsystems.

1. *Piezoresistors.* Silicon piezoresistors are most commonly used in microsensors because of minute size and high sensitivity in signal transduction. Piezoresistors can be produced on substrates other than silicon—in such materials such as GaAs and polymers, for example. Characteristics of silicon piezoresistors were presented in Section 7.6. A major disadvantage of using piezoresistors is the stringent control of the doping process required to achieve good quality, and an even more serious drawback is the strong temperature dependence of resistivity (see Table 7.10). The sensitivity of piezoresistors deteriorates rapidly with increasing temperature. Proper temperature compensation in signal processing is required for applications at elevated temperatures.

2. *Piezoelectric.* Piezoelectric materials are made of crystals as described in Section 7.9. Common piezoelectric crystals are presented in Table 7.14, by which engineers can select the proper crystal for the specific application. For example, the PZT ($PbTi_{1-x}Zr_xO_3$) crystals are used primarily for displacement transducers and for accelerometers. Barium titanate ($BaTiO_3$) is commonly used to transduce signals from microaccelerometers. Quartz crystals, on the other hand, are used for oscillators in ultrasonic transducers.

Most piezoelectric materials are brittle. Special consideration should be given to packaging this material to avoid brittle fracture. Size and machinability are two problems in using piezoelectric materials. Piezoelectrics are suitable for use in accelerometers for measuring dynamic or impact forces that exist in short periods of time, as sustained piezoelectric action will cause overheating of the crystals and thus deteriorate their conversion capability. PZT crystals are the most popular piezoelectric used in industry because of their high piezoelectric coefficient as shown in Table 7.14.

3. *Capacitance.* We learned the working principles of capacitance signal transduction and electrostatic actuation in Chapter 2. While this method is particularly attractive for high-temperature applications, the nonlinear input/output relationships between the change of the gap of the electrodes and the output voltage, as demonstrated in Equation (2.3), require special compensation for errors in interpretation of the output signals. This method also requires larger space in a microsystem, as the output capacitance of a parallel-plate capacitor is directly proportional to the overlapped area of the plate electrodes, as shown in Equation (2.2).

4. *Resonant vibrator.* The working principle of signal transduction using resonant vibration was presented in Section 2.2.5. While this technique can offer higher resolution and accuracy for signal transduction in micropressure sensors, its application to other microdevices is limited by the complexity of fabrication, such as fusion bonding of the silicon beam to the substrate, as in the case of micropressure sensors. The required space for the vibrating members is another drawback of this transduction method.

5. *Electroresistant heating.* This technique is widely used in microactuation, such as in microvalves and pumps in fluidics (see Section 2.3). The technique is simple and straightforward. However, the requirements for precise control of the heating of the actuating element with thermal inertia may affect the timely response of the intended actuation. Also, the contact of the heated element with the working medium can cause local heat transfer, which may in turn alter the flow pattern of the working medium in the case of microvalves. The technique thus has serious drawbacks in liquid-based microfluidic systems. Sealing of heat transmitting fluids often causes problems in the packaging and operations of these microsystems.

6. *Shape memory alloy.* Shape memory alloy (SMA) is a good actuating material when it is used in conjunction with electroresistant heating. A major drawback is the limited availability of SMA, and often the deformation of SMA cannot be accurately predicted because of its sensitivity to temperature.

10.2.5 Electromechanical System

No microsystem can function without electrical power. Electrical circuitry that provides the flow of electric current, or maintains voltage and/or current supply in the case of actuators, is an integrated part of the system. In the case of microsensors, the electronic signals produced by the transducers need to be led to the outside of the device, and be conditioned and processed by a suitable electrical system. Whatever the electrical system for the intended product, a preliminary assessment on the interface between the mechanical actions and the electrical system is needed in order to configure the product. For example, in a micropressure sensor design, the layout of the piezoresistors and the leads for the connecting circuitry are to be deposited on the surface of the silicon die as illustrated in Figure 2.8. These layouts may affect the overall configuration of the die.

10.2.6 Packaging

As mentioned earlier in this book, the cost of packaging a micropressure sensor can be as low as 20 percent of the overall cost of the product with simple plastic encapsulation, or as high as 95 percent of the overall cost with complex passivation and stainless steel or tungsten casings for special-purpose units. The design parameters that affect packaging thus need to be considered at this early stage in the design process. Design engineers need to consider the following major factors that will affect the packaging of the product:

- Die passivation
- Media protection
- System protection
- Electric interconnect
- Electrical interface

- Electromechanical isolation
- Signal conditioning and processing
- Mechanical joints (anodic bonding, TIG welding, adhesion, etc.)
- Processes for tunneling and thin-film lifting
- Strategy and procedures for system assembly
- Product reliability and performance testing

A more detailed description of each of the above items and other packaging design considerations will be presented in Chapter 11.

10.3 | PROCESS DESIGN

Once the design engineer has selected a specific manufacturing technique, whether it is bulk micromanufacturing or surface micromachining or the LIGA process, a search for appropriate microfabrication processes begins. We will classify the various microfabrication processes involved in micromanufacturing in three categories: (1) photolithography, (2) thin-film fabrications, and (3) geometry shaping. The LIGA process includes two additional stages: electroplating and injection molding.

10.3.1 Photolithography

Photolithography is the only viable way to produce micropatterns that depict the three-dimensional structural geometry of microsystems by the current state of the art. It is also used in the process for producing masks (or *mask sets*) for all micromanufacturing techniques that include those masks for etching in bulk manufacturing, for thin-film deposition and for etching in surface micromachining (see, for example, Fig. 9.10), and for micromolds in the LIGA process. Detailed working principles of photolithography were presented in Section 8.2. There are three tasks that need to be carried out in the systems design process: (1) design of patterns for the substrates, (2) design of masks for lithography, and (3) fabrication processes for the mask sets.

Let us look at an example of the silicon die used in a micropressure sensor as illustrated in Figure 10.3. In this illustrative case, the pressure of the medium is applied at the back side, or the cavity side, of the silicon die. The front side of the die has four piezoresistors diffused beneath its surface. The location and orientations of these resistors are shown at the top view (mask for SiO_2) of the die in Figure 10.4.

We can visualize that two masks are required for the silicon die, one for the etching of the cavity and the other for the diffusion of the piezoresistors and the deposition of thin films for electrical conductors connecting these four resistors. As we have learned from Chapter 8, both SiO_2 and Si_3N_4 are suitable candidate materials for this purpose. However, because deep etching is required for the production of many cavities in silicon dies, Si_3N_4 is a more suitable masking material.

As illustrated in Figure 10.4, the patterns for both masks are quite different. The pattern for cavity etching is simply a square opening, whereas the pattern for the doping of piezoresistors and the connecting leads is quite complicated. The white lines for the SiO_2 mask are for the electrical connection of the resistors in a Wheatstone bridge as described in Chapter 2. The two square pads at the lower left corner of the top mask are the leads to the signal conditioning and processing units outside the pressure sensor.

Figure 10.3 | Cross section of a micropressure sensor.

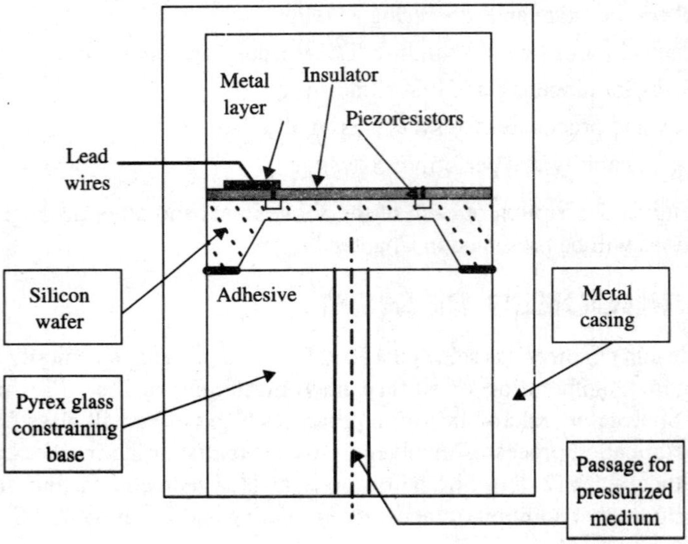

Figure 10.4 | Patterns in masks for a micropressure sensor die

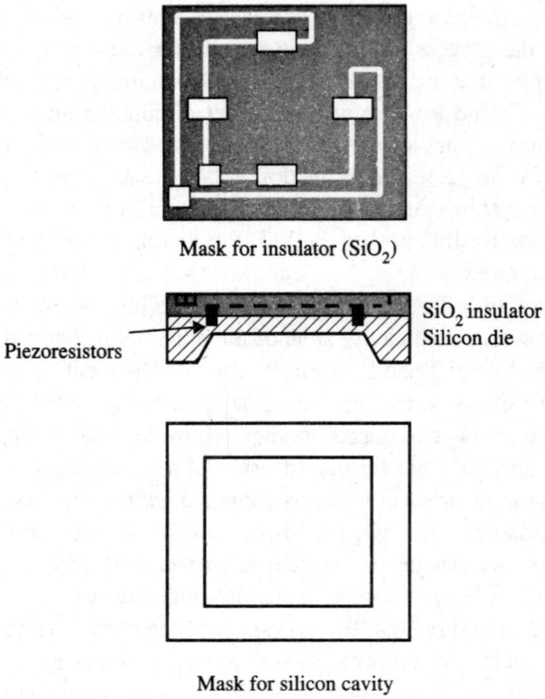

10.3.2 Thin-Film Fabrication

There are several ways to produce thin films over the substrate's surface, as described in Chapter 8. The design engineer may use Table 10.1 for the selection of particular processes for the production of thin films for microsystems. Many of these processes require high-temperature environments, which would likely result in residual stresses and residual strains as described in Section 9.3.3. Every effort should be made to minimize these residual effects.

Table 10.1 | Summary of thin-film production for microsystems

Processes	Principal applications	Building up or building in	High or low temperature	Approx. rate of production
Ion implantation (Section 8.3)	For doping p–n junctions or other impurities	In	Low	Equation (8.1)
Diffusion (Section 8.4)	For doping of p–n junctions or other impurities	In	High	Equation (8.4)
Oxidation (Section 8.5)	For SiO_2 layers using O_2 or steam	In	High	Equations (8.9) and (8.10)
Deposition (Section 8.6)	Physical deposition for metals; chemical deposition (APCVD, LPCVD, PECVD) for SiO_2, Si_3N_4, and polysilicon	Up	Moderate to high	Equation (8.23) for APCVD
Sputtering (Section 8.7)	Thin metal films	Up	High	Madou [1997]
Epitaxial deposition (Section 8.8)	Thin films of the substrate material	Up	High	Table 8.9
Electroplating	Thin metal films over polymer photoresist materials in LIGA or SLIGA manufacturing process	Up	Low	Equation (10.1)

The electroplating process was presented in Section 9.4.3. The rate of deposition for a common metal, i.e., nickel, over photoresist material can be estimated from the following expression [Madou 1997] for 100 percent yield:

$$m = \frac{1}{z}\, iAt\, \frac{1}{F}\, M \tag{10.1}$$

where

A = electrode surface
t = electroplating time
F = Faraday constant (96487 Å/s-mol)
i = current density
z = electrons involved in reaction $Ni^{2+} + 2e^- \rightarrow Ni$
M = atomic weight of Ni

10.3.3 Geometry Shaping

The complex geometry of silicon-based microdevice components can be produced either by depositing thin films of various materials over the substrates as described in Section 10.3.2, or by removing portions of material from the bulk substrates.

An effective process for removing material from substrates is etching, as described in Sections 8.9 and 9.2. Tables 9.1 and 9.2 and the description in Section 9.2.3 for chemical (wet etching) and Section 9.2.5 for plasma-assisted (dry) etching can be used to estimate the rates of etching.

10.4 | MECHANICAL DESIGN

The principal objective of mechanical design is to ensure the structural integrity and the reliability of the microsystem when it is subjected to specified loading at both normal operating and overload conditions. The latter condition relates to possible mishandling or unexpected surges of load due to system malfunction. Design methodologies developed for machines and structures in macro- and mesoscales are used here with provisions to accommodate the necessary modifications according to the scaling laws presented in Chapter 6, as well as the mechanical engineering design principles in Chapters 4 and 5.

10.4.1 Thermomechanical Loading

Most of the loads that a microsensor or actuator is subjected to are common to macrostructures. These can be categorized in the following way:

1. Concentrated forces, such as the contact forces between actuating members and the fluid passages in microvalves as shown in Figures 2.29 to 2.31 and Example 4.6
2. Distributed forces, such as the pressure loads on the diaphragms in micropressure sensors as illustrated in Figure 2.7 and Example 4.4
3. Dynamic or inertia forces, as in the case of microaccelerometers in Figures 2.33 to 2.36 and Example 4.12
4. Thermal stress induced by mismatch of coefficients of thermal expansion in layered structures (see Fig. 2.16 and Examples 4.14 and 4.15)
5. Friction forces between moving components in a microsystem; examples are the bearing of a rotary micromotor (Fig. 2.28), a linear motor (Fig. 2.27), or a micropump

The following forces are unique to microsystems structures:

1. Electrostatic forces for actuation. Evaluation of these forces is available in Section 2.3.4 and Example 2.4.
2. Surface forces due to piezoelectricity. These forces are generated by the mechanical deformation of piezoelectric crystals with the application of

electric voltages. They are used to drive actuators such as those illustrated in Figure 2.19, and also in micropumping as described in Section 5.6.3 and Example 7.5.

3. van der Waals forces that exist between closely spaced surfaces as described in Section 9.3.3. A van der Waals force is a form of electrostatic force but at the molecular level. Accurate estimation of the force is not straightforward as a result of the physical change of atomic cohesion involved in generating these forces.

10.4.2 Thermomechanical Stress Analysis

Stress analysis is a major effort in design analysis. For microsystems that involve either fabrication at high temperature (see Table 10.1) or are expected to operate at elevated temperature, a thermal analysis should precede the stress analysis.

Thermomechanical stress analysis of microsystems can be handled by the formulations provided in Section 4.4.3, or by a finite element method [Hsu 1986]. However, thermomechanical stress analysis of microsystems can significantly differ from that of macro systems. In the case of microsystems, there is a strong presence of intrinsic stresses resulting from the involved microfabrication processes. One of such stresses is the residual stress and strain resulted from these fabrication processes as illustrated in Section 9.3.3. These inherited residual stresses must be evaluated and included in the subsequent stress analysis [Hsu and Sun 1998]. Other possible sources for intrinsic stresses induced in thin films on thick substrates include the following [Madou 1997]:

- Doping of substrates with impurities would cause intrinsic stresses because of lattice mismatch and variation of atomic sizes
- Atomic peening due to ion bombardment by sputtering atoms and working gas densification of the thin film
- Microvoids in the thin film as result of the escape of working gases
- Gas entrapment
- Shrinkage of polymers during cure
- Change of grain boundaries due to change of interatomic spacing during and after deposition or diffusion

An important consideration in using finite element analysis for thermomechanical analysis of MEMS and microsystem structures is that we must ensure that proper finite element formulations are chosen. Many constitutive laws, and thus constitutive equations derived for continua at macroscale, require substantial modifications for structures in submicrometer scale (i.e., structures with dimensions less than 1 μm). One such example is the heat conduction equation given in Equation (5.53), which is significantly different from that for solids of macroscale, Equation (5.38). Properties of most materials at submicrometer scale become size dependent, which invalidates most commercial finite element codes for MEMS and microsystems components.

10.4.3 Dynamic Analysis

Dynamic analysis is conducted for microsystems that involve motion. The primary reason for the analysis is to find (1) the inertia forces that act on the device or the components of the device and (2) the natural frequencies of the moving structures at several modes of vibration.

Inertia forces due to acceleration or deceleration of the components must be accounted for in the stress analysis. The natural frequencies obtained from the modal analysis, on the other hand, will be used to avoid resonant vibration as described in Chapter 4. However, in some microsystem design, resonant vibrations of beams or plates are used to enhance the output signals of sensors, as demonstrated in Section 2.2.5 for certain types of pressure sensors.

In general, excessive vibration can be a major cause for structural failure due to the fatigue of the materials. The finite element method is widely used for dynamic analysis of microsystems.

EXAMPLE 10.1

The following case illustrates the intricacy of the mechanical and electrical coupling aspects in the design of fragile microstructures. The case is not meant to be realistic, as it is unlikely that a simple capacitor with parallel plates is used to generate the necessary actuation forces in a microgripping device, as in the case. This case is used as a comprehensive demonstration of how mechanical behavior of a structure can influence the electrical effects and vice versa.

The microgripper in the example is actuated by a capacitor made of two square plates with an initial gap of 4 μm as illustrated in Figure 10.5. The gripping arms are made of single-crystal silicon. The gripper is designed to pick up a rigid object weighing 3 mg. The capacitor operates in an atmospheric environment, so a value of 1.0 for the relative permittivity is used in the capacitor. Determine the following:

1. The required electric voltage to pick up the weight. Assume a friction coefficient of 1.0 between the gripping points and the object surface.
2. The maximum stress in the gripping arms.
3. The maximum deflection in the gripping arms.
4. The effect of arm deflection on the performance of this microdevice.

Solution

1. To find the necessary voltage required for gripping the weight, we need to first determine the necessary electrostatic force for gripping. This force can be determined by static equilibrium of the forces acting on the gripping arms.

We assume that each of the two arms carries equal weight; i.e., $W/2$ = 1.5 mg. Forces acting on the arm are as shown in the following free-body diagram.

Figure 10.5 | An illustrative microgripper.

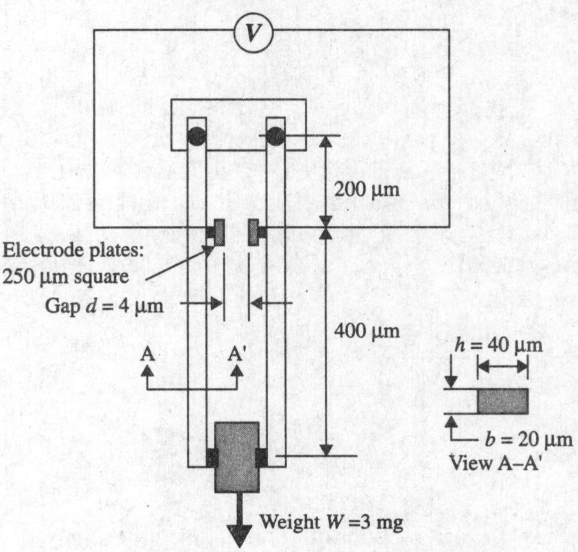

Free-body diagram of forces acting on the gripping arm:

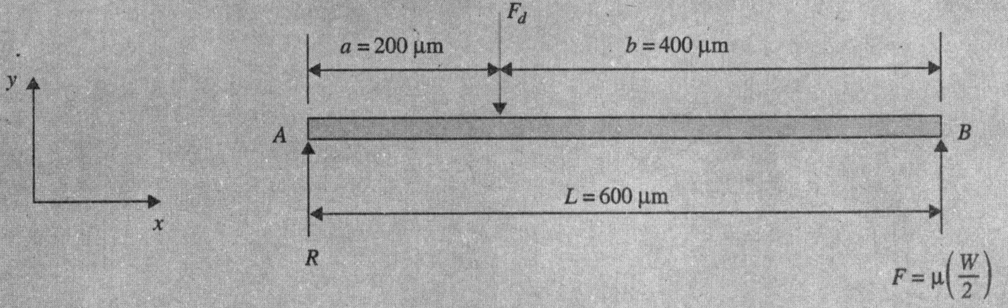

in which

R = reaction at the hinged support
F_d = required electrostatic force generated by the capacitor
F = required gripping force
μ = coefficient of friction (= 1.0)
W = weight of the mass to be picked up (= 3 mg)

From the first static equilibrium condition, i.e., $\Sigma, F_y = 0$:

$$R + F - F_d = 0 \quad \text{(a)}$$

and from the other static equilibrium condition on the moments about A, i.e. $\Sigma M_A = 0$:

$$200F_d - 600F = 0 \tag{b}$$

We solve for $F_d = 3F = 3(W/2) = 1.5W$ from Equations (a) and (b)

Since the design weight to be picked up by the gripper, $W = 3$ mg $= 29.43 \times 10^{-6}$ N, we may compute the required electrostatic force $F_d = 44.15 \times 10^{-6}$ N from the above expression. We have also learned from Section 2.3.4 that the electrostatic force F_d produced by a capacitor with the plate sizes as shown in the figure below can be related to the applied voltage by Equation (2.8):

$$F_d = -\frac{1}{2}\frac{\varepsilon_0 \varepsilon_r WL}{d^2}V^2$$

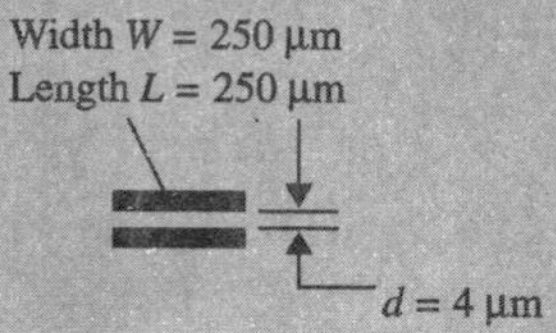

in which W and L are respectively the width and length of the electrode plates, ε_0 = absolute permittivity of free space $= 8.85 \times 10^{-12}$ C/N-m^2, $\varepsilon_r = 1.0$ in air.

By substituting the dimensions of W and L, the given permittivity, and the required $F_d = 44.15 \times 10^{-6}$ N into the above expression, we obtain the required voltage supply to be $V = 50.54$ V.

2. To determine the maximum stress in the gripping arm: The arm is subjected to combined bending and tensile forces as illustrated in the figure below, in which $F = F_w = 0.5W = 14.72 \times 10^{-6}$ N, and from Equation (a), we obtain the reaction $R = F_d - F = 29.43 \times 10^{-6}$ N.

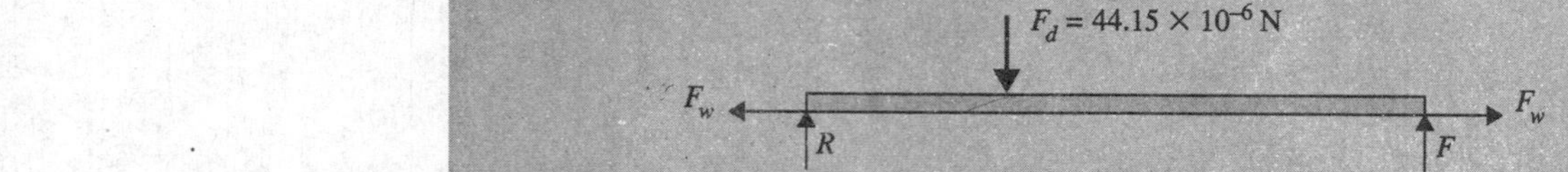

We need to determine the maximum bending moment in the beam for the maximum bending stress. The moment distribution in a simply supported beam subjected to a concentrated force, such as the present case, can be found in many engineering mechanics textbooks. Thus, by referring to the bending moment distribution diagram as shown below, we may come up with the following expressions.

Moment distributions:

$$M_a(x) = \frac{Pbx}{L} \quad \text{for } 0 \le x \le a \tag{c}$$

$$M_b(x) = Pa\left(1 - \frac{x}{L}\right) \quad \text{for } a \le x \le L \tag{d}$$

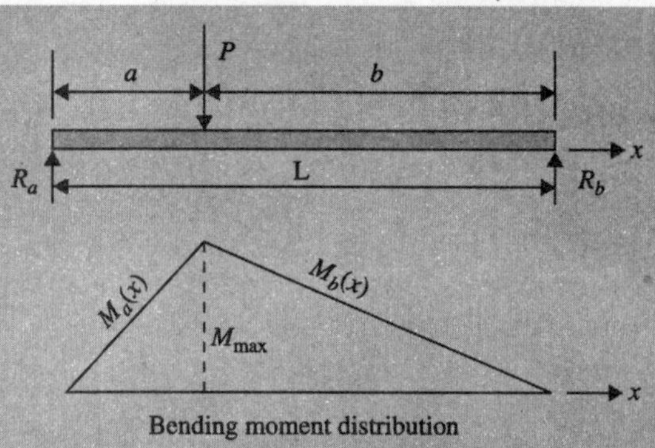

The maximum bending moment in the beam occurs at $x = a$, with $M_{max} = Pba/L$.

Deflections:

$$\delta_a(x) = -\frac{Pbx}{6EIL}(L^2 - b^2 - x^2) \qquad \text{for } 0 \le x \le a \tag{e}$$

$$\delta_b(x) = -\frac{Pbx}{6EIL}(L^2 - b^2 - x^2) - \frac{P(x-a)^3}{6EI} \qquad \text{for } a \le x \le L \tag{f}$$

The maximum deflection of the gripping arm occurs at:

$$x_m = \sqrt{\frac{L^2 - b^2}{3}} \tag{g}$$

The deflection under load P is:

$$\delta_p = -\frac{Pba}{6EIL}(L^2 - a^2 - b^2) \tag{h}$$

We are now ready to calculate the maximum stress in the beam by using the following expression for beams subject to combined bending and tensile loads:

$$\sigma_{max} = \sigma_{m,b} + \sigma_t \tag{i}$$

in which $\sigma_{m,b}$ is the maximum bending stress and σ_t is the tensile stress in the beam.

To determine both stresses, we need to find the cross-sectional area A and the area moment of inertia I of the beam for the cross section of the beam as shown below:

We can thus calculate that

$$A = 8 \times 10^{-10}\ \text{m}^2$$

$$I = 1.0666 \times 10^{-19}\ \text{m}^4$$

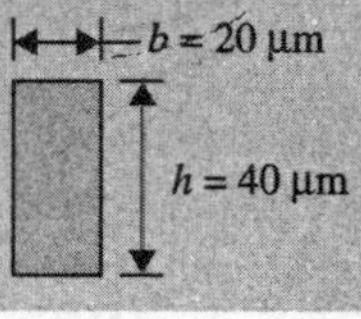

The maximum bending stress is:

$$\sigma_{m,b} = \frac{M_{max}\,C}{I} = \frac{M_{max}\,h}{2I}$$

in which $M_{max} = Pba/L = F_d ba/L = 5.887 \times 10^{-9}$ N-m, and $h = 20 \times 10^{-6}$ m. We have $\sigma_{m,b} = 1{,}103{,}800$ N/m², or 1.1038 MPa.

The tensile stress $= F_w/A = 0.0184 \times 10^6$ N/m² $= 0.0184$ MPa.

Thus the maximum stress in the gripping arm is:

$$\sigma_{max} = \sigma_{m,b} + \sigma_t = 1.1038 + 0.0184 = 1.1222 \text{ MPa}$$

We realize that $\sigma_{max} \ll$ yield strength of silicon (= 7000 MPa as shown in Table 7.3).

3. The maximum deflection in the beam: The maximum deflection of the gripping arm occurs at x_m, given in Equation (g), or $x_m = 258.2 \times 10^{-6}$ m in the present case. The corresponding maximum deflection δ_{max} can be calculated from Equation (f) to be $\delta_{max} = 0.00833$ μm with the Young's modulus of 1.9×10^{11} N/m² used in the computation.
4. As we may realize from Figure 10.5, the effect of the deflection of the gripping arm is to narrow the gap d in the capacitor, which produces the electrostatic force for the gripping. Significant narrowing of the gap will drastically increase the magnitude of F_d, which will in turn further deflect the arm. The situation is clearly a self-accelerating process that can lead to the collapsing of the gripper. It is thus critical to assess the deflection-induced gap change in the capacitor in this particular design case. Consequently, we need to calculate the deflection under the force F_d by using Equation (h). In the present case, the corresponding deflection is computed to be 0.007746 μm, which means a narrowing of the gap d by 0.015 μm, or a 0.375 percent change of the gap. Such a change is negligible for the present case.

   A mechanical stopper, which will prevent the continuous closing of the gap of the capacitor due to the coupled electromechanical effect as described in this design case, may become necessary in order to avoid instability of the gripper structure.

10.4.4 Interfacial Fracture Analysis

Microsystem components often involve inclusions of foreign substances. Many are also made of layers of thin films of a variety of materials. These foreign substance inclusions are either produced in the substrates by diffusion or by ion implantation, whereas thin films are deposited on the surface of the substrate by various deposition techniques. All three micromanufacturing techniques described in Chapter 9 involve multilayer thin films. Interfaces between dissimilar materials are thus common in microsystem structures.

Multilayer structures made of dissimilar materials present serious mechanical problems. In addition to excessive thermal stresses and strains due to mismatch of CTE, they are vulnerable to interfacial fracture. Delamination of interfaces is a principal concern in the mechanical design of microstructures.

In Section 4.5.3, we learned how to analyze the mechanical strength of an interface of two dissimilar materials by computing the coupled stress intensity factors relating to the opening and shearing modes of fracture of the interfaces. A major problem, however, is the lack of available values of the corresponding fracture toughness, which is required for determining whether the computed stress intensity factors are below the safe limits set by the fracture toughness as illustrated in Figure 4.47.

10.5 | MECHANICAL DESIGN USING FINITE ELEMENT METHOD

We have demonstrated that theories and principles of continuum mechanics and heat transfer as presented in Chapters 4 and 5 can be used in the mechanical design analysis of many MEMS and microsystem components. However, there are times when more sophisticated analytical tools are needed to handle microsystems involving components with complex geometry and loading/boundary conditions. Also, as we have learned from Chapters 8 and 9, many of these design analyses require the inclusion of the thermomechanistic effects induced to the finished components by microfabrication and manufacturing processes. In such cases, finite element analysis becomes the only viable tool that is available to design engineers.

An overview of the working principle of finite element analysis was presented in Section 4.7. What will be covered in the following are the basic formulation of this popular and effective method, and how this method can be used to handle mechanical design of microsystems. Detailed derivation of the finite element method is available in several excellent reference books, for example in Zienkiewicz [1971], Segerlind [1976], Bathe and Wilson [1976], Hsu [1986], and Heinrich and Pepper [1999].

10.5.1 Finite Element Formulation

Discretization This process begins with subdividing a continuum in Figure 10.6a into an assembly of finite number of elements interconnected at the nodes as illustrated in Figure 10.6b. We realize that, after such discretization, the original curved boundaries are approximated by straight edges of the elements along the boundary. Once satisfied with the discretized model, the user will relate the *primary unknown quantities* in the elements to those at the associated nodes. Primary unknown quantities are the unknown quantities that are used as the basic quantities in the finite element formulations. The computed values of these unknown quantities will lead to the values of other unknown quantities required in the analyses. For instance, the primary unknown quantities in a stress analysis are the displacement components in the elements. The element displacements can be used to compute the element strains and stresses by using the constitutive equations. The primary unknown quantities vary from application to application. For instance, as shown in Figure 10.6a, $\{\mathbf{U}(\mathbf{r}, t)\}$ designates the element displacement components in a stress analysis, element temperatures, $T(\mathbf{r}, t)$, are for a heat conduction analysis and the element velocity components, $\{\mathbf{V}(\mathbf{r}, t)\}$, are for a fluid dynamic analysis. In these mathematical designations, we represent the spatial coordinates with a position

Figure 10.6 | Discretization of continua for finite element analyses.

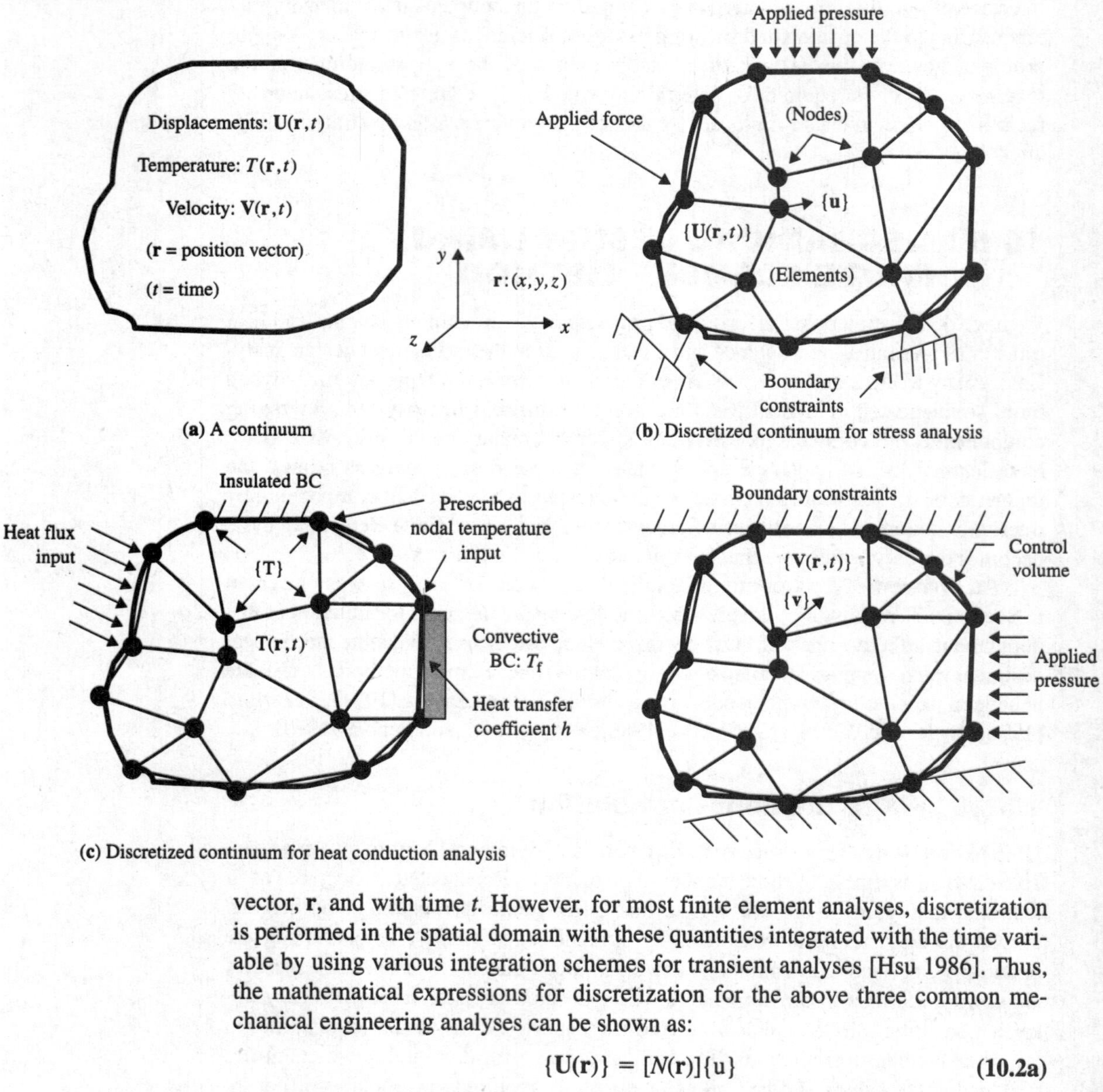

vector, **r**, and with time t. However, for most finite element analyses, discretization is performed in the spatial domain with these quantities integrated with the time variable by using various integration schemes for transient analyses [Hsu 1986]. Thus, the mathematical expressions for discretization for the above three common mechanical engineering analyses can be shown as:

$$\{\mathbf{U}(\mathbf{r})\} = [N(\mathbf{r})]\{\mathbf{u}\} \tag{10.2a}$$

for stress analysis, in which $[N(\mathbf{r})]$ is the *interpolation function,* with **r** denoting the position vector that defines the elements. This position vector represents the spatial coordinates, e.g., (x, y, z) for cartesian coordinates, or the cylindrical polar coordinates (r, z) for an axisymmetric solid. The vector $\{u\}$ represents the corresponding nodal displacement components.

Likewise, as shown in Figure 10.6c, the discretization for the heat conduction and fluid dynamics analyses can be expressed as:

$$\mathrm{T}(\mathbf{r}) = [N(\mathbf{r})]\{\mathrm{T}\} \qquad \textbf{(10.2b)}$$

for heat conduction analysis, with {T} the corresponding nodal temperature, and

$$\mathbf{V}(\mathbf{r}) = [N(\mathbf{r})]\{\mathrm{v}\} \qquad \textbf{(10.2c)}$$

for fluid dynamic analysis with {v} as the corresponding velocity components at the nodes as illustrated in Figure 10.6d.

The values of the primary unknown quantities at the nodes of an element can be related to the applied nodal loads by the following *element equations*:

$$[K_e]\{q\} = \{Q_e\} \qquad \textbf{(10.3)}$$

where

$[K_e]$ = element coefficient matrix
$\{q\}$ = primary unknown quantities at the nodes
$\{Q_e\}$ = applied loads at the nodes

The element coefficient matrix contains constants that are related to the shape of the elements—the interpolation functions used in the discretization, and the material properties of the elements. Material properties used in the finite element analyses include the elastic constants and Poisson's ratio in a stress analysis, or thermal conductivity in a heat conduction analysis.

The applied nodal loads in Equation (10.3) are the applied forces at the nodes, and/or the equivalent nodal forces of the applied pressures on the surface of the elements in a stress analysis (Fig. 10.6b). They can also be the heat flux and/or prescribed nodal temperature in a heat conduction analysis (Fig. 10.6c). The convective boundary condition in Figure 6.10c can be handled by the relationship in Equation (5.45), with a conversion to equivalent thermal forces at the adjacent nodes of the corresponding elements along the solid/fluid boundary [Hsu 1986]. In fluid dynamic analysis, the applied nodal load can be the pressure at the surface of a control volume (Fig. 10.6d).

The primary unknown quantities at all nodes in a discretized continuum $\{q\}$ can be obtained from the overall equilibrium equations as shown below:

$$[K]\{q\} = \{Q\} \qquad \textbf{(10.4)}$$

in which the overall coefficient matrix $[K]$ is constructed by summing up all element coefficient matrices $[K_e]$ in the discretized model:

$$[K] = \sum_{1}^{M} [K_e] \qquad \textbf{(10.5a)}$$

and the overall load matrix $\{Q\}$ is obtained by:

$$\{Q\} = \sum_{1}^{n} \{Q_e\} \qquad \textbf{(10.5b)}$$

where M = total number of elements and n = total number of nodes in the discretized model.

The primary unknown quantities $\{q\}$ at all nodes in the discretized model can be obtained by solving the simultaneous algebraic equations given in Equation (10.4) by using such developed methods as gaussian elimination technique with back substitutions [Allaire 1985].

Derivation of Element Equations It is apparent that the element equation in Equation (10.3) is the key to a finite element analysis. This equation can be derived in several ways, depending on the nature of the analysis. For problems that can be described by differential equations, such as the heat conduction equations in Equation (5.38) or the Navier–Stokes equation in Equation (5.20) for fluid dynamics analysis, the *Galerkin method* is used for such derivations. On the other hand, for problems that cannot be described by distinct differential equations, such as the case of stress analysis, the *Rayleigh-Ritz method* is the viable way to derive the element coefficient matrix. Derivation of element equations using both these methods is available in Hsu [1986]. We will outline these methods here with suggestions as to how they can be modified to account for the submicrometer scale structures.

The Galerkin Method This method is often referred to as the *weighted residuals* method. The principle of this method is based on the fact that a continuum is no longer a continuum after the discretization illustrated in Figure 10.6. The primary quantities obtained from a discretized model in a finite element analysis, in fact, give the approximated values of those in the original continuum.

Consequently, let us consider the "exact" solutions to physical problems to be those for problems that can be described mathematically in the form of differential equations with proper boundary conditions as:

$$D(\phi) = 0 \qquad \text{in domain } V \tag{10.6a}$$

and

$$B(\phi) = 0 \qquad \text{on boundary S} \tag{10.6b}$$

where D and B are the respective differential operators for the differential equation and the boundary conditions. The function ϕ represents the primary unknown quantities as indicated in Figure 10.6a for the three common types of finite element analyses in microsystem design.

Equation (10.6a) represents the differential equations for field problems such as Equation (5.20) for fluid dynamics and Equation (5.38) for the temperature field in a solid. The boundary conditions in Equation (10.6b) represent such expressions in Equations (5.43) to (5.45) for heat conduction analysis.

The above system, with ϕ being the primary unknown quantities in Equation (10.6), can be replaced by an integral equation:

$$\int_V WD(\phi)\, dv + \int_S \overline{W} B(\phi)\, dS = 0 \tag{10.7}$$

in which W and $\overline{W}$ are arbitrary weighting functions.

Equation (10.7) represents the "exact" solution for the continuum represented by Equation (10.6a, b). A similar equation for the discretized continuum of Equation (10.2b, c) with the generic expression $\phi = \Sigma N_i\phi_i$, in which N_i and ϕ_i represent respectively the interpolation function and the primary unknown quantities at the nodes, will lead to the following solution:

$$\int_v W_j D(\Sigma N_i\phi_i)\, dv + \int_S \overline{W}_j B(\Sigma N_i\phi_i)\, dS = R \tag{10.8}$$

where W_j and $\overline{W}_j$ are the respective weighting functions for the discretized continuum and its boundary conditions.

We notice that the substitution of the unknown quantities ϕ in Equation (10.7) with the discretized values $\Sigma N_i\phi_i$, in which i = total number of nodal components in the element, leads to a *residual* value R as shown in Equation (10.8). The existence of the residual value is due to the approximate nature of the solution offered by the discretized continuum. In a physical sense, the residual represents the difference between the exact solution to the problem for a continuum and the approximate solution obtained from the same continuum but with discretization.

A good discretized system, of course, requires the condition that $R \to 0$. By choosing the weighting functions that have identical form as the interpolation function, i.e., by letting $W_j = \overline{W}_j = N_i$, one may arrive at the element equation as presented in Equation (10.3) after performing a variational process on the residual with respect to the variable ϕ_i in Equation (10.8).

For structures that are in the submicrometer size, the differential equations derived for continua in macroscale, such as the heat conduction equation in Equation (5.38), need to be replaced by those for continua in submicrometer and nanoscales in Equation (5.53). Likewise, the Navier–Stokes equation in Equation (5.20) needs to be substituted by the Boltzmann equation [Beskok and Karniadakis 1999]. In the case of heat conduction analysis, the thermal conductivity must also be modified for submicrometer scale. The expression in Equation (5.52) needs to be used for this purpose.

The Rayleigh–Ritz Method This method involves the variation of a *functional* for deriving the element equations in a finite element analysis. It is a viable alternative to the Galerkin method, and is specifically suitable for problems that cannot be described by distinct differential equations, such as the stress analysis of solid structures. A functional is defined as a function of functions.

For stress analysis of solid structures, the functional used in the variation is the *potential energy* $P(\{u\})$, of the deformed structure, in which $\{u\}$ denotes the unknown nodal displacement vector as shown in Figure 10.6b. The derivation of the potential energy for elastically deformed solids is available in many finite element books. Potential energy for elastic-plastically deformed solids is presented in Hsu [1986]. The following relation is used to derive this energy function:

$$P(\{u\}) = U(\{u\}) - W_k(\{u\}) \tag{10.9}$$

where $U(\{u\})$ is the strain energy and $W_k(\{u\})$ is the work done to the solid by the applied body forces and surface tractions.

An element equation in the form of Equation (10.3) is derived by applying the variational principle to the potential energy in Equation (10.9) with respect to the nodal displacements {u}. Consequently, the following expression is obtained:

$$[K_e]\{u\} = \{F\} \tag{10.10}$$

where $[K_e]$ and $\{F\}$ are respectively the element stiffness and nodal force matrices.

The element stiffness matrix can be determined from the following integral:

$$[K_e] = \int_v [B(\vec{r})]^T [C][B(\vec{r})]\, dv \tag{10.11}$$

and the nodal force matrix can be obtained by the following integrals:

$$\{F\} = \int_v [N(\vec{r})]^T \{f\}\, dv + \int_s [N(\vec{r})]^T \{t\}\, ds \tag{10.12}$$

where

v = volume of the element
s = surface of the element
$\{f\}$ = body force acting on the element
$\{t\}$ = surface traction acting on the element surface, s

The matrix $[B(\vec{r})]$ in Equation (10.11) relates the element strain components $\{\varepsilon\}$ to the nodal displacements {u}. The matrix $[C]$ is the elasticity matrix involving Young's modulus and Poisson's ratio of the material. For elastic–plastic stress analysis, an elastic–plasticity matrix $[C_{ep}]$, derived from the incremental plasticity theory, replaces the $[C]$ matrix in Equation (10.11) [Hsu 1986].

Once the displacements {u} are determined at the nodes, one may find the corresponding element displacements by using Equation (10.2a), and the element strains and stresses through appropriate formulations in the theory of elasticity and the constitutive relations, such as the generalized Hooke's law for elastic stress analysis.

As in the thermofluids analyses, the finite element analysis for deformable structures in submicrometer scale requires significant modification to accommodate the geometry and size-dependent anisotropic material properties and the distinct constitutive laws and equations that are required in the finite element formulation. Much research effort is needed in these areas.

10.5.2 Simulation of Microfabrication Processes

We have mentioned in several places in this book the inherent effects of microfabrication on the mechanical behavior of MEMS and microsystem structures. Integration and simulation of microfabrication processes in finite element analyses have thus become a critical requirement. A finite element algorithm was proposed to simulate micromanufacturing of microsystems [Hsu 1998]. The algorithm that uses pseudoelements in the discretization of the continua is involved in the analysis. This simulation technique requires the finite element (FE) mesh in the microstructure to include two parts: (1) the part with real elements for the substrate material and (2) the part with *pseudoelements* for simulating the materials being added or removed in microfabrication processes. In these micromanufacturing processes, materials are added to substrates by "deposition," "electroplating," or "epitaxial growth" of substrate crystals in either surface micromachining or in the LIGA process, whereas portions of the subtracted materials are removed by "etching" in bulk micromachining. By varying the assigned material properties to the pseudoelements before and after microfabrication, one can simulate the three micromanufacturing processes described in Chapter 9. The working principle of this simulation technique is described below.

Simulation of Surface Micromachining and the LIGA Process Surface micromachining is an additive process, and the pseudoelements used in these processes are referred to as the *add elements*. Following is the procedure of the simulation:

Step 1: The add elements are included in the initial FE mesh. The thickness of each add element is made to equal the thickness of the layers of additive materials to be built on the substrate.

Step 2: Initial material properties assigned to the add elements have the following material characteristics: low Young's modulus, $E = 1$ Pa; high yield strength, $\sigma_y = 10^6$ MPa; the same coefficient of thermal expansion, $\alpha_{\text{substrate}}$; and low mass density $\rho = 10^{-6}$ g/m^3.

Step 3: Real material properties are assigned to the add elements after one pass of "deposition" or "electroplating" or "epitaxial growth" is completed.

Step 4: Compute the stress distribution in the overall FE mesh with the current set of add elements, using the real layer material properties, while the remaining add elements still have pseudoproperties. In this case, the element stiffness matrix is calculated according to Equation (10.11).

Step 5: Update stresses, strains, and the nodal coordinates in the FE mesh.

Step 6: Update the stiffness matrix $[K]$.

Step 7: Repeat the computation for the next surface "layer" in the fabrication process.

Simulation of Bulk Micromachining Bulk fabrication is a subtractive process with removal of substrate material at desired localities. The pseudoelements used in this process are called *subtract elements.*

Step 1: Subtract elements are included in the FE mesh for the parts that are to be etched.

Step 2: Initial properties assigned to the subtract elements are identical to those of the substrate. The etching front in subtract elements is identified according to etching rates used in the process.

Step 3: The $[B(\vec{r})]$ matrix in Equation (10.11) is adjusted according to the current nodal position coinciding with the etching front with temporary assigned nodal coordinates. The reduced element stiffness is computed according to Equation (10.11) as follows:

$$[K_e] = \int_v [B(\vec{r})]^T [C][B(\vec{r})]\, dv$$

in which $\vec{r}$ represents the position vectors in the $[B]$ matrix involving temporary nodal coordinates.

Step 4: The entire subtract element is switched to a pseudoelement with pseudomaterial properties once the etching front passes from one end to the other in the subtract element. Pseudomaterial properties in this case are: low element stiffness matrix, $[K_e] = 10^{-6}$; low Young's modulus, $E = 1$ Pa; high yield strength, $\sigma_y = 10^6$ MPa; the same coefficient of thermal expansion, $\alpha_{substrate}$; and low mass density, $\rho = 10^{-6}$ g/m^3.
Step 5: Repeat the same procedure for the subsequent etching process.

Hsu [1986] developed an algorithm based on a similar principle to that presented above for the simulation of crack growth in solids [Kim and Hsu 1982]. A popular commercial FE code, ANSYS was used to simulate the etching process involving a silicon substrate with an SiO_2 layer [Hsu and Sun 1998]. The "birth" and "death" elements in that code, which presumably were constructed using principles similar to the proposed add and subtract elements, were used for the simulation of oxidation of silicon substrates. The residual stresses and strains obtained from the simulation were accounted for as the initial stresses and strains in the subsequent stress and strain analysis of a micro pressure sensor die/diaphragm subjected to specified operating loading conditions.

10.6 | DESIGN OF A SILICON DIE FOR A MICROPRESSURE SENSOR

We will demonstrate the mechanical design of microdevices by a case study relating to the design of silicon dies used in pressure sensors. Silicon dies are key components in micropressure sensors. Proper mechanical design is necessary to ensure the proper functioning of these sensors.

General Description The working principles of various types of micropressure sensors were described in Chapter 2. Here, we will take a closer look at how mechanical design of one type of pressure sensor is carried out. The cross section of a typical square silicon die in such application is illustrated in Figure 10.7. The die is made of a square silicon substrate with a cavity produced by etching from one side of the substrate. The thinned portion of the die is the diaphragm, which deflects when a medium applies pressure at the back face (the bottom face) of the die. Associated with the induced deflection are bending and shearing stresses in the diaphragm.

Figure 10.7 | A silicon die in a typical micropressure sensor.

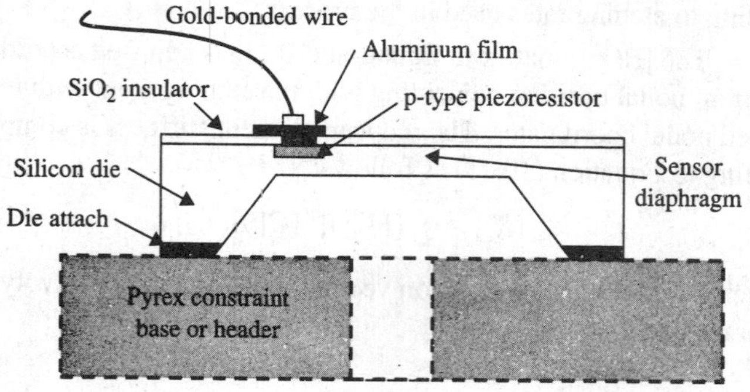

Piezoresistors are planted in the diaphragm to sense the induced stresses. The die is attached to a constraint base or a header if it is made of metal, with die attachment by such means as soldering or epoxy adherents.

Design Considerations There are a number of factors that a design engineer needs to be aware of before embarking on the actual design analysis of the die. These factors may include the following:

1. *Applications.* Different applications will mean different users groups. A sensor to be used in automobiles and machine tools will be required to survive in a tougher environment than those used in high precision instruments in laboratories.
2. *Working medium.* The pressure side of the die is normally in contact with the pressurizing medium. These contacts can be tolerated if the medium is inert and harmless to the die material. Unfortunately, many of these media are hostile and toxic, such as the exhaust gas of an automobile. In such cases, *die passivation* becomes necessary and the mechanical design of the die must provide such protection.
3. *Constraint base.* Sensors are normally attached to a base or a header. A vibrating base can cause the die attach to fail in fatigue. The natural frequency of the die should be designed to be much higher than expected excitation frequencies during operation. The lowest natural frequency for the die of a pressure sensor is 100 Hz to 2 KHz for automobile applications.
4. *Other considerations.* Considerations should be given to unexpected conditions such as mechanical shock due to accidental dropping the sensor from mishandling. Thermal shock may be another unexpected condition that should be considered in the design.

Geometry and Size of the Die Most pressure sensor dies are square in geometry. Rectangular and circular dies were used in early applications but they rarely appear in recent products.

Careful consideration should be given to the size of the die, as it will affect not only the cost of the sensor, but also the sensitivity in the performance. Figure 10.8 indicates the key dimensions of a square silicon die.

Figure 10.8 | Key dimensions of a square die for a pressure sensor.

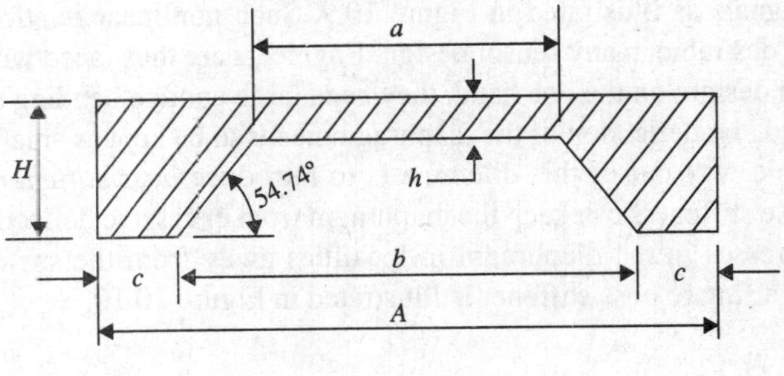

In this figure, the overall size of the die is ($A \times A \times H$). The thickness of the die, H, is limited by the standard thickness of silicon wafers, for example the thickness of 100-mm diameter wafers is 500 μm, and the thickness of other size wafers is available in Section 7.4.2. The height of the die needs to be properly established, as it will provide the necessary isolation for the die from mechanical effects induced by the constraint base. The diaphragm has the size of (a $\times$ a $\times$ h), in which the length a is determined by the dimension b and the etch angle, 54.74°. This angle exists if wet etching is applied to the (100) plane of the substrate. The dimension b, on the other hand, depends on the size of the die "foot print," c. The latter must be large enough so that the stresses in the die attach (see Fig. 10.7) can be kept below plastic yielding strength. It is thus clear that all dimensions of the die are mutually dependent.

Another factor that affects the size of the die is the size of the signal transducers. In general, two types of transducers are used for pressure sensors: piezoresistors and capacitors. In either case, adequate space on the die surface must be provided for these transducers. For best results, these transducers must transmit signals from localized points, which means that either the die is made large enough that the signals can be treated as localized, or the transducers are made as small as possible. The die size is thus also constrained by the transducer size.

One factor that determines the cost of the die, of course, is its overall size. It is desirable to keep the die as small as physically possible for low material cost and also for conservation of space.

Strength of the Die We have shown that the maximum bending stress and deflection of the diaphragm can be estimated from Equations (4.10) and (4.11). These equations are expressed in slightly different forms below:

$$\sigma_{max} = C_1 P\left(\frac{a}{h}\right)^2 \qquad \textbf{(10.13)}$$

$$w_{max} = C_2 P\left(\frac{a^4}{h^3}\right) = C_2 Pa\left(\frac{a}{h}\right)^3 \qquad \textbf{(10.14)}$$

where C_1 and C_2 are constant coefficients and P is the applied pressure.

From the point of view of sensor performance, it is desirable to generate sufficient σ_{max} in the diaphragm for significant signals from the piezoresistors. This would mean thin diaphragms, i.e., low in the h dimension in Equation (10.13). However, a low h dimension will increase the maximum deflection w_{max} much faster than σ_{max} as indicated in Equation (10.14). Large deflections of the diaphragm will result in a nonlinear relationship between the applied pressure and the bending stress and hence the output signals as illustrated in Figure 10.9. Such nonlinear relationships obviously are not desirable in any sensor design. Engineers are thus faced with a dilemma in diaphragm design; on the one hand, they need large enough bending stress but, on the other hand, the deflection of the diaphragm needs to be kept as small as possible.

A possible way out of this dilemma is to introduce *boss stiffeners* in the diaphragm. These stiffeners can keep the diaphragm from excessive deflection but allow maximum stresses in the diaphragm in localities away from the stiffeners. A diaphragm with a square boss stiffener is illustrated in Figure 10.10.

Figure 10.9 | Linearity of output signal in a pressure sensor.

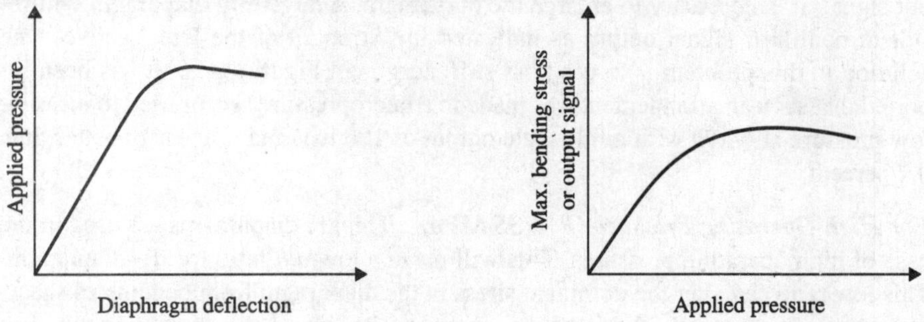

(a) Force–deflection (b) Pressure–output signal

Figure 10.10 | A silicon die with a square boss stiffener.

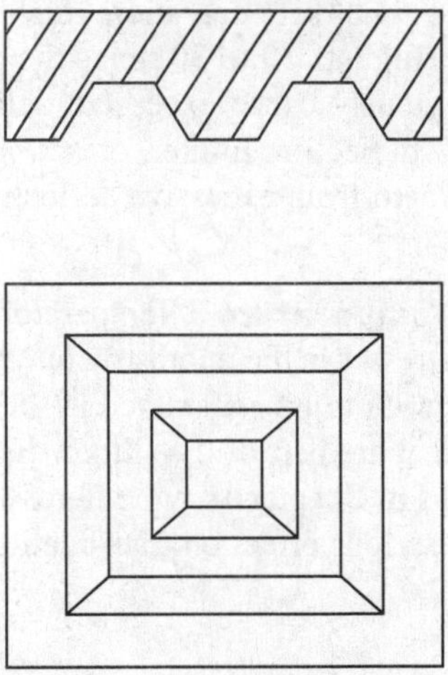

Design for Operating Pressures Depending on the application, micropressure sensors are designed to operate from very low pressure to very high pressure. In the automotive industry, micropressure sensors are typically operated at 10 MPa for gasoline engine cylinders, 100 MPa for diesel engine cylinders, and 20 kPa for brake fluids and hydraulic suspensions.

Flat diaphragms are best for pressure sensors operating in the medium pressure range, i.e., between 35 kPa and 3.5 MPa. A linear relationship between the piezoresistor output and the applied pressure (Fig. 10.9b) exists. However, this desirable situation does not always exist for applied pressures beyond this range.

For Low Operating Pressure ($P < 30$ kPa) The low applied pressure implies the need for a thin diaphragm [i.e., low h in Eqs. (10.13) and (10.14)]. A thin diaphragm also means a small diaphragm size (the dimension a in Fig. 10.8), which leads to low bending stress and hence low signal output. Thus, in order to generate significant output signal, it is necessary to enlarge the diaphragm. A large thin diaphragm will result in nonlinear signal output as indicated in "Strength of the Die," above. One solution to this problem is to use boss stiffeners as in Figure 10.10. It has been reported that such an arrangement was made in a micropressure sensor used to measure low pressure at 7 kPa with a full-scale output of 100 mV and a linearity better than 0.1 percent.

For High Operating Pressure ($P > 35$ MPa) Thicker diaphragms are used in the case of high operating pressures. This will mean a low a/h ratio for the diaphragm. This low ratio can alter the dominant stress in the diaphragm from bending to shearing stress. Consequently the resulting piezoresistive signal versus applied pressure exhibits a nonlinear relationship. One solution to this problem is to carefully locate the piezoresistors in the region where the shearing stress is minimal. Using local boss stiffeners can control the distribution of bending and shearing stresses in the diaphragm. Finite element analyses such as demonstrated in Figure 4.48 can aid such design.

Design for Overpressure For safe operation, micropressure sensors are often designed for overpressure as high as 10 to 30 times the design operating pressure. Such high pressure will obviously cause excessive deflection of the diaphragm. To avoid such occurrence, a deflection limiter, or a *stopper,* must be provided in the die to prevent the diaphragm from excessive deflection under the overpressure conditions.

Design for Operating Temperature Temperature will affect the material strength and induce thermal stresses in the microstructure as described in Section 4.4. Another critical factor that needs to be considered in the design is the temperature effect on the piezoresistors that are part of the silicon die. We will leave the design of signal transduction systems to Chapter 11 on "Microsystem Packaging." The fact that temperature can have a serious effect on transduction, however, should not be overlooked.

10.7 | DESIGN OF MICROFLUIDIC NETWORK SYSTEMS

We have mentioned in several places in this book that bioMEMS is a rapidly growing field of research, with new applications reported frequently by the industry. Microfluidic devices have become principal components in many of these new bioMEMS. In particular, the *capillary electrophoresis (CE) on-a-chip* has gained increasing popularity in the pharmaceutical, genetic, and human genome industries in recent years. CE on-a-chip is a microfluidic system that offers a means of miniaturizing chemical and biomedical analysis systems and achieve the sensing of the constituents in minute samples with almost instant results. As the complete analysis systems are integrated with signal processing in a single chip, they can be produced in batches in large quantities, thereby drastically reducing the production cost. Furthermore, these systems are disposable after use, which eliminates all maintenance costs. Most of these systems are so minute that assemblies with one thousand channels for parallel CE analyses on a single chip are made possible by micromanufacturing technologies. These arrangements have made CE on-a-chip the most effective and efficient analytical tool for the aforementioned purposes.

CE involves the combined applications of electro-osmotic and electrophoretic processes as described in Chapter 3. Both these processes involve fluid movements under the influence of applied electric fields. The electro-osmotic process moves an aqueous solution past a stationary solid surface. It requires the creation of a charged double layer at the solid–liquid interface. The resultant velocity profile across the passage cross section is nearly flat. The electrophoretic process, on the other hand, does not move bulk fluid. Instead, it moves charged species in the buffer solution under the applied electric field.

Typical CEs are made of a network of capillary tubes or channels. Common cross sections of these capillaries are illustrated in Figure 10.11. Typical dimensions of these capillaries are less than 100 μm in diameter for the tubes and 30 to 50 μm wide by 10 to 30 μm deep for the channels.

Figure 10.11 | Typical cross sections of capillaries.

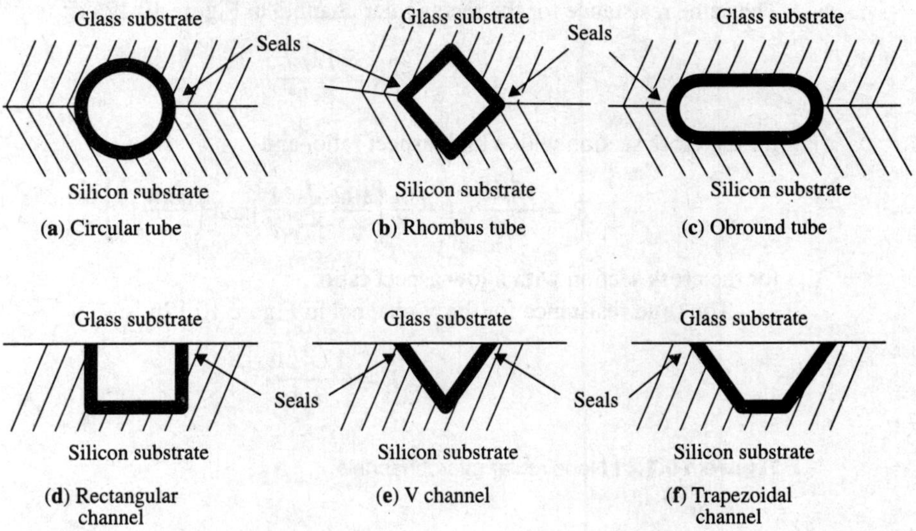

The length of the tubes or channels may vary from a few millimeters to a few centimeters. Tubes and channels are typically fabricated by etching, and epoxy seals are used for closing the conduits for the passage of fluids, as shown in Figure 10.11. The closed capillaries shown in Figure 10.11a, b, and c are produced in halves by microfabrication techniques, and the two halves are then bonded together by various bonding techniques as will be described in Chapter 11. The open channels in Figure 10.11d, e, and f are bonded to glass substrates and sealed at the joints by epoxy [Harrison and Glavina 1993]. These tubes and channels must be electrically insulated from the substrates in order to avoid voltage breakdown in the electrophoretic and electro-osmotic processes during the analysis.

10.7.1 Fluid Resistance in Microchannels

We realize that cross sections of capillary tubes and channels may vary, as shown in Figure 10.11. Channels with noncircular cross sections are used because they can easily be fabricated by microfabrication techniques such as etching. These noncircular cross sections require special formulas for estimating the flow resistance in these channels.

Fluid resistance R is defined as the ratio of pressure drop ΔP to the volumetric flow rate of the medium Q along a specific length of the channel. Mathematically, it can be expressed as [Kovacs 1998]:

$$R = \frac{\Delta P}{Q} \tag{10.15}$$

The fluid resistance for a circular tube with a radius a can be readily obtained from Equation (5.17) as:

$$R = \frac{8\mu L}{\pi a^4} \tag{10.16a}$$

The same resistance for the rectangular channel in Figure 10.12a is:

$$R = \frac{12\mu L}{wh^3} \tag{10.16b}$$

for the cross section with a high aspect ratio, and

$$R = \frac{12\mu L}{wh^3}\left\{1 - \frac{h}{w}\left[\frac{192}{\pi^5}\sum_{n=1}^{\infty}\frac{1}{n^5}\tanh\left(\frac{n\pi w}{h}\right)\right]\right\}^{-1} \tag{10.16c}$$

for the cross section with a low aspect ratio.

The fluid resistance for the V channel in Figure 10.12b is:

$$R = \frac{17.4L\mu}{w^4} \tag{10.16d}$$

Figure 10.12 | Noncircular microchannels.

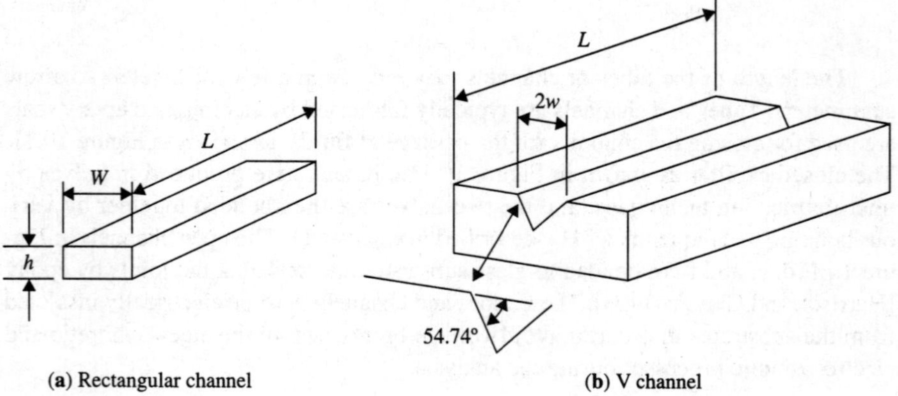

If we use the units N/m^2 for the pressure drop in Equation (10.15) and the units m^3/s for the volumetric flow rate, then the fluid resistance R has units of $N\text{-}s/m^5$. A more revealing unit for the fluid resistance would be $Pa/m^3/s$, which reflects the physical meaning of R as the pressure drop in pascals per unit volumetric fluid flow in m^3/s.

The fluid resistance R is an important indicator that assists engineers in selecting the geometry of flow channels in a microfluidic system. The following example will illustrate such selection.

EXAMPLE 10.2

Evaluate the resistance to water flow in microchannels of three distinct cross sections of the same hydraulic diameter d_h, as illustrated in Figure 10.13. The circular capillary tube has an inside radius $a = 15$ μm. The dynamic viscosity of water μ at room temperature is given in Table 4.3 as 1001.65×10^{-6} $N\text{-}s/m^2$.

Figure 10.13 | Microflow channels.

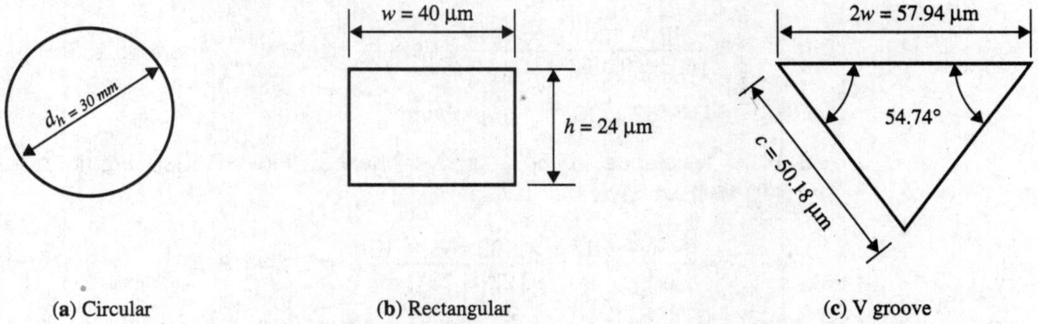

Solution

We will first determine the dimensions of the rectangular channel in Figure 10.13b and the V-groove channel in Figure 10.13c so that both these channels will have same hydraulic diameter d_h as the circular tube in Figure 10.13a.

The hydraulic diameter of noncircular cross sections can be found by using Equation (5.19) in Chapter 5. We can thus compute this diameter for both rectangular and V-groove channel cross sections as:

$$d_h = \frac{4(wh)}{2(w+h)} = \frac{2wh}{w+h}$$

for the rectangular cross section in Figure 10.13b, and

$$d_h = \frac{4wh}{2c + 2w}$$

for the V-groove channel in Figure 10.13c.

It is readily proved that $d_h = 2a$ for the circular tube by Equation (5.19).

By letting $d_h = 2a = 30\ \mu m$ and w for the rectangular channel be 40 μm in the above expressions, we will find h to be 24 μm and w for the V-groove channel to be 28.9 μm, as shown in Figure 10.13.

With the dimensions of the capillary tube and two microchannels determined, we may proceed to calculate the fluid resistance R for these conduits as follows.

1. For the capillary tube (Fig. 5.12): We may use Equation (10.16a) for evaluating the fluid resistance per unit length of the tube, R/L:

$$\frac{R}{L} = \frac{8\mu}{\pi a^4} = \frac{8 \times 1001.65 \times 10^{-6}}{3.14 \times (15 \times 10^{-6})^4} = 5.0416 \times 10^{16} \qquad \text{Pa/m}^3\text{/s/m}$$

2. For the rectangular channel (Fig. 10.13b): We realize that the aspect ratio of the cross section is $W/h = 1.6667$, which is regarded as low. Consequently, Equation (10.16c) is used for the evaluation of the fluid resistance per unit length of the channel.

$$\frac{R}{L} = \frac{12\mu}{wh^3}\left\{1 - \frac{h}{w}\left[\frac{192}{\pi^5}\sum_{n=1}^{\infty}\frac{1}{n^5}\tanh\left(\frac{n\pi w}{h}\right)\right]\right\}^{-1}$$

$$= \frac{12 \times 1001.6 \times 10^{-6}}{(40 \times 10^{-6})(24 \times 10^{-6})^3}\left\{1 - \frac{24}{40}\left[0.6274 \sum_{n=1}^{\infty}\frac{1}{n^5}\tanh(5.233n)\right]\right\}^{-1}$$

$$= 3.5183 \times 10^{16} \qquad \text{Pa/m}^3\text{/s/m}$$

3. Fluid resistance in the V-groove channel (Fig. 10.13c): From Equation (10.16d) we have:

$$\frac{R}{L} = \frac{17.4\mu}{w^4} = \frac{17.4 \times 1001.65 \times 10^{-6}}{(28.97 \times 10^{-6})^4} = 2.4744 \times 10^{16} \qquad \text{Pa/m}^3\text{/s/m}$$

We summarize these results in Table 10.2. It is interesting to note that the circular capillary tubes appear to be the least desirable choice among the three selected geometries for microfluidic systems because of its high fluid resistance. It appears that the V-groove channel generates the least resistance to microflows.

Table 10.2 | Fluid resistance in various microchannel cross sections with the same hydraulic diameters

Channel cross sections	**Fluid resistance per unit length, 10^{16} Pa/m³/s/m**
Circular	5.0416
Rectangular	3.5183
V groove	2.4744

10.7.2 Capillary Electrophoresis Network Systems

As mentioned at the beginning of this section, the capillary electrophoresis (CE) system is an effective (and powerful) tool for fast and accurate analysis of chemical and biological samples. The simplest form of CE network consists of two channels intersecting an angle. Figure 10.14 is a plan view of a CE network. One channel, called the *injection channel*, is for the passage of analyte that contains the species to be separated and measured. The other channel, called the *separation channel*, is where separation of the species occurs. The injection channel is connected to the analyte reservoir at the left and the analyte waste reservoir at the right. The separation channel is connected to a buffer reservoir at the top and a waste reservoir at the bottom.

Figure 10.14 | Plan view of a CE network.

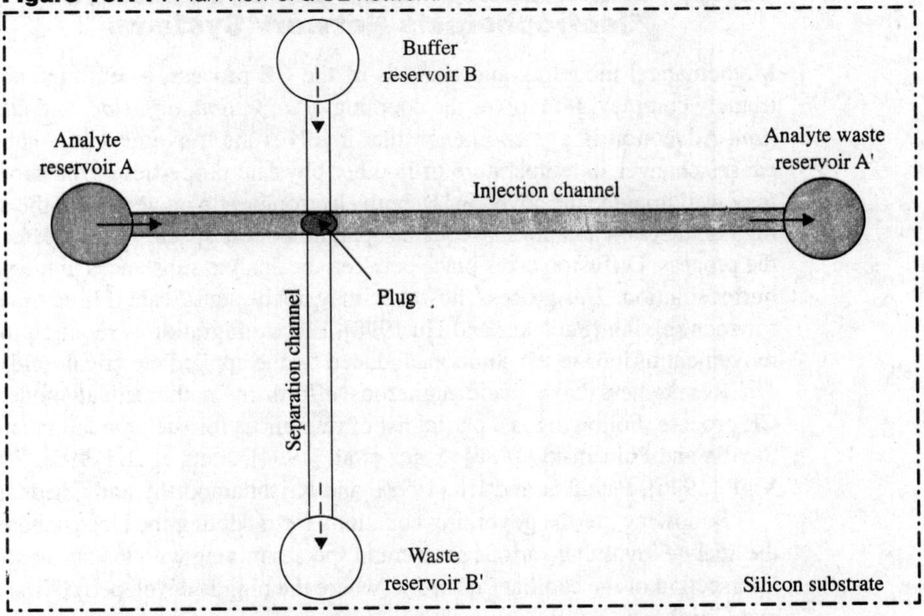

The general procedure for CE analysis is as follows:

1. An analyte solution with measurand species is injected at the injection channel from the analyte reservoir *A* by the application of an electric field ranging from 150 to 1500 V/cm between the analyte reservoir *A* and the analyte waste reservoir *A'*. The applied voltage forces the analyte solution to flow from reservoir *A* to reservoir *A'* by an electro-osmotic process. A *sample plug* is developed at the intersection of the two capillary channels as shown in Figure 10.14. Sample separation does not occur during the injection process.
2. An electric field is then applied between the buffer reservoir *B* and the waste reservoir *B'*. Consequently, the sample plug starts moving in the separation channel. The flow becomes electrophoretic, in which the various species in the fluid will flow at different velocities according to their respective *ion* (or electro-osmostic) *mobilities*. Separation of species in the measurand fluid thus takes place in this portion of the separation channel. The ion mobility ω_i for the ions of species i can be evaluated from the following formula [Manz et al. 1994]:

$$\omega_i = \frac{z_i\, q}{6\pi\, r_i\, \mu} \qquad \textbf{(10.17)}$$

where z_i is the charge of ion i, r_i is the radius of the ion, and μ is the dynamic viscosity of the ion in the solvent. The charge q of an electron is 1.6022×10^{-19} C, as indicated in Equation (3.12).

3. The separated species in the analyte solution can be detected either by an amperometric electrochemical detector or by a fluorescence detector as described by Harrison and Glavina [1993] and Woolley and Mathies [1994].

10.7.3 Mathematical Modeling of Capillary Electrophoresis Network Systems

Mathematical modeling and analysis of the CE process, as outlined above, is extremely complex. It involves the coupling of *advection*, *diffusion*, and *electromigration*. Advection is a phenomenon that involves the movement of a substance that causes changes in temperature or in other physical properties of the substance. One may well imagine the advection in both electrophoretic movement of the analyte during the injection and the subsequent separation of the species in the buffer solution in the process. Diffusion takes place between the analyte substances in the plug and the buffer solution. This process, however, may not be significant if little time is allowed for such mixing [Patankar and Hu 1998]. Electromigration is required to model the movement of ions in the solutions induced by the applied electrical fields.

Researchers have made numerous efforts in mathematical modeling of the CE process. Following is a partial list of references for such modeling and analyses: Saville and Palusinski [1986], Manz et al. [1994], Jiang et al. [1995], Williams and Vigh [1996], Patankar and Hu [1998], and Krishnamoorthy and Giridharan [2000].

Following are the governing equations for modeling the electrophoretic flow of the analyte involving various measurand species mixing with the buffer solvent at the intersection of the capillary network (where the plug is developed) [Krishnamoorthy and Giridharan 2000].

The advection equation for the rate of change of concentration of species i is:

$$\frac{\partial C_i}{\partial t} = -(\nabla \cdot \mathbf{J}_i) + \mathbf{r} \qquad \textbf{(10.18)}$$

in which C_i is the concentration of species i in the solution, t is the time into the process, $\mathbf{r}$ is the rate of production of the species. The flux vector, $\mathbf{J}_i$ has the form:

$$\mathbf{J}_i = \mathbf{V}\, C_i - z_i\, \omega_i\, C_i \nabla\phi - D_i \nabla C_i \qquad \textbf{(10.19)}$$

where

$\mathbf{V}$ = velocity vector of species i in the solution, e.g., with components $V_x(x, y)$ and $V_y(x, y)$ in the respective x and y directions in a flow defined by the x–y plane
z_i = the valence of ion i
ω_i = electro-osmotic mobility of the ith species as shown in Equation (10.17)
ϕ = applied electrical potential
D_i = diffusion coefficient of the ith species in the solution.

The rate of production of species *i*, **r** in Equation (10.18), is usually neglected in the modeling a stable process. The electric field in Equation (10.19) can be obtained by solving the following differential equation:

$$\nabla \cdot (\sigma \nabla \phi) = 0 \qquad \textbf{(10.20)}$$

in which the electrical conductivity σ is defined as:

$$\sigma = F \sum_i z_i^2 \omega_i C_i \qquad \textbf{(10.21)}$$

where F is the Faraday constant with a numerical value of 9.648×10^4 C/mol.

The bulk fluid velocity due to electro-osmotic mobility is:

$$V_0 = \omega_0 \nabla \phi \qquad \textbf{(10.22)}$$

with V_0 being the imposed slip velocity at the channel wall and ω_0 the electro-osmotic mobility of the species.

10.8 | DESIGN CASE: CAPILLARY ELECTROPHORESIS NETWORK SYSTEM

Following is a hypothetical design case, which is used to illustrate the capillary electrophoresis (CE) process. Krishnamoorthy of CFD Research Corporation in Huntsville, Alabama, developed this design case [private communication 2000]. The geometry and dimension of the CE network are shown in Figure 10.15. Both

Figure 10.15 | A capillary electrophoresis for a design case study.

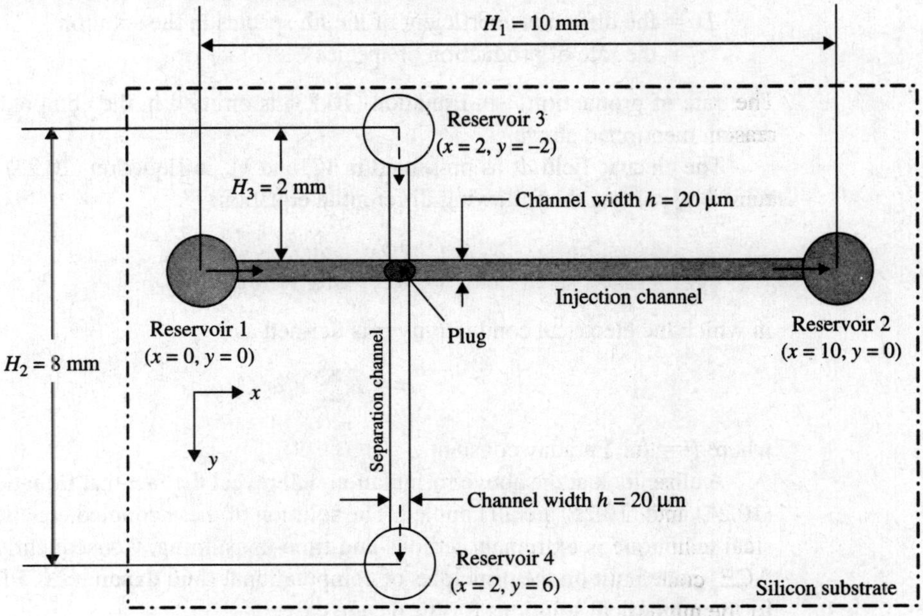

channels have rectangular cross sections 20 μm wide × 15 μm deep. The analyte contains three species with distinct electro-osmotic mobility of $\omega_1 = 2 \times 10^{-8}$ m²/V-s for species A, $\omega_2 = 4 \times 10^{-8}$ m²/V-s for species B, and $\omega_3 = 6 \times 10^{-8}$ m²/V-s for species C. All species are assumed negatively charged. For the sake of simplicity, the flow is assumed to take place in a two-dimensional pattern in the x–y plane as shown in Figure 10.15. The process takes place after an equilibrium condition is reached in the analyte, so that the production rate in Equation (10.18) can be omitted. The CFD-ACE$^+$ code marketed by the CFD Research Corporation was used in this design case.

Since the flow is assumed to be confined in the x–y plane, the velocity vector, **V** in Equation (10.18) consists of two components, $V_x(x, y)$ and $V_y(x, y)$ along the respective x and y coordinates. Consequently the two-dimensional form of Equation (10.18) becomes:

$$\frac{\partial C_i}{\partial t} + (V_x + V_{ex})\frac{\partial C_i}{\partial x} + (V_y + V_{ey})\frac{\partial C_i}{\partial y} = \frac{\partial}{\partial x}\left(D_i\frac{\partial C_i}{\partial x}\right) - \frac{\partial}{\partial y}\left(D_i\frac{\partial C_i}{\partial y}\right) + \dot{r}_i \tag{10.23}$$

where

C_i = concentration of species i in the solution
t = time into the process
$V_{ex} = -\omega_i z_i \dfrac{\partial \phi}{\partial x}$, the x component of the electromigration (the drift velocity)
$V_{ey} = -\omega_i z_i \dfrac{\partial \phi}{\partial y}$, the y component of the electromigration (the drift velocity)
z_i = the valence of ion of species i
ω_i = electro-osmotic mobility of the ith species in Equation (10.17)
ϕ = externally applied electrical potential
D_i = the diffusion coefficient of the ith species in the solution
$\dot{r}_i$ = the rate of production of species i

The rate of production $\dot{r}_i$ in Equation (10.23) is omitted in the computation for the reason mentioned above.

The electric field ϕ, as presented in V_{ex} and V_{ey} in Equation (10.23), can be obtained by solving the following differential equation:

$$\frac{\partial}{\partial x}\left(\sigma\frac{\partial \phi}{\partial x}\right) + \frac{\partial}{\partial y}\left(\sigma\frac{\partial \phi}{\partial y}\right) = 0 \tag{10.24}$$

in which the electrical conductivity σ is defined as:

$$\sigma = F\sum_i z_i^2 \omega_i C_i \tag{10.25}$$

where F = the Faraday constant.

A close look at the above formulation will reveal the fact that Equations (10.23), (10.24), and (10.25) are all coupled. The solution of these coupled equations by classical technique is extremely tedious and time-consuming. Consequently, the CFD-ACE$^+$ code built on the principles of computational fluid dynamics (CFD) was used for the numerical solutions for the design case.

In using the above CFD code, a condition is set to have the separation accomplished with the help of an electrokinetic switching technique as discussed under general procedure for the CE process in Section 10.7.2. This technique consists of two cycles:

1. *Injection cycle:* An electric field is maintained in the injection channel till the sample flows past the intersection.
2. *Separation cycle:* An electric field is maintained in the separation channel for the required length of time, till separation is completed.

The magnitude of the applied electric fields depends on the analyte properties, volume of the sample plug desired at the intersection, and geometry of the system.

In Figure 10.16, the contour map of the sample, predicted by the CFD-ACE$^+$ code, at the end of the injection cycle is shown. During this cycle, analyte reservoir 1 in Figure 10.15 is grounded and analyte waste reservoir 2 is maintained at 250 V.

Figure 10.16 | Contour profiles of a sample that consists of species A, B, and C at the end of injection process.

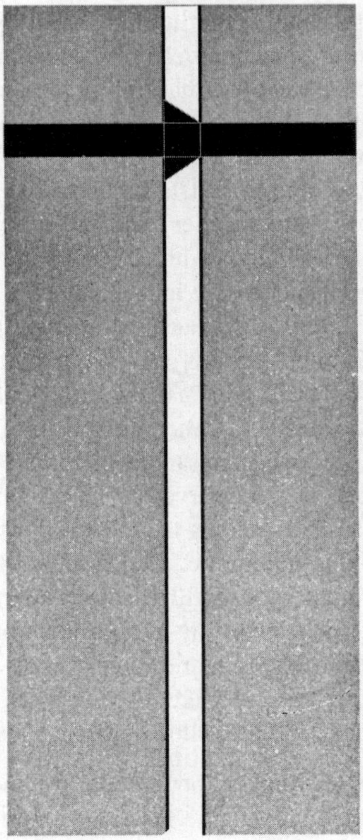

The buffer reservoir 3 and the buffer waste reservoir 4 are maintained at 30 V and 0 V, respectively. The strong electric field in the injection channel causes the sample to flow from reservoir 1 to 2. As it reaches the junction, it is squeezed because of the weak electric field in the separation channel. Consequently, the sample at the injection cross has a trapezoidal shape (see Fig. 10.16). This method of generating a sample of specified volume at the intersection by using a "pinch" in the flow field is known as the *pinched loading technique* [Culbertson et al. 1998]. This process is also known as *electrokinetic focusing*. The pinch in the flow field is caused by the nonuniform electric field at the intersection created by the channel geometry and orientation.

The next cycle in the CE process is the separation of species in the sample. After the sample reaches the intersection, the voltages are switched from the *sample-loading* mode to *sample-separation* mode. A potential of 250 V is applied to buffer waste reservoir 4 while buffer reservoir 3 is grounded. The analyte and analyte waste reservoirs (1 and 2) are maintained at 70 V and 100 V respectively. The reason for maintaining a weak electric field in the injection channel during the separation process is to avoid leakage of sample from the injection channel into the separation channel so that a clean separation may be achieved. The results from the CFD simulation are shown in Figure 10.17a to c for three different time levels, 0.1 s, 0.3 s, and 0.5 s, respectively. These figures show the contour map of species A, B, and C and a line plot of concentration of these species along the center of the separation channel.

During the separation cycle, the sample at the junction starts flowing into the separation channel. As the sample continues its journey, the species A, B, and C migrate with different speeds since their electro-osmotic mobilities are different. As expected, Figure 10.17 clearly shows that, in a given length of time, the distance traveled by a species is directly proportional to its mobility, and species C, which has the highest electro-osmotic mobility, will be detected by the amperometric electrochemical or the fluorescence detector first. The line plots also indicate that the peak values of the concentration of the species decrease with time. This phenomenon is called *band-broadening* or *dispersion*, which refers to the process of dilution of analyte concentration as it moves across the microchannel.

Dispersion/diffusion is the main cause of broadening. A nonuniform electric field may also cause the different regions of the sample to migrate at different velocities [Culbertson et al. 1998] and thus cause band broadening. The nonuniform electric field occurs primarily because of the curvature in the channel geometry. For example, consider a rectangular-shaped sample plug traveling in a curved channel by electrophoresis. If the two ends of the curved channel are maintained at constant voltage, then the higher current density along the inner wall of the bend will cause the inner region of the sample-plug to migrate faster than the outer region. Consequently, the sample may attain a shape of a highly distorted parallelogram and eventually cause band broadening. Krishnamoorthy and Giridharan [2000] have studied this phenomenon in S-shaped and rectangular channels.

A time-accurate CFD calculation like the one presented will help engineers to design and optimize CE devices. The following design issues are noted:

1. One of the parameters defining an optimal electrokinetic injection system is the shape and width of the sample in the separation column. The broader the

Figure 10.17 | Contour profiles of the species A, B, and C are shown at times 0.1 s, 0.3 s, and 0.5 s.

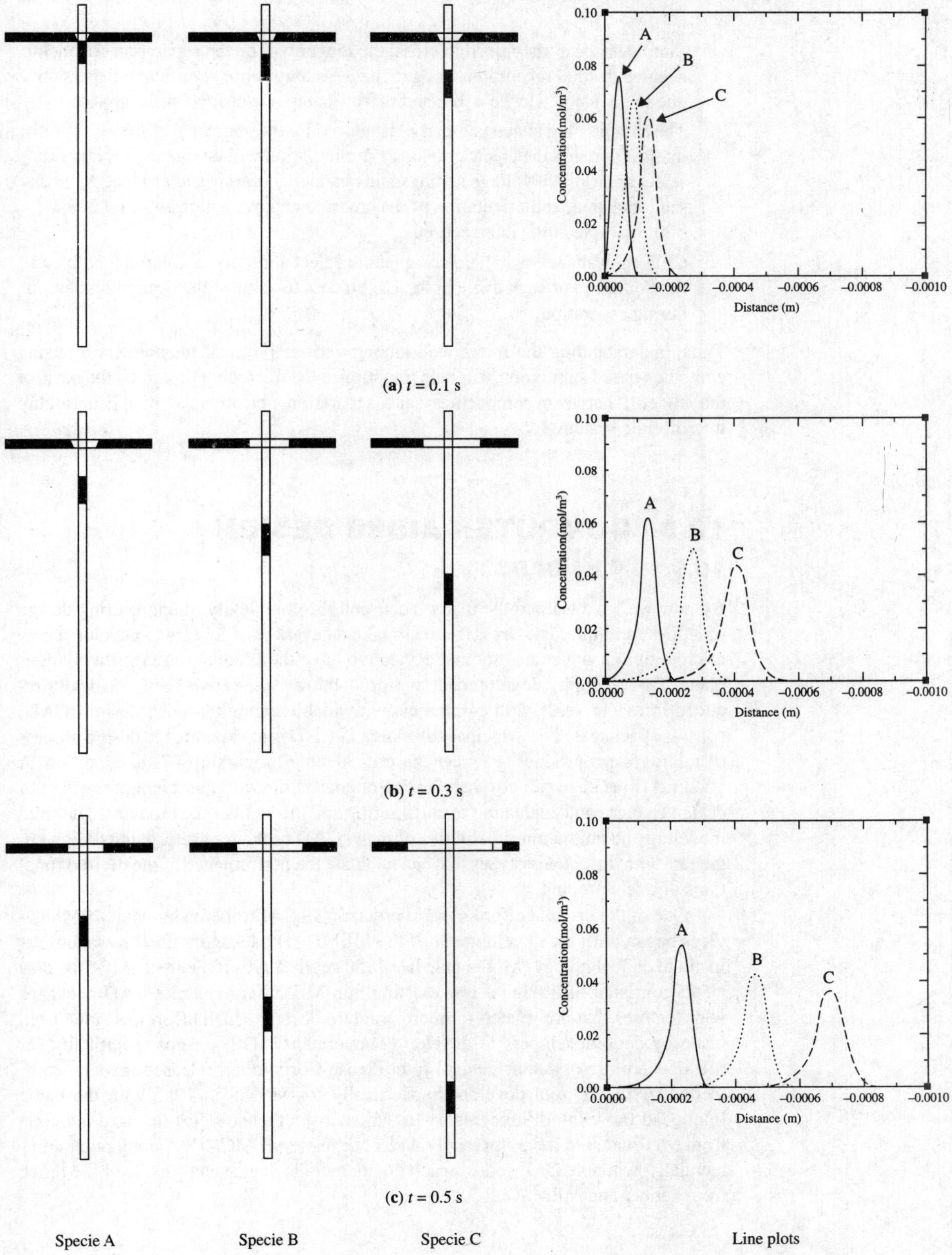

sample shape at the injection cross, the longer will be the separation channel to achieve desired resolution. Longer channels may cause significant dispersion due to diffusion. Hence a thin and narrow sample is desired at the cross.

2. The sample plug shape is greatly influenced by the geometry of the intersection and the electric field. CFD simulation will enable the designer to select appropriate voltage settings and channel geometry such as length, cross-sectional area, and orientation of the cross, to achieve a narrow and thin band of the sample at the intersection.
3. CFD analysis will also help designers to select appropriate column length and time duration of each cycle in the CE process to achieve the desired resolution during separation.

Thus, understanding the interaction among various physical phenomena by using computer-based simulation will help to optimize the microchip design on the basis of the tradeoff between compactness and separation performance of CE-on-a-chip microfluidic system.

10.9 | COMPUTER-AIDED DESIGN

10.9.1 Why CAD?

By now we have learned the many facets and the complexity of engineering design of MEMS and microsystems. It used to take an average of 5 years to develop a new microsystem product and another 5 years to have the product to reach the marketplace. These lengthy development and production cycles have been drastically reduced in recent years with commercially available computer-aided design (CAD) software packages. The principal advantage of CAD is to expedite the design process of microsystem products to meet the critical time-to-market (TTM) condition. A good CAD package can also help design engineers in swiftly assessing the effect of design changes and evaluating manufacturing and the yield of the products. The solid modeling and animation capabilities of many CAD packages can provide design engineers with virtual prototypes that can simulate the performance of the desired functions of a real product.

One of the original efforts toward producing a CAD tool to simulate MEMS devices began with the development of the MEMCAD package at the Massachusetts Institute of Technology (MIT) in the late 1980s/early 1990s [Gilbert et al. 1996]. This effort consisted of linking a few existing non-MEMS commercial CAD packages with features that are related to microstructure design. Much effort has since been made by code developers to develop commercial CAD packages specifically for MEMS and microsystems design. IntelliSense Corporation released the first commercialized CAD tool developed specifically for MEMS in 1995 with the name IntelliCad (now IntelliSuite). In 1996, Microcosm Technologies licensed the code from MIT and sold it commercially under the name MEMCAD. Among other commercially available CAD tools, which do not include device analysis, are MEMSPro from Tanner and MEMSCaP.

10.9.2 What Is in a CAD Package for Microsystems?

While CAD for microsystems is still an emerging effort, a viable CAD package should include at least three major interactive databases: (1) *electromechanical design database*, (2) *material database*, and (3) *fabrication database*. The content of a generic CAD package including the above databases is schematically shown in Figure 10.18.

Figure 10.18 | General structure of CAD for microsystem product design.

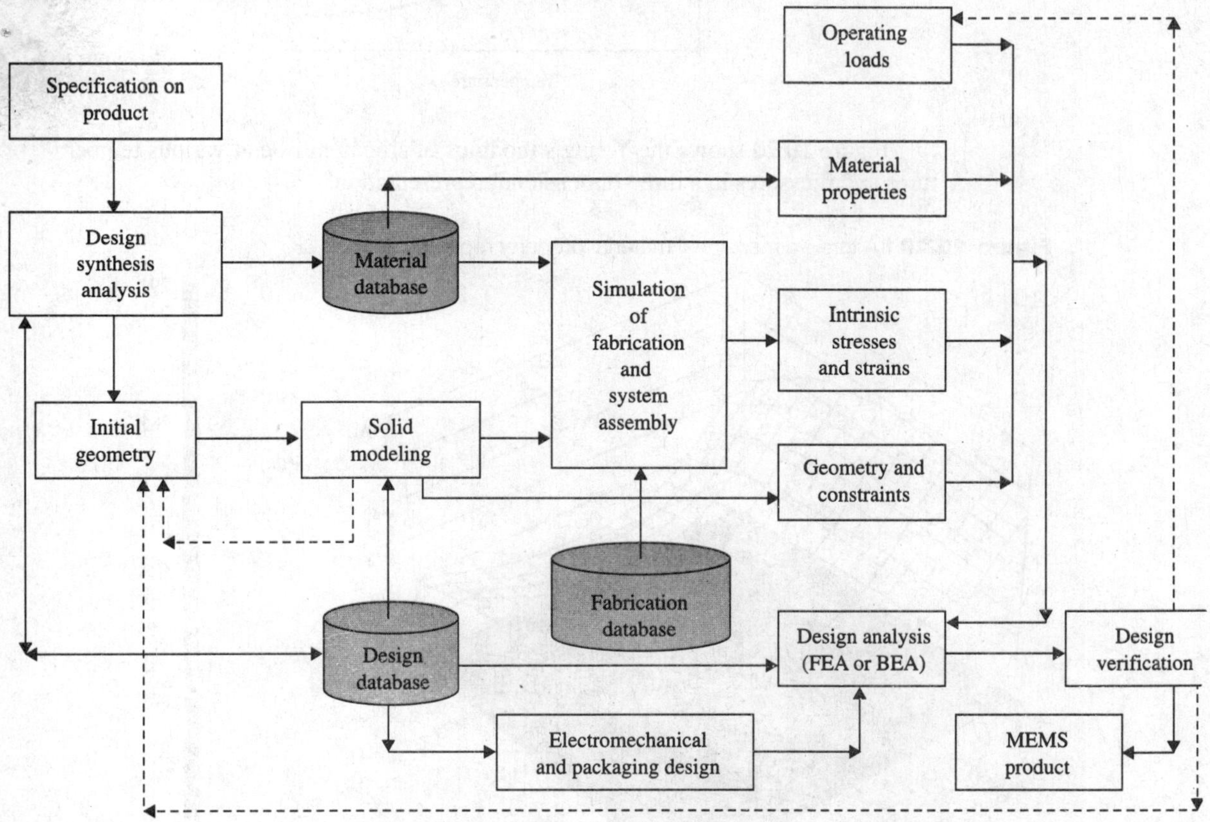

As we can see from Figure 10.18, the design database provides the necessary information and tools such as the inference machine for design synthesis, codes for FEA (finite element analysis) and BEA (boundary element analysis), as well as tables and charts for other design considerations. The need for a material database in a CAD for microsystems is obvious, as the properties of many materials used in microsystems are not available from traditional materials handbooks. This database should contain complete information on material properties such as those presented in Table 7.3. Many of these properties should be in two- or three-dimensional graphs as illustrated in Figures 10.19 and 10.20. It should also include properties for transduction components such as piezoelectric and piezoresistive materials.

Figure 10.19 I A two-dimensional material property representation.

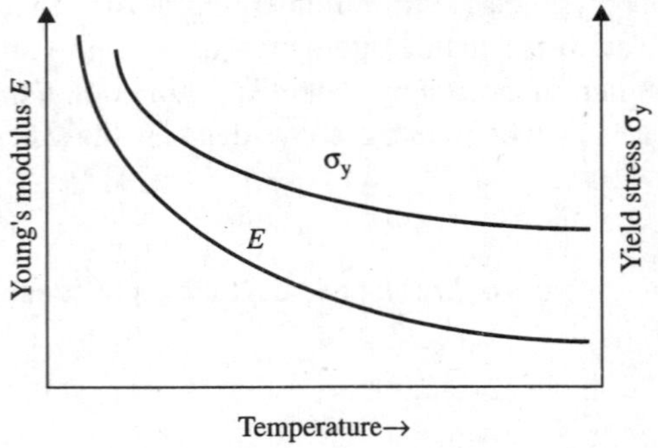

Figure 10.20 shows the Young's modulus of silicon nitride at various temperatures and pressures in a three-dimensional representation.

Figure 10.20 I A three-dimensional material property representation.

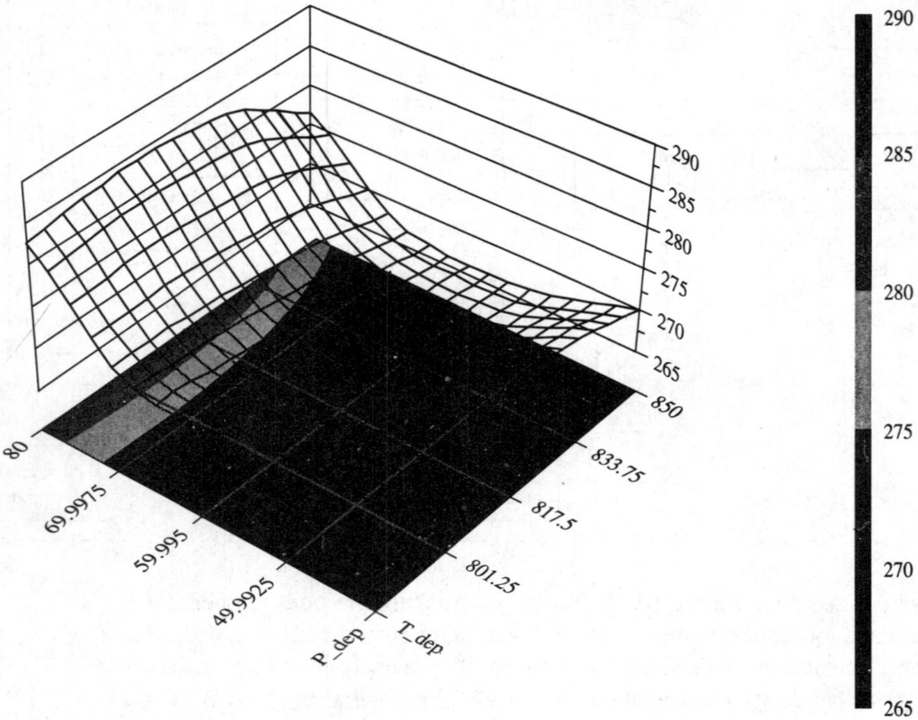

(Courtesy of IntelliSense Corporation, Wilmington, Massachusetts.)

The fabrication database, which is unique for microsystem design, involves all fabrication process simulations required for specifically selected fabrication and manufacturing processes as described in Chapters 8 and 9. This database should also include wafer treatment such as the required cleaning processes for photolithography and thin-film depositions. The results of these fabrication process simulations often include the inherent residual stresses and strains and other intrinsic stresses, which are used as input to the subsequent design analysis under normal operating and overload conditions. Engineers can visualize the designed product in three-dimensional perspectives by using the solid model option provided by the CAD package. Most CAD packages have provisions for animations, which allow engineers to visualize the functions "performed" by the virtual prototype, such as comb-driven actuators.

The flowchart in Figure 10.18 is self-explanatory. Design engineers will first establish a "process flow table" by selecting a substrate material once the product is configured from the design synthesis analysis. The CAD package will offer a possible prephotolithographic substrate treatment process from the fabrication process database. A mask is then either imported from external sources or created by the built-in design database for the subsequent photolithography on the substrate. The same database is then used to determine the appropriate fabrication process flow or steps that may include oxidation, diffusion, ion implantation, etching, deposition, and other processes such as bonding, as selected by the designer. The CAD package offers detailed information on the selected processes, for instance, the etchants for the etching process with an estimated required time for each of such processes. The CAD package also provides automatic flow of information between the material database and the fabrication database. Once the fabrication processes have been established, electromechanical design begins. Here, the design engineer uses the solid model constructed by the package for automatic mesh generation for the electromechanical analysis. Depending on the nature of the product, the CAD package can perform the finite element analysis for thermal conditions and mechanical strength of the structure, as well as electrostatic and electromagnetic analyses in the cases that involve actuation by the products. The latter analyses require the input of electrical potential and current to the finite element analysis. In addition to graphical displays for the analytical results, many CAD packages also offer animation of the designed product for kinematic and dynamic effects. Engineers may either terminate the design at this stage if the outcome of the design process is satisfactory, or make any necessary revisions to the configuration or loading or boundary constraints until all design objectives and criteria are met.

10.9.3 How to Choose a CAD Package

Commercial CAD packages available for the MEMS community differ from one package to another. Following is a general guideline, which engineers may find useful in selecting a suitable package for their specific MEMS needs. In general, a prospective buyer should look for:

1. User friendliness.
2. The adaptability of the package to various computers and peripherals.
3. Easy interfacing of this CAD package with other software, e.g., nonlinear thermomechanical analyses and the integration of electric circuit design.
4. Completeness of the material database in the package.

5. The versatility of the built-in finite element or boundary element codes.
6. Pre- and post-processing of design analyses by the package.
7. Capability of producing masks from solid models.
8. Provision for design optimization.
9. Simulation and animation capability.
10. Cost of purchasing or licensing and maintenance.

10.9.4 Design Case Using CAD

A commercial CAD package with the tradename IntelliSuite was used in the design of a microcell gripper in a student project [Griego et al. 2000]. A general description of the package is available in a published reference [He et al. 1997]. The package consists of three major databases similar to those shown in Figure 10.18: (1) a material database, (2) an electromechanical database, and (3) a fabrication process database. This CAD package offers material selection from the following categories:

1. Substrates:
   Silicon (Si)
   Polysilicon
   Gallium arsenide (GaAs)
   Quartz
   Sapphire
   Alumina
   Doped semiconductors
2. Interconnects and mask materials:
   Aluminum (Al)
   Gold (Au)
   Silver (Ag)
   Chromium (Cr)
   Silicon dioxide (SiO_2)
3. Other substrate and insulating materials:
   Silicon dioxide (SiO_2)
   Silicon nitride (Si_3N_4)
4. Photoresist materials

The overall dimension of the comb-drive-actuated gripper in a plan view is presented in Figure 10.21, and the detailed arrangement of the electrodes is shown in Figure 10.22. The following design case illustrates the steps involved in the design process. The design was a student project, so no effort was made by the designers to optimize the cost and time required for the fabrication of the cell gripper. Only a single gripper was designed for production from a single wafer. In reality, of course, one would seek a maximum yield of MEMS components produced from a single wafer.

The strategy adopted in this design case was to construct all required layers on the substrate wafer by oxidation, deposition, and sputtering processes. These layers

Figure 10.21 | A MEMS cell gripper.

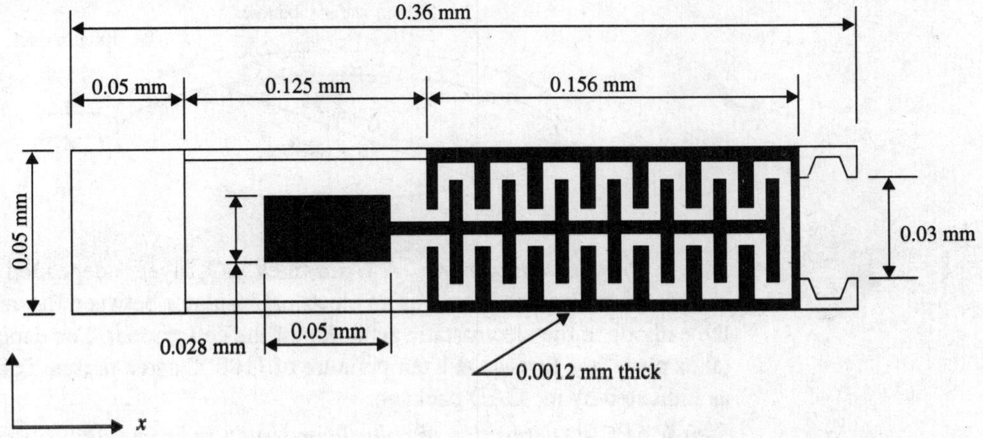

Figure 10.22 | Gaps between electrodes in a MEMS cell gripper.

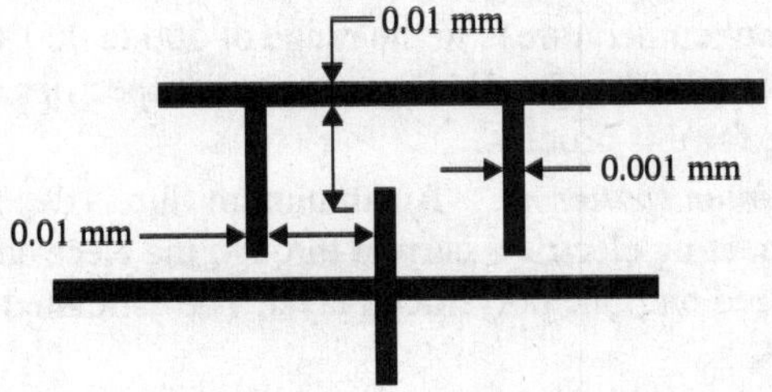

are necessary for the electrical insulation, gripper structures, and electrical terminals. A bulk manufacturing technique with etching was then used to remove materials and thereby shape the gripper as needed. The IntelliSuite package provided the solid modeling of the substrate after each of these steps with estimated required time for each of these processes. The solid model was also used for an electromechanical analysis assessing the reliability of the microstructure.

Major steps involved in this design case are outlined below:

Step 1: Substrate selection. A silicon wafer is chosen to be the substrate material for the gripper. The wafer is of the standard 100-mm diameter, sliced 500 μm thick from a single silicon crystal boule produced by the Czochralski method as described in Chapter 7. The surface of the wafer is normal to the <100> orientation as illustrated in Figure 10.23.

Step 2: Substrate cleaning. The designers choose Pirahna solvent to clean the wafer surface. This is one of several options offered by the CAD package. This solvent contains 75% H_2SO_4 and 25% H_2O_2. The substrate is submerged in the solvent for 10 minutes. The cleaned wafer is ready for oxidation on one of its surfaces.

Figure 10.23 | Silicon wafer for a microgripper.

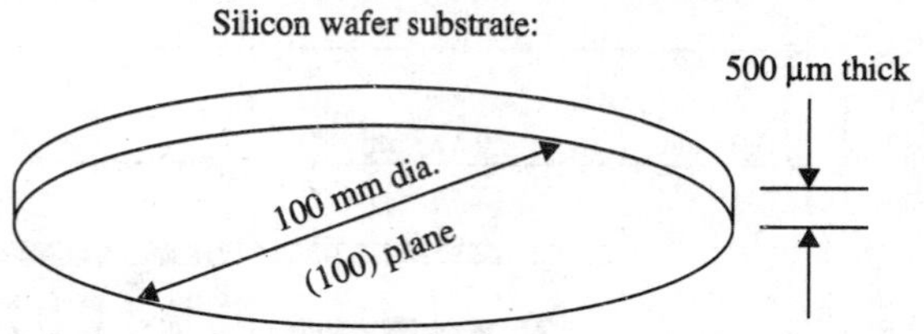

Step 3: Dry SiO_2 deposition. A 1-μm-thick SiO_2 layer is deposited on the surface of the wafer to serve as an electrical insulator between the anode and the cathode in the electrostatic actuation of the cell gripper. The deposition takes place in a furnace at a temperature of 1100°C and a pressure of 101 kPa as indicated by the CAD package.

Step 4: LPCVD deposition of polysilicon structure layer. Polysilicon is chosen to be the material for the cell gripper structure. A 1.2-μm-thick polysilicon layer is deposited over the oxide layer with a medium-temperature LPCVD process, with detail parameters provided by the IntelliSuite package. The deposition temperature is in the range of 500 to 900°C, with an annealing temperature of 1050°C. The CAD package also specifies 60 minutes as the required time for this process.

Step 5: Aluminum sputtering. An aluminum film is deposited for the lead wire for conducting electrical current through the electrodes. A 3-μm-thick film is sputtered onto the polysilicon layer. The estimated time for this process is 10 minutes.

Step 6: Application of photoresist. Positive photoresist is applied to the aluminum layer. A 4000-rpm spinning speed of the chuck, as illustrated in Figure 8.3, is used to spread the photoresist. The photoresist-covered substrate assembly is baked at 115°C, yielding a 3-μm-thick layer. All films, including the photoresist, deposited on the silicon wafer are shown in Figure 10.24.

Figure 10.24 | Thin-film depositions on silicon substrate for a microgripper.

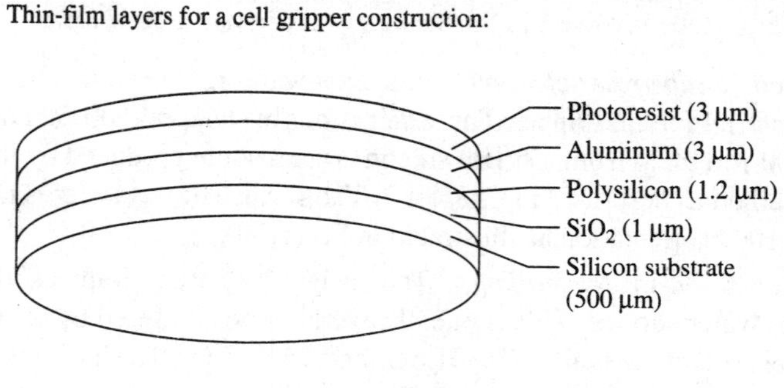

Step 7: Photolithography by UV exposure. A photolithographic process using a UV light source at 250 W with a wavelength $\lambda = 436$ nm is used in the

process over a mask created for the anode and cathode. The exposure time in this case is 10 seconds. This mask is provided by the CAD package.

Step 8: Wet etching to remove photoresist. The solvent KOH, described in Chapter 8, is used as the etchant to remove the exposed photoresist. The unexposed resist stays attached to the aluminum layer.

Step 9: Wet etching on aluminum. A special etchant is selected to remove the unprotected aluminum from the surface. This etchant contains 75% H_2SO_4, 20% $C_2H_4O_2$, and 5% HNO_3. The depth of the aluminum layer to be removed is 3 μm. The estimated time for this process is 15 minutes.

Step 10: Wet etch to remove photoresist from aluminum. Once again, KOH is used to remove the photoresist left on the surface of the aluminum anode and cathode.

Step 11: Photoresist deposition and photolithography of gripper structure. Positive photoresist is applied to the entire surface of the wafer by the same procedure as step 6. Another mask that outlines the gripper structure to comply with what is shown in Figure 10.21 is used for photolithography by the same procedure as in step 7.

Step 12: Remove photoresist by wet etch. The same procedure as described in step 10 is used for this purpose.

Step 13: Etch polysilicon by reactive ion etching (RIE). Because of the relatively high aspect ratio of the gripper structure, RIE is chosen to remove the unprotected region of the polysilicon layer for the net shape of the gripper structure. The reactive chemical species with chlorine or fluorine in plasma are used in this process.

Step 14: Remove the SiO_2 sacrificial layer. This process involves the use of wet etching in conjunction with a laser photochemical etching process. The latter etching process uses a SiH_4 etchant and a KrF laser at 0.3 J/cm^2 intensity. The combined etching provides an etching rate of 40 Å/s. The process in this step releases the gripper arms and tips from the SiO_2 layer.

Step 15: Separation of gripper and the substrate. The net shape of the structure after Step 14 is the gripper structure as shown in Figure 10.21 attached to the silicon substrate of the same structural outline bonded by a thin SiO_2 film. Separation of the gripper structure from the substrate requires the removal of the in-between SiO_2 layer (a sacrifice layer). The removal of this thin layer can be accomplished either by using a thin diamond saw or by using the "etch pit" technique described in a paper by Kim et al. [1991].

Step 16: Electromechanical analysis. The purpose of this analysis is to assess whether the gripper fabricated by the above processes will perform the desired functions.

A *charge density analysis* is performed first with the applied voltage input to the aluminum terminals to assess if the highest charge buildup in the comb drive will produce sufficient electrostatic forces for the gripping function. The CAD package offers a graph of the distribution of such charges with color-coded charge density at various parts of the gripper structure. Engineers are thus assured that the electrostatic charges at the comb fingers have sufficient magnitudes to generate the desired gripping forces.

Finite element strength analysis is followed after the charge density analysis to assure that: (1) The maximum stress in the gripper structure is not excessive in the *x, y,* or in the *z* direction. The coordinate system used in this case is shown in Figure 10.21, with the *z* coordinate being normal to the *x–y* plane. These stresses and the von Mises stress expressed in Equation (4.69) must be kept below the yield strength of the materials. (2) The deformation-induced distortion of the structure is not large enough to affect the function of the comb-drive actuation. The coupled mechanical and electrostatic effect of the actuation is illustrated in Example 10.1. The CAD package offers automatic finite element mesh generation for the required analyses with results indicated in coded colors over the solid models of the gripper structure.

The solid model of the cell gripper as produced by the IntelliSuite CAD package is shown in Figure 10.25.

Figure 10.25 | Solid model of a cell gripper.

PROBLEMS

Part 1. Multiple Choice

1. A major difference between the design of microsystems and the design of traditional products in macroscale is that the design of microsystems requires (1) the integration of design and fabrication, (2) the integration of design and marketing, (3) the integration of chemical and mechanical forces.
2. Microsystem components are fabricated by (1) precision machine tools, (2) physical-chemical means, (3) micromachine tools.
3. Microsystem design involves (1) the single task of design analysis, (2) the coupling of two tasks, (3) the coupling of three tasks.
4. *Process flow* is part of microsystem (1) design, (2) manufacturing, (3) production.
5. An effective way to shorten the microsystem design cycle is to (1) involve more engineers, (2) use better design method, (3) use computer-aided design.
6. One critical design consideration for microsystems is TTM, which stands for (1) total time management, (2) time to market, (3) targeted time management.
7. Three critical design considerations related to environmental conditions are (1) thermal, mechanical, and chemical; (2) electrical, mechanical, and chemical; (3) electrical, mechanical, and materials.
8. There are (1) one, (2) two, (3) three types of substrate materials used in microsystems.
9. The most commonly used substrate material is (1) silicon, (2) GaAs, (3) quartz.
10. The most suitable material for micro-optical components is (1) silicon, (2) GaAs, (3) quartz.
11. The most dimensionally stable material for microcomponents is (1) silicon, (2) GaAs, (3) quartz.
12. The cheapest thermal and electrical insulation material is (1) silicon dioxide, (2) silicon nitride, (3) silicon carbide.
13. The most chemically stable material for microsystems is (1) silicon dioxide, (2) silicon nitride, (3) silicon carbide.
14. The most suitable material for the masks used in deep etching is (1) silicon dioxide, (2) silicon nitride, (3) silicon carbide.
15. In general, the cheapest way to produce MEMS is (1) bulk micro-manufacturing, (2) surface micromachining, (3) the LIGA process.
16. In general, the most expensive way to produce MEMS is (1) bulk micromanufacturing, (2) surface micromachining, (3) the LIGA process.
17. In general, the technique that offers the most flexibility in micromanufacturing is (1) bulk micromanufacturing, (2) surface micromachining, (3) the LIGA process.
18. In general, the micromanufacturing technique that offers a high aspect ratio in geometry is (1) bulk micromanufacturing, (2) surface micromachining, (3) the LIGA process.

19. The principal advantage of using piezoresistors is (1) small size, (2) high sensitivity, (3) low cost in production.
20. The most serious drawback of a piezoresistor transducer is (1) its high cost, (2) its slow responses, (3) its strong temperature dependence on sensitivity.
21. The principal advantage of a piezoelectric transducer is (1) its fast response, (2) its low cost in production, (3) its high sensitivity.
22. A serious drawback of a piezoelectric transducer is (1) its high cost, (2) its vulnerability to brittle fracture, (3) its strong temperature dependence on sensitivity.
23. The principal advantage of a capacitance transducer is (1) its suitability for high-temperature applications, (2) its simplicity, (3) its low cost in production.
24. A serious drawback of a capacitance transducer is (1) its bulky size, (2) its nonlinear input/output relationship, (3) its low sensitivity.
25. The resonant vibration transducer has the advantage of (1) simplicity, (2) reliability, (3) high sensitivity and precision.
26. A serious drawback of resonant vibration transducer is (1) complexity in fabrication, (2) the strong temperature dependence of sensitivity, (3) unreliability in input/output.
27. The problem with actuation using electro-resistant heating is (1) temperature, (2) heat source, (3) the control of heating and cooling rates.
28. SMA stands for (1) smart material actuator, (2) shape memory alloy, (3) shape memory actuator.
29. Microsystem packaging needs to be considered in early stage of design to ensure (1) low packaging costs, (2) high customer demand, (3) acceptable product appearance.
30. Photolithography means to (1) make a photograph of a microsystem, (2) create the pattern of the geometry of a microsystem, (3) produce the label on a microsystem.
31. Selection of thin-film deposition processes in the early stage of microsystem design is important because these processes (1) are expensive, (2) are delicate, (3) may result in adverse consequences in the performance of the microsystem.
32. For closely spaced microstructural components, the force that causes most problems is (1) thermal force, (2) electrostatic force, (3) van der Waals force.
33. Intrinsic stresses in microstructures are induced by (1) microfabrication processes, (2) the applied loads, (3) thermomechanical effects.
34. Resonant vibration in microsystems (1) should always be avoided, (2) is necessary in some microdevices, (3) is irrelevant, as it never happens in microsystems.
35. Interfacial fracture is an important design consideration in microsystems that involve (1) layers of dissimilar materials made by thin-film deposition, (2) high-temperature environment, (3) microstructures with cracks.

36. The finite element method is (1) a universal, (2) viable, (3) irrelevant analytical tool for microstructure analysis.
37. As a rule of thumb, commercial finite element codes developed for macroscale structures can be used for microscaled structures of size greater than (1) 0.1 μm, (2) 1 μm, (3) 100 μm.
38. The interpolation function in a finite element analysis relates (1) element and the corresponding nodal quantities, (2) element quantities in the entire structure, (3) nodal quantities in the entire structure.
39. The Galerkin method used to derive the element equation in a finite element analysis requires (1) the geometry of the element, (2) the governing differential equation, (3) the potential energy in the discretized continuum.
40. The Rayleigh–Ritz method used to derive the element equation in a finite element analysis requires (1) the geometry of the element, (2) the governing differential equation, (3) the potential energy in the discretized continuum.
41. The "birth" or "add" elements are used to simulate the microfabrication process of (1) etching, (2) deposition, (3) molding in a finite element analysis.
42. The "death" or "subtract" elements are used to simulate the microfabrication process of (1) etching, (2) deposition, (3) molding in a finite element analysis.
43. In designing a die for micropressure sensors, one of the critical design considerations is (1) die isolation, (2) mechanical strength, (3) contamination.
44. Better die isolation in a micropressure sensor can be achieved by (1) a thinner structure, (2) passivation by coating protective materials, (3) flexible die attach.
45. Passivation of the die in a pressure sensor is achieved by (1) plastic encapsulation, (2) coating of protective materials, (3) keeping the pressurized medium away from the die.
46. To achieve maximal sensitivity of a pressure sensor, one would maximize the (1) stress, (2) strain, (3) deformation in the diaphragm.
47. Bosses are introduced in the diaphragm of a pressure sensor die to provide (1) extra stiffness, (2) better appearance, (3) uniform stress distribution in the diaphragm.
48. In sensor design, including that for pressure sensors, the (1) magnitude, (2) linearity, (3) fast response of output signals is the primary concern to the design engineer.
49. Capillary electrophoresis (CE) on a chip is primarily used in (1) microfluidic actuator, (2) micropressure measurements, (3) biomedical analysis.
50. The principal advantages of using CE on a chip is that it (1) provides a cheap method of biomedical analysis, (2) involves minute sample size and fast responses, (3) is an easy to use analytical method.
51. Typical capillary electrophoresis involves (1) a network of capillary tubes and microchannels, (2) capillary tubes and microvalves, (3) capillary tubes and micropumps.

52. Fluid flow in capillary electrophoresis is prompted by (1) application of surface forces from the conduit wall, (2) application of volumetric pumping forces, (3) application of electrical fields.
53. Two principal parts in a capillary electrophoresis network are (1) injection and separation channels, (2) injection and sample flow channels, (3) sample separation channels.
54. *Advection* is a physical phenomenon that involves a moving substance that changes its (1) pressure, (2) temperature, (3) phase during the movement.
55. Design analysis for capillary electrophoresis requires the coupling of (1) advection and diffusion, (2) advection and electromigration, (3) advection, diffusion, and electromigration.
56. Separation of various species in a biosample during capillary electrophoresis is due to the difference of (1) density, (2) buoyancy, (3) electro-osmotic mobility of the species.
57. The principal purpose of using computer-aided design (CAD) in microsystem design is to (1) make the design beautiful, (2) make it more accurate, (3) shorten the time required in the design process.
58. Modern CAD for microsystems requires the integration of (1) design and analysis databases and prototyping, (2) design, material, and analysis databases, (3) design, material, and fabrication databases.
59. The design database in a CAD for microsystem design involves (1) mechanical and electrical design analyses, (2) mechanical and chemical analyses, (3) electrical and chemical analyses.
60. One major advantage of using CAD in microsystem design is the capability of (1) graphics representation of the results, (2) animation of the device, (3) obtaining fast results.

Part 2. Descriptive Problems

1. What are the principal sources of intrinsic stresses induced in the microstructures?
2. Why is modal analysis important in the design of microsystems involving motion?
3. Evaluate the resistance to water flow in microchannels of the cross sections in (a) a rhombus, (b) an obround shape, and (c) a trapezoid with the same hydraulic diameter of 30 μm as indicated in Example 10.2. Tabulate the computed resistance to water flow for all these three cross sections and the other three cases as presented in Example 10.2.
4. Formulate the element equation for a finite element analysis using the Galerkin method in conjunction with the modified heat conduction equation for solids in submicrometer scale, i.e., Equation (5.53).
5. Reformulate Equations (10.18) to (10.22) for a capillary electrophoresis network similar to that shown in Figure 10.15 but with a 45° angle between the two channels.

Chapter 11

Microsystem Packaging

CHAPTER OUTLINE

11.1 | INTRODUCTION

Most MEMS and microsystems involve delicate components with sizes in the order of micrometers. These components are vulnerable to malfunctions and/or structural damage if they are not properly packaged. Reliable packaging of these devices and systems is a major challenge to the industry because microsystem packaging technology is far from being as mature as that for microelectronic packaging. Microsystems packaging, in a broader sense, includes the three major tasks of *assembly, packaging,* and *testing,* abbreviated AP&T, as referred to in a special report [National Research Council 1997]. AP&T of MEMS makes up a large portion of the overall cost of production. The cost for packaging micropressure sensors, for example, can be 20 percent of the overall production cost with plastic passivation designed for a friendly environment in mass productions, or as high as 95 percent of the total cost for special pressure sensors for high-temperature applications with toxic pressurized media. AP&T cost for MEMS and microsystems varies from product to product. Currently, this cost represents on average 80 percent of the total production cost. A more serious note is that packaging is usually the source of failure of most microsystems. Packaging technology is thus a key factor in MEMS and microsystem product design and development.

A well-known fact is that the purpose of IC packaging is merely to protect the silicon chip and the associated wire bonds from environmental effects. Microsystem packaging, on the other hand, is expected not only to protect the delicate components such as silicon dies from a hostile environment, but also allows these dies to probe that environment at the same time. For example, microsensors are required to sense such things as the pressure of exhaust gas from an internal combustion engine as well as the composition of the gas. The sensors therefore need to be exposed to these highly corrosive pressurized gases at high temperature. Proper protection of the silicon die and the delicate transducers and the associate wire bonds from excessive heat and corrosive chemicals is absolutely essential. Microsystem packaging is thus much more challenging to engineers than microelectronic packaging.

11.2 | OVERVIEW OF MECHANICAL PACKAGING OF MICROELECTRONICS

The purpose of microelectronic packaging is to provide mechanical support, electrical connections, protection of the delicate integrated circuits from all possible attacks by mechanical and environmental sources, and removal of heat generated by the integrated circuits. Other major roles that packaging plays are power delivery to constituents and signal mapping from outside a package to within and between the constituents [Lyke and Forman 1999]. We have learned from Chapter 1 and several other places in this book that MEMS and microsystems technologies are the result of evolution from microelectronics technology. As such, most microfabrication techniques presented in Chapter 8 for microelectronics can be used in microsystems manufacturing as described in Chapter 9. We will find that certain techniques that were developed for microelectronic packaging can also be used for microsystem packaging. However, microsystem packaging is further complicated by additional requirements such as ensuring light transport and mapping for many optical MEMS, fluidic transport and mapping, and environmental access for chemical and biological sensing in microfluidics.

It is useful to take an overview of microelectronic packaging technology before we look into microsystem packaging in detail. There are generally four levels in the electronic system packaging hierarchy as illustrated in Figure 11.1. The first level of packaging is *chip-and-module* level, in which integrated circuits on silicon chips (L0 as referred to in the microelectronics industry) are packaged into a module (L1). The second level (L2) of packaging is the card level. The modules are packaged on the function cards that perform various specific functions. The next level of packaging (L3) involves the assembly of cards to the board. The final level (L4) of packaging involves the assembly of various boards to make the system.

Figure 11.1 I The four levels of the electronics packaging hierarchy.

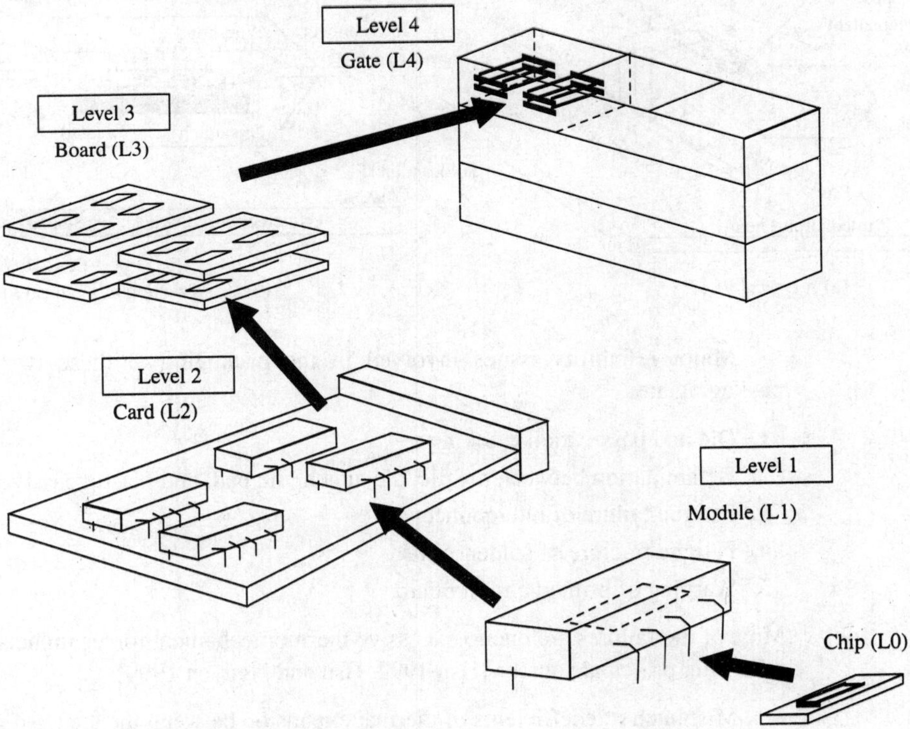

Generally, there are two types of encapsulations for the silicon die in which the integrated circuits are situated: ceramics and plastics. Ceramics offer hermetic sealing, high resistance to environmental effects, and greater durability. However, a ceramic package is costly to produce. Plastics, on the other hand, are less expensive but are vulnerable to degradation due to moisture and temperature effects. Currently, plastic-encapsulated microcircuits (PEMs) make up 95 percent of microelectronic chips in the global market. Figure 11.2 illustrates levels 1 and 2 of microelectronic packaging. A typical configuration of PEMs is presented in Figure 11.2a, and its cross section is illustrated in Figure 11.2b.

The silicon die, in the form of a thin-film containing integrated circuit, is bonded to a metal die pad by die-attach materials. Electrical leads from the IC in the silicon die to the interconnects, are made by fine wire bonds. These components are encapsulated in an epoxy molding compound by means of injection molding. There are generally two types of interconnect: J leads and gull-wing leads, as shown in Figure 11.2b. The interconnects are attached to the printed (or wire-bonded) circuit board using solders in the case of surface mount packaging.

Figure 11.2 | A typical plastic-encapsulated microcircuit.

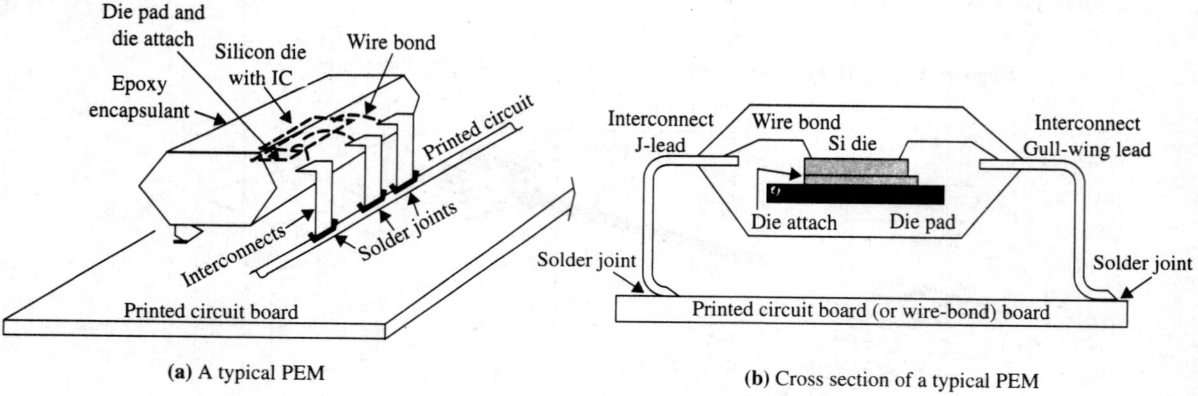

Major reliability issues involved in the packaging at these two levels of packaging are:

1. Die and passivation cracking
2. Delamination between the die, die attach, die pad, and plastic passivation
3. Fatigue failure of interconnects
4. Fatigue fracture of solder joints
5. Warping of printed circuit board

Most of the failures are due to excessive thermomechanical forces induced from the following principal sources [Hsu 1992, Hsu and Nguyen 1999]:

- Mismatch of coefficients of thermal expansion between the attached materials
- Fatigue fracture of materials due to thermal cycling and mechanical vibration
- Deterioration of material strength due to environmental effects such as moisture

- Intrinsic stresses and strains from microfabrication processes as described in Chapter 8

All the above sources of problems that appear in microelectronic packaging are also present in microsystem packaging.

11.3 | MICROSYSTEM PACKAGING

We presented the many differences between microelectronics and microsystems in Table 1.1 in Chapter 1. For this reason, the assembly and packaging of MEMS and microsystems often require significantly different approaches than those used in microelectronic packaging. There are no generic standard packaging materials and methodologies adopted by the industry at the present time. Most MEMS and microsystem packaging has been carried out on the basis of specific applications by the industry. Little has been reported in the public domain on the strategies and methodologies used in packaging MEMS and microsystem products.

A major challenge to MEMS and microsystem packaging is that the core elements of these products, such as microsensors and actuators, usually involve delicate, complex three-dimensional geometry made of layers of dissimilar materials. They are often required to interface with environmentally unfriendly working media such as hot pressurized fluids or toxic chemicals. Yet they are expected to generate a variety of signals—e.g., mechanical, optical, biological, chemical—as shown in Table 11.1 [National Research Council 1997]. These incoming signals need to be converted into electronic signals for speedy and accurate interpretation of the results. In the case of actuators, the interface of the actuating elements with electrostatic power or thermal power sources presents another major challenge to design engineers.

Table 11.1 | Input and output signals in microsystems

Signals	Input	Output
Chemical	Yes	Yes
Electrical	Yes	Yes
Fluid/hydraulic	Yes	Yes
Magnetic		Yes
Mechanical	Yes	Yes
Optical	Yes	Yes

11.3.1 General Considerations in Packaging Design

Proper packaging of MEMS and microsystems products is a critical factor in the overall product development cycle. There are indeed many aspects to be considered in such an endeavor. Following are the principal design requirements that engineers should consider before embarking on the detailed design analyses:

1. The required costs in manufacturing, assembly, and packaging of the components
2. The expected environmental effects, such as temperature, humidity, and chemical toxicity that the product is designed for

3. Adequate overcapacity in the packaging design for mishandling and accidents
4. Proper choice of materials for the reliability of the package
5. Achieving minimum electrical feed-through and bonds in order to minimize the probability of wire breakage and malfunctioning

Because of the great variety of market demands for MEMS and microsystems, structural geometry, materials, and configurations for MEMS and microsystems are far from being standardized. We will thus focus our attention on the packaging of some more common devices and systems, such as micropressure sensors and actuators as presented in Chapter 2. We will further assume that substrate materials are made of silicon unless otherwise specified. The packaging of most components produced by the LIGA process is thus excluded.

11.3.2 The Three Levels of Microsystem Packaging

Unlike electronic packaging with a hierarchy of four levels, microsystem packaging can be categorized in three levels:

Level 1: die level
Level 2: device level
Level 3: system level

The relationships among these three levels of packaging are illustrated in Figure 11.3. We may view the level 1 microsystem packaging as parallel to levels 1 and 2 in microelectronic packaging as illustrated in Figure 11.1, whereas the other two levels of microsystem packaging are similar to levels 3 and 4 in microelectronic packaging.

Figure 11.3 I The three levels of microsystem packaging.

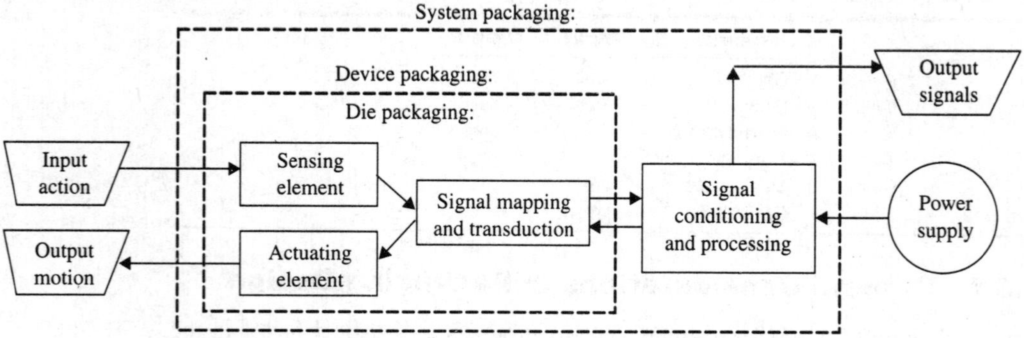

11.3.3 Die-Level Packaging

Die-level packaging involves the assembly and protection of many delicate components in microdevices. Examples are:

The die/diaphragm/constraint base assembly of the pressure sensor in Figures 2.7 and 2.8

The cantilever of the actuator in Figures 2.17 and 2.19

The diaphragm–seismic mass assembly in the accelerometer in Figures 2.33 to 2.36

The electrodes in micromotors in Figures 2.27 and 2.28

The microfluidics components that include microvalves in Figures 2.29 and 2.30, micropumps in Figure 2.31, and microchannels in Figures 10.11 and 10.14

The primary objectives of this level packaging are:

1. To protect the die or other core elements from plastic deformation and cracking
2. To protect the active circuitry for signal transduction of the system
3. To provide necessary electrical and mechanical isolation of these elements
4. To ensure that the system functions at both normal operating and overload conditions

Die-level packaging for many MEMS and microsystems also involves wire bonds for electronic signal transmission and transduction, such as the embedded piezoresistors in a pressure sensor die and the circuits that connect them (see Figs. 2.8 and 10.4). In some cases, that includes bonding the lead wires to the interconnects as illustrated in Figures 1.2 and 11.4, for micropressure sensors, and Figure 11.5, for microaccelerometers.

Figure 11.4 | Die-level packaging of micro pressure sensors.

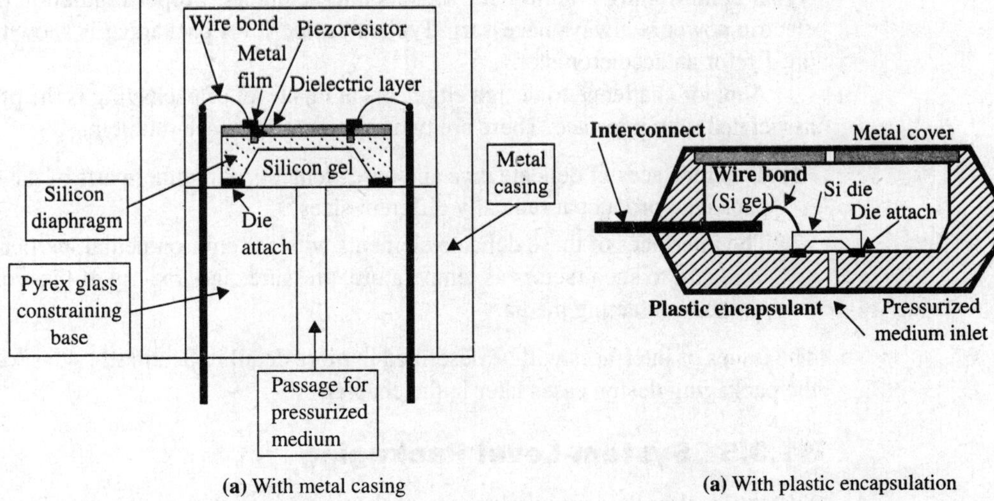

By comparing of Figure 11.2 with Figures 11.4 and 11.5, one will realize that there are indeed common features between microelectronics and MEMS packaging at this level. The similarities of the two packaging technologies include (1) both use silicon dies, (2) die attaches are involved, and (3) there are wire bonds between the die and interconnect. Consequently, the technologies developed and used in microelectronic packaging in these three areas can be used for the MEMS packaging.

Figure 11.5 I Die-level packaging of microaccelerometers.

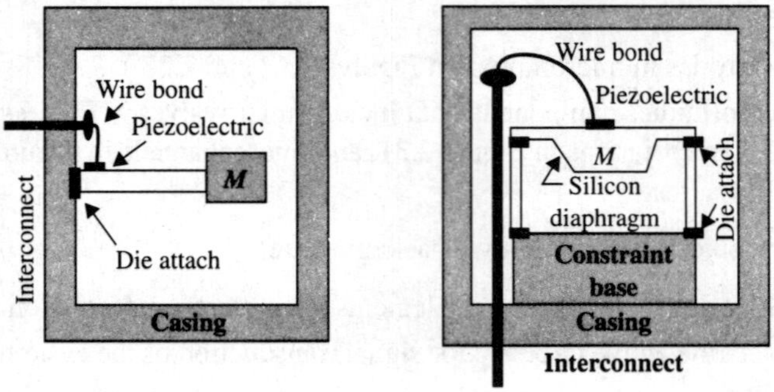

Encapsulation of dies and wire bonds in microsystem packaging can vary from plastic to stainless steel casings (Figure 11.4). The latter encapsulation material is not used in microelectronic packaging.

There are several other issues relating to die isolation and protection in this level of packaging. We will deal with these issues in specific packaging design cases later in the chapter.

11.3.4 Device-Level Packaging

Device packaging as categorized in Figure 11.3 requires the inclusion of proper signal conditioning and processing, which in most cases involves electric bridges and signal conditioning circuitry for sensors and actuators. Proper regulation of input electric power is always necessary. Typical device-level packaging is shown in Figure 1.6 for an accelerometer.

A major challenge to design engineers at this level of packaging is the problems associated with interface. There are two aspects to interface problems:

1. The interfaces of delicate dice and core elements with other parts of the packaged product at radically different sizes
2. The interfaces of these delicate elements with the environmental, particularly in regard to such factors as temperature, pressure, and toxicity of the working and the contacting media

The issues of interfaces will be described in more detail in Section 11.4, as well as in the packaging design cases later in the chapter.

11.3.5 System-Level Packaging

System-level packaging involves the packaging of primary signal circuitry with the die or core element unit. System packaging requires proper mechanical and thermal isolation as well as electromagnetic shielding of the circuitry. Metal housings usually give excellent protection from mechanical and electromagnetic influences.

The interface issue at this level of packaging is primarily the fitting of components of radically different sizes. Assembly tolerance is a more serious problem at this level of packaging than at the device level. Figures 1.2 and 1.22b show system-level packaging of pressure sensors, whereas Figure 1.8b shows system-level packaging for an accelerometer.

11.4 | INTERFACES IN MICROSYSTEM PACKAGING

As we pointed out in Section 11.3, interfaces between MEMS components and their operating environment often present major challenges to design engineers. These interfaces often make the selection of proper materials for the components, the signal transduction, and in some cases, the sealing of working media and electromagnetic fields a critical factor in the design of a successful microsystem. We will take an in-depth view of the interfaces problems associated with various kinds of microsystems in the following paragraphs [National Research Council 1997].

Biomedical Interfaces Interfaces in packaging of biomedical sensors and biosensors as described in Section 2.2.2 are much more critical than in other microsystems. Microbiosystems are subject to vigorous government regulation, being treated as medical products. The packaged systems need to be biologically compatible with human systems and they are expected to function for a specific lifetime. Every microbiosystem must be built to satisfy the following requirements related to interfaces:

1. Be inert to chemical attack during the useful lifetime of the unit
2. Allow mixing with biological materials as described in Section 2.2.2 in a well-controlled manner when used as a biosensor
3. Cause no damage or harm to the surrounding biological cells in instrumented catheters such as pacemakers
4. Cause no undesirable chemical reactions such as corrosion between the packaged device and the contacting cells.

Optical Interfaces There are two principal types of optical MEMS used at the present time. One type relates to the direction of light in such devices as microswitches involving mirrors and reflectors, as illustrated in Figure 1.12, and the other type is optical sensors, as described in Section 2.2.4. Both these optical MEMS require:

1. Proper passages for the light beams to be received and reflected
2. Proper coating of the surface on which light beams are received and reflected
3. Enduring quality of the coated surface during the lifetime of the device
4. Freedom from contamination of foreign substances on the exposing surfaces
5. Freedom from moisture in the enclosure; moisture in the environment may cause stiction of minute delicate optomechanical components

All these requirements must be adequately met to assure a quality optical MEMS.

Mechanical Interfaces Mechanical interfacing is a design issue with moving parts in MEMS. These parts need to be interfaced with their driving mechanisms, which may be thermal, fluidic or magnetic, such as the microvalves and pumps in Figures 2.29 and 2.31 and other microactuators and robots. Improper handling of the interfaces may cause serious malfunctions or damage to these device components. An obvious example of this is thermally actuated microvalves or pumps, in which the diaphragms are expected to be in contact with the working media. Thermal contact conditions at the interfaces can result in negative effects on performance or overstress of the diaphragm structure. Mechanical sealing at the interfaces is another major problem to be overcome in this type of MEMS.

Electromechanical Interfaces Electrical insulation, grounding, and shielding are typical problems associated with MEMS and microsystems. These problems are more obvious in the systems that operate at low voltage levels. The intimate electromechanical interfaces in design are illustrated in Example 10.1. Selection of materials for electrical terminals and the shielding of electrical conductors for microdevices is another major consideration in the packaging.

Interfaces in Microfluidics The prominence of microfluidics in chemical and biomedical applications has been described in several places in this book. The particular value of capillary electrophoresis (CE) in these applications is presented in Section 10.7. These systems require precise fluid delivery, thermal and environmental isolation, and mixing. Material compatibility for applications, such as the compatibility of materials of the microchannel and contained solvent, is another critical requirement. The sealing of the fluid flow and the interfaces between the contacting channel walls and the fluid are two major packaging issues relating to the interfaces.

11.5 | ESSENTIAL PACKAGING TECHNOLOGIES

As mentioned at the beginning of this chapter, there is a lack of industry standards in the assembly and packaging of MEMS and microsystems. New techniques and processes as well as materials, are being developed continuously by researchers and engineers. It is not possible to cover all these newly developed packaging technologies in this chapter. What we will cover in this section are the essential technologies that are necessary for packaging most of these products.

11.5.1 Die Preparation

It is a rare practice to use an entire silicon wafer to produce just one die of a MEMS or microsystem, or use one wafer to produce one device. In reality, hundreds of tiny dies required for various parts of microdevices are produced from a single wafer, as illustrated in Figure 11.6. These dies may all be of the same type and size, as in batch production, or have different sizes and shapes (as indicated by the solid black rectangles) to be cut along the dotted lines shown in the figure.

Sawing the wafer (or, in industrial terminology, *wafer dicing*) along the dotted lines as shown in Figure 11.6 is required for the production of individual dies. A common practice for wafer dicing is to mount the thin wafer on a sticky tape. The saw

Figure 11.6 | Silicon wafer with dies.

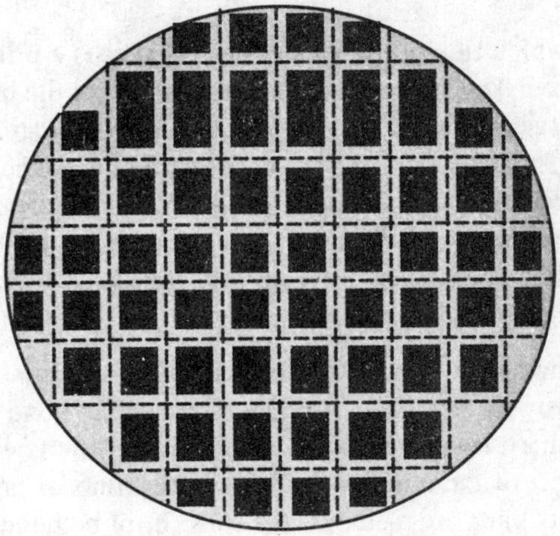

blade is made of a diamond/resin or diamond/nickel composite material. The standard procedure adopted by the microelectronics industry is cutting between the dies imprinted on the silicon wafer. The dies are spaced according to the thickness of the saw blades. For example, the space may be set to about 50 μm with a typical saw blade thickness of 20 μm. The cutting wheel is 75 to 100 mm in diameter. Wafer dicing is normally carried out at a cutting speed of 30,000 to 40,000 rpm.

11.5.2 Surface Bonding

Bonding of microsystem components is a great deal more challenging than that of microelectronics. It is one of the most challenging issues in microsystem packaging because MEMS and microsystems are of three-dimensional geometry and made up of layers of dissimilar materials. Another factor is that many of these systems contain fluids or environmentally hostile substances. Hermetic sealing of these media is required for many of the bonding surfaces. In many cases, bonded dissimilar materials are expected to simultaneously achieve hermetic sealing and provide flexibility at the sealed surface for die isolation, as will be described in more detail in the case of micropressure sensor packaging.

A relentless effort is being made by the microsystem industry to develop new and more effective bonding techniques and procedures. Following is an overview of some of the available techniques.

Adhesives Adhesives are primarily used for attaching dies onto the supporting constraint bases. Typically, the silicon die in a pressure sensor is attached to the glass constraint base by adhesion. There are two common adherents used for this type of bonding: epoxy resins and silicone rubbers.

Epoxy resin bonds provide flexibility for the bonded dies as well as good sealing. Good bonding depends on proper surface treatment and control of the curing process. Unfortunately, epoxy resins are also vulnerable to thermal environments.

The bond should be kept below the glass transition temperature, which is normally around 150 to 175°C.

In the cases in which flexibility of bonding surfaces is a primary requirement, soft adhesives are used. The softest commercially available die bonding material is silicone rubber. One such adhesive material is the room-temperature vulcanizing silicone rubber (RTV). This material cures at room temperature. The soft nature of this type of adherent makes it the most flexible for the die bonding and thus provides the best die isolation. Unfortunately, the chemical resistance of this type of material is not as good, and peeling and flaking develop when it gets in contact with air. It is not suitable for high-pressure applications.

Soldering Bonding of a silicon die to the base, or other similar component bonding, can be achieved by using eutectic solders. This type of bond has the advantage of being chemically inert. It also provides stable and hermetic seals. A good bond requires intimate contact of the bonding surfaces. Thin films of gold in the order of fractional micrometers thick are plated on the surfaces of both the mating parts by a sputtering process, as described in Chapter 8. The normal melting temperature of the gold–silicon system is 370°C. One candidate solder alloy is 60Sn–40Pb. It is nearly eutectic, with a solidus temperature of 183°C. A major shortcoming of solder bonding is its vulnerability to creep at elevated temperature. As most materials, including solder alloys, creep takes place at a temperature above half of the homologous melting temperature. For the silicon–gold solder bond systems, the corresponding homologous melting temperature is about 370°C or 643K. The corresponding half melting temperature for solder alloys is 322K, or 72°C. This means that a soldered silicon bond operating at 72°C or above will be subjected to creep deformation.

Anodic Bonding This process is reliable and effective for attaching silicon wafers to thin glass or quartz substrates. It provides a hermetic seal and is an inexpensive method of die bonding. The procedure is illustrated in Figure 11.7. The silicon wafer is placed on the top of a thin glass constraint base. A high dc voltage source (about 1000 V) is applied across the set at a temperature between 450 and 900°C. A major drawback of this technique is the low aspect ratio of the bonded compounds. A silicon wafer 500 μm thick was successfully bonded to 750 to 3000-μm-thick Pyrex glass substrates.

Figure 11.7 | Anodic bonding of silicon to glass.

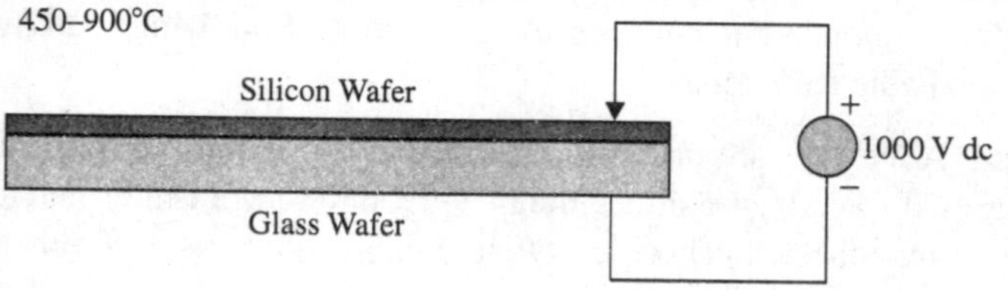

Silicon Fusion Bonding Silicon fusion bonding (SFB) is an effective and reliable technique for bonding two silicon wafers or substrates without the use of intermediate adhesives. The concept of joining two silicon wafers was first developed in the early 1960s for bonding discrete transistor chips [Barth 1990]. It was not until

1988 that the application of this technique in MEMS was first reported [Petersen et al. 1988].

The SFB process begins with thorough cleaning of the bonding surfaces. These surfaces must be polished, then made hydrophilic by exposure to boiling nitric acid. These two surfaces are naturally bonded even at room temperature. However, strong bonding occurs at high temperature in the neighborhood of 1100 to 1400°C. The SFB process is thus considered a chemical bonding process, and the surface treatment is critical to the success of such bonding. Several applications of SFB in micropressure sensors and accelerometers have been reported. It is also a useful process in surface micromachining as reported in the two references cited above.

Silicon-on-Insulator (SOI) We have learned from Table 7.1 in Chapter 7 that silicon is a semiconductor. As such, it has the ability to conduct electricity when it is subjected to high electric potentials or at elevated temperature. For example, silicon becomes increasingly electrical conductive at a temperature above 125°C [Maluf 2000]. This transformation of silicon from a semiconductor to a conductor limits silicon sense elements to being effective in elevated temperature applications. The process of *silicon-on-insulator* offers a viable solution to this problem.

SOI is a process that is used in microelectronics to avoid leakage of charges in p–n junctions [Sze 1985]. It involves bonding the silicon with an amorphous material such as SiO_2 by way of epitaxial crystal growth as described in Section 8.8. An amorphous material does not have long-range order or crystal structure [Askeland 1994].

Figure 11.8 illustrates an SOI process [Maluf 2000]. The process uses two silicon substrates, one with one of its surfaces heavily doped with boron atoms to produce a layer of p-silicon, and the other with a thin silicon oxide film on one of its faces, as shown in Figure 11.8a. The two substrates are then mounted one on the top of the other as shown in Figure 11.8b. The process of silicon fusion bonding joins the two substrates together. The bonded substrates are then exposed to etching to etch the exposed surfaces of the bonded substrates. The heavily p-doped region can act as an etch stop as described in Section 9.2.4. Consequently, one may obtain either a p-silicon layer on the SiO_2 insulator, or sandwiched silicon substrates with SiO_2

Figure 11.8 | A silicon-on-insulator process.

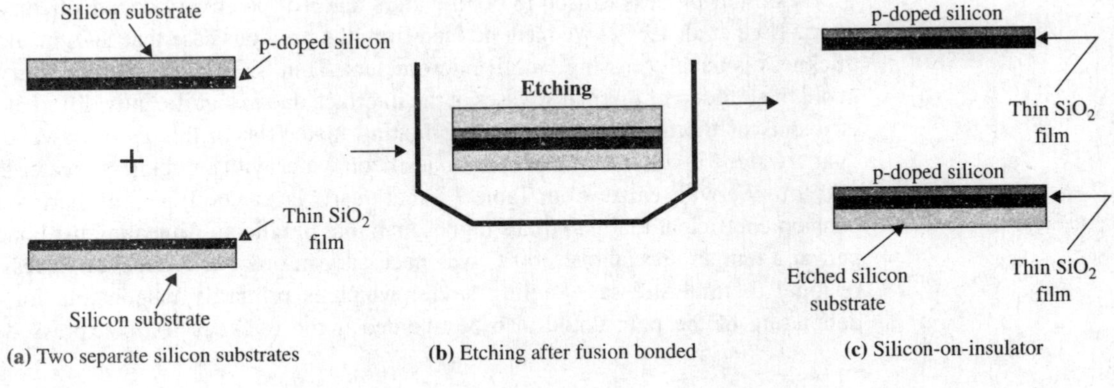

insulator in between as shown in Figure 11.8c. This technique, which involves bonding of substrates followed by etching the bonded substrates is termed the *bonding-and-etchback* technique.

Instead of doping with boron to form a p-silicon layer in the so-called donor wafer, one may implant H_2 ions into the contacting surface. The more active hydrogen ions allow the donor wafer to be bonded to the receptor wafer with SiO_2 grown on its surface at a lower temperature. There is no loss of materials in etching in this case compared with the bonding-and-etchback process described above.

This technique was used to introduce a SiO_2 insulation layer over a microchannel [Yun and Cheung 1998] as illustrated in Figure 11.9. It illustrates the process of introducing a SiO_2 insulator over the top of a silicon substrate. The substrate, called the receptor wafer, contains an embedded microchannel of a depth of 1 μm. A 60-nm-thick oxide film is thermally grown over the surface of the donor wafer. It is then implanted with H^+ ions with a dose of $8 \times 10^{16}/cm^2$ at 40 keV of energy. The two wafers are then bonded face-to-face at room temperature after thorough cleaning of the surfaces. The bonded pair is then heated to a temperature of 470°C briefly. This temperature is sufficient to cause a crack at the peak of the implanted hydrogen region as shown in the figure. The cracking at this face results in the bonded insulator adhering to the receptor wafer after being separated from the donor wafer as illustrated. Consequently, a SiO_2 film insulator is produced to cover the silicon substrate, creating an embedded channel.

Figure 11.9 | SOI for an embedded channel.

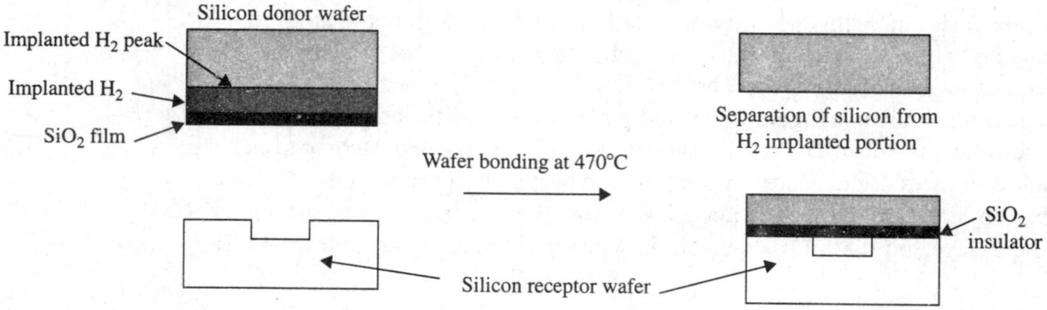

A similar process is used to bond wafers made of positively doped silicon and quartz [Lee et al. 1997]. We have noticed from the previous case that SiO_2 insulator thickness is usually in single-digit micrometers. Thin SiO_2 layers are necessary to avoid the buildup of thermal stresses at the interface due to significantly different coefficients of thermal expansion of the mating materials. In this case, however, a quartz wafer 525 μm thick is bonded to a silicon wafer with a thickness greater than 300 μm. We will realize from Table 7.3 that quartz has a coefficient of thermal expansion coefficient that is 3 times higher than that of silicon. Annealing the bonded pair at a temperature above 200°C was necessary in order to reduce the excessive residual thermal stresses. Peeling stress, which is primarily responsible for the debonding of the pair, could also be avoided if the thickness ratio expressed by

Equation (11.1) is followed. This equation is useful in determining the thickness of the bonding wafers:

$$\frac{t_1}{t_2} = \sqrt{\frac{E_2(1 - \nu_1^2)}{E_1(1 - \nu_2^2)}} \tag{11.1}$$

in which t_1 and t_2 are the respective thicknesses of the mating substrates, E_1 and E_2 are the Young's moduli, and ν_1 and ν_2 are respectively the Poisson's ratios of the mating materials.

Low-Temperature Surface Bonding with Lift-Off Processes Effective bonding techniques such as silicon fusion bonding and the SOI technique involve processes that require high temperatures above 500°C, which may result in adverse effects on the bonded materials. It is desirable to develop bonding processes that can be carried out at low or moderate temperatures. One such technique involves the adhering of a heterostructure of thin-film on a special substrate, or growing the thin film on the substrate by epitaxial deposition as described in Section 8.8. This film is then bonded to the desired substrate (the receptor substrate) by applying mechanical pressure. A lift-off process is used to separate the thin-film from the substrate by passing a UV laser beam through the donor substrate. This technique was successfully used to bond gallium nitride (GaN) to silicon wafer at 200°C [Wong et al. 1999]. This bonding process is outlined below.

The heterostructure film used in this process is gallium nitride (GaN) commonly used in many micro-optical devices. A GaN film in the order of 3 μm thick is deposited on a donor substrate made of sapphire alumina (Al_2O_3). A thin palladium (Pd) film (100 nm thick) is deposited onto the surface of the GaN, to be followed by yet another deposition, indium (In) film 1 μm thick, as illustrated in Figure 11.10a. These two metal film coatings are chosen because Pd adheres well to most semiconductor materials and In has a low melting point, 100 to 200°C, so they can be readily bonded to most semiconductors and polymers. The PdIn alloy has a melting point at 664 to 710°C. The receptor wafer, which is silicon in this case, is deposited with Pd film. Both the receptor and donor substrates are placed one on the top of the other with an applied pressure in the order of 3 MPa for bonding at 200°C (see Fig. 11.10b). The bonding takes about 30 minutes. The bonded unit is then ready for lift-off by a low power excimer KrF pulse laser with 600 mJ/cm^2 energy output at a wavelength of 248 nm and a pulse width of 38 ns as illustrated in Figure 11.10c. The laser radiation results in fracture at the interface of the GaN and the Al_2O_3 sapphire. Consequently, the end result of a solid bond of GaN to a silicon wafer is successfully accomplished. The bonding agent is the palladium–indium ($PdIn_3$) film. This bonding technique can also be used to bond GaN with other substrates such as gallium arsenide (GaAs) or polymers.

11.5.3 Wire Bonding

Wire bonding provides electrical connection to or from the core elements, such as the silicon diaphragm in the pressure sensor illustrated in Figure 11.4, or the actuating members in the microaccelerometer in Figure 11.5. Common wire materials are

Figure 11.10 | Low-temperature bonding with laser lift-off.

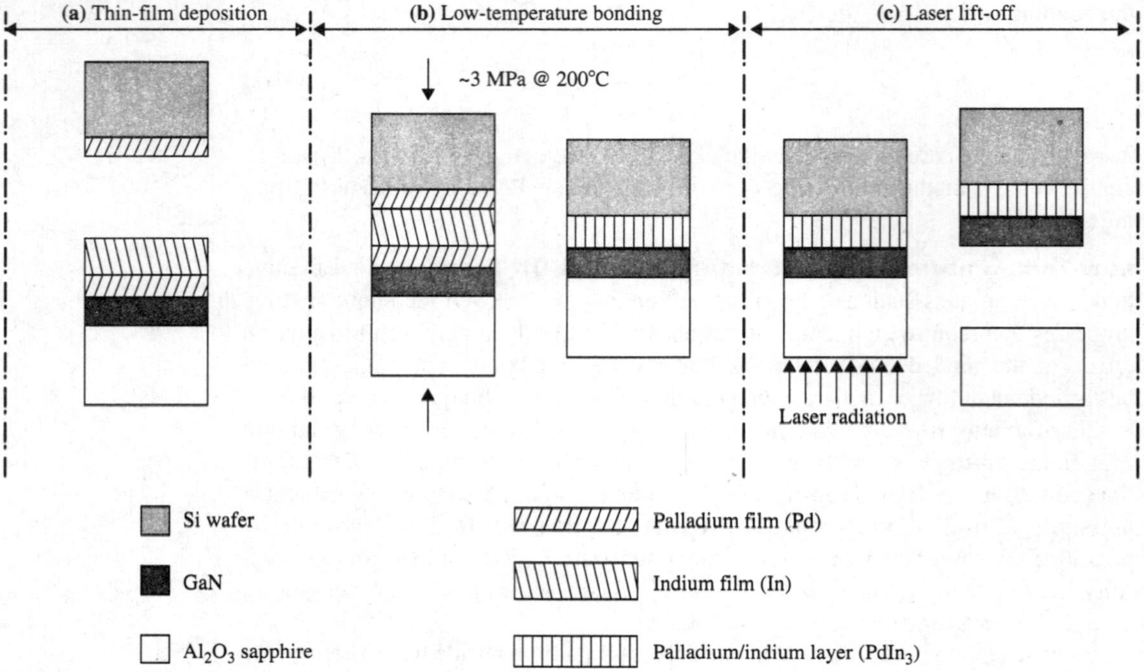

(After Wong et al. [1999].)

gold and aluminum. Other wire materials include copper, silver, and palladium. Wire sizes are in the order of 20 to 80 μm in diameter. Wire bonding techniques that are used in the IC industry are used for microsystems. Three wire bonding techniques are commonly used by the industry: (1) thermocompression, (2) wedge–wedge ultrasonic wire bonding, and (3) thermosonic.

Ultrasonic bonding is used for very fragile structures. Working principles of each of these three techniques are outlined in the subsequent paragraphs. Detailed descriptions of the equipment and the procedures used for wire bonding can be found in a reference [Pecht et al. 1995].

Thermocompression Wire Bonding The principle of thermocompression wire bonding is to press heated metal balls onto metal pads as illustrated in Figure 11.11. In Figure 11.11a, the metal wire is fed through a capillary bonding tool with the tip of the wire above the metal pad. A torch heats the wire tip to about 400°C, at which temperature the wire tip turns into the shape of a ball. At this point, the tool, which has a semispherical mold shape at the end, is lowered to press the ball onto the pad (Fig. 11.11b). After typically 40 ms of pressing, the tool is retracted and the wire is bonded to the pad, as illustrated in Figure 11.11c. The solid bonding of the wire ball to the flat metal pad is accomplished by two physical actions: the plastic deformation of the ball onto the pad by the tool and the atomic interdiffusion of the two bonded materials.

Figure 11.11 | Wire bonding by thermocompression.

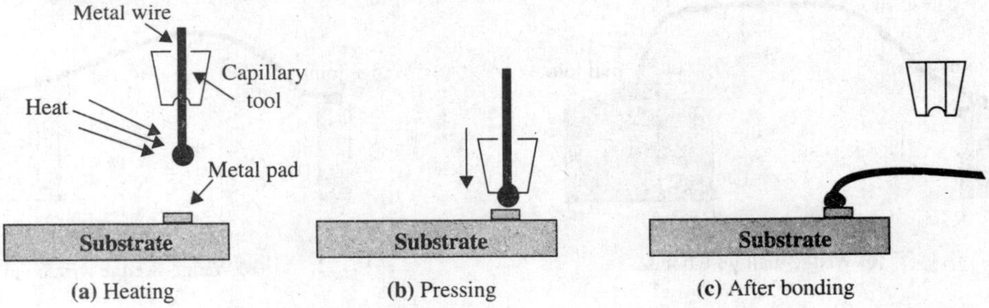

Wedge–wedge Ultrasonic Bonding Unlike thermocompression bonding, ultrasonic wire bonding is a low-temperature process. The bonding tool is in the shape of a wedge, through which the wire is fed. The energy supplied for the bonding is an ultrasonic wave that is generated by a transducer that vibrates the tool at a frequency from 20 to 60 kHz. The bonding tool travels in parallel to the bonding pad as illustrated in Figure 11.12 a. After moving overtop the desired bonding pad, the wedge tool is lowered to the pad's surface. A compressive force is applied and a burst of ultrasonic energy is released to break down the contact surfaces and achieve the desired surface bonding (Fig. 11.12b). Typical time required for bonding is about 20 ms.

Figure 11.12 | Wedge–wedge wire bonding.

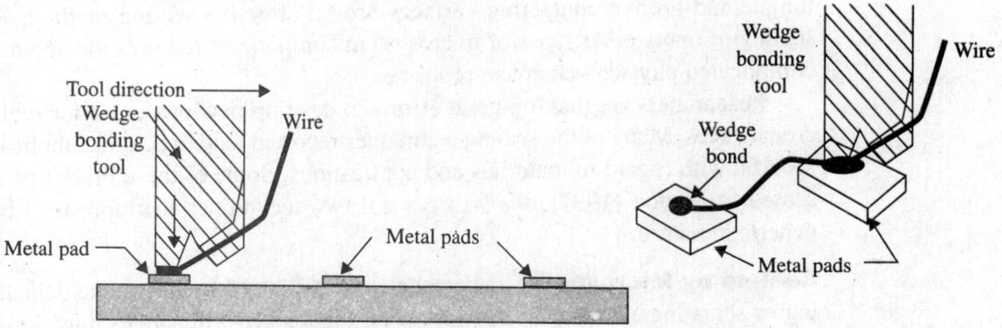

Thermosonic Bonding This technique combines the previous two wire bonding techniques. Ultrasonic energy is used with thermocompression at moderate temperatures of 100 to 150°C. A capillary tool is used to provide ball joints.

For MEMS application, wire bonding is often combined with ball–wedge joints or wedge–wedge joints as illustrated in Figure 11.13.

11.5.4 Sealing

As described in Section 11.4, the interfacing of key components in microdevices is a critical problem in microsystems packaging. Devices, whether they are microsensors or accelerometers with damping, need to be interfaced with the measurand media.

Figure 11.13 | Wire bonding for microsystems.

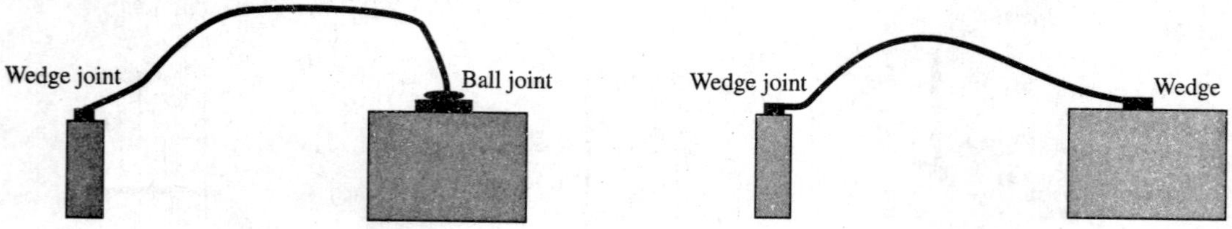

(a) Wedge–ball wire bond (b) Wedge–wedge wire bond

Many other types of microdevices such as microchannels, valves, and pumps in microfluidic systems need to be in contact with working fluids. All these devices and systems require that the contacting fluids be hermetically sealed off from the package, either for the purpose of protecting the delicate core components such as sensing elements, check valves, or actuating beams or for ensuring the proper functioning of the device. Sealing is thus a critical challenge to design engineers.

Figure 11.14 illustrates situations in which sealing of fluid is necessary. Hermetic seals for microchannels are indicated in Figure 10.11.

Mechanical sealing of the interfaces of mating components by epoxy resins is common in microchannels in fluidic systems, as illustrated in Figure 10.11. This method of sealing is usually adequate for systems that do not operate at elevated temperature and/or with toxic working media, and they are not expected to last for a long period of time. In microvalves and pumps, micro O-rings in conjunction with tongue-and-groove contacting surfaces are an effective sealing method. Reliable sealing of many other types of microsystem components requires the application of complicated physical–chemical processes.

Researchers are making great efforts to develop new processes for sealing micropackages. Many of the sealing techniques reported tend to be too complicated and specific with regard to materials and applications. Some of these processes are presented in Madou [1997]. We will present two techniques that appear to be more generic in nature.

Sealing by Microshells Microshells are produced to protect the delicate sensing or actuating elements in microdevices. Figure 11.15 illustrates how these shells are produced.

A surface micromachining technique is used to produce microshells. The procedure involves depositing a sacrificial layer over the die to be protected, as shown in Figure 11.15a. A shell material is then deposited over the sacrificial layer. An etching process then follows to remove the sacrificial layer. Consequently, a gap space between the die and the microshell is created as shown in Figure 11.15b. These gaps can be as small as 100 nm.

Reactive Sealing Technique This technique relies on specific chemical reactions to produce the necessary sealing of the mating components. As illustrated in Figure 11.16, the encapsulant cover for the die is initially placed on the top of the

Figure 11.14 | Required sealing in micropressure sensors, microvalves, and micropumps.

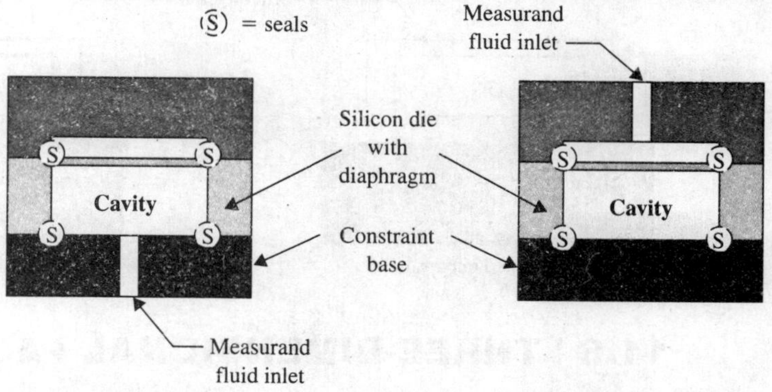

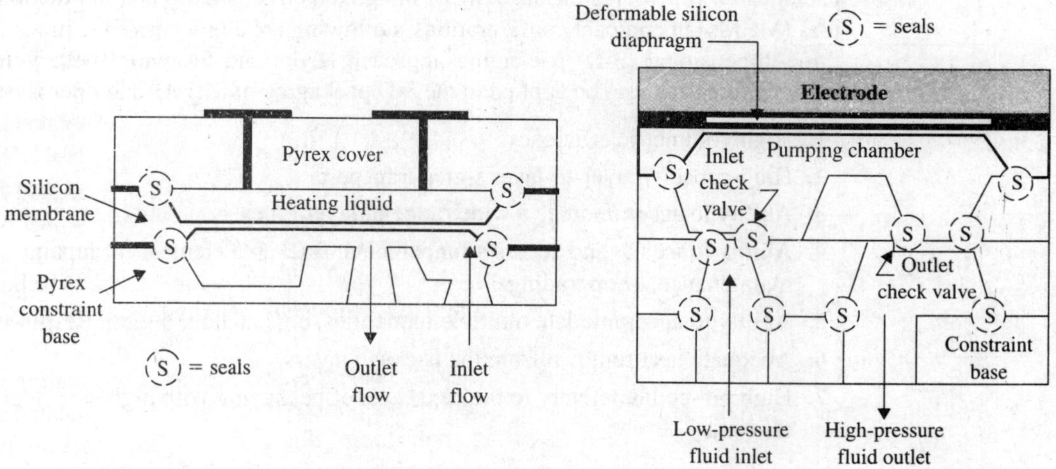

Figure 11.15 | Sealing by microshell.

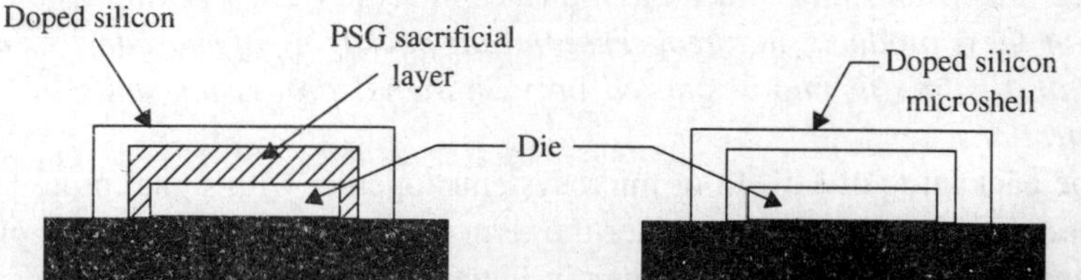

(a) With sacrificial layer (b) With sacrificial layer etched away

die/constraint base with small gaps (Fig. 11.16a). The unit is subject to a chemical reaction such as the thermal oxidation process described in Section 8.5. The growth of SiO_2 from both ends of the silicon encapsulant and the constraint base can provide a reliable and effective seal for the encapsulated die.

Figure 11.16 | Reactive sealing.

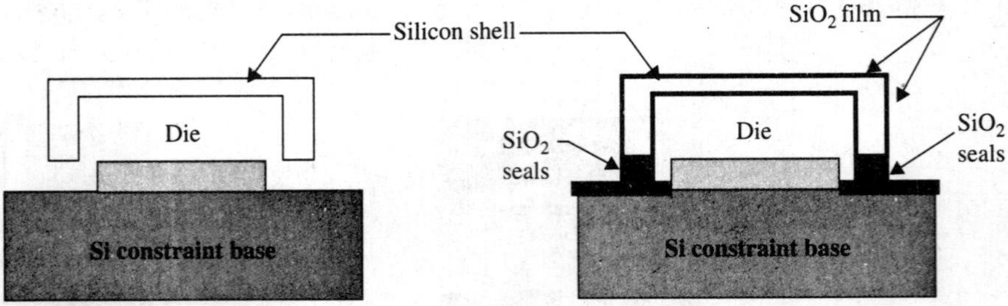

(a) Unsealed encapsulant **(b)** Encapsulation by oxide seals

11.6 | THREE-DIMENSIONAL PACKAGING

Three-dimensional packaging is a relatively mature technology in the microelectronics industry. It involves the stacking of integrated circuits (ICs) and multichip modules (MCMs) in compact configurations. Following are a few desired features in the three-dimensional (3-D) packaging approach [Lyke and Froman 1999]. Some of these features can also be applied to the 3-D packaging of MEMS and microsystems.

1. High volumetric efficiency
2. High-capacity layer-to-layer signal transport
3. Ability to accommodate a wide range of layer types
4. Ability to isolate and access a fundamental stackable element for repair, maintenance, or upgrading
5. Ability to accommodate multiple modalities, e.g., analog, digital, RF power
6. Adequate heat removal from the package layers
7. High pin-count delivery to the next level of packaging with high electrical efficiency

The term *three-dimensional microsystem packaging* may appear ambiguous to many engineers, as we have long recognized the fact that MEMS and most microsystems are three-dimensional in geometry as indicated in Table 1.1. *Three-dimensional packaging for these products, in a real sense, means packaging of microdevices and systems with distinct functions stacked up with signal processing units in compact configurations.*

The packaging of MEMS or microsystems together with signal processing has been practiced by the industry under the term of *lab-on-a-chip*. An example is the packaged microaccelerometers shown in Figure 1.6. A 3-D packaged microsystem would package the system in more compact forms, e.g., by placing the two force-balanced accelerometers in x- and y-axis orientations, one on the top of the other. The stacked accelerometers can further be placed on the top of the signal-processing chip. This concept is illustrated in Figure 11.17. Most of the microfluidics devices using capillary electrophoresis for biological analyses are built as lab-on-a-chip structures.

Stacking layers of MEMS and microsystems of various functions and the signal processing unit requires the application of many of the essential packaging technologies described in Section 11.5. However, shielding of electromagnetic and thermal

Figure 11.17 | Conceptual three-dimensional packaging.

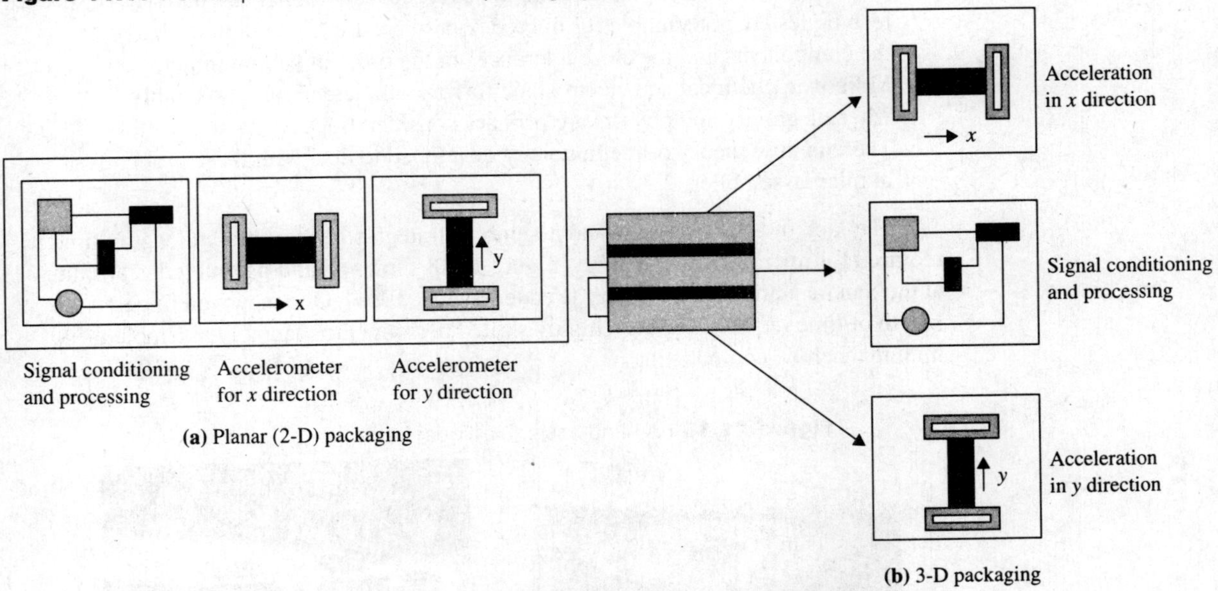

effects, and the hermeticity of moving fluids in many of these systems, become even more critical issues in such packaging. Electromechanical leads must be properly designed, and proper fabrication processes for the interfaces must be selected.

11.7 | ASSEMBLY OF MICROSYSTEMS

Assembly of microsystems is regarded as a part of microsystem packaging, as described in Section 11.1. Assembly of microsystems with components whose sizes range from 1 μm to a few millimeters presents major challenges to engineers in terms of reliability and cost. There are many reasons for the high cost associated with microassembly. Following are just a few of the obvious ones:

1. There is lack of standard procedures and rules for such assemblies. Products are assembled according to the specific procedures chosen on the basis of either individual customer requirements or the personal experience of the design engineer.
2. There is lack of effective tools for microassembly. Tools such as microgrippers, manipulators, and robots are still being developed. Microassemblies also require reliable visual and alignment equipment such as stereo electron microscopes and electron-beam, UV-stimulated beam, or ion beam imaging systems specially designed for microsystem assembly.
3. There is a lack of established methodology in setting proper tolerances of parts in insertion and assembly. There are many sets of tolerances involved in the assembly. The strategies for setting tolerances for parts feeder, grasping surface to mating surface, fixtured surface to mating surface, etc. have not been established for microassembly at the present time.

4. Microcomponents to be assembled are mostly made by physical–chemical processes that have strong material dependence. Traditional assembly techniques are not suitable for microdevices because of the minute size of the components and the close tolerances in the order of submicrometers. Moreover, chemical and electrostatic forces dominate in microassembly, whereas gravity and physics are primary considerations in macroassembly. There is little theory or methodology developed to deal with these problems in microassemblies.

The lack of effective tools and assembly strategies has resulted in lengthy time required for microsystem assembly. Figure 11.18 shows a hand-assembled gear train at the Sandia National Laboratory [Feddema et al. 1999]. One may well imagine the length of time required for even highly skilled personnel to manually assemble these minute machine components.

Figure 11.18 | A hand-assembled gear train.

Press-fit pin transmission:

Tool steel gauge pins, 0.0059 in dia.
Aluminum substrate
Ni gears

(Courtesy of Sandia National Laboratory.)

Despite the many differences between microassembly and traditional assembly in methodologies and tools, major steps are quite similar between the two scales in automated assembly. These steps include:

1. *Part feeding:* Some of the part feeding techniques such as the common tape-bonded feeders used in microelectronics can be adopted in microsystem assembly.
2. *Part grasping:* Micro grippers, manipulators, and robots are desirable tools for this task. However, these tools cannot function properly for handling minute parts without intelligent end effectors. An intelligent end effector requires the integration of gripping, positioning, sensing, and orientation for accurate alignment in microscale of mating parts.
3. *Part mating:* As mentioned earlier, at the micrometer scale, electrostatic and chemical forces dominate the interaction between grippers and parts, as well as between the mating parts. Special design of grippers that can discharge these

forces for possible stiction is necessary for easy releasing of parts from the gripper, and for the mating of small components.

4. *Part bonding and fastening:* Various bonding techniques are available, as described in Section 11.5. Most of these bonding techniques involve microfabrication processes. Automated assembly is possible with batch fabrication of the bonding parts. Other methods of joining parts, including pulse laser deposition, welding, soldering, "snap-fits" based on surface chemistry, and thin-film chemistry can be used for fastening microparts.
5. *Sensing and verification:* Three-dimensional machine vision systems such as stereo microscopy are effective for visual identification of parts and for part alignment. Other types of microsensors, e.g., tactile and thermal sensors, are also required in assembly and inspection. Near- and far-field infrared (IR) sensors can be used for microthermal feedback for monitoring welds or solder joints. Ideally these sensors should be integrated with the grippers and/or other process tooling. One common problem with most of these sensors is the short depth of field near the wavelength of the light source in optical sensing systems. Another problem is the requirement for the sensing head to access the parts that need to be sensed.

As in traditional automated assembly, both serial and parallel assembly processes have been used in microassembly. Self-assembly is an attractive option for free-form fabrication of certain component groups of a microsystem. One such example is the work reported on the self-orienting fluidic transport (SOFT) technique [Smith 1999]. It allows the placement of millions of microscopic objects with $\pm 1\ \mu m$ accuracy each minute. The working principle of this assembly technique involves objects that are micromachined with a trapezoidal cross section and held in a slurry. Matching recesses are micromachined in a target substrate, and assembly of the objects and the recesses takes place randomly but with high yield and accuracy. The objects can be as small as 30 μm in size. This assembly technique provides self-alignment of mating parts at high yield.

Parallel assembly is attractive for mass production of microsystems. A parallel assembly workcell was developed at Sandia National Laboratory [Feddema et al. 1999]. The workcell was originally intended to assemble microdevices made of high-aspect-ratio components. As shown in Figure 11.19, the cell consists of a cartesian robot with a microgripper, a visual servo, and a microassembly planning unit.

The cartesian robot is used to press 385- and 485-μm pins into a substrate (see Fig. 11.20) and then place a 3-in diameter wafer with gears made from the LIGA process onto the pins. Upward- and downward-looking microscopes were used to locate holes in the substrate, the pins to be pressed into the holes, and the gears to be placed on the pins. This vision system can locate parts within 3 μm, whereas the robot can place the parts within 0.4 μm.

11.8 | SELECTION OF PACKAGING MATERIALS

Section 7.11 presented an overview of packaging materials. We will take a closer look at each of these materials in this section. Properties of many of these materials

Figure 11.19 | A microassembly workcell.

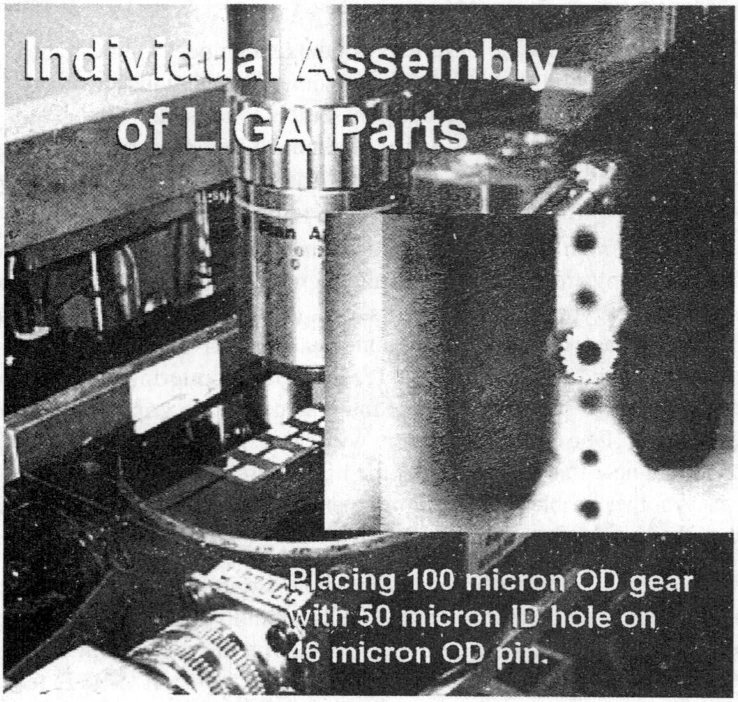

(Courtesy of Sandia National Laboratory.)

Figure 11.20 | Parallel assembly of LIGA components.

Close-up view:

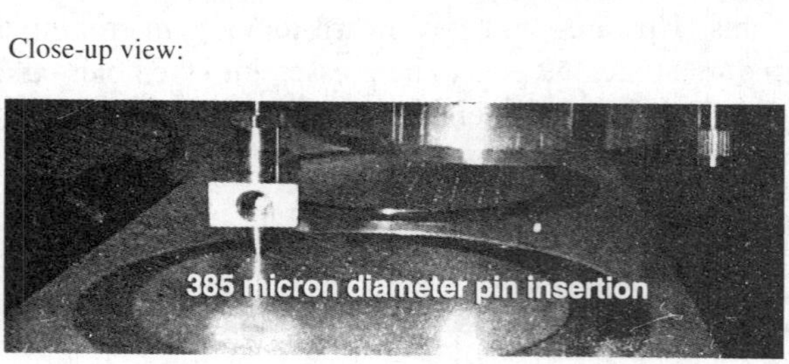

(Courtesy of Sandia National Laboratory.)

may be used for design purposes. Table 11.2 presents a summary of materials that can be used for packaging various parts of a microsystem. Table 11.3 lists properties of some of the frequently used materials for die packaging.

Table 11.2 I Common materials for microsystem packaging

Microsystem components	Available materials	Remarks
Die	Silicon, polycrystalline silicon, GaAs, ceramics, quartz, polymers	Refer to Chapter 7 and Section 10.2.2 for selection
Insulators	SiO_2, Si_3N_4, quartz, polymers	Materials are in order of increasing quality and cost
Constraint base	Glass (Pyrex), quartz, alumina, silicon carbide	Pyrex and alumina are more commonly used materials
Die bonding	Solder alloys, epoxy resins, silicone rubber	Solder for better seal, silicone rubber for better die isolation
Wire bonds	Gold, silver, copper, aluminum, and tungsten	Gold and aluminum are popular choices
Interconnect pins	Copper and aluminum	
Headers and casings	Plastic, aluminum, and stainless steel	

Table 11.3 I Summary of die packaging material properties

Materials	Young's modulus, MPa	Poisson's ratio	Thermal expansion coefficient, ppm/K
Silicon	190,000	0.29	2.33
Alumina	344,830–408,990 (20°C) 344,830–395,010 (500°C)	0.27	6.0–7.0 (25-300°C)
Solder (60Sn40Pb)	31,000	0.44	26
Epoxy (Ablebond 789-3)	4,100		63 below 126° 140 above 126°C
Silicone rubber, RTV (Dow Corning 730)	1.2	0.49	370

Source: Schulze [1998].

Design engineers should be aware of the strong temperature dependent properties of die bonding materials when performing design analyses. Tables 11.4 and 11.5 provide temperature-dependent material properties of solder and epoxy resins listed in Table 11.2. Interpolations and extrapolations can be used to determine properties within and beyond given temperature and strain ranges.

Table 11.4 | Temperature-dependent properties of 60Sn–40Pb solder alloy

Strain range, 10^{-6}	Young's modulus, MPa	Poisson's ratio	Yield strength, MPa
0–500	−40°C: 46,100	0.32	60
	25°C: 27,700	0.43	38
	125°C: 17,000	0.43	14
500–1500	−40°C: 27,800	Same as above	Same as above
	25°C: 16,200		
	125°C: 4,670		
1500–3000	−40°C: 5,600	Same as above	Same as above
	25°C: 5,290		
	125°C: 1,140		
3000–10,000	−40°C: 1,490	Same as above	Same as above
	25°C: 700		
	125°C: 210		

Source: Schulze [1998].

Table 11.5 | Temperature-dependent properties of epoxy resin (Ablebond 789-3)

Strain range, 10^{-6}	Young's modulus, MPa	Poisson's ratio	Fracture strength, MPa
0–500	−40°C: 7990	0.42	55
	25°C: 5930	0.42	60
	125°C: 200	0.49	1.5
500–2000	−40°C: 4680	Same as above	Same as above
	25°C: 4360		
	125°C: 110		
2000–10,000	−40°C: 3830	Same as above	Same as above
	25°C: 3620		
	125°C: 60		
10,000–20,000	−40°C: 3610	Same as above	Same as above
	25°C: 2650		
	125°C: 40		
20,000–30,000	25°C: 1790		
	125°C: 30		

Measured stress versus strain relations for silicone rubber with the trade name RTV show significant scattering. The mean value of Young's modulus for this material is about 1 Pa at all temperatures.

11.9 | SIGNAL MAPPING AND TRANSDUCTION

11.9.1 Typical Electrical Signals in Microsystems

Signal mapping relates to developing and establishing strategies in selecting both the types and positioning of transducers for a microsystem. It is one of the most critical issues in packaging. As we have seen from Section 11.3.4, signal transduction is a major part in the device level packaging of a microsystem. Electrical or mechanical signals generated from the die of a sensor or input to an actuator must be converted into a desired form that can be readily measured and processed. Various transducers have been used for that purpose, as illustrated in Figure 10.2. The electrical signals take the forms shown in Table 11.6.

Voltmeters and ammeters can be used for the measurements of the electrical voltage and current listed in Table 11.6. We will present ways for measuring resistance and capacitance with electric bridges.

Table 11.6 | Common transducers for microsystems

Transducers	Electric signals	Input or output	Typical applications
Piezoresistors	Resistance *R*	Output	Pressure sensors
Piezoelectric	Voltage *V*	Input or output	Actuators, accelerometers
Capacitors	Capacitance *C*	Input or output	Actuators (by electrostatic force), pressure sensors
Electro-resistant heating/Shape memory alloys	Current *i*	Input	Actuators

11.9.2 Measurements of Resistance

A popular way to measure the change of electrical resistance from a strain gage or piezoresistor is to use a Wheatstone bridge. A typical bridge circuit is shown in Figure 11.21. The bridge consists of four resistors connected as shown in the figure. A constant voltage supply V_{in} is applied to the circuit. In this bridge circuit, a strain gage or piezoresistor whose resistance is to be measured after being deformed by a strain replaces the resistor R_1. Let us assign R_g to be the resistance of that strain gage

Figure 11.21 | A typical Wheatstone bridge circuit.

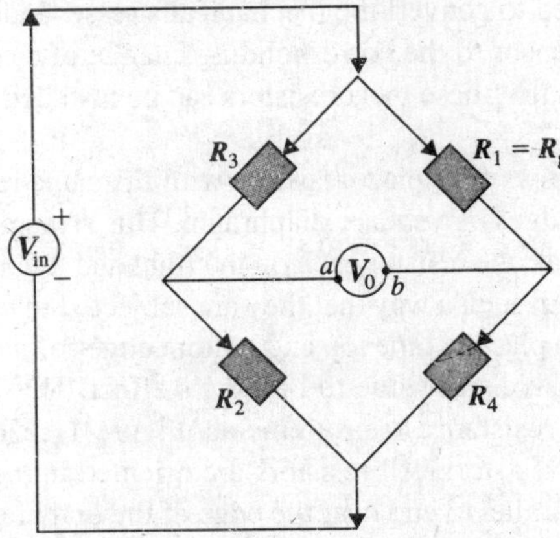

or piezoresistor. There are two measurement modes for the variable resistor R_g: the static balance mode and the dynamic deflection operation mode [Histand and Alciatore 1999]:

- *The static balanced mode:* The voltage V_0 between terminal *a* and *b* is adjusted to zero to achieve a balanced bridge circuit. The following relation can be obtained in such a balanced condition:

$$\frac{R_g}{R_4} = \frac{R_3}{R_2} \tag{11.2}$$

from which we can evaluate the resistance R_g as:

$$R_g = \frac{R_3 R_4}{R_2} \tag{11.3}$$

- *The dynamic deflection operation mode:* For the case in which the resistance in the strain gage or piezoresistor undergoes continuous variation due to changing loads, the dynamic deflection mode is used. In such a case, the output voltage V_0 measured in a continuous mode can be expressed as:

$$V_0 = V_{\text{in}}\left(\frac{R_g}{R_g + R_4} - \frac{R_3}{R_2 + R_3}\right) \tag{11.4}$$

Thus, the change of the resistance in resistor R_g can be determined by the following formula:

$$\frac{\Delta R_g}{R_1} = \frac{\dfrac{R_4}{R_1}\left(\dfrac{\Delta V_0}{V_{\text{in}}} + \dfrac{R_3}{R_2 + R_3}\right)}{1 - \dfrac{\Delta V_0}{V_{\text{in}}} - \dfrac{R_3}{R_2 + R_3}} - 1 \tag{11.5}$$

in which R_1 is the original resistance of R_g.

11.9.3 Signal Mapping and Transduction in Pressure Sensors

Piezoresistors are used to convert the mechanical stresses induced in the diaphragm of a micropressure sensor to the corresponding change of electric resistance. There are a number of ways that these piezoresistors can be installed beneath the surface of the diaphragm.

Figure 11.22a shows four piezoresistors with the same resistivity placed in the midspan of the four edges of a square diaphragm. This arrangement is the most popular design of all. It has one resistor each on the right and left sides of the diaphragm, and they are oriented in such a way that they are subjected to tensile stress. The other two piezoresistors are placed at the top and bottom edges of the diaphragm. They are subjected to compressive stress due to Poisson's effect. This arrangement provides the optimum gain in resistance measurements. Figure 11.22b shows a rectangular diaphragm with four resistors. All resistors are oriented in the same direction. Two resistors are placed parallel to and near the edge of the diaphragm, whereas the other

Figure 11.22 | Arrangements of piezoresistors in pressure sensors.

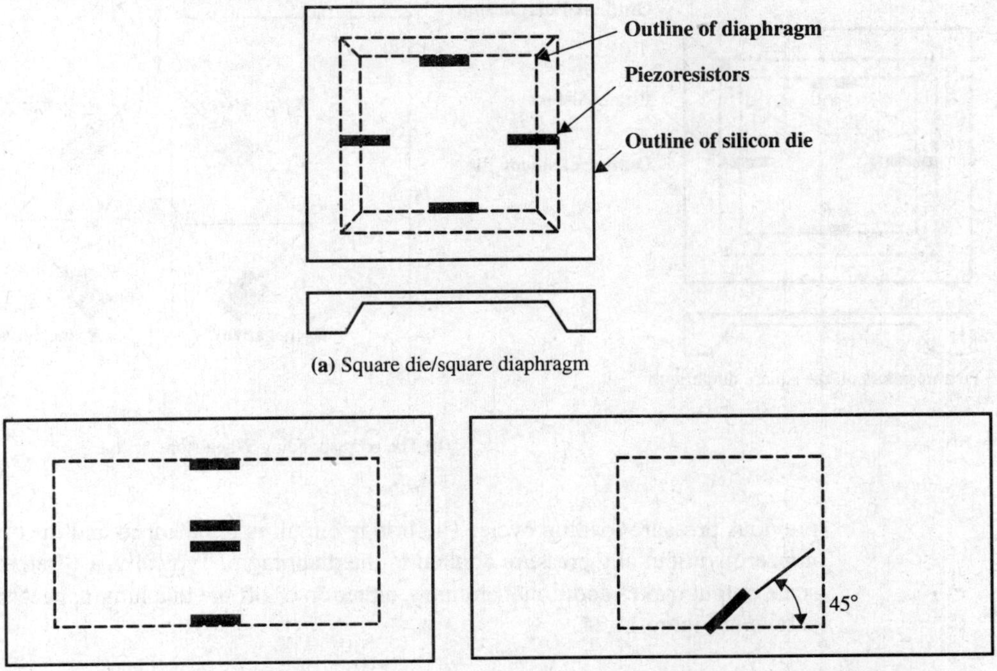

(a) Square die/square diaphragm

(b) Rectangular die/diaphragm

(c) For shear deformation measurements

two are placed near the center. This arrangement can increase the sensitivity of the resistors and allows the use of a single center boss or double bosses to concentrate the stresses and minimize the nonlinearity of the output. The arrangement in Figure 11.22c is for the measurement of the shearing stress. However, because of nonsymmetry of the piezoresistor, this arrangement is more vulnerable to thermal compensation than the other arrangements.

In the case of pressure sensors with a square diaphragm containing four piezoresistors placed as shown in Figure 11.22a, the proper wiring to the Wheatstone bridge is shown in Figure 11.23b, with the piezoresistor designations shown in Figure 11.23a.

Correct measurements of the resistance change in piezoresistors (due to the application of pressure on the diaphragm) by the Wheatstone bridge circuitry require proper conditioning. Two situations warrant special attention:

1. Offset balancing and temperature compensation for the resistivity changes in piezoresistors due to change of temperature coefficient of resistance (TCR) and temperature coefficient of piezoresistance (TCP) as described in Table 7.10 of Section 7.6.
2. Sensitivity normalization, linearization, and temperature compensation due to thermal expansion of piezoresistors at elevated temperature.

The offset voltage results from a mismatch of the Wheatstone bridge circuit, or by possible residual stresses in either the diaphragm or the piezoresistors from a

Figure 11.23 | Electric bridge for pressure sensors with square diaphragms.

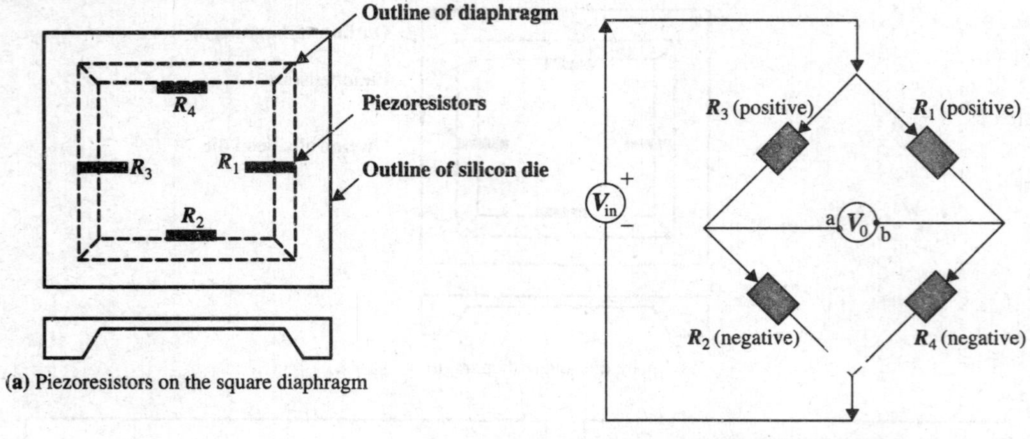

(a) Piezoresistors on the square diaphragm

(b) The corresponding Wheatstone bridge

previous pressure loading cycle. The bridge circuit is unbalanced and the output is nonzero without any pressure applied to the diaphragm. Typically, a 1 percent full-scale output makes additional circuitry, either on or off the diaphragm, necessary, as shown in Figure 11.24.

Figure 11.24 | Typical circuitry for piezoresistors in pressure sensors.

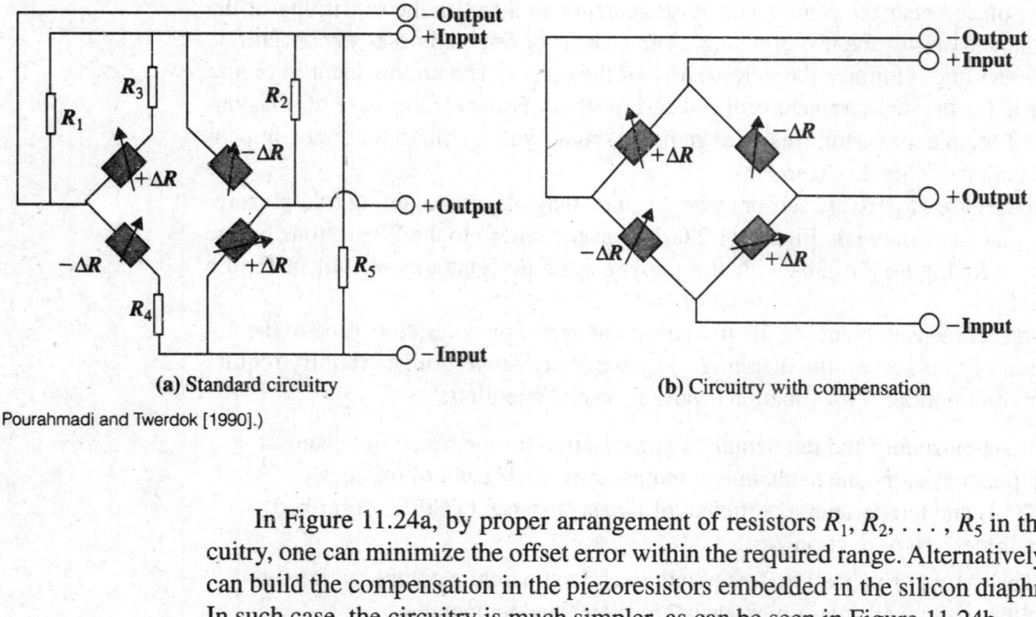

(a) Standard circuitry

(b) Circuitry with compensation

(From Pourahmadi and Twerdok [1990].)

In Figure 11.24a, by proper arrangement of resistors $R_1, R_2, \ldots, R_5$ in the circuitry, one can minimize the offset error within the required range. Alternatively, one can build the compensation in the piezoresistors embedded in the silicon diaphragm. In such case, the circuitry is much simpler, as can be seen in Figure 11.24b.

11.9.4 Capacitance Measurements

Capacitors are common transducers as well as actuators used in microsystems. The use of capacitance transducers in micropressure sensors is illustrated in Figures 2.9 and 2.11 in Chapter 2. The capacitance variation in a capacitor can be measured by simple circuits such as illustrated in Figure 11.25.

Figure 11.25 | A typical bridge for capacitance measurements.

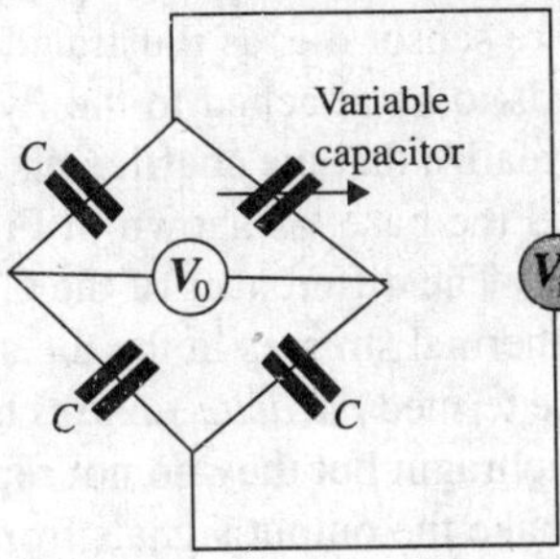

(From Bradley et al. [1991].)

The electrical bridge in Figure 11.25 is similar to that of the Wheatstone bridge for resistance measurements. The variable capacitance can be measured by measuring the output voltage V_0. The following relationship can be used to determine the variable capacitance in the circuit:

$$\Delta C = \frac{4C\, V_0}{V_{in} - 2V_0} \tag{11.6}$$

where ΔC is the capacitance change in the capacitor in the microsystem and C is the capacitance of the other capacitors in the bridge. The bridge is subjected to a constant voltage supply designated by V_{in}.

11.10 | DESIGN CASE: PRESSURE SENSOR PACKAGING

The following packaging design case for a micropressure sensor is presented to illustrate the major steps involved in such design.

Dies and Substrate Preparations This level of packaging begins with the configuration of silicon dies and the production of these dies from wafers. The design engineer must bear in mind the constraints imposed by using the standard wafer sizes and dimensions as presented in Section 7.4.2. The thickness of silicon substrates or dies is limited by whatever wafer size is chosen. A careful assessment of the maximum number of dies or substrates that one can cut from a standard wafer is thus necessary. Wafer dicing is described in Section 11.5.1. Many companies have their own established procedures in cutting thin wafers into small dies and substrates.

Primary Packaging Consideration Dies are the most critical component for proper functioning of a MEMS device or microsystem. As illustrated in Chapter 2, dies are the components that generate or receive signals. It is essential that these dies are packaged in such a way that the surroundings and the environment conditions will not affect their intended mechanical or electrical performance. *Die isolation* is thus the primary consideration in packaging a micropressure sensor and other similar MEMS devices or microsystems.

Die isolation often becomes a critical part in packaging design. Let us take a look at a typical micropressure sensor die, as illustrated in Figure 11.26. The silicon die of a pressure sensor needs to be attached to the Pyrex glass constraint base as shown in Figure 11.26a. We realize that the coefficients of thermal expansion (CTE) of the die, the die attach, and the base, as shown in Figure 11.26b, can be significantly different in magnitudes. The differences of the CTE of the individual materials will result in substantial thermal stresses in the die as indicated in Section 4.4.3. These undesirable stresses are termed *parasite stresses* because the thermal environment induces them in the diaphragm but they do not represent the applied pressure. These parasite stresses can make the output signals from the transducers inaccurate and introduce significant error in the measured results.

Figure 11.26 | Schemes for die isolation.

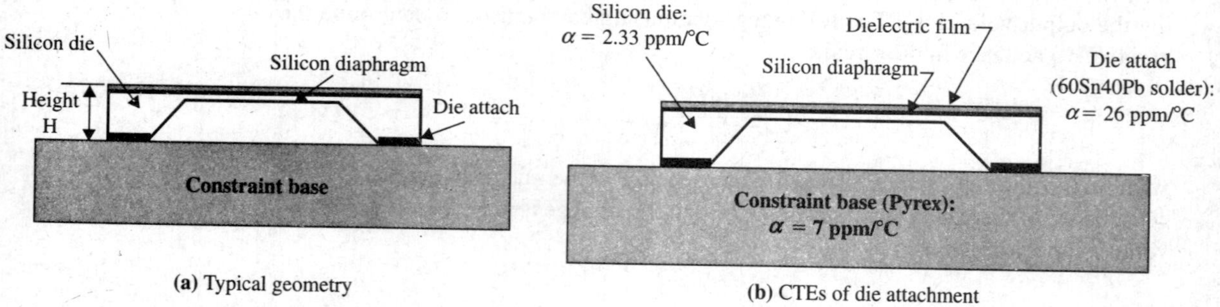

Die isolation may be accomplished by proper design of the die configuration and also the die attach. The height H in Figure 11.26a is an important parameter in isolating the die from the mechanical and thermal effects on the base. A larger value of H will provide more flexibility for the diaphragm. However, the dimension H in a pressure sensor die is constrained by the standard thickness of silicon wafers, from which the cavity is created. A spacer, such as shown in Figure 11.27, can provide additional flexibility of the die support in order to improve die isolation. Unfortunately, this extra component obviously increases the cost of packaging, as well as the overall dimensions of the die.

Die Down This procedure relates to the bonding of the die to the constraint base. The latter can be glasses or ceramics or metals. Die bonding requires the use of proper die attach as illustrated in Figure 11.4. There are generally three ways to attach the die to the constraint bases. These are: (1) anodic bonding, (2) soldering, and

Figure 11.27 | Spacer for die isolation.

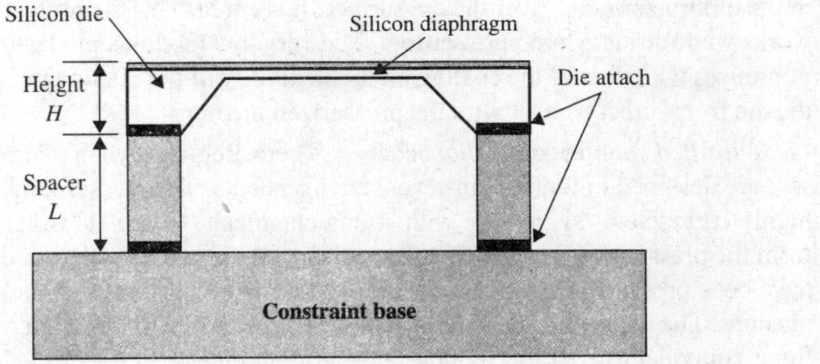

(3) adhesion, as described in Section 11.5.2. Each of these die bonding methods has its advantages and disadvantages, as described in Section 11.5.2.

Die Protection The need for protecting the pressure sensor die from the contacting pressurized medium was described in Section 11.4. Since the dies in micropressure sensors are most likely to be in the situation described above, we may consider three common ways of protecting the dies in this type of sensor [Brysek et al. 1991, Schulze 1998].

1. *Vapor-deposited organic.* A material called Parylene is used for this purpose. This material adds a uniform layer of passivation over the die surface by an LPCVD process but at near room temperature. Annealing takes place after the deposition. The coating is usually thinner than 3.5 μm for sensors operating up to 105°C. The coating has a negative effect on the performance of the device, as it stiffens the diaphragm. Thus, the effect of the thickness of the coating on the diaphragm flexibility is a limiting factor in this practice. However, the technique is useful in protecting the die and the wire bonds of the sensor.

2. *Coating with silicone gel.* A common practice for protecting the die is to apply a coating of silicone gel over the die surface, as shown in Figures 2.11 and 11.28. The

Figure 11.28 | Die protection with silicone gel.

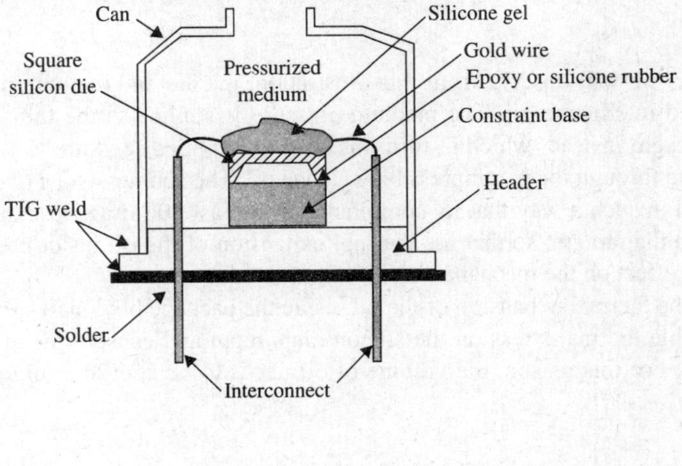

gel consists of one or two parts of siloxanes. After the gel is dispensed as a thick film in the millimeter range over the die surface, it is cured. This die protection technique works well for many pressure sensors. The very low modulus of elasticity of the gel minimizes the effect of the coating on the flexibility of the diaphragm, yet it protects the die from direct contact with the pressurized medium.

3. *Oil-filled stainless-steel diaphragm.* There are times when micropressure sensors are designed to function in severe environments. The pressurizing media can be highly corrosive or aggressive with strong chemicals. Complete isolation of the die from the pressure medium becomes necessary. For example, micro pressure sensors have been produced to measure the air blast pressure in simulated nuclear bomb explosions. The measured pressure is in the range of 70 kPa to 350 MPa with an impact force equivalent to 10 to 20,000g (g = gravitational acceleration). In addition to these extreme mechanical conditions, the measurements needed to be made in harsh environments with high-velocity dust particles, high intensity of light, and above all, a temperature rise to 5000°F in a few milliseconds. In such cases, robust packaging is the only way to ensure the survival and proper functioning of these sensors. Lucas NovaSensor in California produced an unusual pressure sensor packaging for such extreme environment. It involved the use of an oil-filled stainless-steel diaphragm, as schematically shown in Figure 11.29.

Figure 11.29 | Oil-filled stainless-steel diaphragm for die protection.

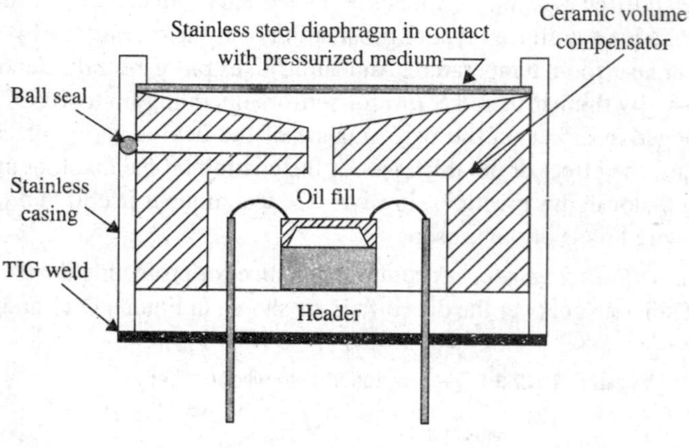

As we will observe from this illustration, the die and the wire bonds are submerged in silicone oil. The medium pressure is applied to the thin stainless-steel diaphragm instead, which in turn transmits the applied pressure to the silicon diaphragm through the incompressible silicone oil. The stainless-steel diaphragm is designed in such a way that its compliance is at least 100 times less than that of the silicon diaphragm, so that the thermal expansion of the oil inside the package has minor effect on the mechanical response of the die.

The thermal expansion of the oil inside the package obviously can generate undesirable thermal stress on the silicon diaphragm and cause error in the measurements. For this reason, the volume of oil needs to be kept to a minimum. Special volume compensators made of low-thermal-expansion materials (e.g., ceramics) are inserted into the package to fill the space between the die and the stainless diaphragm.

The oil used to transmit the pressure has to be baked out and outgassed before it is injected into the package. This type of pressure sensor is used to measure pressures up to 70 MPa. However, the cost of the packaging is very high because of the stringent requirements and the low demand for these special sensors.

Device-Level Packaging Once the critical die and wire bonds are properly secured in the constraint bases, they are assembled for passivation with plastics (see the package at the right of Fig. 1.2), ceramics (the package at the middle of Fig. 1.2), or metallic casings (the package at the left of Fig. 1.2). Major tasks involved in this level of packaging are the design and packaging of signal transduction and the interconnection of the die and the required user connectors. Plastic encapsulation of micropressure sensors is common for those operating at moderate temperature environment. On the other hand, metal encapsulation is used for high-temperature environments. Injection molding is used for plastic-encapsulated packages as in IC manufacturing. The fabrication techniques used for metal packaging involve soldering and welding. There is no standard practice established for microdevice packaging at this time. Virtually every microsystem requires a special design in the final product packaging.

System-Level Packaging In many cases, the devices are shipped as shown in Figure 1.2, with internal arrangements such as illustrated in Figure 11.4. The customer normally includes the sensor and the conditioning electronics in a new package. Although this is a common practice, it is not always economical, as the microsystem package is usually built around existing engineering systems. A better approach is to include only the primary signal-conditioning circuitry in the package of the sensor. In such cases, the package has to provide the necessary mechanical and thermal insulation as well as electromagnetic shielding of the circuitry such as shown in Figures 1.22a and 1.22b. The accuracy of the measurements will be compromised if this protection is not sufficient. A metal housing usually gives adequate protection for mechanical and electrical influences. However, plastic encapsulation of the sensor and the conditioning circuitry is better suited for mass production purposes. In this case, thin metal layers for the necessary electromagnetic shielding are usually required.

PROBLEMS

Multiple Choice

Select the correct answers from the following questions.

1. There are (1) two, (2) three, (3) four levels of packaging in electronic systems.
2. There are (1) two, (2) three, (3) four levels packaging in microsystems.
3. The softest die-attach (bonding) material is (1) solder alloy, (2) epoxy resin, (3) silicone rubber.
4. Silicone gel is used to (1) protect the die, (2) strengthen the die, (3) isolate the die.
5. A spacer is often used in the silicon die of pressure sensors for (1) extra strength, (2) extra size, (3) better die isolation.
6. Anodic bonding of a silicon/glass substrate takes place under (1) high temperature, (2) high temperature and pressure, (3) high temperature and high electric voltage.
7. Silicon fusion bonding takes place under (1) high temperature, (2) high temperature and pressure, (3) high temperature and high electric voltage.
8. Silicon fusion bonding is used to bond (1) silicon to silicon, (2) silicon to glass, (3) silicon to quartz.
9. Bonding followed by lift-off takes place under (1) high temperature, (2) moderate temperature, (3) high temperature and high electric voltage.
10. SOI stands for (1) splitting of ions, (2) silicon on insulator, (3) substrate on insulator.
11. The purpose of SOI is to prevent (1) leakage of electric charge, (2) leakage of thermal effect, (3) spreading of corrosion across the silicon substrates.
12. The SOI process normally takes place at (1) high temperature around 1000°C, (2) medium temperature around 500°C, (3) low temperature below 200°C.
13. The lift-off process takes place at (1) high temperature around 1000°C, (2) medium temperature around 500°C, (3) low temperature below 200°C.
14. The silicon fusion bonding process takes place at (1) high temperature around 1000°C, (2) medium temperature around 500°C, (3) low temperature below 200°C.
15. The wire bonding technique that operates at room temperature is (1) thermocompression, (2) wedge–wedge ultrasonic, (3) thermosonic bonding.
16. Reliable hermetic sealing of microsystems can be accomplished by (1) mechanical means only, (2) electromechanical means, (3) physical–chemical processes.
17. Microshells can be created by (1) bulk micromanufacturing, (2) surface micromachining, (3) the LIGA process.
18. Packaging technologies for microelectronics and microsystems are (1) the same, (2) different, (3) interchangeable.
19. Packaging cost in microsystems is (1) trivial, (2) very significant, (3) somewhat significant in the cost of production.

20. A major challenge in microsystem packaging is to (1) protect the core elements from the contacting environment, (2) prevent these core elements from disintegrating, (3) wire these elements correctly.
21. Packages for bioMEMS must be (1) inert to biological attack of human systems, (2) inert to the body temperature, (3) inert to mishandling by the user.
22. Packaging of optical MEMS requires (1) sensitivity to light, (2) adequate access to light beams, (3) reflection of light beams.
23. One serious packaging problem in microvalves is (1) speed, (2) sensitivity, (3) sealing of fluid.
24. Epoxy resins are (1) suitable, (2) not suitable, (3) not a matter of concern for high-temperature applications
25. Solder bonds are vulnerable to (1) creep, (2) thermal stresses, (3) melting when they are used for high-temperature applications.
26. A reactive sealing process involves (1) chemical reactions, (2) physical treatment, (3) mechanical adherence.
27. Three-dimensional microsystem packaging involves stacking up layers of (1) microcircuits, (2) devices for different functions, (3) different materials.
28. The primary objective of three-dimensional microsystems packaging is to achieve (1) volume efficiency, (2) material efficiency, (3) functional efficiency.
29. One major problem in microsystem assembly is lack of (1) standard, (2) practice, (3) funds.
30. One major problem in automated microassembly is lack of (1) strategy on parts and gripper tolerances, (2) interest, (3) demand.
31. One problem in gripping microcomponents is (1) achieving the required gripping pressure, (2) the mitigation of chemical and electrostatic forces, (3) high gripping sensitivity.
32. Signal mapping is related to (1) mapping out the transducers used in the system, (2) setting the strategy for selecting and positioning the transducers, (3) tracing transducers in the system.
33. A Wheatstone bridge is used to measure (1) electric current, (2) electric voltage, (3) electric resistance.
34. A Wheatstone bridge can be used to measure electric resistance in (1) static, (2) dynamic, (3) both static and dynamic situations.
35. Sensitivity normalization in electric resistance measurements is to compensate for (1) thermal expansion of the device, (2) the change of resistivity of the gage due to temperature, (3) thermal expansion of the gage.
36. Offset balance in electric resistance measurements is required to compensate for (1) thermal expansion of the device, (2) the change of resistivity of the gage due to temperature, (3) thermal expansion of the gage.
37. Silicone gel is a die protective material for (1) high temperature, (2) moderate temperature, (3) all temperatures of operation.

38. A piezoresistor is implanted in a rectangular silicon diaphragm at a 45° angle with the longer edge to measure (1) the normal bending strain, (2) the shearing strain, (3) both normal and shearing strains of the diaphragm.
39. Hydrogen ions are implanted in a silicon substrate in the SOI process to (1) act as a fracture initiator, (2) initiate a chemical process, (3) introduce an insulation layer.
40. Microsystem packaging takes up on average (1) 60, (2) 70, (3) 80 percent of total production cost with current technology.

BIBLIOGRAPHY

Abramowitz, M., and Stegun, I., *Handbook of Mathematical Functions*, Dover Publications, New York, 1964.

Allaire, P. E., *Basics of the Finite Element Method*, Wm. C. Brown Publishers, Dubuque, Iowa, 1985.

Angell, J. B., Terry, S. C. and Barth, P. W., "Silicon Micromechanical Devices," *Scientific American*, vol. 248, no. 4, April 1983, pp. 44–55.

Askeland, D. *The Science and Engineering of Materials*, 3rd ed., PWS Publishing, Boston, 1994.

ASTM, *Manual on the Use of Thermocouples in Temperature Measurement*, American Society for Testing and Materials, ASTM Special Technical Publication 470A, 1974.

Bart, S. F., Tavrow, L. S., Mehregany, M., and Lang, J., "Microfabricated Electrohydrodynamic Pumps," *Sensors and Actuators*, vol. A21–A23, 1990, pp. 193–197.

Barth, P. W., "Silicon Fusion Bonding for Fabrication of Sensors, Actuators and Microstructures," *Transducers '89*, Proceedings of the 5th International Conference on Solid-State Sensors and Actuators and Eurosensors III, vol. 2, 1990, pp. 919–926.

Bathe, K–J., and Wilson, E. L., *Numerical Methods in Finite Element Analysis*, Prentice Hall, Englewood Cliffs, New Jersey, 1976.

Bean, K., "Anisotropic Etching of Silicon," *IEEE Transactions on Electron Devices*, vol. ED-25, no. 10, October 1978, pp. 1185–1193.

Beckwith, T. G., Marangoni, R. D., and Lienhard V. J. H., *Mechanical Measurements*, 5th ed., Addison–Wesley, Reading, Massachusetts, 1993.

Beer, F. P., and Johnston, Jr., E. R., *Vector Mechanics for Engineers*, 5th ed., 1988.

Beskok, A., and Karniadakis, G. E., "A Model for Flows in Channels, Pipes, and Ducts at Micro and Nano Scales," *Microscale Thermophysical Engineering*, vol. 3, no. 1, 1999, pp. 43–77.

Bley, P., "The LIGA Process for Fabrication of Three-Dimensional Microscale Structures," *Interdisciplinary Science Reviews*, vol. 18, no. 3, 1993, pp. 267–272.

Bley, P., "Polymers—An Excellent and Increasingly Used Material for Microsystems," *SPIE 1999 Symposium on Micromachining and Microfabrication,* Santa Clara, California, September 20–22, 1999.

Boley, B. A., and Weiner, J. H., *Theory of Thermal Stresses*, John Wiley & Sons, New York, 1960.

Bouwstra, S., and Geijselaers, B. "On the Resonance Frequencies of Micro Bridges," *1991 International Conference on Solid-State Sensors and Actuators (Transducers '91),* San Francisco, 1991, pp. 538–542.

Bowen, C. R., Stevens, R., Perry, A., and Mahon, S. W., "Optimization of Materials and Microstructure in 3–3 Piezocomposites," *Proceedings of Engineering Design Conference 2000 on Design for Excellence*, Sivaloganathan, S., and Andrew, P. T. J., eds., June 27–29, 2000, Professional Engineering Publishing, London, 2000, pp. 361–370.

Bradley, D. A., Dawson, D., Burd, N. C., and Loader, A. J., *Mechatronics*, Chapman & Hall, London, 1993.

Brown, T. L., and LeMay, Jr., H. E., *Chemistry*, 2nd ed., Prentice-Hall, Englewood Cliffs, New Jersey, 1981.

Bryzek, J., Petersen, K., Mallon, Jr., J. R., Christel, L., and Pourahmadi, F., *Silicon Sensors and Microstructures,* Lucas NovaSensors, Fremont, California, 1991.

Bryzek, J., Petersen, K., Christel, L., and Pourahmadi, F., "New Technologies for Silicon Accelerometers Enable Automotive Application," Sensors and Actuators 1992, sp-903, SAE International, 1992, pp. 25–32.

Bryzek, J., Petersen, K., and McCulley, W., "Micromachines on the March," *IEEE Spectrum*, May 1994, pp. 20–31.

Buerk, D. G., *Biosensors, Theory and Applications*, Technomic Publishing, Lancaster, Pennsylvania, 1993.

Chau, K. H.–L., Lewis, S. R., Zhao, Y., Howe, R. T., Bart, S. F., and Marcheselli, R. G., "An Integrated Force-Balanced Capacitive Accelerometer for Low-G Applications," *Transducers '95, Proceedings, 8th International Conference on Solid-State Sensors and Actuators,* Stockholm, June 25–29, 1995, pp. 593–596.

Chilton, J. A., and Goosey, M. T., eds., *Special Polymers for Electronics and Optoelectronics*, Chapman & Hall, London, 1995.

Chiou, J., "Pressure Sensors in Automotive Applications and Future Challenges," *IMECE'99*, Nashville, November 17, 1999.

CRC Handbook of Mechanical Engineering, F. Kreith, ed., CRC Press, Boca Raton, 1997.

Choi, I. H., and Wise, K.D., "A Silicon-Thermopile-Based Infrared Sensing Array for Use in Automated Manufacturing," *IEEE Transactions on Electron Devices*, vol. ED-33, no. 1, January 1986, pp. 72–79.

Culbertson, C. T., Jacobson, S. C., and Ramsey, J.M., "Dispersion Sources for Compact Geometrics on Microchips," *Analytical Chemistry*, vol. 70, no. 18, 1998, pp. 3781–3789.

Desai, C. S., *Elementary Finite Element Method*, Prentice Hall, Englewood Cliffs, New Jersey, 1979.

Doscher, J., "Accelerometer Design and Applications," Analog Devices Inc., Norwood, Massachusetts, private communication, 1999.

Dove, R. C., and Adams, P. H. *Experimental Stress Analysis and Motion Measurement*, Charles E. Merrill Publishing, Columbus, Ohio, 1964.

Fan, L.–S., Tai, Y.–C., and Muller, R. S., "IC-Processes for Electrostatic Micro-motors," *IEEE International Electronic Devices Meeting*, December 1988, pp. 666–669.

Feddema, J., Christenson, T., and Polosky, M., "Parallel Assembly of High Aspect Ratio Microstructures," *SPIE Proceedings on Microrobotics and Microassembly*, vol. 3834, Boston, September 21–22, 1999, pp. 153–164.

Flik, M. I., and Tien, C. L., "Size Effect on Thermal Conductivity of High-T_c Thin-Film Superconductors," *Journal of Heat Transfer, ASME Transactions*, vol. 112, 1990, pp. 873–881.

Flik, M. I., Choi, B. I., and Goodson, K. E., "Heat Transfer Regimes in Microstructures," *Journal of Heat Transfer, ASME Transactions*, vol. 114, August 1992, pp. 666–674.

French, P. J., and Evans, A. G. R., "Piezoresistance in Single Crystal and Polycrystalline Si," sec. 3.4 in *Properties of Silicon*, INSPEC, 1988, pp. 94–103.

Gabriel, K. J., Trimmer, W. S. N., and Walker, J. A., "A Micro Rotary Actuator Using Shape Memory Alloys," *Sensors and Actuators*, vol. 15, no. 1, 1988, pp. 95–102.

Giachino, J. M., and Miree, T. J., "The Challenge of Automotive Sensors," *Proceedings of SPIE on Microlithography and Metrology in Micromachining*, vol. 2640, October 1995.

Gilbert, J. R., Ananthasuresh, G. K., and Senturia, S. D., "3D Modeling of Contact Problems and Hysteresis in Coupled Electro-Mechanics," *9th Annual International Workshop on Micro Electro Mechanical Systems, MEMS '96*, San Diego, 1996, pp. 127–132.

Gise, P., and Blanchard, R., *Modern Semiconductor Fabrication Technology*, Prentice Hall, Englewood Cliffs, New Jersey, 1986.

Griego, J., Smith, M., and Wood, J., "MEMS Cell Gripper," senior project report, San Jose State University, San Jose, California, May 2000.

Halliday, D., Resnick, R., and Krane, K. S., *Physics*, vol. 2, extended version, 4th ed., John Wiley & Sons, New York, 1992.

Harrison, D. J., and Glavina, P. G., "Towards Miniaturized Electrophoresis and Chemical Analysis Systems on Silicon: An Alternative to Chemical Sensors," *Sensors and Actuators B*, vol. 10, 1993, pp. 107–116.

He, Y., Marchetti, J., and Maseeh, F., "MEMS Computer-aided Design," *European Design & Test Conference and Exhibition on Microfabrication*, Paris, March 17–20, 1997.

Heinrich, J. C., and Pepper, D. W., *Intermediate Finite Element Method*, Taylor and Francis, Philadelphia, 1999.

Helvajian, H., and Janson, S.W., "Microengineering Space Systems," *Microengineering Aerospace Systems*, Henry Helvajian, ed., American Institute of Aeronautics and Astronautics, Reston, Virginia, 1999, Chapter 2, pp. 29–72.

Henning, A. K., Fitch, J., Hopkins, D., Lilly, L., Faeth, R., Falsken, E., and Zdeblick, M., "A Thermopneumatically Actuated Microvalve for Liquid Expansion and Proportional Control," *Proceedings, Transducers '97*, IEEE Conference on Sensors and Actuators, June 1997, pp. 825–828.

Henning, A. K., "Microfluidic MEMS," Paper 4.906, *IEEE Aerospace Conference*, Snowmass, Colorado, March 25, 1998.

Higuchi, T., Yamagata, Y., Furutani, K., and Kudoh, K., "Precise Positioning Mechanism Utilizing Rapid Deformations of Piezoelectric Elements," *Proceedings of IEEE, Micro Electro Mechanical Systems*, February 1990, pp. 222–226.

Histand, M. B., and Alciatore, D. G., *Introduction to Mechatronics and Measurement Systems*, McGraw-Hill, New York, 1999.

Howe, R., "Resonant Microsensors," *4th International Conference on Solid-State Sensors and Actuators*, 1987, pp. 843–848.

Hsu, T. R., *The Finite Element Method in Thermomechanics*, Allen & Unwin, London, 1986.

Hsu, T. R., "On Nonlinear Thermomechanical Analysis of IC Packages," Advances in Electronic Paging, ASME, 1992, pp. 325–326.

Hsu, T. R., and Sinha, D. K., *Computer-Aided Design—An Integrated Approach*, West Publication Co., St. Paul, 1992.

Hsu, T. R., and Zheng, X. M., "Tensile Creep Strength of Eutectic Solder Cores," *ASME Winter Annual*

Meeting, New Orleans, November 28–December 3, 1993, paper 93–WA/EEP-11.

Hsu, T. R., Chen, G. G., and Sun, B. K., "A Continuum Damage Mechanics Model Approach for Cyclic Creep Fracture Analysis of Solder Joints," *Advances in Electronic Packaging, Proceedings of the 1993 ASME International Electronics Packaging Conference*, Binghamton, New York, September 29–October 2, 1993, pp. 127–138.

Hsu, T. R., and Sun, N. S., "Residual Stresses/Strains Analysis of MEMS," *Proceedings of MSM '98*, Santa Clara, April 6–8, 1998, pp. 82–87.

Hsu, T. R., and Nguyen, L., "On the Use of Fracture Mechanics in Plastic-Encapsulated Microcircuits Packaging," Proceedings of ASME International Mechanical Engineering Congress & 11th Symposium on Mechanics of Surface Mount Assemblies, November 1999 pp. 1–8.

Janna, W. S., *Introduction to Fluid Mechanics*, 3rd ed., PWS Publishing, Boston, 1993.

Janson, S., Helvajian, H., and Breuer, K., "Micropropulsion Systems for Aircraft and Spacecraft," *Microengineering Aerospace Systems*, Henry Helvajian, ed., American Institute of Aeronautics and Astronautics, Reston, Virginia, 1999, Chapter 17, pp. 657–696.

Janusz, B., Petersen, K., Christel, L., and Pourahmadi, F., "New Technologies for Silicon Accelerometers Enable Automotive Applications," SAE Technical Paper Series 920474, also *Sensors and Actuators*, 1992.

Jerman, H., "Electrically-Activated, Micromachined Diaphragm Valves," *Technical Digest, IEEE Solid-State Sensor and Actuator Workshop*, June 1990, pp. 65–69.

Jerman, H., "Electrically-Activated, Normally-Closed Diaphragm Valves," *Technical Digest, 6th International Conference on Solid-State Sensors and Actuators*, San Francisco, June 1991, pp. 1045–1048.

Jiang, X. N., Zhou, Z. Y., Yao, J., Li, Y., and Ye, X.Y., "Micro-Fluid Flow in Microchannel," *Proceedings of Transducers '95, 8th International Conference on Solid-State Sensors & Actuators & Eurosensors IX*, Stockholm, June 25–29, 1995, vol. 2, pp. 317–320.

Kasap, S. O., *Principles of Electrical Engineering Materials and Devices*, Irwin, Chicago, 1997.

Keneyasu, M., Kurihara, N., Katogi, K., and Tabuchi, K., "An Advanced Engine Knock Detection Module Performance Higher Accurate MBT Control and Fuel Consumption Improvement," *Proceedings of Transducers '95, Eurosensors IX*, 1995, pp. 111–114.

Kim, C. J., Pisano, A. P., and Muller, R. S., "Overhung Electrostatic Microgripper," *Technical Digest, IEEE International Conference on Solid-State Sensors and Actuators*, San Francisco, 1991, pp. 610–613.

Kim, Y. J., and Hsu, T. R., "A Numerical Analysis on Stable Crack Growth Under Increasing Load," *International Journal of Fracture*, vol. 20, 1982, pp. 17–32.

Kovacs, G. T. A., *Micromachined Transducers Sourcebook*, McGraw-Hill, New York, 1998.

Kreith, F., and Bohn, M. S., *Principles of Heat Transfer*, PWS Publishing, Boston, 1997.

Krishnamoorthy, S., and Griridharan, M. G., "Analysis of Sample Injection and Band-Broadening in Capillary Electrophoresis Microchips," *Proceedings of 2000 International Conference on Modeling and Simulation of Microsystems*, San Diego, March 27–29, 2000, pp. 528–531.

Kumar, S., and Cho, D., "Electric Levitation Bearings for Micromotors," *Technical Digest, IEEE International Conference on Solid-State Sensors and Actuators*, San Francisco, 1991, pp. 882–885.

Kwok, H. L., *Electronic Materials*, PWS Publishing, Boston, 1997.

Lee, T. H., Tong, Q. Y., Chao, Y. L., Huang, L. J., and Gosele, U., "Silicon on Quartz by a Smarter Cut Process," *Proceedings of the 8th International Symposium on Silicon-On-Insulator Technology and Devices*, S. Crisstoloveanu, ed., The Electrochemical Society, New Jersey, 1997, pp. 27–32.

Lipman, J., "Microfluidics Puts Big Labs on Small Chips," *EDN Magazine*, December 1999, pp. 79–86.

Lober, T. A., and Howe, R. T., "Surface-Micromachining Processes for Electrostatic Microactuator Fabrication," *Technical Digest, IEEE Solid-State Sensor and Actuator Workshop*, June 1988, pp. 59–62.

Lyke, J., and Forman, G., "Space Electronics Packaging Research and Engineering," *Microengineering Aerospace Systems*, H. Helvajian, ed.. Aerospace Corporation, El Segundo, California, 1999, Chapter 8, pp. 259–346.

MacDonald, G. A., "A Review of Low Cost Accelerometers for Vehicle Dynamics," *Sensors and Actuators*, A21–A23, 1990, pp. 303–307.

Madou, M., *Fundamentals of Microfabrication*, CRC Press, Boca Raton, 1997.

Maluf, N., *An Introduction to Microelectromechanical Systems Engineering*, Artech House, Boston, 2000.

Manz, A., Effenhauser, C. S., Burggraf, N., Harrison, D. J., Seller, K., and Fluri, K., "Electroosmotic Pumping and Electrophoretic Separations for Miniaturized Chemical Analysis Systems," *Journal of Micromechanics and Micro Engineering*, vol. 4, 1994, pp. 257–265.

Marks' Standard Handbook for Mechanical Engineers, 13th ed., E. A. Avallon and T. Baumeister III, eds., McGraw-Hill, New York, 1996.

Mehregany, M., Gabriel, K. J., and Trimmer, W. S. N., "Integrated Fabrication of Polysilicon Mechanisms,"

IEEE Transactions on Electron Devices, vol. ED-35, no. 6, 1988, pp. 719–723.

Mehregany, M., Senturia, S. D., and Lang, J. H., "Friction and Wear in Microfabricated Harmonic Side-Drive Motors, " *Technical Digest, IEEE Solid-State Sensor and Actuator Workshop,* June 1990, pp. 17–22.

Merriam Webster's Collegiate Dictionary, 10th ed., F. C. Mish, ed.-in-chief, Merriam-Webster, Springfield, Massachusetts, 1995.

Moroney, R. M., White, R. M., and Howe, R. T., "Fluid Motion Produced by Ultrasonic Lamb Waves," *IEEE 1990 Ultrasonics Symposium Proceedings*, Honolulu, 1990, pp. 355–358.

Moroney, R. M., White, R. M., and Howe, R. T., "Microtransport Induced by Ultrasonic Waves," *Applied Physics Letters*, vol. 59, 1991, pp. 774–776.

National Research Council, "Microelectromechanical Systems—Advanced Materials and Fabrication Methods, 5: Assembly, Packaging, and Testing," no. NMAB-483, National Academy Press, Washington, D.C., 1997, pp. 38–49.

Newell, W. E., "Miniaturization of Tuning Forks," *Science*, vol. 161, 1968, p. 1320.

Nguyen, L. T., Hsu, T. R., and Kuo, A.Y., "Interfacial Fracture Toughness in Plastic Packages," *Application of Fracture Mechanics in Electronic Packaging*, AMD–vol. 222/EEP–vol. 20, ASME 1997, pp. 15–24.

Nyborg, W. L. M., "Acoustic Streaming," *Physical Acoustics*, W. P. Mason, ed., Academic Press, New York, 1965, pp. 265–331.

O'Connor, L., "MEMS: Microelectromechanical Systems," *Mechanical Engineering*, American Society of Mechanical Engineers, February 1992, pp. 40–47.

Ohnstein, T., Fukiura, T., Ridley, J., and Bonne, U., "Micromachined Silicon Microvalve," *IEEE Micro Electro Mechanical Systems*, February 1990, pp. 95–98.

Ozisik, M. N., *Boundary Value Problems of Heat Conduction*, International Textbook Co., Scranton, Pennsylvania, 1968.

Patankar, N. A., and Hu, H. H., "Numerical Simulation of Electroosmotic Flow," *Analytical Chemistry*, vol. 70, 1998, pp. 1870–1881.

Patterson, G. N. *Introduction to the Kinetic Theory of Gas Flow*, University of Toronto Press, 1971.

Paulsen, J., and Giachino, J., "Powertrain Sensors and Actuators: Driving toward Optimized Vehicle Performance," *39th IEEE Vehicular Conference*, vol. II, 1989, pp. 574–594.

Pecht, M. G., Nguyen, L. T., and Hakim, E. B., eds., Plastic-Encapsulated Microelectronics, John Wiley & Sons, Inc., New York, 1995.

Petersen, K. E., "Silicon as a Mechanical Material," *Proceedings of IEEE*, vol. 70, no. 5, May 1982, pp. 420–457.

Petersen, K., Barth, P., Poydock, J., Brown, J., Mallon Jr., J., and Bryzek, J., "Silicon Fusion Bonding for Pressure Sensors," *Technical Digest, IEEE Solid-State Sensor and Actuator Workshop,* June 1988.

Petersen, K., Pourahmadi, F., Brown, J., Parsons, P., Skinner, M., and Tudor, J., "Resonant Beam Pressure Sensor Fabricated with Silicon Fusion Bonding," Proceedings of Transducers '91, the 1991 International Conference on Solid-State Sensors and Actuators, San Francisco, CA, June 24–27, 1991, pp. 177–180.

Pfahler, J., Harley, J., and Bau, H. H., "Liquid and Gas Transport in Small Channels," *Microstructures, Sensors and Actuators*, DSC–vol. 19, ASME, November 1990, pp. 149–157.

Pottenger, M., Eyre, B., Kruglick, E., and Lin, G., "MEMS: The Maturing of a New Technology," *Solid State Technology*, September 1997, pp. 89–96.

Pourahmadi, F., Christ, L. and Petersen, K., "Variable-Flow Micro-Valve Structure Fabricated with Silicon Fusion Bonding," *Technical Digest, IEEE Solid-State Sensor and Actuator Workshop,* June 1990, pp. 78–81.

Pourahmadi, F., and Twerdok, J.W., "Modeling Micromachined Sensors with Finite Elements," Engineering & Technology Guide, *Machine Design,* July 26, 1990, pp. 44–60.

Powers, W. F., and Nicastri, P. R., "Automotive Vehicle Control Challenges in the Twenty-First Century," IFAC, Beijing, 1999.

Putty, M. W., and Chang, S.–C., "Process Integration for Active Polysilicon Resonant Microstructures," *Sensors and Actuators*, vol. 20, 1989, pp. 143–151.

Riethmuller, W., Benecke, W., Schnakenberg, U., and Heuberger, A., "Micromechanical Silicon Actuators Based on Thermal Expansion Effects," *Transducer '87*, 4th International Conference on Solid-State Sensors and Actuators, June 1987, pp. 834–837.

Rice, J. R., "Elastic Fracture Mechanics Concept for Interfacial Cracks," *Journal of Applied Mechanics, ASME Transactions*, vol. 55, March 1988, pp. 98–103.

Roark, R. J., *Formulas for Stress and Strain*, 4th ed., McGraw-Hill, New York, 1965.

Rohsenow, W. M., and Choi, H. Y., *Heat, Mass and Momentum Transfer*, Prentice Hall, Englewood Cliffs, N.J., 1961, Chapter 20.

Ruska, W.S., *Microelectronic Processing*, McGraw-Hill, New York, 1987.

Saville, D. A., and Palusinski, O. A., "Theory of Electrophoretic Separations," *AIChE Journal*, American Institute of Chemical Engineers, vol. 32, no. 2, February 1986, part I, pp. 207–214; part II: pp. 215–223.

Schulze, V., "Design and Analysis of Typical Micro Pressure Sensors for Automotive Application," MS thesis, San Jose State University, San Jose, California, 1998.

Segerlind, L. J., *Applied Finite Element Analysis*, John Wiley & Sons, New York, 1976.

"Smart Cars," cover story, *Business Week*, June 13, 1988, pp.67–77.

Smith, J. S., "Massively Parallel Assembly Using SOFT (Self Oriented, Fluidic Transport)," *Proceedings of the Integration of Nano- to Millimeter Sized Technologies Workshop*, DARPA/NIST, Arlington, Virginia, March 11–12, 1999.

Starr, J. B., "Squeeze-Film Damping in Solid-State Accelerometers," *Technical Digest, IEEE Solid-State Sensor and Actuator Workshop,* June 1990, pp. 44–47.

Sulouff, Jr., R. E., "Silicon Sensors for Automotive Applications," *Proceedings of Transducers '91, 1991 International Conference on Solid-State Sensors & Actuators,* San Francisco, June 24–27, 1991, pp. 170–176.

Sze, S. M., *Semiconductor Devices—Physics and Technology*, John Wiley & Sons, New York, 1985.

Teshigahara, A., Watnabe, M., Kawahara, N., Ohtsuka, Y., and Hattori, T., "Performance of a 7-mm Microfabricated Car," *Journal of Microelectromechanical Systems*, vol. 4, no. 2, June 1995, pp. 76–80.

Tien, C. L., and Chen, G., "Challenges in Microscale Conductive and Radiative Heat Transfer," *Journal of Heat Transfer, ASME Transactions*, vol. 116, November 1994, pp. 799–807.

Tien, C. L., Majumdar, A., and Gerner, F.M., eds. *Microscale Energy Transport*, Taylor & Francis, Washington, D.C., 1998.

Timoshenko, S., and Woinowsky-Krieger, S., *Theory of Plates and Shells*, 2nd ed., McGraw-Hill, New York, 1959.

Trim, D. W., *Applied Partial Differential Equations*, PWS-Kent Publishing, Boston, 1990.

Trimmer, W. S. N., and Gabriel, K. J., "Design Considerations for a Practical Electrostatic Micro-Motor," *Sensors & Actuators*, vol. 11, 1987, pp. 189–206.

Trimmer, W. S. N., "Microrobots and Micromechanical Systems," *Sensors & Actuators*, vol. 19, no. 3, 1989, pp. 267–287.

Trimmer, W. S. N., ed., *Micromechanics and MEMS—Classic and Seminal Papers to 1990*, IEEE Press, New York, 1997.

Tzou, D.Y., *Macro- to Microscale Heat Transfer*, Taylor & Francis, Washington, D.C., 1997.

Van der Spiegel, J., Tau, J. F., Ala'ima, T. F., and Lin, P. A., "The ENIAC-History, Operation and Reconstruction in VLSI," *International Conference on the History of Computing,* HNF Paderborn, Germany, August 14–16, 1988.

Van Zant, P. *Microchip Fabrication*, 3rd ed., McGraw-Hill, New York, 1997.

Waanders, J. W., "Piezoelectric Ceramics, Properties and Applications," *Philips Components*, N.V. Philips Gloeilampeenfabrieken, Eindhoven, The Netherlands, 1991.

White, F. M., *Fluid Mechanics*, 3rd ed., McGraw-Hill, New York, 1994.

White, R. M., "A Sensor Classification Scheme," *IEEE Transactions on Ultrasonic Ferroelectric Frequency Control*, UFFC-34, 1987, pp. 124–126.

Williams, B. A., and Vigh, G., "Fast, Accurate Mobility Determination Method for Capillary Electrophoresis," *Analytical Chemistry*, vol. 68, no. 7, 1996, pp. 1174–1180.

Williams, K., "New Technology & Applications at Lucas NovaSensor," *Tribology Issues and Opportunities in MEMS*, B. Bhusshan, ed., Kluwer Academic Publishers, 1998, pp. 121–135.

Wise, K. D., "Integrated Microelectromechanical Systems: A Perspective on MEMS on the 90s," *IEEE MEMS '91*, 1991, pp. 33 –38.

Wong, W. S., Wengrow, A. B., Cho, Y., Salleo, A., Quitoriano, N. J., Cheung, N. W., and Sands, T., "Integration of GaN Thin Films with Dissimilar Substrate Materials by Pd-In Metal Bonding and Laser Lift-off," *Journal of Electronics Materials*, vol. 28, no. 12, 1999, pp. 1409–1413.

Woolley, A. T., and Mathies, R. A., "Ultra-High-Speed DNA Fragment Separations Using Microfabricated Capillary Arrays Electrophoresis Chips," *Proceedings of National Academy of Science*, U.S.A, vol. 91, November 1994, pp. 11348–11352.

Yun, C. H., and Cheung, N. W., "SOI on Buried Cavity Patterns Using Ion-Cut Layer Transfer," *Proceedings of 1998 IEEE International SOI Conference*, October 1998, pp. 165–166.

Zengerle, R., Richter, M., and H. Sandmaier, "A Micro Membrane Pump with Electrostatic Actuation," *Proceedings of IEEE Micro Electro Mechanical Systems Workshop*, Travemunde, Germany, February 1992, pp. 19–24.

Zienkiewicz, O. C., *The Finite Element Method in Engineering Science*, McGraw-Hill, New York, 1971.

Zum Gahr, K.–H., "Microtribology," *Interdisciplinary Science Reviews*, vol. 18, no. 3, 1993, pp. 259–266.

INDEX